AF553276

SP
Pvt. Ltd.

Text Book of

Tractor Engine And Systems

Concepts, Objectives & Numericals

Prof. Manoj Kumar Ghosal
Department of Farm Machinery and Power
Orissa University of Agriculture and Technology
Bhubaneswar, Odisha - 751 003

2019

Studium Press (India) Pvt. Ltd.

Text Book of

Tractor Engine And Systems

Concepts, Objectives & Numericals

ISBN: 978-93-85046-33-9 HB

ISBN: 978-93-85046-35-3 PB

Published by:

Studium Press (India) Pvt. Ltd.
4735/22, 2nd Floor, Prakash Deep Building
(Near Delhi Medical Association),
Ansari Road, Darya Ganj, New Delhi-110 002
Tel.: + 91-11-43240200-15 (15 lines); Fax: 91-11-43240215
E-mail: pubdir@studiumpress.in

Printed at India

About Author

Dr. M. K. Ghosal, presently the Professor in the Department of Farm Machinery and Power, Orissa University of Agriculture and Technology (OUAT), Bhubaneswar, Odisha. His areas of expertise are Renewable Energy, Farm Power and Greenhouse Technology. Author of five books (four in Renewable Energy Technology and one in Farm Power). He did his B.Sc. (Agril. Engg.) from OUAT, Bhubaneswar, Odisha, M.Tech. (Farm Power and Machinery) from IIT, Kharagpur, West Bengal, Ph.D. (Energy Engineering) from IIT, Delhi. To his credit, he has published 110 research articles in various national and international scientific journals. He is at present the Associate Editor of two agricultural based Journals *i.e.* International Journal of Tropical Agriculture and International Journal of Agricultural Engineering. He has been actively involved in teaching Renewable Energy and Farm Power subjects both in UG and PG level of Agricultural Engineering discipline since last 20 years.

Dr. Manoj Kumar Ghosal
B.Sc. (Agril. Engineering) (O.U.A.T, Bhubaneswar)
M.Tech (IIT, Kharagpur), Ph.D. (Energy Engineering) (IIT, Delhi)
Professor, Department of Farm Machinery and Power
Orissa University of Agriculture and Technology (O.U.A.T.), Bhubaneswar-751003, Odisha, India

Preface

Adequacy in the availability of power in the agricultural sector is one of the critical parameters for increasing crop productivity.The increasing level of mechanization in the present scenario of farming, demands more use of power sources for operating improved implements and machinery. There has been substantial increase in use of mechanical and electrical power in order to enhance productivity and profitability in agriculture. During last decades, the use of mechanical power (through tractor, power tiller, oil engines etc.) and electrical power (through irrigation pumps, operation of various stationary machines etc.) has increased manifolds because of more demand in timeliness of farm operations. It is observed that the farm power availability and productivity have increased from 0.32 kW/ha to 2.02 kW/ha and 0.636 t/ha to 2.11 t/ha respectively over the years from 1965-66 to 2013-14.

At present, the tractors and power tillers are major sources of farm power for tractive work and electric motor for stationary operations. Their density is also rapidly increasing to meet the power requirements of more improved and matching implements in the agriculture sector. The effective repair, maintenance and periodical check-up of tractors and power tillers depend upon the skill and technical know-how of the technician, mechanic and operator. This book aims at providing the concepts and practical information regarding the engine and systems of tractors and power tillers. The details on the working principle, construction of various components of systems, mechanisms, hydraulic system, test and performance, maintenance etc. of tractors and power tillers along with the basic concepts of electrical energy, electric motor, generator etc. have been discussed.

The book has also been designed as a text book for the branch of Agricultural Engineering at under graduate and post graduate level in Farm Machinery and Power specialization to acquire a thorough understanding and proficiency in their field of specialization. The contents of the book have been framed to meet the syllabi prescribed by the Agricultural Universities of India for the courses "Tractor and Automotive Engines" and "Tractor Systems and Controls" in Agricultural Engineering discipline. Inclusion of objective type of questions, viva voce questions and solved numericals would help the students in appearing various competitive examinations. This book can also be used as a reference material in other disciplines mainly Mechanical Engineering/Electrical Engineering/Polytechnics/ITIs etc.Important definitions and glossary of related terms have been given towards the end of the book for their clear and appropriate meaning. Above all, the book is intended to be used by field workers, technicians, manufacturers, students and professionals who are engaged in the profession of Agricultural Engineering.

The author feels highly indebted to all those who have directly and indirectly helped in bringing the book to this shape and would also like to express appreciations to my family members for continued patience, understanding and support throughout the preparation of the manuscript.

Special thanks are to M/S Studium Press (India) Pvt. Ltd. New Delhi for taking interest in publishing the book.

Despite careful scrutiny of the proofs, there may still be some errors. The author would feel obliged if the readers point out the errors for rectification in the successive editions.

Suggestions for further improvement of the book by the readers will be appreciated.

September 2018 **Manoj Kumar Ghosal**

Table of Contents

Chapter 1

Sources of Farm Power

1.1 INTRODUCTION

Sources from which power is derived for doing various farm operations are called sources of farm power. Adequacy in the availability of power in the agricultural sector is one of the critical parameters for increasing crop productivity. It has been reported that there is a strong correlation between the inputs of power and productivity in farming (Anon, 2013). The increasing level of mechanization in agriculture demands more use of power sources for operating improved implements and machinery. With the introduction and adoption of intensive agriculture and multiple cropping systems in the present scenario of farming in Indian for achieving food security of the growing population, the time gap between the harvest of the previous crop and sowing of the subsequent crop is very short. Within the availability of very short period of hardly 2-3 weeks, it is not possible to complete the pre-sowing operations with the help of traditional tools and implements. The farmers are therefore forced either to own the power-driven machinery or to use the high capacity and labour saving machines on hiring basis for achieving timeliness of farm operations. Shortage of agricultural labourers during peak seasons of crop cultivation, increase in cropping intensity due to shrinking of cultivable land and more use of costly inputs like hybrid seeds, high dose of fertilizer, pesticides etc. are some other limiting factors for increased availability of power sources in view of quality, precision and timeliness of operations. Enhanced use of power is therefore indispensable for increasing yield, bringing more land under farming, improving utilization efficiency of water and other scarce agricultural inputs. Indian agriculture is thus gradually shifting from traditional farming to the mechanized farming since last decade by using mainly in animate power sources such as tractor, power tiller, diesel and petrol engine, self propelled agricultural machinery, electric motors etc. Over the years, the attempts for rural electrification are also rising and the use of electrical power mainly through motors of various capacities is becoming the prime source of power for performing different stationary jobs like operating irrigation equipment, threshers, shellers, cleaner and graders and other post harvest machinery. The larger sizes of electric motor are also increasing day by day and being used in the pump sets due to decline of water table in many regions. The farm power availability (kW/ha) in Indian farm is still very low compared to USA (6 kW/ha), South Korea (7 kW/ha) and Japan (14 kW/ha). The farm power input of India was 0.22 kW/ha in 1960-61, increased to 0.73 kW/ha in 1990-91 and further increased to 2.02 kW/ha in 2014-15. The highest farm power availability among the states of India is Punjab *i.e.* 3.5 kW/ha followed by Haryana, Uttar Pradesh, Andhra Pradesh and Tamil Nadu. As agriculture is a significant contributor of Indian economy, more power is needed in the farming and by 2022, the availability of power has to be increased to about 2.2 kW/ha in order to meet the rising demands of food grains of the burgeoning population of our country. The sources of power utilized at present in Indian agriculture are mainly from fossil fuels (electricity and petroleum oils) whose emissions during their use are now the major and emerging concerns of global warming and climate change, causing adverse effects on the productivity of crop. It has been reported that agriculture alone is now contributing to about 15 percent of green house gases emissions in India (Ghosal, 2017). Looking into the current trends of increasing input power in farming and thus the rising use of fossil fuels in Indian agriculture, it is the high time to think of alternate and environment–friendly energy sources for achieving sustainability in the farming. Power from non-conventional energy sources needs to be supplemented with the fossil fuels in order to reduce their uses gradually. There is therefore now a call for shifting from green revolution to the ever-green revolution. This necessitates the gradual transition from fossil fuels to renewable energy sources in the agricultural sector. Renewable sources of energy from solar, wind and biomass are now-a-days gaining momentum as the reliable and sustainable sources of power not only in agricultural sector but also in other power consuming sectors to curtail the harmful emissions of green house gases or atleast to stabilize their concentrations in the atmosphere for the well beings of organisms living in the earth.

1.2 CROPPING INTENSITY AND PRODUCTIVITY WITH POWER AVAILABILITY ON INDIAN FARMS

India is an agrarian country with more than 60 % of its population being dependent directly and indirectly on agriculture. It supports about 21 % of the nation's GDP. It not only provides employment and livelihood to the considerable section of India's population but also a source of raw materials for a large number of industries established in the country. Rising population and gradual decline of the cultivable land are the major factors for adopting multiple cropping systems and thus increase in cropping intensity in order to become self sufficiency in food grain production. During eighties, there was the acute shortage of food grain in our country. Adoption of intensive agriculture, multiple cropping systems, use of improved practices in crop cultivation, more inputs of power in farming etc. resulted into becoming not only self-sufficient but also a net exporter of food grains. The trends in farm power availability, cropping intensity and food grain productivity on Indian farms have been mentioned in Table-1. It has also been observed that the states with higher farm power availability have more productivity. The variation of farm power with productivity is shown in Fig. 1. The farm power availability and productivity increased from 0.32 kW/ha to 2.02 kW/ha and 0.636 t/ha to 2.11 t/ha respectively over the years from 1965-66 to 2013-14. Thus, food grains productivity is positively associated with unit power availability in Indian agriculture (Fig. 2). The relationship between food grains productivity and unit farm power availability for the period 1960-61 to 2013-14 was estimated by linear function, with highly significant value of coefficient of determination (R^2) as mentioned below.

$Y = 0.5512 + 0.8195 X$; $R^2 = 0.990$ where, Y = food grains productivity, t/ha, and X = power availability, kW/ha. This indicates that productivity and unit power availability is associated linearly. It is therefore evident that farm power input has to be increased further to achieve higher food grains production for the growing population of the country.

Table 1: Cropping Intensity and Power Availability on Indian Farms (Singh *et al*., 2014)

Year	*Cropping Intensity (%)*	*Food grain productivity (t/ha)*	*Power available (kW/ha)*	*Power per unit produc-tion (kW/t)*	*Net sown area per tractor (ha)*
1965-66	114.00	0.636	0.32	0.50	2162
1975-76	120.30	0.944	0.48	0.51	487
1985-86	126.80	1.184	0.73	0.62	174
1995-96	130.80	1.499	1.05	0.70	82
2005-06	135.90	1.715	1.49	0.87	45
2010-11	140.50	1.930	1.78	0.92	34
2011-12	141.50	2.079	1.87	0.90	31
2012-13	140.90	2.129	1.94	0.91	29
2013-14	142.00	2.111	2.02	0.96	27

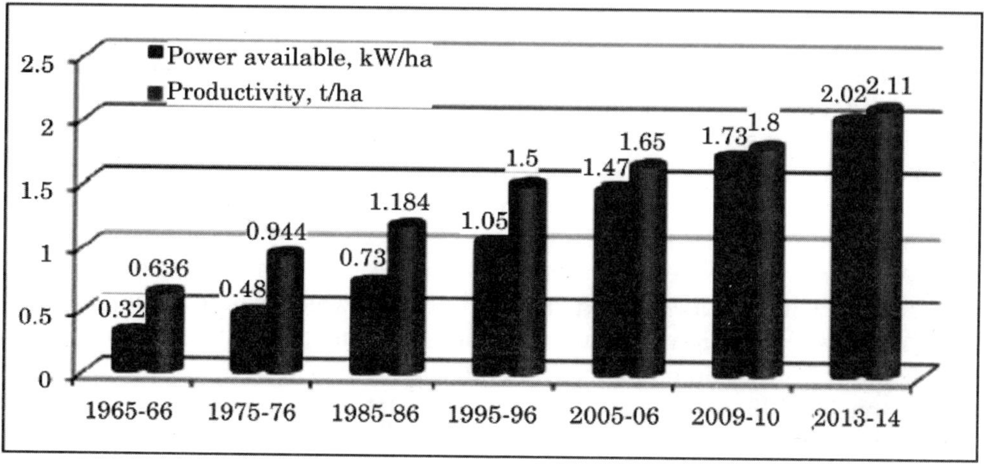

Fig. 1: Power Availability and Food Grain Productivity over the Years

1.3 FARM AND FARM ACTIVITIES

A farm is an area of land along with buildings in it which is used for growing crops and rearing animals. Farm activity is categorized into two types; one is the tractive work and the other is the stationary work. Tractive works are those which require

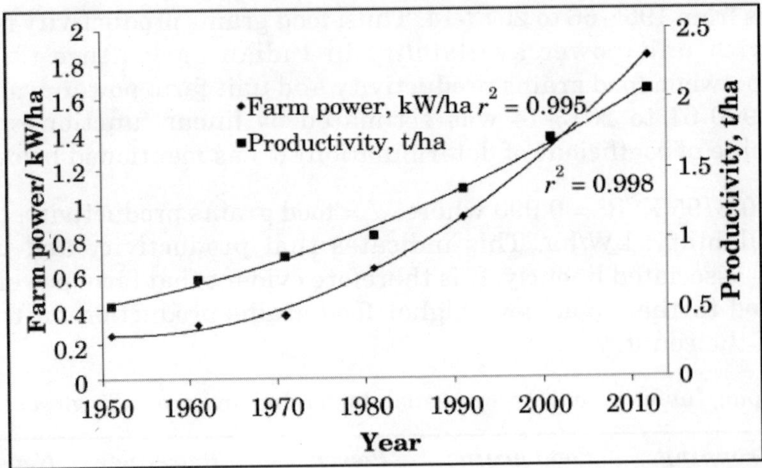

Fig. 2: Food Grain Productivity and Farm Power Availability Relationship in Indian Agriculture

pulling force for doing tillage, sowing, weeding, spraying, harvesting, transportation work etc. Stationary works are those which are accomplished by belt, pulley, gears, P.T.O. or direct drive to operate stationary machines like threshers, irrigation pumps, winnowers, silage cutters, feed grinders, fruit processing units, saws, cane crusher etc. The important sources of power for performing the above operations are from human labor, draught animal, diesel engine, petrol engine, tractor, power tiller, self propelled agricultural machines, electricity and renewable energy.

Increased availability of power from the above sources helps in enhancing production and productivity, because of timely farm operations, maximizing efficiency of agricultural inputs by their precise and judicious applications and reducing drudgery as well as cost of operations. The cost towards power and machinery is as high as 60 per cent of the total input cost in raising a crop. Even with the slow growth of mechanization, the Indian agriculture has achieved major success by increasing food grain production from 51 million tones in 1950-51 to 211 million tons in 2000 and 272 million tons in 2016-17 and productivity from about 0.636 t/ha in year 1965-66 to 2.111 t/ha in 2013-14 with surplus of food grain for export. The increase in food grain production was possible through adoption of better inputs such as high yielding varieties, increased dose of inorganic/organic fertilizers, plant protection measures and agricultural practices besides increase in cropping intensity. Efficiency of these inputs was further increased through adoption of appropriate farm machinery powered especially by mechanical and electrical sources of farm power.

India is predominantly an agricultural country where about 74 % of population from 1200 million people live in rural areas and depend directly or indirectly on agriculture for their livelihood. Of its total geographical area of 329 million hectare (m-ha), about 195 million hectare is the cultivable land and net area sown is about 141 million hectare (Anon, 2014). Due to ever-increasing population, there is a gradual reduction in average size of farm holdings. The agricultural land is also gradually converted into non-agricultural uses due to rising demands of houses, infrastructure, urbanization and industrialization. As per agriculture census, 2010-11, small and marginal holdings less than 2 ha account for about 85% of the total operational holdings and 44% of the total operational area. The number of operational holdings was estimated to be 71 million in 1970-71 which increased to 81.6 million in 1976-77, 88.9 million in 1980-81, 97.2 million in 1985-86 and 106 million in 1990-91 and 138 millions in 2010-11 (Table 2).

Table 2: Farm Holdings Break-up in India (Singh *et al.* 2015)

Category of Holdings	*Number of holding (Million)*			*Area (Million hectare)*		
	2000-01	*2005-06*	*2010-11*	*2000-01*	*2005-06*	*2010-11*
Marginal (less than 1 ha)	75.4	83.7	92.4	29.8	32	35.4
Small (1-2 ha)	22.7	23.9	24.7	32.1	33.1	35.1
Semi-Medium (2-4 ha)	14	14.1	13.8	38.2	37.9	37.5
Medium (4-10 ha)	6.6	6.4	5.9	38.2	36.6	33.7
Large (More than 10 ha)	1.2	1.1	1	21.1	18.7	17.4
All holdings	119.9	129.2	137.8	159.4	158.3	159.1
Hectare/holding				1.33	1.23	1.15
Holdings (%)	81.8 %	83.3 %	85.0 %			

The farm holdings in India are classified as marginal (<1 ha), small (1-2 ha), semi-medium (2-4 ha), medium (4-10 ha) and large (> 10 ha). Since the total cultivable land remained about 195 million ha, average size of farm holdings gradually reduced from 2.28 ha in 1971 to 1.84 in 1981, 1.57 ha in 1991, 1.33 ha in 2001 and 1.16 ha in 2011 (Table 3), due to small and scattered land holdings. This is because of inheritance law in which land gets divided among the children, resulting into very small plot size. This small land holding is suitable for cultivating by animate (human + draught animal) power only. But, due to increased cropping intensity, animate power was not adequate, and these sources of power have been supplemented by electro-mechanical power sources (tractor, oil engines, power tiller, electric motor etc.). India has therefore 195 million hectares of operated land owned by more than 138 million farm holders with the present average land holding size of 1.16 ha. The details of sources of farm power and their present status in Indian agriculture have been discussed in the subsequent section.

Table 3: Land Holdings in India (Singh *et al.*, 2015)

Category	*Percentage number of holdings in each category*				*Area under each category*			
	1971	*1991*	*2001*	*2011*	*Percentage*			*Average (ha)*
					1991	*2001*	*2011*	*2011*
Marginal (less than 1 ha)	50.6	59.2	62.4	67.0	15.0	18.7	22.2	0.38
Small (1-2 ha)	19.0	18.7	19.1	17.9	17.4	20.2	22.1	1.42
Semi-Medium (2-4 ha)	15.2	13.6	11.9	10.1	23.2	23.9	23.6	2.71
Medium (4-10 ha)	11.3	7.0	5.6	4.3	27.1	24.0	21.2	5.76
Large (More than 10 ha)	3.9	1.5	1.0	0.7	17.3	13.2	10.9	17.37
Average holding size (ha)	2.28	1.57	1.33	1.16				
All holdings (million)	70.5	106.6	119.9	137.8				

1.4 SOURCES OF FARM POWER

Energy is the basic requirement for human and animal beings, agriculture, industry, transportation, communication and all other economic activities of the present civilization. The quantum of energy input in agricultural sector *i.e.* for the farming operations determine the production and productively of a nation. The inability to complete farm operations in time is one of the key factors that results into reduced productivity and loss of production. Research results reveal that the percentage loss in yield for a few major crops like paddy, wheat and cotton due to delay in sowing for a fortnight are respectively 4, 5 and 20. Therefore, mechanization of agriculture through different sources of power assures timely completion of farm operations and hence maintains productivity and prevents loss of production. The mechanization level in Indian agriculture has been increased manifold due to gradual increase in use of power in various farm operations. This has also been revealed from the present status of manufacturers of agricultural implements and equipment in India

Table 4: Status of Agricultural Implements and Equipment in India during 2014-15 (Singh 2015).

Equipment manufacturers	*Number of units*
Agricultural tractors	22
Power tillers	5
Irrigation pumps	600
Plant protection equipment	300
Combine harvesters	48
Reapers	60
Threshers	6,000
Seed drills and planters	2,500
Diesel engines	200
Plough, cultivators, harrows	5,000
Chaff cutters	50
Rural artisans	More than 1 million

The sources of power for different farm operations are as follows:

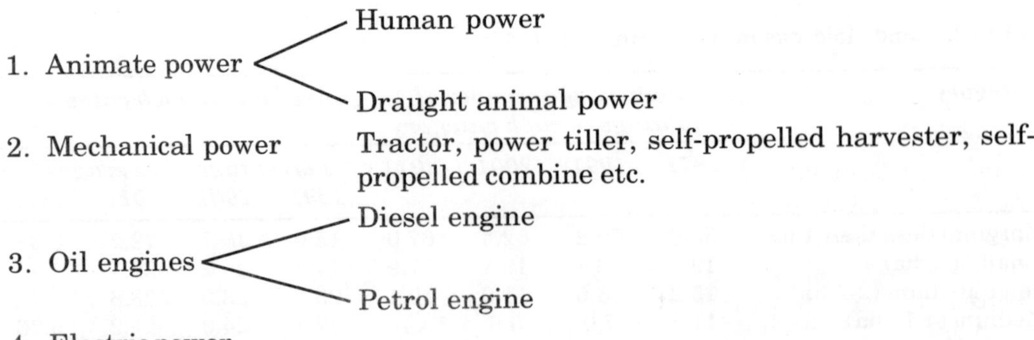

1.4.1 Human Power

Human beings are the main source of power in agriculture for operating small tools and implements. They are also employed for doing stationary work like threshing, winnowing, chaff cutting, lifting irrigation water etc. Several farm operations are still performed with hand tools in the developing countries like India as this source of power is easily available in rural areas where about 70 % of the population is engaged in farming operations. It is estimated that the agricultural workers' population in India during 2013-14 is 272 million. The agricultural workers are comprised of small cultivators and agricultural laborers. Of the total agricultural workers, about 30 % (81.6 million) is the women agricultural workers. On an average, a man develops nearly 75 watts power for a period of 6-8 hours of work per day. Therefore, the total power available through human source may be about 20.4 million kW. Similarly, on an average a woman can develop 60 watts power. But higher rates can be obtained while working for a short period. However, at present, there is a steady decline in the number of landless agricultural laborers for doing farm work in rural areas. With urbanization, industrialization and good transport system, these laborers daily go and work in industries due to payment of higher wages than farm work. The gradual declines of agricultural laborer have now resulted in the introduction of labor saving devices and mechanical power in Indian agriculture. The percentage of agricultural workers to the total workers was 59.1 in 1990-91, which reduced to 54.6 % in 2011-12. (Table-5, Mehta *et al.* 2014). But, in absolute terms, due to increase in overall population, the number of agricultural workers available in rural areas has increased from 131.1 million in 1960-61 to 272 million in 2013-14 and thereby registered an annual compound growth rate of 1.38% during the last 53 years, however, the percent of female agricultural workers has been increased from 11.4 to 37.2 % during 1981 to 2011. The population trend of agricultural workers in India is given below,

Table 5: Time Series Population of Agricultural Workers (Mehta *et al.* 2014)

Particulars	***1991***	***2001***	***2011***	***2020 (Projected)***
Country's population (million)	846.4	1028.7	1210.7	1323.0
Total no. of workers (million)	313.7	402.2	481.7	566.0
No. of workers as % of population	37.1	39.1	39.8	42.8
No. of agricultural workers (million)	185.3	234.1	263.0	230.0
Cultivators (million)	110.7	127.3	118.7	110.0
Agricultural labourers (million)	74.6	106.8	144.3	120.0
Percentage of agricultural workers to total workers	59.1	58.2	54.6	40.6
Percentage of females in agricultural work force	35.1	39.0	37.2	45.0

Advantages and Disadvantages of Human Power

Advantages	Disadvantages
1. Suitable for all kinds of farm work	1. Costliest power compared to other forms of power
2. Easy source of farm power	2. Efficiency is significantly low
3. Work does not hamper if one man falls sick	3. Cannot be continuously engaged for longer duration
4. Does not require any investment	4. Needs rest at frequent intervals
5. Flexible source of power according to demand	5. Mostly affected by adverse weather
6. Safe source of power for emergency work	6. Requires full maintenance during idle period.

Power developed by a human worker

According to Campbell (1990), the useful power developed by a human worker is given by

Power (hp) = 0.35 – 0.092 log t; where t = time in minutes. Now for 3 hours of work, the power developed by a person would become 0.15 hp or 0.11 kW. Similarly, for 4 hours of work, the power developed would be 0.13 hp and for 6 hours of work, it would become 0.11 hp or 0.085 watts. Hence more the hours of continuous work with small period of rest in between, less is the power developed.

Example 1: Calculate the power developed by a worker with 2 hours of continuous work and compare the same with similar kind of work if it is performed continuously for 4 hours with one time of rest in between for small period?

Solution: According to Campbell (1990), the power developed by the operator is given by P(hp) = 0.35 – 0.092 log t; where t = time in minutes. Working time in minutes = 2 hours = 120 minutes. P = 0.35–0.092 log 120 = 0.15 hp = 0.11 kW.

For 4 hours of work, P = 0.13 hp = 0.09 kW. This clearly indicates that power developed, is decreased with the increase in duration of work and not related linearly.

Example 2: Calculate the force (push/pull) developed by a worker with 2 hours of continuous work while operating an agricultural machine in the field at a speed of 2.5 kmph?

Solution: Power developed (P) kW = [Push (N) × speed (m/s)] /1000. In Example 1, the power developed by a worker for 2 hours of continuous work = 0.11 kW. The speed of the operator = 2.5 kmph = 0.70 m/s. Therefore, Push = [0.11 × 1000] / 0.70 = 157 N = 16.0 kgf.

Example 3: Calculate the work done by a labourer while calibrating in a tread mill when he walks at a linear speed of 4.0 kmph on an inclined running belt of 15^0? The weight of the worker may be assumed to be 60 kg.

Solution: The power developed by the worker while moving in a tread mill is given by

P (kW) = [W × V × sin θ] / 1000 where W = weight of the worker in (N); V = linear speed of belt (m/s), θ = inclination of belt , degrees.

P = [60 × 9.81 × 1.11 × sin 15^0] / 1000 = 0.17 kW. For 1 minute of work, the work done = 10.2 kJ/min.

1.4.2 Draught Animal Power

Draught animal power refers to the muscle power of working animals used for the various tasks like pulling agricultural implements, hauling carts, giving motive power to devices such as water pumps, cane and seed crushers, and electricity generation equipment, carrying loads on the back as pack animals, handling, dragging and stacking timber logs in forests. Draught animal power is used in agriculture for ploughing, harrowing, planting, ridging, weeding, mowing and harvesting; in transport, for pulling carts and loads over a surface, logging and carrying loads (pack animals); in irrigation for driving water-pumps and pulling water from wells; in the building industry, for assisting in earthmoving for road works, for carrying bricks, etc and to provide power for the operation of stationary implements such as threshing machines, grain mills and food processing machines through rotary gear systems.

Draught animal is mostly used for tractive works and is also a major source of farm power in Indian Agriculture. Bullocks and he-buffaloes are generally used for field operations. The pack animals like horses, elephants, mules and donkeys are also used in some agricultural operations, transport work as well as for commercial activities. The importance of this source of power has been well established in Indian agriculture and will continue be so for small and marginal farmers as well as for hill agriculture in future also. It is estimated that about 50 million draft animals (in 2011 census) supply power for major farm works in India. The average force a bullock can exert is nearly equal to one-tenth of its body weight. But for a very short period, the average force exerted is more than that. Generally, a medium size bullock can develop power between 360 and 550 watts during sustained working. The power developed by them depends primarily on the characteristics of animals (breed, species, sex, age, training and temperament etc.). Also, it depends upon how they are tamed, trained and harnessed. According to work condition, bullocks can develop draft up to about 10-12 % of their body weight, 16-18 % for camels and about 25-30 % for donkeys. Generally, 90 % of the bullocks and buffaloes above 3 years of old are employed for draft purpose on the farm. The total power output from the draught animals is therefore ($50 \times 10^6 \times 0.3$) = 15 million kW. In addition to bullocks and buffaloes, other animals like camels, mules and elephants are also frequently used in some places of India for transportation purpose.

1.4.3 Renewability of Animal Energy

Animal energy is a renewable and sustainable source of energy. It is renewable because the draught animals can be reproduced by breeding and rearing the required number of animals. It is sustainable because the animals derive their energy for work from feed and fodder made available from agricultural production, indeed largely from agricultural by- products. In addition, there are other environmental contributions of the working animal stock, mainly through the replacement of fossil-fuel run agricultural machinery. It saves natural resources, fossil fuels and prevents emission of greenhouse gases, mostly released to the atmosphere by use of petroleum fuels in the agricultural tractors, power tillers and self-propelled machinery. The by-products of livestock can be recycled through biogas production. The fossil-fuel equivalent of the animal energy used in the Indian agriculture has been found pretty large, as much as 19 million tons of diesel as per 2003 livestock census. If this much amount of fuel were to be burnt through combustion to run tractors in the absence of the working animal stock of over 50 million, it would have released about 5 million tonnes of carbon dioxide to the environment. Hence, interdependence of livestock and crop production for energy is a distinguishing characteristic of the mixed crop-livestock systems. Crops supply feed and fodder to livestock—the source of their energy and they return that energy in the form of food (milk and meat), dung (manure and fuel) and draught power. In India, despite increasing mechanization of agricultural operations, animals remain an important source of energy for crop production. India at present has a stock of about 50 million working animals used for various agricultural operations, and thus saving fossil fuel (diesel) worth of Rs 150 billion, annually (GoI, 2007). Animals are therefore the consumers and producers of biomass (Fig. 3). In both its raw and digested forms, biomass production is always included in assessments of renewable energy potential.

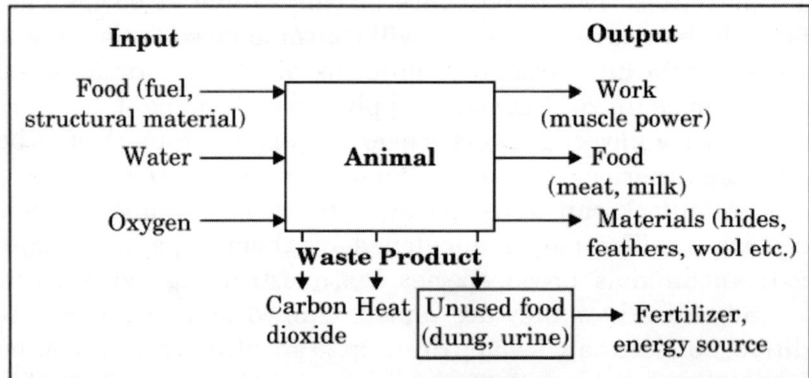

Fig 3: Input-output block diagram for working animals (source FAO, 2010).

Animals are produced and maintained locally and don't require any special infrastructure. Moreover, when the value of machines is likely to be depreciated over time, animals may appreciate because of growth. The principal environmental advantage of draught animal power (DAP) compared with mechanization is that DAP relies on bio-energy for its creation, maintenance and functioning instead of fossil energy. So, DAP provides the renewable source of energy and itself is indirectly a renewable source of energy.

1.4.4 Contribution of Draught Animal Power in Indian Agriculture

Draught animals, particularly bullocks and he-buffaloes, are still the predominant source of mobile power on about 55 % of the cultivated area consisting of about 87 million ha. They are very versatile and dependable source of power and are used in sun and rain under muddy and rough field conditions. They are born and reared in the village system and maintained on the feed and fodder available locally. They are ideal for rural transport where proper roads are not available. They reduce dependence on mechanical sources of power and save scarce petroleum products. Their dung and urine are also used as indirect source of energy, farmyard manure, biogas. They also help in maintaining ecological balance. Draught animal system is in perfect harmony with the village ecosystem. Draught animals, besides being a source of power are also a major source of by-products such as manure and urine, which enhance the fertility of the soil. Despite rapid increase in tractors and power operated machines, animal power is still prevalent for small and marginal farmers and for hilly regions of the country. Similarly, under Indian conditions where majority of the people are vegetarian and even amongst non-vegetarians, majority of them don't eat beef, draught animals as by-product of milch animals, will continue to be available for draught purposes in future also. About 5-6 decades back, most of the farm operations, water lifting, rural transport, oil extraction, sugarcane crushing, chaff cutting etc, were being done using draught animals only. But with the modernization of agriculture, development of pucca roads connecting village and availability of electricity in those villages, most of the jobs earlier being done using draught animals, except field operations, are now being done using other convenient and cheaper options. Over the years, the annual use of draught animals is going down. While earlier, a pair of animals is being used for about 1000-1500 hours annually, their average annual use has now come down to about 250-400 h only, that too for tillage, sowing, weeding and a little bit of rural transport on kuchha roads. The time series population of draught animals during 1971 to 2012 is given in Table 6 which shows that the population of draught animals during the last 15 years has been going down as the demand of tractors and power tillers have subsequently gone high.

Advantages and Disadvantages of Draught Animal Power

Advantages	Disadvantages
1. Mostly used for tractive works and limited use for stationary work	1. Cost of maintenance is high
2. Easy source of power for farm work	2. Requires constant care to keep animal in good health
3. Initial investment is less	3. Cannot be continuously engaged for longer period
4. Cheaper than human power	4. Needs rest at frequent intervals

5. Supplies manure to the soil and fuel (in form of cow dung cake or biogas) to the farmers	5. Hot and cold weather reduces the working capacity.
6. Uses farm produce as feed	6. Requires longer space for shelter
7. Utilizes also for short and medium distance transport work	7. Creates unhygienic condition near the farm house
	8. Working speed is very slow.

Table 6: Time Series Population of Draught Animals in India during 1997-98 to 2011-12 (in millions)

Type of draught animal	*1997-98*	*2003-04*	*% change*	*2007-08*	*% change*	*2011-12*	*% change*
Cattle	55.76	54.32	–2.58	49.70	–8.50	44.48	–10.50
Buffalo	6.80	5.84	–14.12	4.93	–15.58	4.25	–13.79
Total bovine	62.56	60.16	–3.84	54.63	–9.19	48.73	–10.80
Camels	0.91	0.03	–96.70	0.39	1200	0.29	–25.64
Yak	0.03	0.15	400	0.06	–60	0.05	–16.67
Mithun	0.08	0.75	837	0.15	–80	0.17	13.33
Horse and poney	0.83	0.18	78.31	0.21	16.67	0.25	19.04
Mules	0.22	0.65	195.45	0.10	–84.62	0.15	50
Donkeys	0.88	0.63	–28.41	0.30	–52.38	0.21	–30
Total	65.51	62.55	–4.52	55.85	–10.71	49.85	–10.74
Draft animal density, animal/ha	0.461	0.445	–3.47	0.396	–11.01	0.356	–10.10
Draft animal power (million kW)	15.64	15.04	–3.84	13.66	–9.17	12.18	–10.83

Example 4: A bullock drawn MB plough is operated at a forward speed of 2.5 km/h. Determine the horsepower developed by the bullocks. Assume that the weight of pair of bullocks is 600 kg.

Solution:

Given: Weight of each bullock = 300 kg.; operating speed of plough = 2.5 km/h = 0.7 m/s. It is assumed that a draught bullock can develop a pull of 1/10th of its body weight. Hence pull developed by a bullock = 300 kg/10 = 30.0 kg. Pull developed by a pair of bullocks = 30 × 2 = 60.00 kg.= 588.6 N. So, power developed by a pair of bullocks (watt) = Pull (N) × speed (m/s) = 588.6 × 0.7 = 412.02 watt = 0.412 kW.

Example 5: Calculate the draft developed by a pair of bullocks operating MB plough at a forward speed of 2.5 km/h. Assume that the weight of pair of bullocks is 600 kg and pull makes 60 ^{0} with horizontal.

Solution:

Given: Weight of each bullock = 300 kg.; operating speed of plough = 2.5 km/h = 0.7 m/s. It is assumed that on an average a medium size bullock develops 0.4 kW power for 3-4 hours of work in a day. Hence power developed by a pair of bullocks = 0.8 kW.

Pull (N) = power (watt)/ speed of operation (m/s)

Pull (N) = 800/ 0.7 = 1142.85

Draft = 1142.85 × Cos 60^0 = 571.42 N = 58.24 kgf.

1.4.5 Mechanical Power

The third important source of farm power is mechanical power that is obtained from tractor, power tiller, oil engines etc. Oil engine having an internal combustion engine, is basically of two types, *i.e.*, diesel engine and petrol engine. The efficiency of diesel engine varies from 32 to 38 per cent where as that of petrol engine varies from 25 to 32 per cent. In recent years, diesel engine is used in both tractor and power tiller. The major adoption of mechanical power is for irrigation by pump sets. The present population of irrigation pumps (in 2015-16) in India is estimated to be 26.00 million with 60 % electric operated and the remaining diesel, tractor or power tiller operated. Similarly, the total number of tractors, power tillers and diesel engines available for Indian agriculture during 2013-14 were about 5.237 million, 0.45 million and 8.45 million respectively. Stationary oil engines are generally used for pumping water, thresher, winnower, chaff cutter, sugar cane crusher, flour mill, oil ghanis etc. The mechanical power is primarily derived through internal combustion (IC) engine. An IC engine is a device to convert fuels (diesel and petrol) into mechanical power. The combustible fuel is burnt inside the engine cylinders in the gaseous form after proper compression and ignition in order to operate piston and crank shaft mechanism. The rotary power is generated at flywheel for use in different ways and applications. The mechanical power obtained through IC engine can be used for tractive works with the help of tractors and power tillers and stationary works with the rotary motion of the fly wheel.

1.4.6 Tractor Power

Tractor power is at present considered to be a power house in the agricultural sector, because a tractor can be used for multiple nature of jobs like tractive power and rotary power. It is becoming popular on the farms even in hilly and difficult regions and are available in various ranges. Tractor power is described in following ways.

Drawbar Power: It is the power available for pulling the loads. It is available at the end of the drawbar (the bar which is used between the two lower hydraulic links). This is the most widely used mode of tractor power on the farm. This is least efficient as its utilization for pulling the loads depends on traction developed by the rear drive wheels of the tractor. It is calculated as

Drawbar power (kW) = drawbar pull (kN) × speed (km/h)/(3.6)

Brake Power: It is a measure of an engine's horsepower without the loss in power caused by the gearbox, alternator, differential, water pump, and other auxiliary components such as power steering pump, muffled exhaust system, etc. It is the power generated at the belt pulley and available for useful work. The output delivered to the driving wheels is less than that obtainable at the engine's crankshaft. It is measured by a suitable dynamometer.

Brake Power (kW) = $(2 \pi N T)/(6 \times 10^4)$; where T= torque (N-m) and N = engine speed, rpm

Power Take Off (PTO) Shaft Power: It is the most efficient method of use of power in a tractor. This is primarily used for positive rotary drives to the mechanisms of the machines such as rotavator, vertical conveyor reaper, threshers, power sprayers, pump sets etc. The shaft transfers engine power to the driven machines attached to the tractor.

PTO Shaft Power (kW) = $(2 \pi N T)/(6 \times 10^4)$; where T= PTO shaft torque (N-m) and N = engine speed, rpm

1.4.7 Power Tiller Power

Power tiller is used many states of India as an important prime mover both for tractive works in small farm and stationary works related to agricultural operations. It is a multipurpose two wheeled hand tractor designed primarily for rotary tilling and other farm operations. It is a low horse power tractor usually fitted with two wheels (pneumatic or steel) in which direction of travel and its control for field operations are performed by operator walking behind it. In view of its high maneuverability and versatility, it is ideally suited for agricultural operations on small fields and farms where larger conventional four wheeled tractor are difficult and uneconomic to use. Two wheeled tractors have a variety of names such as single axle tractor, hand tractor and walking tractor. Power tiller was introduced in India with import of two units from Japan in 1961. The import continued till 1974 while domestic production started from 1965. Although six manufacturers were licensed to manufacture power tillers, only two established firms continued their production releasing about 6 models in the range of 6-9 kW (8-12 hp). Besides, there are many others who are importing Chinese make of power tillers and selling in the country. The major sales of power tillers are in West Bengal, Tamil Nadu, Karnataka, Assam, Kerala, Odisha and Maharashtra. At present, most of the power tillers are fitted with diesel engines with water cooling system.

India presently is the largest manufacturer of tractor in the world. There are more than 20 manufacturers of tractors in the country producing about 60 models of tractors in different hp ranges. Tractor population in India has grown from 0.037 million in 1960-61 to 5.237 million units in the year 2013-14 at an annual compound growth rate of about 10 per cent during the last 53 years (Table 7). Farm power availability from tractor has consequently increased from 0.007 kW/ha in 1960 to 0.218 kW/ha in 1990 at an annual compound growth rate of 12.14%. The growth rate in the next decade decreased to 8%. Sale of tractors and power tillers has also constantly increased during last 10 years. Time series population of tractors and power tillers is given below.

Table-7: Time Series Population of Tractors and Power Tillers in India (in millions)

Year	*Tractor*	*Power tiller*
1960-61	0.037	0
1970-71	0.168	0.0096
1980-81	0.531	0.0162
1990-91	1.192	0.0323
2000-01	2.531	0.1147
2010-11	4.207	0.3213
2011-12	4.553	0.3621
2012-13	4.858	0.4021
2013-14	5.237	0.4409
CAGR (%)		
1960-61 to 1990-91	12.27	6.25
1991-92 to 2013-14	6.65	12.03
1960-61 to 2013-14	9.79	9.30

(CAGR = Compound Annual Growth Rate)

Advantages and Disadvantages of Mechanical Power

Advantages	Disadvantages
1. Suitable for all tractive and stationary works	1. Initial cost is high
2. Higher efficiency compared to human and animal power	2. Requires costly fuel
3. Space requirement is less	3. Requires technical know-how to operate
4. No maintenance during idle period	4. Repair and maintenance cost is high
5. Operating cost per hectare is less compared to human and animal power	5. Risky if not handled properly.
6. Can run for a longer period	
7. Unaffected by weather.	

1.4.8 Stationary Power Sources (Diesel/Petrol Engine)

Diesel and petrol engines are generally used in stationary condition for performing various agricultural operations mainly in pump sets, portable sprayers, threshers, chaff cutters, sugar cane crushers, dal mill, cleaner and graders etc. Diesel and petrol engines are heat engines having internal combustion engine. A heat engine is a device which converts heat energy into mechanical energy. Heat generated by the combustion of a fuel creates high pressure or explosion, resulting into the movement of piston in the cylinder, which in turn rotates the crank shaft. The crankshaft delivers power to the desired load. Diesel (Compression Ignition) engine is designed to convert chemical energy of heavier fuel into mechanical energy. The injected fuel is ignited by the heat of the air which is compressed by the piston within the cylinder head. In this engine, only air is sucked during suction stoke. However, in case of petrol (Spark Ignition) engine, the petrol or gasoline fuel is atomized, vaporized and mixed in air in correct proportion before entering the engine cylinder. The fuel is ignited in the cylinder by the electric spark. In such engine, the mixture of air and fuel is drawn into the cylinder instead of only air as in diesel engine. The basic elements of diesel engine are same as spark ignition engine, but the method of fuel introduction and ignition are different to a great extent. The engine of diesel has high compression ratio, hence the air in the cylinder attains very high temperature and pressure at the end of the compression stroke. At the end of the compression stroke, the fuel is sprayed into the cylinder through injectors. The air inside cylinder at high pressure and temperature helps in igniting the fuel in atomized form. Such engines are called compression ignition engines because the ignition of fuel takes place due to heat of compression. Diesel engine is equipped with fuel injection pump and injectors. For both SI and CI engines, four strokes include intake, compression, power and exhaust. Valves are provided for admitting air to CI engines or air/fuel mixture to SI engines and for expelling exhaust gases. Two revolutions of crankshaft are required for each engine cycle and timing gears are arranged to allow two revolutions of crankshaft for each revolution of the cam shaft. The CI combustion chamber designs fall into two categories: direct injection (DI) and indirect injection (IDI). The IDI design is more fuel-tolerant *i.e.* it accommodates fuels with a wider range of viscosities and cetane ratings than the DI design, but the latter has about 8-10 % better fuel economy. Thus the trend is towards the use of DI design in newer CI engines.

Diesel engines are widely used by the farmers for operating various stationary machines. The population of this prime mover has increased significantly since the Green Revolution. Diesel engine population in the country increased about 37 times between 1960-61 and 2013-14 (Table 8), while the annual compound growth rate had been 10.66% during the period 1960-61 to 1990-91, with increased availability of electricity it reduced to 7.04% during the period of 1990-91 to 2013-14. Farm power from diesel engines increased from 0.009 kW/ha in 1960-61 to 0.247 kW/ha in 2000-01and 0.335 kW/ha in 2013-14, registered an annual compound growth rate of about 7% during the last 53 years. The time series population diesel engines in Indian farm are mentioned below.

Table-8: Time Series Population of Diesel Engines and Electric Motors in Indian Agriculture (in millions)

Year	*Diesel Engines*	*Electric Motors*
1960-61	0.23	0.20
1970-71	1.70	1.60
1980-81	2.88	3.35
1990-91	4.80	8.07
2000-01	5.90	13.25
2010-11	8.20	16.50
2011-12	8.30	16.70
2012-13	8.35	16.80
2013-14	8.45	17.00
CAGR (%)		
1960-61 to 1990-91	10.66	13.12
1991-92 to 2013-14	2.50	3.29
1960-61 to 2013-14	7.04	8.74

For adoption of higher level of technology to perform complex operations within time constraints and with comfort and dignity to the operators, mechanical power becomes essential. Thus, the extent of use of mechanical power serves as an indicator of acceptance of higher level of mechanization on farms

1.4.9 Electrical Power

Now-a-days electricity has become a very important source of power in the farms especially for stationary power. Its use in Indian farming has significantly been increased during past few years. The farmers generally prefer electric motors over the diesel engine as motors are easy to operate and require less maintenance and care. Motors are now therefore widely used in farm for stationary power in operating pump set, power thresher, winnower, chaff cutter, cane crusher etc. and using in dairy industry, cold storage, farm product processing, fruit industry, poultry industry, cattle feed grinding and many similar things for farmers. The share of total electricity consumption by the agriculture sector increased from 81673 GWh in 2001-02 to 153116 GWh in 2012-13 (1G = 10^9) and electric energy consumption in agriculture was recorded to be about 17.89% of the total electrical consumption during 2015-16. Towards the end of 2010, the number of electric motors for use in farms increased to 15 million and further to 17 million. The increased use of electricity in the farm is due to its lowest operating cost compared to other sources of farm power. The rural electrification programme launched by the Government of India during eighties has helped in making electricity available to 18.5 % villages during 1970-71 and increased to about 90 % villages during 2014-15. Electric motor population thus increased 85 times between 1960-61 and 2013-14 at an impressive annual compound growth rate of 8.7% (Table 8). Farm power availability from electric motors consequently increased exponentially from 0.041 kW/ha during 1971-72 to 0.494 kW/ha by 2012-13 with an annual compound growth rate of about 8.94% during the same period (Table-9).

Table 9: Sources of Farm Power Availability in Indian Agriculture

Year	*Farm Power, kW/ha*						*Total*
	Agril. workers	*Draught animals*	*Tractors*	*Power tillers*	*Diesel engines*	*Electric motors*	*power (kW/ha)*
1971-72	0.045	0.133	0.020	0.001	0.053	0.041	0.293
1975-76	0.045	0.135	0.040	0.001	0.078	0.056	0.358
1981-82	0.051	0.128	0.090	0.002	0.112	0.084	0.467
1985-86	0.057	0.129	0.140	0.002	0.139	0.111	0.578
1991-92	0.065	0.126	0.230	0.003	0.177	0.159	0.760
1995-96	0.071	0.124	0.320	0.004	0.203	0.196	0.918
2001-02	0.079	0.122	0.480	0.006	0.238	0.250	1.175
2005-06	0.087	0.120	0.700	0.009	0.273	0.311	1.500
2011-12	0.100	0.119	0.804	0.014	0.295	0.366	1.698
2012-13	0.093	0.094	0.844	0.015	0.300	0.494	1.841

Advantages and Disadvantages of Electrical Power

Advantages	Disadvantages
i. Cost of operation is very cheap	i. Initial cost is high
ii. Highly efficient	ii. Requires costly transmission system
iii. Can be run continuously	iii Needs sound technical knowledge
iv. Requires less space for installation	iv It is fatal and dangerous if proper care is not taken
v. No maintenance during idle period	v. Erratic supply causes problem
vi. Unaffected by weather	
vii. Weight per horse power is less	

Example 6: Find out the pulling capacity of a tractor having 32 kW PTO power if tractor is operating a disc harrow with draft of 20 kN at 4.0 kmph in clayey soil. Assume the transmission efficiency of PTO as 75 %.

Solution:

Given: Draft of harrow, D = 20 kN; forward speed of tractor = 4.0 kmph; PTO power of tractor = 32 kW.

Drawbar power of tractor (kW) = Draft (kN) × Speed (kmph)/(3.6) = 20 × 4/3.6 = 22.22

Drawbar power available from tractor = PTO power × transmission efficiency of PTO
= 32 × 0.75 = 24 kW

Since the available power from tractor is more than the power required for disc harrow, the tractor will be able to pull the disc harrow.

Example 7: A 4-bottom 35 cm disc plough is being pulled at a speed of 5 kmph by a tractor in a soil which offers a resistance of 0.7 kg/cm^2. How much power is being developed at the tractor drawbar when the plough is going 20 cm deep?

Solution:

Given: Number of bottoms = 4; width of each bottom = 35 cm; speed = 5kmph; soil resistance = 0.7 kg/cm^2; depth of ploughing = 20 cm. Total width of ploughing = 35 × 4 =140 cm

Depth of ploughing = 20 cm; furrow cross section = 140 × 20 cm^2

Total resistance offered by the soil = 140 × 20 × 0.7 = 1960 kgf = 19.22 kN

Draft requirement = 19.22 kN

Drawbar power of tractor (kW) = Draft (kN) × Speed (kmph)/(3.6) = 19.22 × 5/3.6
= 26.70

Example 8: Find out the size of a tractor to pull a 4 × 40 cm MB plough at a working depth of 20 cm. The soil resistance is assumed to be 0.5 kg/cm^2 and forward speed to be 5.0 kmph. The transmission efficiency and tractive efficiency of tractor is 80 % and 50 % respectively.

Solution:

Given: Number of bottoms = 4; width of each bottom = 40 cm; speed = 5 kmph; soil resistance = 0.5 kg/cm^2; depth of ploughing = 20 cm. Total width of ploughing = 40 × 4 =160 cm

Depth of ploughing = 20 cm; furrow cross section = 160 × 20 cm^2

Total draft of plough = cross section of furrow (cm^2) × soil resistance (kg/cm^2) = 160 × 20 × 0.5 = 1600 kgf = 1600 × 9.81/1000 = 15.69 kN

Drawbar power of tractor (kW) = Draft (kN) × Speed (kmph)/(3.6) = 15.69 × 5/3.6 = 21.80

Power of tractor required (kW) = Drawbar power (kW)/(transmission efficiency × tractive efficiency) = 21.80 /(0.8 × 0.5) = 54.5 kW.

Example 9: An I.C. engine consumes high speed diesel at the rate of 0.5 kg/h. The heat value of fuel is 44,100 kJ/kg. Calculate the power of engine?

Solution:

Heat value = quantity (kg/h) × calorific value (kJ/kg) = 0.5 × 44,100 = 22,050 kJ/h
= 22,050/3600 = 6.12 kW.

Example 10: A diesel engine transmits 24 kW power at the end of the crankshaft moving at 1600 rpm. Find the torque exerted by the piston?

Solution:

Power (W) = 2 π N T/60 where N = rpm and T= torque (N-m).

24 × 1000 (W) = 2 π × 1600 × T/60 $\rightarrow$ T = 143.31 N-m.

Example 11: A diesel engine consumes 4.0 litres fuel per hour and develops 10 kW brake power at 1600 rpm. The specific gravity of fuel and its heat value are respectively 0.82 and 44,100 kJ/kg. Calculate the fuel power and the thermal efficiency of the engine?

Solution:

Fuel consumed per hour = vol. of fuel (litre/h) × sp. Gravity of fuel × density of water = 4.0 lit/h × 0.82 × 1 kg/litre = 3.28 kg

Fuel power = 44100 × 3.28/3600 = 40.18 kW

Engine develops 10 kW brake power

Thermal efficiency = (Engine brake power/fuel power) × 100 = 24.88 %

Example 12: A petrol engine consumes 8.0 litres fuel per hour and develops 22 kW brake power at 1400 rpm. The specific gravity of fuel and its heat value are respectively 0.76 and 46,600 kJ/kg. Calculate the fuel power and the thermal efficiency of the engine?

Solution:

Fuel consumed per hour = vol. of fuel (litre/h) × sp. Gravity of fuel × density of water = 8.0 lit/h × 0.76 × 1 kg/litre = 6.08 kg

Fuel power = 46600 × 6.08/3600 = 78.70 kW

Engine develops 22 kW brake power

Thermal efficiency = (Engine brake power/fuel power) × 100 = 27.95 %

Example 13: A current of 2.5 amperes from a 230-volt source flows through an electric iron for one hour. Calculate the energy consumed?

Solution:

Current I = 2.5 amperes; voltage = 230 volts; time of operation = 1 hour

Energy = V × I × t = 230 × 2.5 × 1/1000 = 0.57 kWh

Example 14: A house is fitted with 8 lamps (each 40 watts) and 2 fans (each taking 1 ampere current). The energy is supplied at 230 volts. The lamps are lighted for 6 hours a day and fan works 8 hours a day. What will be electric bill for 30 days if the rate of supply of electricity is at Rs. 5.00 per kWh?

Solution:

Energy used by the lamps in a month = 40 × 8 × 6 × 30/1000 = 57.6 kWh

Energy used by fans in a month = 2 × 1 × 230 × 8 × 30/1000 = 110.4 kWh

Total energy consumed = 57.6 +110.4 = 168 kWh (units)

Amount of electric bill = 168 × 5 = Rs. 840

Example 15: Calculate the water power which is required to discharge liquid @ 30 litres/min at 30 kg/cm^2 pressure?

Solution:

Given: Discharge (Q) = 30 litres/min = (30 × 10^{-3}/60) m^3/s

Pressure = 30 kg/cm^2 = 30 × 9.81 × 10^4 N/m^2

Water power (W) = Discharge (m^3/s) × Pressure (N/m^2) = Q P

Water power (W) = Discharge (m^3/s) × Density of liquid to be pumped (kg/m^3) × Acceleration due to gravity, g (m/s^2) × Head of pumping (m) = Q ρ g H

1 kg/cm^2 = 9.81 × 10^4 = N/m^2

1 litre of water = 10^{-3} m^3

Water power is the power required to deliver liquid from the pump.

Water power (kW) = Discharge (m^3/s) × Pressure (N/m^2) = Q P = {30 × 10^{-3}/60} × {30 × 9.81 × 10^4}/1000 = 1.47 kW

Example 16: Find out the size (kW) of an electric motor required to operate a centrifugal pump running at a speed of 1400 rpm and pumping 1000 litres of water per minute at a total head of 20 meter. Assume the efficiency of the pump 70%.

Solution:

Given: Discharge (Q) = 1000 litres/min = (1000 × 10^{-3}/60) m^3/s

Total head = 20 metre

Water power (kW) = {(1000 × 10^{-3})/60} × 1000 kg/m^3 × 9.81 × 20 m/1000 = 3.27

Power required to drive the pump = 3.27 × 100/70 = 4.67 kW

Example 17: Determine the power of the motor required to drive a centrifugal pump which is installed 10 metres above the pumping water level for delivering 250 litres of water per minute against a height of 30 metre above the pump. Losses due to friction in pipe are 0.5 metre per 40 metres of length of pipe and total length of pipe is 40 metres. Losses in bends and other fittings are equal to 3 metres. Assume the efficiency of pump as 60 %.

Solution:

Given: Discharge = 250 litres/min = (250 × 10^{-3}/60) m^3/s, Total length of pipe = 40 m

Suction head = 10 m, Delivery head = 30 m, Total friction head loss = 0.5 × 40/40 = 0.5 m

Losses in bends and fittings = 3 m

Total head = suction head + delivery head + losses = 10 + 30 + 0.5 + 3 = 43.5 m

Water power (kW) = {(250 × 10^{-3})/60} × 1000 kg/m^3 × 9.81 × 43.5 m/1000 = 1.77

Power of motor = 1.77 × 100/60 = 2.95 kW

Example 18: An electric pump set is lifting water at the rate of 1700 litres per minute against a total head of 27 m. Calculate the water power of pump and power input to motor if pump and motor efficiencies are respectively 70 % and 80 %?

Solution:

Given: Discharge (Q) = 1700 litres/min = $(1700 \times 10^{-3}/60)$ m^3/s, Total head = 27 m

Pump efficiency = 70 %, Motor efficiency = 80 %

Water power (kW) = $\{(1700 \times 10^{-3})/60\} \times 1000$ kg/m^3 $\times 9.81 \times 27$ m/1000 = 7.5

Water power of pump = water power/pump efficiency = 7.5/0.7 = 10.71 kW

Power input to motor = Pump power/motor efficiency = 10.71/0.8 = 13.38 kW

Example 19: A 10 kW electric motor is used to operate a centrifugal pump. The pump is used to irrigate a field crop for 10 hours. Calculate the cost of irrigation if electrical energy costs Rs. 5.00 per unit and the cost of use of the pumping set is Rs. 30.00 per hour?

Solution:

Given: Size of motor = 10 kW; working time = 10 hours; cost of use of pumping set = Rs. 30.00 per hour.

Electrical energy consumed, units = size of motor × operating hours = 10 × 10 = 100 kWh = 100 units.

Cost of electrical energy = 100 × 5 = Rs. 500

Cost of use of pumping set = 30 × 10 = Rs. 300 for 10 hours

Cost of irrigation for 10 hours = 500 + 300 = Rs. 800

Cost of irrigation per hour = 800/10 = Rs. 80.00

Example 20: A 3 phase motor rated 440 V, 500 kVA is operated at 0.8 power factor for 10 hours in a day and 24 days in a month. The cost of electrical energy is Rs. 5/- per kWh. Calculate the total cost of electrical energy taken from the supply point in 1 month?

Solution:

Power (3 phase) = $\sqrt{3}$ VI cos ∅ Kw

where ∅ is the phase angle between V and I, cos ∅ is the power factor.

$\sqrt{3}$ VI = 500 kVA and cos ∅ = 0.8

Power (P) = 500 × 0.8 = 400 kW

Energy consumed in 10 hours in a day = 400 × 10 = 4000 kWh

Cost of energy = 4000 × 5 =Rs 20,000

For one month= 24 × 20,000 = Rs. 4,80,000

1.5 POWER FROM RENEWABLE ENERGY SOURCES

The requirements of diesel and electricity are rising day by day due to fast growth of mechanization in Indian agriculture and is expected to grow in the future because of the necessity of more inputs of power in the farming sector. The rising price of diesel and electrical energy would thus threaten the sustainability of agricultural development for the majority of small and marginal farmers existing in India. The irregular and erratic supply of electricity in remote areas is also becoming a major concern for its use in the farm. In this context, the use of alternative sources of energy is at present the need of hour and renewable energy sources are therefore now gaining importance not only in agricultural sector but also in other energy dependent sectors such as industries, transportation, residential houses, infrastructure, manufacturing units etc. Besides the energy reliability, these sources of energy are environment-friendly and can be harnessed with the upcoming technologies through intensive research and development in the relevant areas. Ministry of New and Renewable Energy (MNRE), Government of India, is making all efforts for the accelerated adoption of renewable energy technologies in the farm by developing need based and suitable devices for the small and marginal farmers. The energy used in the farm, can especially be derived from sun, biomass and wind. The energy from solar thermal devices, solar photovoltaic systems, biomass, biogas, biodiesel, biofuel cell, wind mill etc may be obtained for fulfilling the requirements of farm and for their uses in a decentralized manner, as renewable energy is inexhaustible and available profusely in the nature. The devices from the above renewable energy sources can be used for lighting, cooking, water heating, water distillation, water pumping, food processing, crop drying, diesel engine running, electricity generation etc. Hence, the renewable energy in the form of solar, biomass (producer gas, biogas, biodiesel, ethanol etc.) and wind can contribute significantly in achieving the energy security and sustainability in the agricultural sector for 21st century if devices based on those sources of energy can be promoted widely among the farming community.

Advantages of Renewable Sources of Energy

i. Renewable energy sources are available in considerable quantities and available continuously in the nature.

ii. Renewable energy sources are financially and economically competitive for certain applications such as in remote locations where the cost of transmitting electrical power or transporting conventional fuels are high. They favour power system decentralization.

iii. Renewable energy sources cannot be depleted unlike fossil fuels. They can provide a reliable and sustainable supply of energy indefinitely. In contrast, the non-renewable sources of energy are finite and can be diminished by extraction and consumption.

iv. Renewable energy sources provide less environmental impacts as compared to other sources of energy. The non-renewable energy sources cause water/ air pollution, acid rain, global warming, climate change, ozone depletion which affect human and animal life and vegetation in the earth.

Disadvantages of Renewable Sources of Energy

i. The availability of renewable energy is very intermittent, irregular and most erratic.
ii. Introduction or use of energy storage system makes the energy generation system costly.
iii. They are site and location specific
iv. Area required for considerable energy extraction from the renewable energy sources is vast as compared to non-renewable energy sources.
v. The efficiency of renewable energy conversion devices is very less as compared to non-renewable systems.
vi. All renewable energy devices are installed far away from residence
vii. Implementation of renewable energy devices is not so easy due to lack of efficient devices, public awareness and sufficient information.

1.6 STATUS OF FARM POWER IN INDIA

Indian agriculture since fifties started using improved implements to achieve timeliness of operations and thus increasing productivity. With the onset of green revolution, adoption of agricultural machinery with higher capacity was highly necessary for timely completion of farm operations for promoting multiple cropping systems. Government policies on extension of subsidies on animal operated implements assisted the small farmers in faster adoption of improved implements. The state Agro-Industries Development Corporation played a catalytic role in providing logistic support for manufacturing, marketing, after sale services of the prime movers and machinery as well as custom hiring of capital-intensive items. The use of these farm machinery necessitated the introduction of comparatively compatible as well as efficient power sources in the farm for performing various tractive and stationary operations. Looking into the timeliness of operation with multiple cropping system, precise and efficient uses of costly inputs as well as comforts to the agricultural workers, attention was diverted towards the more use of mechanical and electrical power in the farm, compared to human labor and draft animal. Table 10 gives the population statistics of different sources of farm power from 1950 to 2014. It is clear from the table that except for draught animals, the population of all other power units is increasing with time. The decline in the use of draft animals is due to their high cost per unit energy as compared to mechanical and electrical energy sources and some expenditure for their upkeep and maintenance.

Table 10: Trends on Farm Power Sources in Indian Agriculture

Year	*Draft animals, million*	*Agricultural workers, millions*	*Tractor, millions*	*Power tillers, 000 nos.*	*Diesel engines, millions*	*Electric motors, millions*
1950	65.0	95.6	0.008	0.0	0.07	0.02
1955	72.0	112.4	0.020	0.0	0.12	0.05
1960	80.4	116.0	0.037	0.0	0.23	0.20
1965	81.4	120.0	0.063	1.5	0.50	0.50
1970	82.6	124.2	0.168	9.6	1.70	1.60
1975	83.4	136.2	0.292	17.9	2.32	2.28
1980	73.4	149.3	0.531	16.2	2.88	3.35
1985	72.6	165.9	0.810	19.6	5.40	4.33
1990	70.9	183.5	1.192	31.2	4.80	8.07
1995	65.2	199.0	1.707	55.2	5.20	11.13
1997	62.6	205.0	2.032	65.9	5.55	11.99
2000	60.3	234.10	2.531	114.7	5.90	13.25
2010	53.50	263.00	4.207	321.3	8.20	16.50
2011	53.0	266.08	4.553	362.1	8.30	16.70
2012	52.8	269.20	4.858	402.1	8.35	16.80
2013-14	52.0	272.00	5.237	440.00	8.45	17.00

Source: Agricultural Engineering Today, vol. 38(4), 2014, p.36

The share of animate, mechanical and electrical power in the use of total farm power from 1970 to 2013 has been presented in Table 11. From the table, it is clear that animate power contributed about 60 % of the total farm power in 1971-72 and mechanical and electrical power together contributed only about 40 %. In 2000-01, the contribution from animate power reduced to about 17 % and mechanical as well as electrical power increased to nearly 83 %. Similarly, in 2012-13, the percentage shares of animate and inanimate (mechanical as well as electrical power together) were respectively 10 and 90. It is however seen that the use of mechanical and electrical power is more for stationary operations than tractive field operations. The availability of farm power per hectare in Indian agriculture has been presented in Table 12. For timely operation, the desirable farm power under the present situation is as follows. The area under tractorization decreased from 2162 ha in 1965-66, 45 ha in 2005-06 and 27 ha in 2013-14 per tractor (Table-1) indicating the increase in the use of tractor per hectare in performing the farming operations in the country. The availability of farm power in a country depends on its farm machinery industries. The net work of agricultural machinery manufactures in India is quite complex ranging from village artisans, small scale industries to state Agro-Industrial Development Corporation and organized tractor, oil engine and processing equipment industries. Table 4 indicates the number of such manufactures in the country.

Table 11: Percentage share of animate, mechanical and electrical power sources in total farm power in India

Year	*Agril. workers*	*Draught animals*	*Tractor*	*Power tiller*	*Diesel engine*	*Electric motor*
1971-72	15.4	45.4	6.8	0.3	18.1	14.0
1975-76	13.4	37.2	11.2	0.3	21.8	15.6
1981-82	10.9	27.4	19.3	0.3	24.0	18.0
1985-86	9.9	22.3	24.2	0.3	24.0	19.2
1991-92	8.6	16.6	30.3	0.4	23.3	20.9
1995-96	7.7	13.5	34.9	0.4	22.1	21.3
2000-01	6.7	10.4	40.8	0.5	20.3	21.3
2005-06	5.8	8.0	46.7	0.6	18.2	20.7
2012-13	5.0	5.1	45.8	0.8	16.3	26.8

Source: Agricultural situation in India, 2015.

Table 12: Availability of farm power per hectare in India

Year	*Total power available in kw/ha*	*Net area sown per tractor (ha)*	*Source wise (%)*		
			Animate	*Mechanical*	*Electrical*
1951	0.25	–	97.4	2.1	0.5
1961	0.31	3600	94.9	3.7	1.4
1971	0.32	960	60.80	25.20	14.0
1981	0.50	260	38.30	43.60	18.0
1991	0.85	120	25.2	54.0	20.9
2000	1.17	55	17.1	61.6	21.3
2005	1.50	45	13.8	65.5	20.7
2012	1.84	29	10.1	62.9	26.8
2014	2.02	27	17.03	52.97	29.60

Source: Agricultural statistics at a glance, 2015, Govt. of India

1.7 POTENTIAL OF FARM POWER FROM RENEWABLE SOURCES OF ENERGY

The Indian agriculture has experienced a phenomenal growth in mechanization that cannot be sustained without inputs of energy in the farm. The 21st century will also experience a further rise in mechanization due to the necessity of more use of tractors and electric motors leading to rise in demand of diesel and electricity. In this century, the contribution of diesel and electrical energy will rise while contribution of animate power will decline. Under the above scenario of rising diesel and electricity demand and increasing oil import bill coupled with dwindling growth of electricity production, sustainability of agricultural production will be seriously threatened if appropriate steps to augment the increasing demands of diesel and electricity are not taken in time through alternative sources of energy. Renewable energy which is inexhaustible and abundantly available in the nature may appear to be a very potential source of alternative energy for replacing the increased use of fossil fuels not only in agriculture but also in the other energy consuming sectors. The country has already taken a lead in introducing the following technologies in rural areas for fulfilling substantially the energy needs of Indian agriculture in future;

i. Producer gas for thermal application and running engines for shaft power
ii. Anaerobic decomposition of surplus crop residues for methane gas and compost production
iii. Use of surplus agro-residues for decentralized electric generation
iv. Biodiesel from plant oil/oil seeds for running tractors and stationary engines
v. Alcohol from sugarcane and other cellulose material for running gasoline consuming engines
vi. Charring and briquetting of biomass for domestic fuel in rural areas and for agro-industries
vii. Efficient utilization of animate sources of energy with matching tools and equipment to increase their efficiency
viii. Efficient utilization of solar energy for crop drying, water lifting, solar refrigeration, solar distillation etc.
ix. Efficient utilization of wind energy for water lifting and shaft power in high wind velocity areas.
x. Introduction of solar-wind hybrid power systems in Indian farm

1.7.1 Status of Renewable Energy in India

The national programmes in different areas of renewable energy sector have resulted not only in generation of public awareness about the advantages of renewable sources of energy but also in a significant increase in the deployment of renewable energy systems and devices for varied applications in the society. The contribution of renewable energy to total installed capacity of power generation has been progressively rising. As on December 2016, the contribution of renewables has reached 50,000 MW, representing about 16 % of total grid capacity of 3,00,000 MW in India. Almost all the areas namely solar, wind, biomass, small hydro and urban as well as industrial waste have contributed to the satisfactory achievement of renewable energy sources in the country. The achievements of various renewable energy sources in India during 2016-17 have been shown in Table.13. The potential utilization of renewable energy technologies in rural areas and particularly in agricultural sector is discussed below.

Solar Energy: India is blessed with abundant supply of solar energy for more than 200 clear sunny days in a year. It is a clean source of energy. It is eco-friendly and inexhaustible. The energy from sun is utilized either as thermal energy or photovoltaic energy.

Solar Thermal Energy: Solar thermal technologies basically comprise of a solar collector where solar energy is collected, a system to use collected solar energy. Solar cooker, solar water heater, solar still and solar crop dryers have been commercialized and are available in different designs and sizes. Solar cooker, solar water heater and solar stills do not have much potential in rural areas but small or medium size solar crop dryers are especially designed to improve quality and to increase the shelf life and value addition of agricultural commodities like resins, chilies, ginger, spices, vegetables etc.

Table 13: Renewable Energy Achievements in India as on December 2016

Sectors	*Achievement during 2016-17 (as on December 2016)*	*Cumulative achievements (31-12-2016)*
I. GRID_INTERACTIVE POWER (CAPACITIES In MW)		
Wind Power	1922.99	28700.44
Solar Power	2249.81	9012.66
Small Hydro Power	59.92	4333.85
Bio-Power (Biomass, Gasification and Bagasse Cogeneration)	151.40	7907.34
Waste to Power	7.50	114.08
Total	4391.62	50068.37
II. OFF-GRID/CAPTIVE POWER (CAPACITIES In MW)		
Waste to Energy	4.47	163.35
Biomass (non-bagasse) Cogeneration	0.00	651.91
Biomass Gasifiers		
Rural	0.00	18.34
Industrial	4.30	168.54
Aero-Generator/Hybrid systems	0.38	2.97
SPV systems	98.50	405.54
Water mills/micro hydel	0.10 MW + 100 water mills	18.81
Total	81.99	1403.70
III. OTHER RENEWABLE ENERGY SYSTEMS		
Family Biogas Plants (in Lakhs)	0.35	49.40

Source: https://mnre.gov.in (Ministry of New and Renewable Energy, New Delhi, India)

Solar Photovoltaic Power: Electricity can be produced from the solar energy by photo voltaic solar cells; which convert the solar energy directly to electricity (direct current). The electricity generated can be stored in batteries and is used for various lighting or shaft power applications.

Electricity is directly generated by utilizing solar energy by photo voltaic process. When photons from the sun light are absorbed in a semi-conductor, they create free electrons with higher energies than the electrons which provide the bonding in the base crystal. Once these free electrons are created, there must be an electric field to induce these highly energized electrons to flow out of the semiconductor to do useful work. The electric field in most of the solar cells is provided by a junction of materials which have different electrical properties. The photovoltaic effect can be described easily with p-n junction in semi-conductor materials of solar cells which are made from p-type and n-type semiconductors A p-type semiconductor is formed by doping Germanium (Ge) or Silicon (Si) with a trivalent (number of valence electrons=3) elements like Indium, Boron or Aluminium. An n-type semiconductor is formed by doping Ge or Si with a pentavalent (number of valence electrons=5) elements like Arsenic or Antimony.

The most significant applications of photovoltaic cell in India, are the energization of pump sets for irrigations, drinking water supply, solar lantern, solar home lighting system, solar street light, community TV sets, solar fans etc. These devices are being marketed by MNRE through its nodal agencies like state energy development agencies. The major impediments to widespread use of photovoltaic devices are their high cost, although there is considerable optimism about reducing their costs. The average factory price for photovoltaic modules has considerably been reduced from almost US $ 90,000/KW in 1975 to around US $ 900/KW now. A further cost reduction of about 50 to 70 % is needed and is expected to achieve in order to make it cost competitive with other electrical energy generation.

Power from Biomass: Materials obtained form plants and animals are called biomass. Plant matter created by the process of photosynthesis is also called biomass. It is one of the potential sources of energy. India has plenty of agricultural and forest resources for production of biomass. Biomass can be used to produce solid fuel like briquettes, liquid fuels like alcohol and biodiesel, gaseous fuels like biogas and producer gas and electrical power through biomass fuelled power plants and biofuel cells.

The method of physical conversion of biomass through the compression of combustible material is known as briquetting. The alcohol fuel or ethyl alcohol is prepared from the fermentation of biomass like molasses, sugar cane and agricultural residues in the presence of yeast. Biodiesel is a liquid fuel obtained from plant oils by transesterification process. Biogas is produced by the anaerobic fermentation of organic materials with the help of bacteria. Methane gas (60 %) is the important constituent of biogas, besides CO_2 (38 %) and some trace gasses (2 %). Producer gas is obtained from the thermal decomposition of biomass in the presence of controlled and limited supply of air. The producer gas is mainly composed of CO and H_2. Biofuel cells are the electro-chemical devices that convert the chemical energy of biomass directly to electrical energy without an intermediate combustion or thermal cycle. With no internal moving parts, fuel cells operate similar to batteries. An important difference between the batteries and fuel cell is that batteries store energy, while fuel cells produce electricity continuously as long as fuel (hydrogen, producer gas) and air are supplied. The fuel cell uses hydrogen produced from biomass through thermo chemical conversion of methane gas from biogas or carbon dioxide from producer gas. Similarly, biomass fueled power plants use the direct burning of biomass with excess air to produce hot flue gases that are used to produce steam in the boilers. The steam generated is usually captured by a turbine and a generator to convert it into electricity. The power generated from the biomass can be used in the form for running of oil engines, tractor, power tiller, self propelled machines, electric motor etc.

1.7.2 Surplus Crop Residues as Energy Source for Farm

Crop residues are of great economic values as livestock feed, fuel and industrial raw material. India being an agriculture-dominant country produces more than 500 million tons of crop residues annually. These residues are used as animal feed, for thatching of houses and as a source of domestic and industrial fuel. A large portion of unused crop residues are burnt in the fields primarily to clear the left-over straw and stubbles after the harvest. Non-availability of labour, high cost of residue removal from the field and increasing use of combines in harvesting the crops are main reasons behind burning of crop residues in the fields. Burning of crop residues causes environmental pollution, is hazardous to human health, produces greenhouse gases causing global warming and results in loss of plant nutrients like N, P, K and S. Therefore, appropriate management of crop residues is highly essential in the farming sector. Now there is therefore a call countrywide, "Earn, Don't burn initiative" for the management and utilization of crop residues.

The utilization of crop residues varies across different states of the country. The residues of cereal crops are mainly used as cattle feed. Rice straw and husk are used as domestic fuel or in boilers for parboiling rice. Farmers use crop residues either themselves or sell it to landless households or intermediaries, who further sell them to industries. The remaining residues are left unused or burnt on-farm. In some states of India, crop residues of cereal crops are not used as cattle feed, a large amount is burnt on-farm. Sugarcane tops are either used for feeding of dairy animals or burnt on-farm for growing a ratoon crop in most parts of the country. Residues of groundnut are burnt as fuel in brick kilns and lime kilns. The residues of cotton, chilli, pulses and oilseed crops are mainly used as fuel for household needs. The shells of coconut, stalks of rapeseed and mustard, pigeon pea and jute and mesta, and sunflower are used as domestic fuel. The surplus residues *i.e.*, total residues generated minus residues used for various purposes, are typically burnt on farm. Estimated total amount of the surplus crop residues in India is 91-141 Mt (million tons). Cereals and fibre crops contribute 58% and 23%, respectively (Figure below) and remaining 19% is from sugarcane, pulses, oilseeds and other crops. Out of 82 Mt surplus residues from the cereal crops, 44 Mt is from rice followed by 24.5 Mt from wheat, which is mostly burnt on-farm. In case of fibre crops (33 Mt of surplus residue) approximately 80% of the residues are from cotton and are subjected to on-farm burning.

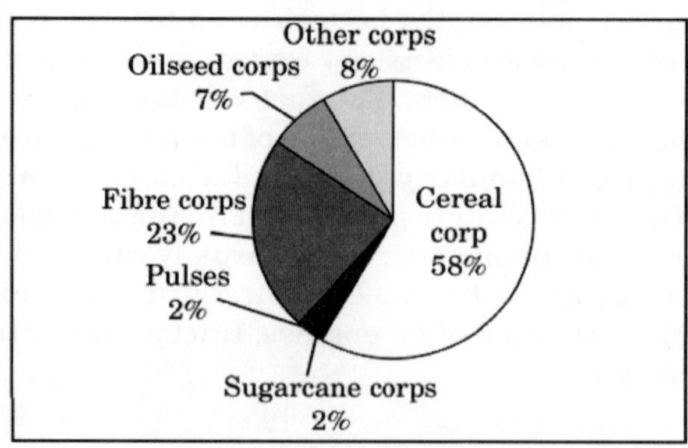

The share of unutilized residues in total residues generated by different crops in India (2012-13)

Now, it is high time to promote utilization of surplus crop residues for various gainful purposes as follows:

- Biomass based power generation
- Paper/board/panel making
- Briquette making for use in brick kilns and furnaces
- Producing bio-ethanol
- Producing producer gas through gasification
- Producing biogas through bio-methanation
- Promote mushroom cultivation and recycling of compost.

1.7.3 Energy Crops/Energy Plantation/Energy Farming

Out of various sources of energy production, energy plantations are indirectly the sustainable sources of energy generation. These are not only renewable, but also eco-friendly. In marginal and wastelands where agricultural crop production is not economic, these are more profitable. Besides giving more economic return from such lands, the plantations consistently improve the site and production system which create congenial micro-climate of the area. A huge parts of land area is available in different countries which come under wastelands and degraded lands and those lands can be put under energy plantations. In India there is more than 47.23 million hectares area under wastelands. There is a wide range of species which can be raised in energy plantations. Some 1200 species have been identified of which 700were highly ranked.

To mitigate global climate change and to act as a substitute for fossil fuels, bioenergy is becoming an important component of national portfolio. Bioenergy plantations are also being considered as a viable option to meet rural energy demands and simultaneously reduce emissions of greenhouse gases (GHGs). Biomass based renewable energy, when produced in an efficient and sustainable manner, has various environmental and social benefits. The ultimate use of bioenergy plantations can be for direct utilization of biomass as fuel, conversion of biomass to energy through gasification or extraction of biofuels to produce convenient and less polluting transportation fuels such as biodiesel whilst continuing to provide biomass for carbon mitigation. The plantations thus sequester carbon in biomass throughout their rotation and abate carbon emissions through replacement of fossil fuels when converted to energy. However, for bioenergy to take advantage of these opportunities, the attitudes towards this potential resource needs to change from the existing perception of being a "poor man's fuel." Energy crops are specifically grown to be used as energy commodities or to offer high output per unit area with lower input. In general, the characteristics of the ideal energy crop are high yield (maximum production of dry matter per hectare), low energy input with higher output, low cost, low nutrient requirements and composition with least contaminants. Plant species that are efficient users of solar energy for converting

CO_2 into biomass, and which can be used as a source of energy, are called energy crops/ energy plantation or energy farming. The nature of biomass obtained from these crops is of the following five types: (1) wood (lingo-cellulose) (2) sugar (3) starch (4) oils and (5) hydrocarbons. Biofuels obtained from such crops that accumulate sugar, starch or oil and are also used as human food or animal feed are called first generation biofuels. The second-generation biofuels are produced from lingo-cellulosic biomass that is not edible and from oil produced from non-food plants like jatropha (*Jatropha curcas*) Karanj, simarua, neem etc. The emerging third generation biofuels are derived from microalgae, which give high biomass yields and do not compete with agricultural production systems. Hence, attempts for raising energy plantation mostly in waste and degraded lands in a large scale would not only favour climate change mitigation approach but also substitute the fast depleting and rising cost of fossil fuels.

1.7.4 Wind Power

Air in motion is called wind. Differences in temperature of air cause wind. Energy derived from wind velocity is called the wind energy. Wind mill extracts energy from the wind and produces mechanical energy. There is a rotor in the wind mill which rotates with the help of flowing air. When the rotating shaft of the rotor is attached to the generator, then mechanical energy can be converted into electrical energy. The threshold (minimum) and cut-out velocity of the wind for which windmill can be used to generate power are 10 km/hour and 35 km/hour respectively. Cut-out velocity means beyond that velocity of wind, the wind energy device cannot produce power and causes damage to it.

The utilization of wind energy in India has been growing at a steady rate over the past few years. The present wind power installed capacity in the country is over 32.7 GW and wind energy constitutes around 55% of the total renewable capacity (about 58 GW) in the country during 2016-17. Wind energy is intermittent and highly site-specific and, therefore, an extensive Wind Resource Assessment Programme is essential for selecting the potential sites. Therefore, MNRE, Government of India, placed emphasis on Wind Resource Assessment since the beginning and today, India has an abundance of data, collected from over 800 wind monitoring stations installed all over India. The recent assessment conducted by National Institute of Wind Energy, Chennai indicates a gross wind power potential of about 302 GW @ 100 m height in the country. Most of this potential exists in seven windy states. The state-wise wind power potential at 100 m height is described in Table 14.

The wind electric generator technology has evolved very rapidly in the country. State-of-the-art technologies are now available for manufacture of wind turbines and all major global players in the field have made their presence in the country. The unit size of machines has gone up to 3.00 MW. Over 50 different models of wind turbines are being manufactured by more than 20 different companies in India. The current annual production capacity of domestic wind turbines is about 10,000 MW. The focus is to promote a technology suitable for low wind regimes of India.

Table 14: Wind power potential in India at 100 m above ground level.

Sl. No.	*States*	*Wind energy potential at 100 m height (GW)*
1	Andhra Pradesh	44.23
2	Gujarat	84.43
3	Karnataka	55.86
4	Madhya Pradesh	10.48
5	Maharashtra	45.39
6	Rajasthan	18.77
7	Tamil Nadu	30.80
Total of 7 windy states		289.96
Other states		12.29
All India total		302.25

Source: https://mnre.gov.in (Ministry of New and Renewable Energy, New Delhi, India)

Water pumping is one of the main applications of wind energy besides electricity generation. Hence the mechanical and electrical energy derived from wind can be utilized for doing various works in the farm. Two designs of wind turbines are available. These are: horizontal axis wind turbine and vertical axis wind turbine. The sixties and seventies saw considerable development activities in the vertical axis wind turbines in India. The vertical axis wind turbines have two main advantages over horizontal axis wind turbines. Firstly, they do not need yaw control to keep rotor pointed towards wind, secondly, the drive train, generator and electrical systems can be located at ground level where maintenance and accessibility is easier. Despite these advantages, the vertical axis wind turbines have not become popular mainly due to high cost and narrow working band of wind velocity. The horizontal axis wind turbines are therefore, very common throughout the world and in our country.

1.7.5 Water Power (Mini hydropower)

Water power is developed by allowing water to fall under the force of gravity. It is used almost exclusively for electric power generation. Potential energy of water is converted into mechanical energy by using the prime mover, known as hydraulic turbine. In hilly areas, where there are natural streams, water power can be tapped through water turbine. The range of mini hydro power is between 2.5 MW to 10 MW. Electric power generation from falling water depends on the volume of water flowing per minute and head or vertical distance of water storage. The scope and adoption of technology for hydropower is huge at national level but is limited at farm level. Only the farms in hilly regions, having water streams can use mini hydropower technology.

1.7.6 Solar Photovoltaic-Wind Hybrid System

As the wind does not blow all the time nor does the sun shine all the time, solar and wind power alone are poor power sources. Hybridizing solar and wind power sources together with storage batteries to cover the periods of time without sun or wind provides a realistic form of power generation. The system creates a stand-alone energy source that is both dependable and consistent. The hybrid solar wind turbine generator uses solar panels that collect light and convert it to energy along with wind turbines that collect energy from the wind. Hybrid solar PV and wind generation system become very attractive solution in particular for stand-alone applications. Combining the two sources of solar and wind can provide better reliability and their hybrid system becomes more economical to run since the weakness of one system can be complemented by the strength of the other one. The integration of hybrid solar and wind power systems into the grid can further help in improving the overall economy and reliability of renewable power generation to supply its load. Similarly, the integration of hybrid solar and wind power in a stand-alone system can reduce the size of energy storage needed to supply continuous power.

Hybrid solar-wind systems can be classified into two types: grid-connected and stand-alone. The integration of combined solar and wind power systems into the grid can help in reducing the overall cost and improving reliability of renewable power generation to supply its load. The grid takes excess renewable power from renewable energy site and supplies power to the site' loads when required. Figures below show the common DC and common AC bus grid-connected to solar PV and wind hybrid system, respectively.

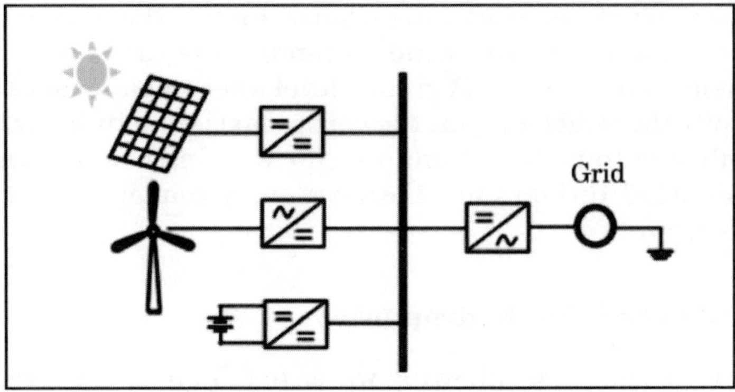

Grid-connected hybrid system at common DC bus

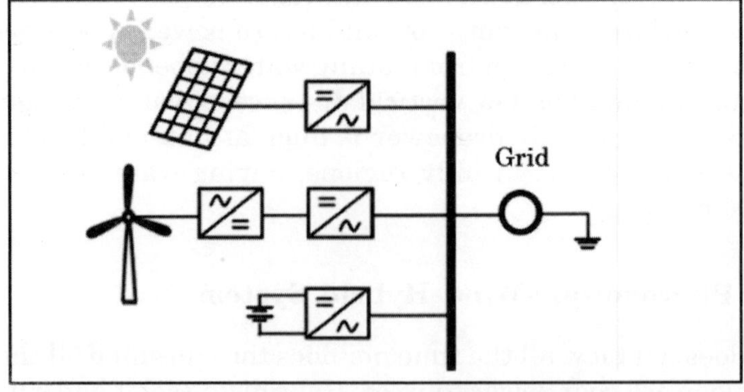

Grid-connected hybrid system at common AC bus

This type of set up with batteries back up can also be established in a farm to obtain reliable source of power throughout the year.

Solar Energy

Advantages	Disadvantages
1. Plentily available in nature	1. Dilute source of energy
2. Non-polluting and has no adverse influence on the environment	2. Availability varies widely with time
3. Low maintenance cost	3. Initial cost is high
4. Decentralized or dispersed power generation at the point of power consumption can save power transmission and distribution costs	4. Efficiency is low compared to conventional power sources
5. Solar energy devices with long life	5. The insolation is unreliable and therefore storage batteries are needed
6. Reliable source of energy	6. Electrical generation cost is high
7. Mostly suitable for remote areas	

Wind Energy

Advantages	Disadvantages
1. Plentily available in nature	1. Initial cost is high
2. Non-polluting and has no adverse influence on the environment	2. Efficiency is low compared to conventional power sources
3. Repair and maintenance is easy	3. Affected by climate and weather conditions
4. Not much technical knowledge is required to operate the wind energy conversion devices	4. It is variable, unsteady, irregular, erratic, intermittent and sometimes, dangerous.
5. Does not need much space	5. Requires large storage batteries for storing electricity and use in the future
6. Operating cost is less compared to conventional energy sources	6. Wind energy devices are located far away from the farm house in vast open areas
7. Mostly suitable for remote areas	7. It is site specific is nature and favorable in geographical locations away from citics.

Example 21: Estimate the amount of solar energy available per year in India? Geographical area of India is 3287×10^3 square kilometre and annual average

global solar radiation on the horizontal surface, incident over India is about 5.5 kWh per square metre per day.

Solution:

The daily average solar energy incident in India varies from 4-7 kWh per square metre depending upon the location. The annual average global solar radiation on the horizontal surface, incident over India is about 5.5 kWh per square metre per day. This makes solar energy in India to be a viable option to generate a clean and cheap source of electricity.

Geographical area = 3287×10^3 km^2 = $3287 \times 10^3 \times 10^6$ m^2

Average solar radiation available per day per square metre = 5.5 kWh

Solar energy available per year in India = $3287 \times 10^3 \times 10^6$ m^2 × 5.5 kWh/m^2-day × 365 days/year = 6598652.5×10^9 kWh/year ... 6500×10^{12} = 6500 trillion kWh/ year.

Example 22: Solar radiation of 700 w/m^2 incident on a flat surface of 100 m^2 area is making an angle of 60 ^{0} with the normal to the surface. Calculate the power collected by the surface?

Solution:

The component of solar radiation incident normal to the surface is **700** × cos 60^0 = 350 W/m^2

Power = solar radiation incident normal to the surface × Area of the surface; P = 350 × 100 = 35 kW

Example 23: Calculate the energy associated with photons that have a wavelength, ë of 0.55 micrometre (the middle of visible portion of the light from the sun)?

Solution:

$E = hf = h\frac{c}{\lambda}$ where h is the Planck's constant and c is the velocity of light

$$E = \frac{6.63 \times 10^{-34} \times 3 \times 10^{8}}{0.55 \times 10^{-6}} = 3.62 \times 10^{-19}\ \text{J} = \frac{3.62 \times 10^{-19}}{1.60 \times 10^{-19}\ \text{J/eV}} = 2.26\ \text{eV}$$

Example 24: A PV system supplies power to a dc motor to produce 1 hp power at the shaft. The motor efficiency is 85 %. Each module has 36 multi-crystalline silicon solar cells arranged in a 9 × 4 matrix. The cell size is 120 mm × 120 mm and cell efficiency is 13 %. Calculate the number of modules required in the PV array? Assume total (global) solar radiation incident normally to the modules as 1 kW/m^2.

Solution:

Motor output power = 1 hp = 746 W

Electric power required by the motor = 746/0.85 = 877.64 W

Cell area in one module = $9 \times 4 \times 120 \times 120 \times 10^{-6} = 0.52\ m^2$

Let N number of modules be required. Solar radiation incident on module = 1 kW/m^2

Cell efficiency = 13 % = 0.13

Output from the solar panel = $1000 \times 0.52 \times N \times 0.13 = 67.6 \times N$

The output of solar array is the input to the motor = $67.6 \times N = 877.64 \rightarrow N = 12.98 \approx 13$

Therefore, 13 modules are required in the panel.

Example 25: Calculate the power produced per square metre of windmill disk area for a windmill operating at 70 % of the theoretical maximum efficiency when the wind velocity is 10 m/s.

Solution:

The power contained in the flowing wind depends on its velocity (V) and density of air (ρ). It is given by the formula;

$$P(\text{watt}) = \frac{1}{2}\rho V^3 A$$

Where A is the area covered by the windmill disk of windmill (swept area). The density of air = 1.21 kg/m^3. Hence P (watt) = $0.6\ V^3\ A$

Theoretical maximum efficiency of windmill is 59 %. Hence power produced per m^2 of windmill disk area of windmill, $P = 0.7 \times 0.59 \times 0.6 \times 10^3 \times 1 = 0.25$ kW/m^2.

Example 26: Calculate the theoretical power available for a windmill having 5 m rotor diameter and receiving wind with an average speed of 12 kmph?

Solution:

The power contained in the flowing wind is formulated as

$$P(\text{watt}) = \frac{1}{2}\rho V^3 A$$

Where A is the area covered by the rotor of windmill (swept area) m^2, V is the velocity of wind in m/s. density of air (ρ) in kg/m^3 = 1.21 kg/m^3. Hence theoretical power available at the windmill is

P (watt) = $(½) \times 1.21$ kg/m^3 $\times (12 \times 1000/3600)^3 \times \pi/4 \times 5^2 = 424$ watts.

Example 27: Calculate the flow rate of water required for generating 1 kW of electrical power if the water falls from a vertical distance of 90 m? Assume 80 % conversion efficiency.

Solution:

The potential energy in joules of a mass, *m*, at a height, h, is *mgh*, where $g = 9.8$ m/s^2, *m* is in kg and *h* in metres. If the flow rate is 1kg/s, the power in the stream of water after a fall from 90 m is

Power = energy/time = mgh/s = (1 kg/s) × (9.8 m/s^2) × 90 m = 882 J/s = 882 W. If this is converted to electricity at 80 % efficiency, the electrical power produced will be 0.8 × 882 = 706 W at a flow of 1 kg/s. To generate 1 kW, the flow rate becomes (1000 W) / (706 W per kg/s) = 1.41 kg/s. This is equal to 1.41 litres/s.

Example 28: Calculate the area of flat plate collector required for a solar water heater to get hot water of 250 litres per day at 70 °C. The temperature of inlet water is 25 °C, daily solar insolation = 5.5 kWh/m^2-day and efficiency of solar water heater = 45 %.

Solution:

Heat required for getting water at 70 °C from 25 °C = $m \times C_w \times \Delta T$ where m is the mass of water in kg, C_w specific heat of water (4.18 kJ/kg per degree rise in temperature), ΔT temperature rise. Mass, *m* of water in the present case = 250 kg, $\Delta T = 70 - 25 = 45$ °C.

Required heat energy = 250 × 4.18 × 45 = 47025 kJ per day

Useful energy obtained = Daily solar insolation × solar water efficiency = 5.5 × 0.45 = 2.47 kWh/m^2-day = 8892 kJ/m^2-day (1 Wh = 3600 joules)

Collector area required = (Required heat energy) / (Useful heat energy) = 47025/8892 = 5.28 m^2

Example 29: Calculate the number of module (each module of 75w_p) required for supplying power to operate a pump of 60% efficiency in order to lift 60 m^{3}of water at a height of 5 m for a period of 4 hours.

Solution:

The potential energy to be given to the lifting water = mass of water × acceleration due to gravity × total vertical lift of water = density of water × volume of water to be lifted × acceleration due to gravity × total vertical lift $= \rho\, v\, g\, h$

Hydraulic power required = hydraulic energy /time

$$= \frac{\left(\frac{1000\text{ kg}}{\text{m}^3}\right)\left(60\text{ m}^3\right)\left(9.8\frac{\text{m}}{\text{s}^2}\right)(5\text{m})}{4\text{ hour} \times 3600\text{ s}} = 204\text{ W}$$

Pump efficiency = 60%, hence actual wattage required for PV module = $\frac{204}{0.6}$ = 340 W

No. of modules required=340/75 = 4.5 H ≈ 5

Hence the no. of modules required for the above purpose is 5.

Example 30: Estimate the biomass production in the earth surface per year. The earth and its atmosphere receive continuously 1.7×10^{17} watts of solar power from the sun, out of which 1.2×10^{14} watts power is utilized in photosynthesis process. The biomass production resulting from photosynthesis is basically glucose (carbohydrate) with an energy content of 3744 cal/g.

Solution:

Biomass production per year in earth surface = $(1.2 \times 10^{14}$ J/s) × (1cal/4.184 J) × $(3.15 \times 10^7$ s/year) × (1 g/3744 cal) = 24×10^{16} g/year … 250×10^9 ton/year.

Example 31: Calculate the power developed by a biogas digester with dry mass input of 20 kg per day, retention time of 30 days, burner efficiency of 70% and methane proportion of 80%. The heat of combustion of methane is 23MJ/m^3.

Solution:

Dry mass input per day = 20 kg; Retention time = 30 kg; biogas yield = 0.22 m^3/kg of dry mass; burner efficiency = 70%, methane proportion = 80%, heating value of methane = 23 MJ/m^3.

Gas produced = 20 × 0.22 =4.4 m^3/day

Thermal energy available = 4.4×23×0.7 = 70.84 MJ/day

Thermal power developed = $\frac{70.84 \times 10^6 \text{J}}{24 \times 3600 \text{ s}}$ = 820 watt

Example 32: Calculate the volume of a fixed dome type biogas digester from the excreta of four cows. Also calculate the thermal energy available from biogas per day. Use the following data, Retention time = 30days, Cow dung produced from one cow = 10 kg/day, Collection efficiency of cow dung = 70 %, Biogas yield = 0.04 m^3/kg of fresh cow dung, Density of slurry = 1050 kg/m^3 Burner efficiency = 70%, Heating value of biogas = 23MJ/m^3

Solution:

Cow dung available per day = Number of cows × dung produced per cow per day × collection efficiency = 4 × 10 × 0.7 = 28 kg. Preparing slurry in the ratio of 1:1 of cow dung and water, the daily input to the biogas digester = 28 + 28 = 56 kg. The retention period is 30 days. Total input material to the digester = 56 × 30 = 1680 kg. = 1.6 m^3 (density of slurry =1050 kg/m^3). Hence the volume of biogas digester = 1.6 m^3

Biogas yield = 0.04 × 28 = 1.12 m^3 per day.

Thermal energy per day = volume × burner efficiency × heating value of biogas = 1.12 × 0.7 × 23 = 18 MJ/day.

Example 33: Calculate the amount of thermal energy required for drying of 1000 kg paddy from the moisture content of 24 % (d.b.) to 12 % (d.b.). Also calculate the amount of moisture evaporated and final weight of paddy after drying.

Solution:

Moisture content indicates the amount of moisture present in a material or produce. It can be expressed either in percentage or decimal on wet or dry basis. The moisture content in wet basis is generally used for commercial purposes. For engineering research purposes, the moisture content on dry basis is preferred and used. This is so because as the moisture is removed or evaporated, the total weight of produce changes while the weight of dry material remains constant and hence the weight change associated with each percentage point of moisture reduction on dry basis becomes constant.

% moisture content (wet basis) = [weight of water in product] / [weight of product sample] × 100

% moisture content (dry basis) = [weight of water in product] / [weight of dry matter of product sample] × 100

The relationship between moisture content of dry and wet basis is given by

$M_d (\%) = \{(M_w) / (100 - M_w)\} \times 100$; also M_d (decimal) = $\{M_w$ (decimal)$\} / \{(1 - M_w$ (decimal)$\}$

$M_w (\%) = \{(M_d) / (100 + M_d)\} \times 100$; also M_w (decimal) = $\{M_d$ (decimal)$\} / \{(1 + M_d$ (decimal)$\}$

where M_d is moisture content on dry basis and M_w is moisture content on wet basis

Dry weight of the produce = Initial weight × $\{(1- M_w$ (decimal)$\}$

Dry weight of the produce = (Initial weight) / $\{(1+ M_d$ (decimal)$\}$

Latent heat of water evaporation = 2260 kJ/kg

Weight of product = 1000 kg, initial mc = 24 % (db) and final mc = 12 % (db)

24 % (db) can be converted to wet basis as (24/124) × 100 = 19.35 % (wb)

Weight of dry matter = 1000/1.24 or 1000 × (1- 0.1935) = 806.5 kg

Amount of moisture to be evaporated = {(weight of dry matter) × (initial mc (d.b.) – final mc (d. b))} /100 = 806.5 (24-12)/100 = 96.78 kg

Final weight of paddy = 1000 – 96.78 = 903.22 kg

Thermal energy required for removal of 96.78 kg moisture = 2260 × 96.78 = 218722.8 kJ.

Example 34: Calculate the amount of thermal energy required for drying of paddy with moisture content of 38 % (w.b.) to produce 1500 kg of final product with moisture content of 14 % (w.b.). Also calculate the initial weight of the product.

Solution:

The dry weight of the product will remain same irrespective of changes in moisture content. The dry weight for the final product of 1500 kg with the moisture content of 14 % (w.b.) = 1500 × (1-0.14) = 1290 kg.

Initial weight of the product with moisture content of 38 % (w.b.) = (Dry weight) / {(*1*- M_w (decimal)} = (1290) / (1-0.38) = 2080.65 kg

Amount of moisture removed is equal to the product of initial dry matter and difference in dry basis moisture content.

Moisture content of 38 % (w.b.) = 61.29 % (d.b.)

Moisture content of 14 % (w.b.) = 16.27 % (d.b.)

The amount of moisture removed after drying to 14 % m.c (w.b.) = 1290 × (61.29 – 16.27)/100 = 580.75 kg.

Thermal energy required for removal of 580.75 kg moisture = 2260 × 96.78 = 1312495 kJ.

1.8 ESTIMATING COST OF USING FARM POWER DEVICES

The cost of using farm power devices consists of expenses for ownership, operation and overhead charges. Ownership costs are independent of the use of the machines and are often called fixed costs. Costs for operating a machine vary directly with use and are referred to as variable costs. The items under fixed cost are to be charged irrespective of the use of machinery whereas the operating cost items are charged only on the basis of working hours. A summary of cost items is given below.

(a) Fixed cost: i) Depreciation, ii) Interest on investment, iii) Insurance and taxes (property, registration and road), iv) Housing.

(b) Variable cost: i) Fuel, ii) Lubricating oil, iii) Repair and maintenance, iv) Wages and labor charges.

(c) Overheads

Fixed Costs

Depreciation: This cost reflects the reduction in the value of a machine with use (wear) and time (obsolescence). The value of a machine is reduced due to wear, weathering, accidental damage, obsolescence etc. It is charged so that the owner may have some capital in his pocket to replace the old one at the end of its life for buying the new machine. The depreciation also depends on the sale value of the machine after its use. Depreciation can be estimated on the basis of different computational methods. The method given below is based on straight-line method.

$$D = \frac{C - S}{L \times H}$$

Where D = Depreciation cost per hour; C = Capital investment; S = Salvage or junk or residual or resale value; L = Useful life of machine in years; H = Number of working hours per year. Salvage value is the approximate value of the machine after the expiry of its useful life. It is approximately taken 5 % of purchase price.

Interest on investment: Annual charges of interest should be calculated on the basis of the actual rate of interest payable. If this information is not available, interest is charged at the rate of 10 % of the average value of machine in the first and last years.

$$\text{Interest I} = \frac{C + S}{2} \times \frac{i}{H} \text{ where i = \% rate of interest per year}$$

Insurance and taxes: Machines are insured to avoid risk. Insurance and taxes are charged at the rate of 5 % of the purchase price of machine.

Housing: Housing charges depend on the size of the machine and its necessity to provide a shed for its shelter. Machines should be kept in a shed in order to protect it from rains, sun, for which charge for housing needs to be added. Housing cost is calculated on the basis of 1.0 per cent of the purchase price of machine.

Variable cost (Operating cost)

Fuel consumption: Consumption of fuel is recorded by observing daily fuel consumption during actual working of the machine. It is common practice to consider average fuel consumption from the varying load test. The cost of fuel is charged on the basis of the prevailing price.

Lubricating oil: The actual oil consumption should be recorded while the machine is working. In case, oil consumption data are not available, the cost towards oil consumption may be charged as 20 % of the cost of fuel consumption per unit working time.

Repair and maintenance: Repair and maintenance expenditures are necessary to keep a machine operable due to wear, parts failure, renewal of tyres, accidents etc. The costs of restoring a machine are highly variable. Good machinery management may keep cost low. The repair and maintenance cost may be charged at the rate of 6% of the purchase price of the machine.

Labor charges: Labor charges are calculated on the basis of the actual prevailing wages of the operator.

Overhead: Overhead charges include supervision, establishment (farm office, management staff). It is assumed to be 20 % of the sum of the fixed and variable costs. In many situations, the overhead charges are neglected.

Adding all those costs, an approximate cost of operating a particular farm power device can be found out.

Example 35: Calculate the cost of operating a 42 hp tractor with the following information;

Sl. No.	*Item*	*Information*
1.	Cost of tractor (Mahindra and Mahindra)	Rs. 6,00,000
2.	Life of tractor (L)	10 years
3.	Annual working hour (H)	1000
4.	Interest rate (i)	10 %
5.	Insurance and taxes	5 % of purchase value of tractor
6.	Housing	1 % of purchase price of tractor
7.	Fuel consumption	4.5 litres/hour
8.	Price of diesel	Rs. 55/litre
9.	Wages of the operator	Rs. 250/day

Solution:

A. Fixed Cost

Sl.No.	*Parameters*	*Cost (Rs./hour)*
1	**Depreciation** $D = \frac{(C - S)}{L \times H}$; Capital cost C = Rs. 6,00,000; Salvage value S = 5 % of purchase cost (Rs. 30000); Life of tractor L = 10 years; Annual working hours, H = 1000 hours	57
2	**Interest** $I = \frac{C + S}{2} \times \frac{i}{100} \times \frac{1}{H}$; (i is percentage rate of interest per year (10 %)	31.5
3	**Insurance and taxes** $= C \times \frac{5}{100} \times \frac{1}{H}$	30
4	**Housing** $= C \times \frac{1}{100} \times \frac{1}{H}$	6.0
Total Fixed cost (Rs. Per hour)		**124.50**

B. Variable Cost

Sl. no.	*Parameters*	*Cost (Rs./hour)*
1	**Fuel** @ Rs. 55 per litre for 4.5 litres per hour	247.50
2	**Lubricants** (20 % of the cost of fuel consumption per unit working time).	49.5
3	**Repair and maintenance** = $C \times \frac{6}{100} \times \frac{1}{H}$ (6 % of purchase price of tractor)	36
4	**Wages** (@ Rs. 250 per day of 8 hours	31.25
Total Variable Cost (Rs. Per hour)		**364.25**

Total cost of operating tractor per hour = Fixed cost per hour + variable cost per hour = Rs. 124.50 + Rs.364.25 = Rs. 488.75 ... **Rs. 488/hour**

Example 36: Calculate the cost of operating a 12 hp power tiller with the following information;

Sl. no.	*Item*	*Information*
1	Cost of power tiller (VST)	Rs. 1,80,000
2	Life of tractor (L)	8 years
3	Annual working hour (H)	800
4	Interest rate (i)	10 %
5	Insurance and taxes	5% of purchase price of power tiller
6	Housing	1 % of purchase price of power tiller
7	Fuel consumption	1.5 litres/hour
8	Price of diesel	Rs. 55/litre
9	Wages of the operator	Rs. 250/day

Solution:

A. Fixed Cost

Sl. no.	*Parameters*	*Cost (Rs./hour)*
1	**Depreciation** $D = \frac{(C - S)}{L \times H}$; Capital cost C = Rs. 1,80,000; Salvage value S = 5 % of purchase cost (Rs. 9000); Life of power tiller L = 8 years; Annual working hours, H = 800 hours	26.71
2	**Interest** $I = \frac{C + S}{2} \times \frac{i}{100} \times \frac{1}{H}$; (i is percentage rate of interest per year (10 %)	11.81
3	**Insurance and taxes** = $C \times \frac{5}{100} \times \frac{1}{H}$	11.25
4	**Housing** = $C \times \frac{1}{100} \times \frac{1}{H}$	2.25
Total Fixed Cost (Rs. Per hour)		**52.03**

B. Variable Cost

Sl. no.	Parameters	Cost (Rs./hour)
1	**Fuel** @ Rs. 55 per litre for 1.5 litres per hour	82.50
2	**Lubricants** (20 % of the cost of fuel consumption per unit working time).	16.5
3	**Repair and maintenance** = $C \times \frac{6}{100} \times \frac{1}{H}$ (6 % of purchase price of power tiller)	13.5
4	**Wages** (@ Rs. 250 per day of 8 hours	31.25
Total Variable Cost (Rs. Per hour)		**143.75**

Total cost of operation per hour = Fixed cost per hour + variable cost per hour = Rs. 52.03 + Rs.143.75 = Rs. 195.78 ≈ **Rs. 195/hour**

Example 37: Calculate the cost of operating a 42 hp tractor operated rotavator with the following information of rotavator and refer Example 30

Sl. no.	Item	Information
1	Cost of rotavator	Rs. 1,00,000
2	Life of rotavator (L)	8 years
3	Annual working hour (H)	400
4	Interest rate (i)	10 %

Solution:

A. Fixed Cost

Sl. no.	Parameters	Cost (Rs./hour)
1	**Depreciation** $D = \frac{(C - S)}{L \times H}$; Capital cost C = Rs. 1,00,000; Salvage value S = 5 % of purchase cost (Rs. 5000); Life of rotavator L = 8 years; Annual working hours, H = 400 hours	29.68
2	**Interest** $I = \frac{C + S}{2} \times \frac{i}{100} \times \frac{1}{H}$; (i is percentage rate of interest per year (10 %)	13.12
Total Fixed cost (Rs. Per hour)		**42.81**

B. Variable Cost

Sl. no.	Parameters	Cost (Rs./hour)
1	**Repair and maintenance** = $C \times \frac{6}{100} \times \frac{1}{H}$ (6 % of purchase price of rotavator)	15
Total Variable Cost (Rs. Per hour)		**15**

Total cost of operating a rotavator per hour = Fixed cost per hour + variable cost per hour = Rs. 42.81 + Rs.15 = Rs. 57.81 ≈ **Rs. 57/hour**

Total cost of operating tractor operated rotavator per hour = Total cost of operating tractor per hour + Total cost of operating rotavator per hour = Rs. 488 + Rs.57 = Rs. 545

Capacity of tractor operated rotavator = 1 acre/hour (0.4 ha/hour)

Total cost of operating a tractor operated rotavator = Rs. 545/Acre (Rs. 1362/hectare)

Example 38: Determine the break-even point and pay-back period of operating a 42 hp tractor as per Example 30, if it is used on custom hiring @ Rs. 550 per hour.

Solution:

Break- even point (hours/annum) =

$$\frac{\text{Total annual fixed cost (Rs. / annum)}}{\text{Custom hiring rate (Rs. / hr)} - \text{Variable cost (Rs. / hr)}}$$

Fixed cost of tractor = Rs.124.50/hour

Variable cost of tractor = Rs.365/hour

Annual working hour of tractor = 1000 hours

Custom hiring rate of tractor = Rs. 550/hour

$$\textit{Break- even point (hours/annum)} = \frac{124.50 \times 1000}{550 - 365} = 672.97 \text{ H} \approx 675 \text{ hours}$$

Pay Back Period

Purchase cost of tractor = Rs. 6, 00,000

Net benefit per hour = custom hiring rate (Rs./hour) – variable cost (Rs./hour)

Net benefit per annum = Annual working hour × net benefit per hour

$$\textit{Pay- back period (years)} = \frac{\text{Investment Cost (Rs.)}}{\text{Net benefit (Rs. / year)}} = (6,\ 00{,}000)/(1000)\ (550 - 365)$$

$$= 3.25 \text{ years}$$

Example 39: Calculate the comparative cost per hour and per kW-h in using various power sources such as human, draught bullock, tractor, power tiller, diesel engine, electric motor, solar PV system and wind aero-generator.

Solution:

(i) **Calculation of cost of operation of draught bullock pair**

Assumptions:

Annual hours of usage	= 240 h
C= Cost of bullock pair	= Rs.40, 000
S = Salvage or junk value	= 10%
Operational hours per day	= 6 h
Life span	= 12 yrs

Calculation of Fixed Cost

Depreciation $D = \frac{(C - S)}{L \times H}$ = Rs. 12.5/h

Interest $I = \frac{C + S}{2} \times \frac{i}{100} \times \frac{1}{H}$; (i is percentage rate of interest per year (10 %) = Rs. 9.16/h

Housing (10 % of the C) $= C \times \frac{10}{100} \times \frac{1}{H}$ = Rs. 16.66 /h

Operational cost

Health and maintenance cost @5% of C =Rs. 8.33/h

Labour cost @ Rs.250/person/ day of 6 hours = Rs. 41.66/h

Feeding cost per day

8 kg paddy straw @ Rs.90/quintal	= Rs. 7.2
Cattle feed 2 kg @ Rs.2000/quintal	= Rs. 40.00
Total feed cost per day	= Rs. 47.20
Cost of feeding per hour	= Rs. 7.86
Total cost of operation for bullock pair	= 12.5 + 9.16 + 16.66 + 8.33 + 41.66 + 7.86
	= Rs. 96.17/h

(ii) **Calculation of cost of operation of diesel engine (1 hp or 746 watt)**

Assumptions:

Annual hours of usage (H) = 500 h

C= Cost of 1 hp diesel engine = Rs.7, 000
S = Salvage or junk value = 10%
I = Interest rate = 10 %
Operational hours per day = 6 h
Life span = 8 yrs
Fuel consumption = 250 ml /h
Engine efficiency = 30 %
Device efficiency = 70 %

Calculation of Fixed Cost

Depreciation $D = \frac{(C - S)}{L \times H}$ = Rs. 1.57/h

Interest $I = \frac{C + S}{2} \times \frac{i}{100} \times \frac{1}{H}$; (i is percentage rate of interest per year (10 %) = Rs. 1.54/h

Total fixed cost = 1.57 + 1.54 = Rs. 3.11/h

Operational cost

Fuel cost per hour = (BHP) × fuel consumed in litres/h/bhp × cost of fuel/litre = 1 × 0.25 × 60 = Rs. 15

Lubricants = 20 % of cost of fuel per unit time = Rs 3/h

Repair and maintenance = 10 % of C = (C) × (10/100) × (1/H) = Rs. 1.4/h

Operator's wages = Rs250/6 = Rs.41.66/h

Total operational cost/h = 15 + 3 + 1.4 + 41.66 = Rs. 61.06

Total cost of operation/h = Fixed cost per hour + Operational cost per hour = 3.11 + 61.06 = Rs.64.17/h.

Cost of operation per kW-h = Rs. 86.01

(iii) **Calculation of cost of operation of electric motor (1 hp or 746 watt)**

Assumptions:

Annual hours of usage (H) = 500 h
C= Cost of 1 hp electric motor = Rs.5, 000
S = Salvage or junk value = 10 %
I = Interest rate = 10 %

Operational hours per day	= 6 h
Life span	= 8 yrs
Motor efficiency	= 70 %
Device efficiency	= 70 %

Calculation of Fixed Cost

Depreciation $D = \dfrac{(C - S)}{L \times H}$ = Rs. 1.12/h

Interest $I = \dfrac{C + S}{2} \times \dfrac{i}{100} \times \dfrac{1}{H}$; (i is percentage rate of interest per year (10 %) = Rs. 0.55/h

Total fixed cost = 1.12 + 0.55 = Rs. 1.67/h

Operational cost

Energy consumption (kWh) cost per hour = [(BHP) / (motor efficiency)] × 0.746 × 1 hour × unit cost of electricity = [1/(0.7)] × 0.746 × 1 × 5 = Rs. 5.32/h

Lubricants = 20 % of cost of electricity per unit time = Rs 1.06/h

Repair and maintenance = 10 % of C = (C) × (10/100) × (1/H) = Rs. 1.0/h

Operator's wages = Rs250/6 = Rs.41.66/h

Total operational cost/h = 5.32 + 1.06 + 1.0 + 41.66 = Rs. 49.04

Total cost of operation/h = Fixed cost per hour + Operational cost per hour

= 1.67 + 49.04 = Rs.50.71/h

Cost of operation per kW-h = Rs. 67.97

(iv) **Calculation of cost of operation of solar PV system (1 kW capacity)**
The cost of modules of 1 kW capacity = Rs. 50,000 @ Rs. 50/- per watt. Cost of other accessories like attached device, battery, charge controller, wires and fitting charges =Rs. 30,000. Total cost of the PV system C = Rs. 80,000/-, Life of system= 25 years, Annual working hours of the attached device (H) = 500 hours, Salvage value = 5 % of initial cost, Interest rate = 10%, Repair and maintenance = 0.5 % of initial cost.

Calculation of Fixed Cost

Depreciation $D = \dfrac{(C - S)}{L \times H}$ = Rs. 6.08/h

Interest $I = \frac{C+S}{2} \times \frac{i}{100} \times \frac{1}{H}$; (i is percentage rate of interest per year (10 %) = Rs. 8.4/h

Total fixed cost = 6.08 + 8.4 = Rs. 14.48/h

Operational Cost

(i) Fuel cost = Nil

(ii) Lubricants = Nil

(iii) Repair and maintenance = (C) × (0.5/100) × (1/H) = Rs. 0.8/hour

(iv) Operator's wages Rs. 250/6 = Rs. 41.66/hour

Total variable cost = 0.8 + 41.66 = Rs. 42.46/hour

Total operation cost per hour of solar PV system with an attached device = Total fixed cost/hour + total variable cost/hour = 14.48 + 42.46 = Rs. 56.94/h

(v) **Calculation of cost of operation of wind aero-generator (1 kW capacity)**

The cost of wind aero-generator of 1 kW capacity = Rs. 1, 50,000, Life of system= 20 years, Annual working hours of the attached device (H) = 500 hrs, Salvage value = 5 % of initial cost, Interest rate = 10%, Repair and maintenance = 2 % of initial cost.

Calculation of Fixed Cost

Depreciation $D = \frac{(C-S)}{L \times H}$ = Rs. 14.25/h

Interest $I = \frac{C+S}{2} \times \frac{i}{100} \times \frac{1}{H}$; (i is percentage rate of interest per year (10 %) = Rs. 15.75/h

Total fixed cost = 14.25 + 15.75 = Rs. 30/h

Operational Cost

(i) Fuel cost = Nil

(ii) Lubricants = Nil

(iii) Repair and maintenance = (C) × (2/100) × (1/H) = Rs. 6/hour

(iv) Operator's wages Rs. 250/6 = Rs. 41.66/hour

Total variable cost = 6 + 41.66 = Rs. 47.66/hour

Total operation cost per hour of solar PV system with an attached device = Total fixed cost/hour + total variable cost/hour = 30 + 47.66 = Rs. 77.66/h

Comparative Cost of Operating Various Sources of Power

Sl.no.	*Sources of Power*	*Number/units equivalent to 1 kW*	*Cost of using per hour (Rs.)*	*Cost of using per kW-hour (Rs.)*
1	Tractor (32 kW)	______	488.00 (Example-1.30)	15.00
2	Power tiller (9 kW)	______	195.00 (Example-1.31)	22.00
3	Electric motor (0.75 kW) without load	______	51.00 (Example-1.34)	68.00
4	Solar PV system (peak capacity 1 kW), (actual power generation = 0.75 of peak capacity = 0.75 kW)	1	57.00 (Example-1.34)	76.00
5	Diesel engine (0.75 kW) without load	______	64.00 (Example-1.34)	86.00
6	Wind aero-generator (1 kW) (Efficiency to be 60 %)	1	78.00 (Example-1.34)	130
7	Draught bullock	A pair of bullocks (developing 0.3 kW power per bullock)	96.00/pair (Example-1.34)	160.00
8	Human	13 @ 75 watt per person for 6 hours of operation per day	42.00 @ Rs. 250 labour charge per person per day of 6 hours	542.00

1.9 ENERGY SAVING APPROACHES FOR MACHINERY USE IN FARM

With the rise in the input of power in Indian agriculture, the necessity of farm equipment in performing various farm operations has also increased. Various types of tools and implements such as rotavators, seed drill/planters, hoes and weeders, sprayers/dusters, reapers, threshers and combine harvesters have been introduced. Many manufacturers in the country have been engaged in popularizing big size machines to meet the needs of the farming community with respect to quantity and quality of work without paying much attention to the conservation and saving of energy. Availability of energy is now-a-days a scarce commodity and its sustainability is a major concern. The demand of energy in the agricultural sector has been increased many folds in the last decade, due to the rapid growth of power driven machinery. Due to the increase in power availability on Indian farms (2.02 kW/ha, 2014-15), cropping intensity has increased to 142 % and side by side cropping pattern has also been changed. Now, there is the need of the availability of secured energy sources and the expected improvements in the design of efficient farm equipment to save and conserve energy if the agricultural mechanization to become sustainable.

1.9.1 Some Aspects on Energy Saving Measures

In the present scenario of Indian agriculture, various types of power driven machinery have been introduced in the farming and most of them use power mainly from mechanical or electrical sources. In some cases, increasing the efficiency of a single piece of equipment would result into saving and conservation of power to a greater extent. In other situations, many small improvements in efficiency and conservation across the farm can add up to meaningful reductions in energy use and operating costs. Tractors and related field equipment can use a lot of energy on the farm, so it is high time to take necessary steps to optimize their efficiency. These include.

- Selecting the proper sizes and makes of tractor and equipment, travel and engine speed
- Reducing the number of field operations (*i.e.*, reduced-till or zero-till farming etc.)
- Reducing tillage depth
- Properly adjusting and maintaining tractors and other self driven machines.
- Simple adjustments such as keeping tractor tires properly inflated/ballasted to improve tractive efficiency and reducing turning/down time can go a long way towards saving fuel and improving overall field efficiency.
- Efficient field equipment uses
- Adjustments, cleaning, and maintenance of equipment and
- Thoughtful planning to select new or replacement of equipment for the most efficient energy use.

Properly maintaining equipment: Replacing clogged air/fuel filters and cleaning injectors can also reduce fuel use. Timely replacement of air and fuel filters and lubricants can reduce fuel use while increasing horsepower. Repair of leaking valves and piston rings will equally improve engine performance and therefore energy efficiency.

Reducing tillage to save fuel: Conservation tillage generally uses less fuel than full tillage systems because the soil is tilled less intensely and less often. These tillage practices may also allow seedbed preparation, fertilizer application, and seeding in fewer passes. Such practices are also eligible for carbon credits.

Gearing up and throttling back: Increasing the gear and lowering the throttle speed can lead to fuel savings. Make sure not to overload the engine, excessive black smoke indicates overloading.

Optimizing wheel slippage: Some wheel slippage is needed to reduce excess wear on the tires. The optimal level is generally 10%, but the actual level depends on the type of tractor, speed, and the implement being used.

Matching implements size to the tractor: Using a large tractor for light loads is inefficient because extra horsepower is used to move the larger tractor. Producers should consider using a smaller tractor if possible. On the other hand, using a smaller tractor to perform operations that require more horsepower can overload a smaller tractor, reducing its efficiency.

Selecting the optimal engine and travel speeds: Most tractor engines have the highest fuel efficiency when operated at or near rated speed and load, or maximum power. For primary tillage implements properly matched to the tractor, the best fuel efficiency in the field is achieved by pulling loads at the fastest speed possible within the acceptable speed range for the implement.

Optimizing efficiency of field operations: Field efficiency refers to the time the operation takes *versus* turning and other non-productive time. Spending an inordinate amount of time in turning around at the ends of short, wide fields or overlapping tillage operations within a field can result in higher fuel consumption and thus higher energy requirement.

All the above approaches discussed relates to different methods of energy saving techniques and applications. The following examples show a little bit of consciousness and judicious use of power driven farm machinery (particularly engine operated prime movers) will save energy and cost to a considerable extent.

Example 40: Calculate the quantity of diesel fuel saved annually if energy saving tools and implements are used in tractor and power tiller and also calculate the total amount in rupees saved by their use for farming operations of India.

Solution:

(A) Tractor

The total number of tractors in India is 5.27 million (as on March 2014)

Average power of tractor = 40 hp

Average fuel consumption of 40 hp tractor = 4.0 lit / h in load condition.

Average annual use of a tractor is 1000 hours.

Total fuel consumption is = 5.27 × 4 × 1000 = 21080 million litres.

Total diesel cost = 13855.52 × Rs 70/ lit = 1475600 million rupees

If energy saving tools and implements are used, there may be 10-20 % reduction in fuel consumption (average 15%), then total diesel cost will be reduced by 1475600 × 0.15 = **221340 million rupees.**

(B) Power Tiller

The total number of power tillers in India is 0.45 million (as on March 2014)

Average fuel consumption of power tillers = 1.5 lit./ h. in load condition.

Average annual use of a tractor is 800 hours.

Total fuel consumption is = 0.45 × 1.5 × 800 = 540 million litres.

Total diesel cost = 540 × Rs 70/ lit = 37800 million rupees.

If energy saving tools and implements are used and there may be reduction of 10% fuel consumption, then total diesel cost will be reduced by 37800 × 0.10 = **3780 million rupees**.

Thus, total saving of cost on diesel in tractor and power tiller per annum comes to 221340 + 3780 = **225120 million rupees. = Rs. 22,512 crores.**

REFERENCES

1. Anonymous. 2007. Agricultural statistics at a glance. Directorate of Economics and Statistics. Ministry of Agriculture, Govt. of India.
2. Anonymous. 2013. State of Indian Agriculture 2012-13. Agricultural statistics at a glance. Directorate of Economics and Statistics. Department of Agriculture and Cooperation, Ministry of Agriculture, Govt. of India.
3. Anonymous. 2014. Annual Report 2013-14. Ministry of New and Renewable Energy (MNRE), Govt. of India, New Delhi.
4. Campbell, J. P. (1990). Modeling the performance prediction problem in industrial and organizational psychology. In M. D. Dunnette & L. M. Hough (Eds.), Handbook of industrial and organizational psychology (Vol. 1, pp. 687–732). Palo Alto: Consulting Psychologists Press.
5. FAO (Food and Agriculture Organization of the United Nations). 2010. Agriculture scenario in globe.
6. Ghosal M.K. 2017. Renewable Energy Technologies. Narosa Publishing House PVT. LTD. New Delhi, India.
7. Mehta C. R., Chandel N. S., Senthilkumar T. and Singh Kanchan K. 2014. Trends of Agricultural Mechanization in India. CSAM Policy Brief, June 2014.
8. Singh R.S., Singh Surendra and Singh S.P. 2015. Farm Power and Machinery Availability on Indian Farms. *Agricultural Engineering Today*. Vol. 39(1), 2015.
9. Surendra, Singh R.S. and Singh S.P. 2014. Farm Power Availability on Indian Farms. *Agricultural Engineering Today*. Vol. 38(4), 2014.

Chapter 2

Developments and Types of Tractor

2.1 INTRODUCTION

Tractor is basically a traction machine. It is used for traction purpose. The literal meaning of traction is the action of pulling or hauling. Hauling means pulling or dragging with effort. At the end of 19th century, machines for traction purpose were known as 'Traction Motor' and later on, in the beginning of 20th century i.e. towards 1905, these machines were designated as 'tractor' by taking half the word from, traction *i.e.* trac and the other half from motor *i.e.* tor. Hence the machines used for pulling purposes are now a days known as tractor. It is now used to pull or push the agricultural implements in the fields.

The automobile (car, bus or truck) is a self propelled vehicles intended for transporting goods, carrying people etc. But tractor is a wheeled or tracked self-propelled vehicle used as a power source for pulling agricultural implements, constructing roads, moving earth, clearing land, pulling trailer etc. Tractor can be used as the prime mover for active (moving) tools or stationary farm machinery through the intermediary of power take off (PTO) shaft, drawbar or belt pulley. Drawbar is a flat crossbar having a number of 20 mm diameter holes and is attached to the implement or trailer for pulling with the help of tractor.

The term traction usually refers to the pulling force a tractor can develop on the ground. Traction is achieved by good grip between wheels and the surface over which tractor moves". Hence, traction is achieved by friction and grip between the wheel and the ground. The modern wheel type tractor is provided with steel rims and pneumatic tyres as the ground drive component. Pneumatic tyres on the rear axle of the tractor are so designed that they grip the ground by penetration as well as by adhesion. The tractor tyre is also provided with tread bars which penetrate in the soil and result in better traction.

2.2 TYPES OF TRACTORS

Tractors are classified in three different types. These are

1. Type of construction

2. Type of drive,
3. The purpose for which it is used.

1. **Type of construction:** According to the type of construction, the tractor may have seat arrangement for operator to sit and the tractor with which operator walks behind (without seat arrangement for operator).
2. **Type of drive:** There are two types of drives (i) track type (ii) wheel type, Track consists of two heavy endless metal chains moving on two iron wheels. These chains are called tracks. Track is used in crawler tractor or chain type tractor. The track provides large area of contact with the ground and have a good track adhesion; they crush and compact the soil well. It is most suited for heavy work, specially earth moving work and land reclamation work. Wheel type of tractor has pneumatic wheel or cage wheel.
3. **Purpose for which it is used:** This is again subdivided into the following categories (i) Utility tractor (ii) Row-crop tractor (iii) Orchard tractor (iu) Garden tractor (v) Special-purpose tractor.

Utility tractor: It is a general purpose tractor and is designed to do major farm operations like tillage, disking, harrowing, sowing, harvesting and to drive many other stationary equipments through its R.T.O. shaft.

Row crop tractor: It is used for row crop work as well as for many other field tasks. For this purpose, these tractors are provided with replaceable driving wheels of different trend widths i.e., wide for general farm work and narrow for row-crop work. In order to prevent plant damage, the tractors have a high ground clearance and a wide wheel track that can be adjusted to suit the particular inter-row distance.

Orchard type tractor: These are special type of tractors only used in orchard. These are made very high in height so that while sitting on the tractor, operator can easily pluck the fruit and trim the tree. No part of the tractor is protruded so that tractor can easily go in between trees safely.

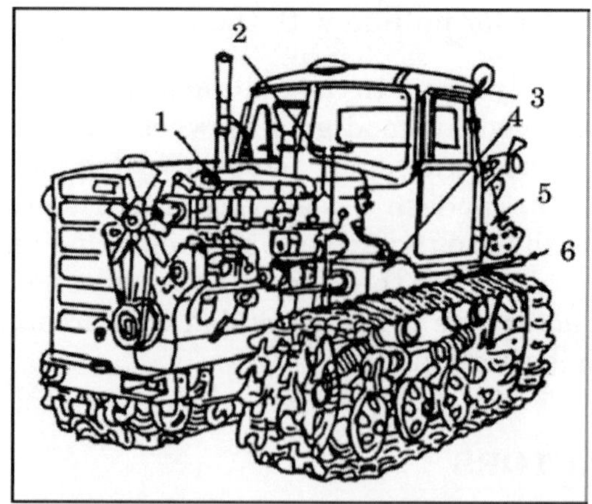

Fig. 2.1: Crawier Tractor

Garden tractors: These tractors are in the range of 1 to 10 H.P and are very small in construction. These are mostly used for grass cutting or making flower beds in the garden. The wheels are also smaller size and like the tyre of the scooter. Power to the rear wheel is transmitted by the belt.

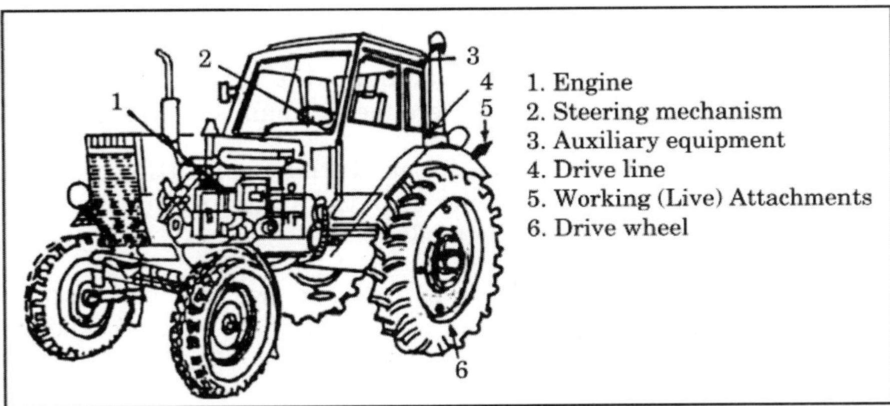

Fig. 2.2: Wheeled Tractor arrangement of three Main Component Parts of the Tractor.

Special purpose tractor: These tractors are the modifications of utility or row-crop tractor models and are used for definite jobs. (e.g. in vineyards, cotton field), or for various jobs under certain conditions (e.g. on forestry, marshy soils, hill sides etc.). Thus special tractors used to mechanize the cultivations of cotton have a single front (steerable) wheel, swamp tractors are equipped with wide tracks enabling them to operate on marshy soils and hillside tractors are designed to work on hillsides sloping at up to 16°.

2.3 PARTS OF TRACTOR

The parts of a tractor are as follows:

(1) Engine
(2) Clutch
(3) Transmission system
(4) Differential
(5) Wheels
(6) Tyres
(7) Steering system
(8) Brake system
(9) Electrical system
(10) Hydraulic lift
(11) Drawbar hitch
(12) P.T.O. shaft
(13) Three point linkage
(14) Control panel.

Engine: A tractor can have petrol or diesel engine fitted in it. But now a day all the tractors are fitted with the diesel engine. Engine provides the source of power to drive the tractors.

Clutch: Clutch is fitted between engine and gearbox and is used to connect or disconnect the engine drive to the gearbox. Engine is disconnected from the transmission for a short period of time while the driver is shifting gears and also to connect smoothly the flow of power from the engine to the driving wheels when starting the tractor from rest.

Transmission: Gearbox is fitted in the tractor to provide the variable speed and torque to the rear wheels. Higher torque is given to the rear wheels in order to enable the tractor to pull more load. According to the field conditions and land requirements, a number of gears are provided.

Control panel: In control panel, hour meter, light switch, horn button, oil pressure indicator, water temperature gauge, ampere meter, main switch are provided.

Differential: Differential helps tractor to take turn by permitting one of the rear wheels of tractor to rotate slower or faster than the other. It also transmits power through right angle drive to two rear wheels.

Wheels: Two wheels are fitted to the front axle and two rear wheels to the rear axle. Engine drive is transmitted to rear axle. That's why rear wheels are called drive wheel. For better traction, rear tyres are provided with lugs.

Tyres: Pneumatic tyres are generally used in tractors. Tyres are like a rubber bag fitted around the rim. Tube is fitted inside tyre and filled with compressed air. They carry load of the vehicles.

Steering system: Steering system enables the operator to steer the tractor to right or left whenever necessary.

Brake system: Brake system is provided in the tractor so as to enable the operator to stop the moving tractor when required.

Electrical system: Electrical system provided in the tractor enables it to work in the night through the electricity produced by the dynamo. Moreover engine is also started through self starter which draws its electric current from the batteries.

Hydraulic lift: With the hydraulic power, the implements are raised up and lowered down or their depths are controlled.

Drawbar hitch: Drawbar is a device by which the pulling power of the tractor is transmitted to the trailing implement. It consists of a crossbar with suitable holes, attached to the lower hitch links. It is fitted at the rear part of the tractor.

Three point linkage: It is a combination of three links, one is the upper link (compression link) and two lower links (tension links). Generally the compression link and one of the tension links are made adjustable to vary the hitch and to connect the implement to the tractor.

P.T.O. (Power take off) Shaft: It is a part of tractor transmission system. It consists of a shaft, a shield and a cover. The shaft is externally splined to transmit torsional power to another machine. A rigid guard fitted on a tractor covers the PTO shaft as a safety device. This guard is called a power take off shield. Agricultural machines are coupled with this shaft at the rear part of the tractor.

Technical terms: Certain technical terms for tractor need to be acquainted and are as follows (Fig. 2.3)

(i) **Track:** It is the distance between centres of front two tyres,

(ii) **Track for dual wheel:** It is the distance between centres of dual wheels.

(iii) **Ground clearance:** It is the smallest distance from the bottom most point of the axle to the ground level.

(iv) **Wheel base:** It is the distance between centres of hubs of front and rear wheel.

(v) **Height of frame:** It is the distance from the upper edge of frame to the ground level.

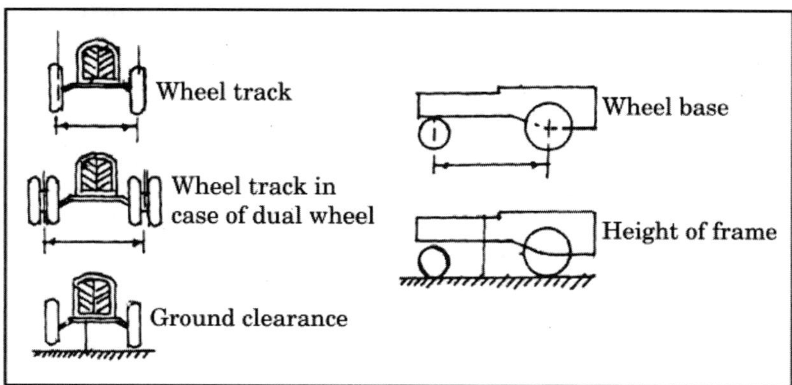

Fig. 2.3: Technical terminologies of a tractor

2.4 DEVELOPMENT OF TRACTOR

Evolution of tractor was thought of with the changes in farm technology and sizes of farm. The tractor has progressed in different stages from its original primary use as a substitute for animal power to the present day requirements of multiple uses. The concept of developing a tractor started in 1855 with the introduction of steam engine for the purpose of operating agricultural threshers in U.S.A. Later on, these engines were made self-propelled for pulling ploughs and were known as self propelled steam traction engines. However, some of the important stages through which tractor has come into its existence are as follows.

1890: Self propelled steam engine with clutch, gear was available under the name of steam ploughs. The term tractor appeared first on record in U.S. patent for steam traction engine.

1906: Successful gasoline tractor was introduced by Charles W. Hart and Charles H. Parr of Charles city, Iowa.

1908: Though steam ploughs and tractor with I.C. engine were in use simultaneously, but during an exhibition in 1908, organized by Winnipeg Industrial Exhibition, field performance of gasoline tractor were compared with these of steam plough.

1911: The first tractor demonstration was held in the United States at Omaha, Nebraska.

1915–1919: The power takeoff was first introduced in tractor.

1920–1924: All - purpose farm tractor was developed.

1930–1937: Diesel engine was used in tractor; pneumatic tyres and higher speeds were introduced.

1937–1941: Three point implement hitch and linkage were introduced; Hydraulic control was also introduced.

1941–1944: Number of lawn and garden tractors were developed.

1950–1960: Manufacturing of diesel tractors on large scale was taken up throughout the world.

1960–1961: Tractor manufacturing was started first in India by M/s Eicher Good Earth Ltd., Faridabad.

1962–1970: The manufacturers like M/s TAPE Madras, Escorts tractor at Faridabad, Hindustan Tractors at Baroda started work during this period.

1971: Escort tractor Ltd. manufactured the tractor in the name of 'Ford Tractor.

1973: Manufacture of HMT tractor was started.

1975: Harsha Tractor Company was established.

1982: Universal tractors were established.

Between 1983 to 2004, 12 tractor manufacturers in India are manufacturing more than 50 model in the range of 15 to 75 HR.

2.5 SELECTION OF TRACTOR

The selection of tractor with the matching implements is an important and difficult job for a buyer. In the midst of so many factors like influence of tractor dealers, bankers, friends, the buyer gets confused in selecting a right tractor for his farm. Hence following points need to be considered while purchasing a tractor.

(i) **Land holding and type of operation:** Among the primary criteria of selection of tractor, the size of land holding is very important. If high horse power tractor is purchased, it remains under-utilized for most of the period in a year causing financial loss in terms of depreciation and interest. Similarly with low horse power tractor, the farm operation cannot be performed in time resulting in the low yield of the crop. But land holding is not the only deciding factor, selection of tractor has also impact on the type of operations to be done in the farm. Tractor is used for various farm operations starting from tillage, sowing, harvesting, threshing and transporting work. Hence while selecting a tractor in terms of hp, it is required to consider the field operation which is toughest and time consuming. Normally, it is the primary tillage operations *i.e.*, ploughing.

The power requirements for different types of farm operations are computed separately and the horse power of tractor is selected on the basis of the highest power requirement of any operations carried out in the farm.

However, it is normally recommended to consider 1 hp for every 2 hectares of land. In other words, one tractor of 20-25 hp is suitable for 40 hectares of farm.

(ii) **Cropping pattern:** Number of crops grown in the farm in a year is considered for selecting a tractor in the farm.. If more than one crop is grown in a year, the time available from the harvest of 1st crop to the sowing of second crop is a very crucial factor for the preparation of land. If a small horse power tractor is available, it may not be possible to prepare the seedbed in time. Generally 1.5 hectare/hp is recommended for the farm where adequate irrigation facilities are available and more than one crop is taken. Hence a 30-35 hp tractor is suitable for 40 hectares farm.

(iii) **Types of soils:** The soil resistance for different soils is different. In heavier soil, it is high and in light soil, it is low. An implement which can give a desired depth of operation in light soil at certain speed will not be able to do the same quantum of work in the same period in heavier soil. In such a case, either speed is to be reduced to allow tractor to be capable of pulling the load at the desired depth or the depth is to be sacrificed. In other words, a tractor capable of performing all operations timely for a specified size of field in light soil, will not be able to do so in heavy soil for the same size of field. Therefore, for a heavier soil, a higher horsepower tractor will be more suitable compared to a tractor needed for same size field of lighter soil.

(iv) **Climatic condition:** In very hot zones and desert area, air cooled engines are preferred over water cooled engines. Similarly for higher altitude, air cooled engines are preferred because water is liable to be frozen at higher altitude.

(v) **Operating cost:** Tractors with less specific fuel consumption should be preferred over others so that operating cost maybe less.

(vi) **Availability of spare parts:** Any mechanical device may be subjected to its parts failure while in running condition. Hence it needs quick replacement of its broken parts, without affecting the farm operations. Good manufacturing companies maintain their sales and service branches throughout the country from where all the parts 'maybe replaced within time. So before buying, the availability of parts and quick services need to be considered. It is also required that the tractor to be purchased has a dealer at nearby place with all technical skills for repair and maintenance of machines.

(uii) **Initial cost and resale value:** The cost of tractor has also a great influence on its selection. While keeping the resale value in mind, the initial cost should not be very high, otherwise the higher amount of interest will have to be paid.

(viii) **Design:** Indian farmers are lacking in technical knowledge. Tractor to be purchased should have easy operating mechanism and have very few adjustments. The controls should be easily understandable. The wearing parts should be as few as possible.

(ix) **Test report:** Test report of tractors released from farm machinery testing stations should be consulted for guidance.

Rules for Safe Tractor Operation

The following attentions need to be paid for safe operation of tractor,

(i) Never change the gear while tractor is in motion.

(ii) Tractor engine should be brought to its operating temperature before the load is applied.

(iii) When backing or hitching, avoid standing between tractor and the hitched implement.

(iv) Never ride on the drawbar of the tractor or on the implement. Always ride on seat, stand or platform of tractor

(v) For safety, tractor should be driven slowly.

(vi) Avoid sharp turnings, reduce speed before taking a turn.

(vii) When going down the slope, always keep the tractor in gear to have full control on it.

(viii) Fuel should be filled in the tank, when tractor comes to normal temperature.

2.6 TRENDS IN TRACTOR POPULATION IN INDIA

Today in India, tractor technology has reached a reasonably advanced level. More emphasis has been concentrated on reducing the total weight of the tractor implement system and improvement in the components for the drivers' comfort and safety. Also world research and development is going on for computer operated tractor. It is expected that after a decade, such type of tractors may be available for agricultural works. About 12 tractor manufacturers are at present engaged in producing different models of tractors for their suitability in various sizes of farm in India. If we compare the production level of tractor 25 years back and today, the increase of the production of tractor in India has been increased to about 50 per cent (92, 000 in 1987-88 and 1, 90, 000 in 2003-04). It has also been estimated that more than 20 lakh tractors are now available in Indian farms (Table 2.1). The importance of tractor lies mainly in the improvement of soil tillage, development of various matching implements and using for transport purpose in the field of agriculture. Timeliness in farm operation in multi cropping system with the help of tractor is also an important factor for its population in the present agricultural scenario.

Table 2.1: Production and sale of tractors in India

Year	*Production (Numbers)*	*Sale (Numbers)*
1987-88	92092	93157
1988-89	109987	110323
1989-90	121624	122098
1990-91	139233	139831
1991-92	151759	150582
1992-93	147016	144336
1993-94	136971	138879
1994-95	164024	164841
1995-96	191311	191329
1996-97	221689	220937
1997-98	257339	251198
1998-99	261609	262251
1999-2000	278556	273181
2000-01	255690	254825
2001-02	219620	225280
2002-03	166889	173098
2003-04	190687	190336

Table 2.2: Sale of tractor in different states

Sl. no.	*States*	*2000-01 (Numbers)*	*2001-02 (Numbers)*	*2002-03 (Numbers)*	*2003-04 (Numbers)*
1	Andhra Pradesh	17958	12703	11370	10479
2	Assam	531	603	497	488
3	Bihar	18031	14949	13137	12181
4	Gujarat	12383	14616	8473	10067
5	Haryana	17978	15704	12003	11528
6	Karnataka	11817	9669	6626	8480
7	Kerala	669	347	135	145
8	Madhya Pradesh	23099	32496	24953	29359
9	Maharashtra	16714	9870	5009	7071
10	Orissa	5033	2822	3082	2930
11	Punjab	23983	23039	16237	13646
12	Rajasthan	15445	17666	11084	18228
13	Tamil Nadu	10205	6523	5078	6150
14	Uttar Pradesh	68354	51965	40185	38789
15	West Bengal	3533	2682	2171	1985

2.7 OPERATING A TRACTOR

2.7.1 Pre-starting Check-up of a Tractor

1. Fuel should be checked in the fuel tank. If it is not adequate, fuel should be added to the tank.
2. Lubricating oil should be checked by a dipstick and if necessary it should be topped up.
3. Water in the radiator should be checked, if necessary, it should be topped up.

4. Air cleaner should be checked to see weather it is clean or blocked. If blocked, it should be cleaned.
5. Transmission oil (gear oil) should be checked by a dipstick, if necessary it should be topped up.
6. Air pressure in the tyres should be checked and if necessary it should be topped up.
7. Fan belt should be checked by hand, if necessary, it may be tightened or loosened.
8. Grease points should be checked, to see whether they have been greased or not.
9. Important nuts and bolts should be checked If any of them are loose, it should be tightened.
10. The water level of the battery should be checked. If it is below the partition wall, it should be filled up with distilled water.

2.7.2 Methods to Stop a Tractor

Stopping of any automobile vehicle/tractor is as important as its starting. If the vehicle is not stopped in proper time and right moment, there may be the chances of accidents. Hence the following steps need to be followed for stopping of a running tractor.

(i) The throttle lever is pulled and engine speed is allowed to decrease to the lowest possible limit,
(ii) Clutch pedal is depressed and brake pedal is pressed to stop the motion of the tractor,
(iii) Gear shift lever is placed in neutral position,
(iv) If an implement is attached to the tractor, hydraulic control lever is moved slowly to the lower position,
(v) Main switch is made off,
(vi) Parking brake is used if it is required.

2.7.3 Rules for Safe Tractor Operation

The following attentions need to be paid for safe operation of the tractor.

(i) Never change the gear while tractor in motion
(ii) Tractor engine should be brought to its operating temperature before the load is applied
(iii) When backing or hitching is done, avoid standing between the tractor and the hitched implement
(iv) Never ride on the drawbar of the tractor or on the implement. Always ride on the seat, stand or platform of tractor

(v) For safety, tractor should be driven slowly *(vi)* Avoid sharp turnings, reduce speed before taking a turn

(vi) When going down a slope, always keep the tractor in gear to have full control on it

(vii) Fuel should be filled in the tank, when tractor comes to normal temperature

Looking into the increase in the power availability per hectare and inadequacy of human power and bullock power during peak period of farm operation, the demand for tractors in India is rising. The production and sale of tractors in India and different states along with the list of manufacturers have been mentioned in the following tables.

Chapter 3

Internal Combustion Engine and its Components

3.1 INTRODUCTION

The power unit of an automobile or tractor is an internal combustion engine. It works on the basic principle of a heat engine, converting heat energy into mechanical energy. The mechanical energy or the tractive effort at the crankshaft is utilized to propel the automobile vehicle or tractor. Engine is a device which converts heat energy into mechanical work. Internal combustion engines are those heat engines that burn their fuel inside the engine cylinder. On the other side, external combustion engines (Fig. 3.1) are those heat engines that burn their fuel outside the engine cylinder. These are steam engines. The energy developed during the combustion of fuel is transmitted to steam which acts on the piston inside the cylinder. In Internal combustion engine, the gases produced after combustion attain high pressure and temperature. This high pressure moves the piston downwards and thus rotates the crankshaft. In this way, the engine produces continuous power to propel the automobile. The classification of engine has been mentioned below.

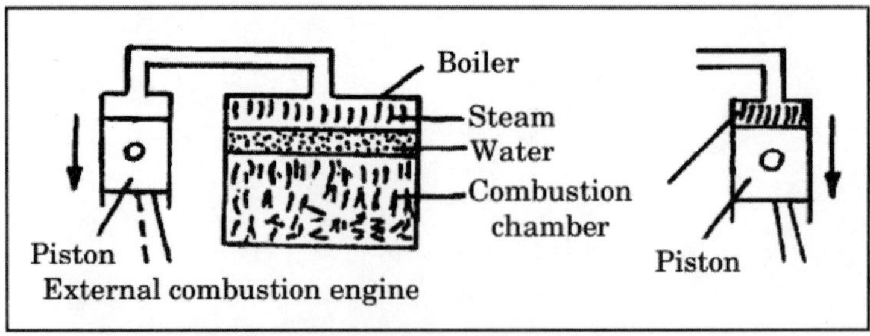

Fig. 3.1: External Combustion Engine

3.2 BASICS OF FUEL BURNING INSIDE ENGINE CYLINDER

It is the oxygen in the air which combines with the fuel for combustion. But the important thing to note is to see how fast we can burn the fuel so that it can exert full force on the piston to get full power from the engine.

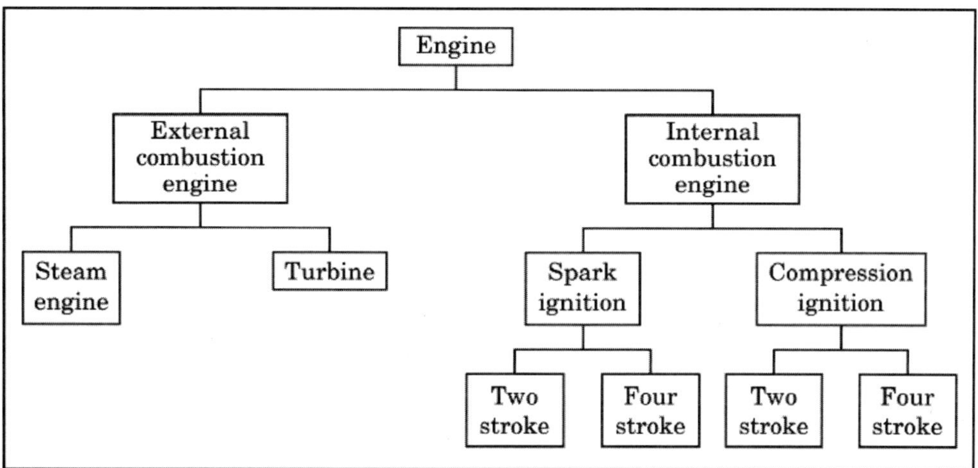

If we burn the fuel in a container, it burns slowly or lazily. This is because the oxygen comes in contact with the surface of the fuel on the top and burning process occurs slowly. But we require a sort of explosive burning of fuel to get full power. However, too powerful explosion is not required as it will destroy the engine.

To make explosive burning or to burn the fuel completely and quickly, we cannot heat the fuel but instead, heat the compressed air with vaporized or atomized fuel to get a big thrust or explosion. This situation actually occurs in the engine cylinder of petrol and diesel engine. Again to understand the basic principle of internal combustion engine (IC engine), consider how a gun propels a bullet from its barrel by rapid burning (explosive power) of a combustible mixture (Fuel charge). The cylinder of the IC engine corresponds to the gun barrel while the piston corresponds to the bullet.

3.2.1 I.C. Engine and steam engines

Both I.C. engine and steam engines are basically heat engines and convert heat energy into mechanical energy. The I.C. engines are now widely used for this purpose due to some reasons. The class of service is the main factor in many cases. One of the outstanding uses is in the field of transportation. Simplicity and low cost operation are the determining factors in the adoption of I.C. engines. The main differences between the I.C. engine and steam engine are as follows:

(i) In I.C. engines, the combustion of fuel takes place inside the engine cylinder, while in steam engines; the combustion of fuel takes place outside the cylinder.

(ii) The working temperatures and pressures in I.C. engines are much higher than those in steam engines.

(iii) Due to high temperature and pressure, materials having better resistance are required for I.C. engines.

(iv) I.C. engines are single acting. Piston is directly connected to connecting rod.

(v) I.C. engines have high efficiency as compared to that of steam engine

(vi) Steam engine requires a big boiler to store water for converting it into steam and a condenser to condense exhaust steam into water. But I.C. engine requires only a small tank to store fuel, and there is no steam condenser in it.

(vii) I.C. engines are easy and quick to start and stop. Steam engines take time in lighting the fire and generating the steam

(viii) I.C. engines are lighter and cheaper as compared to steam engine.

3.2.2 Materials and their properties

The durability and strength of engine component depends on the quality of the material used for construction.

(a) Metal: Iron and steels are the most commonly used materials in engine components.

(b) Pig iron: Pig iron which is obtained from the blast furnace is refined to produce wrought pure iron, cast iron and steel. Wrought iron is almost pure iron and contains very low carbon content.

- **(i) Cast iron:** It is an alloy of carbon and iron and carbon percentage ranges from 2.5 to 4.5 %. It is very hard and brittle.
- **(ii) Steel:** Steel is also an alloy of carbon and iron but lower percentage of carbon as compared to cast iron. Steels are classified as

Mild steel: It is an alloy with up to about 0.25 % carbon content.

Medium carbon steel: Alloy with upto about 0.25 % to 0.5 % carbon content

High carbon steel: Alloy with upto about 0.6 to 1.5 % carbon

Brass: Alloy of copper (60 %) and zinc (40 %)

Bronze: Alloy of copper and tin. (88 % copper, **10** % tin and 2 % zinc)

3.3 CLASSIFICATION OF IC ENGINE

The internal combustion engine are classified on the following basis

(i) Type of fuel used: *(a)* petrol engine (or gasoline engine), *(b)* diesel engine, *(c)* gas engine by using LPG, natural gas, biogas producer gas etc.

(ii) Cycle of operations: *(a)* Otto cycle engine, *(b)* diesel cycle engine, *(c)* dual combustion cycle engine.

(iii) Number of strokes per cycle: *(a)* two stroke cycle engine, *(b)* four stroke cycle engine.

(iv) Type of ignition: *(a)* compression ignition (Cl) engine, *(b)* spark ignition (Sl), engine.

(v) Number of cylinders: *(a)* single cylinder engine, *(b)* Two cylinder engine, *(c)* three cylinder engine, *(d)* four cylinder engine, (e) six cylinder engine, *(f)* eight cylinder engine.

(vi) Arrangement of cylinders: *(a)* In-line type, *(b)* V-type, *(c)* opposed cylinders type, *(d)* radial type.

(vii) Valve arrangement: *(a)* L-head engine, *(b)* I-head engine, *(c)* F-head engine, *(d)* T-head engine.

(viii) Engine speed: *(a)* Low speed engine, *(b)* High sped engine, *(c)* Medium speed engine.

(ix) Method of cooling: *(a)* air cooled engine, *(b)* water cooled engine.

3.3.1 Classification of engine by cylinder arrangement

In-line type: The cylinders are arranged in one row, side by and parallel to each other. The cylinders arranged in one line transmit power to a single crankshaft as shown in Fig.3.2. The cylinders are counted from the radiator side to the flywheel side. A few manufacturers however, specify the cylinders number from the flywheel side to the radiator side. The cylinder nearest to the radiator is designated as No.1 cylinder while counting the number from the radiator side to the flywheel side.

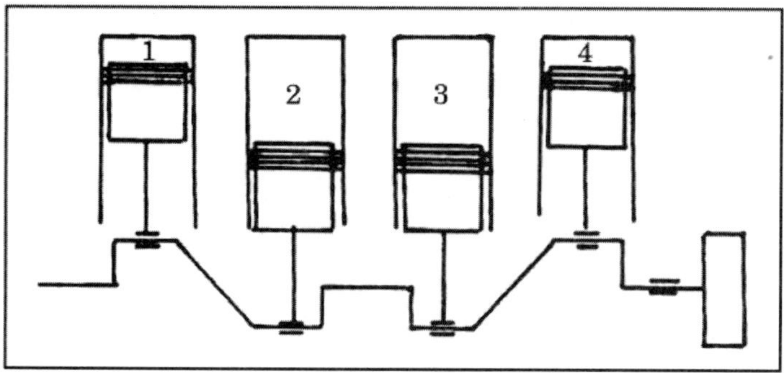

Fig. 3.2: Four Cylinder in line engine

V-type: Two rows of cylinders are arranged such that the cylinders are at an angle, usually 60° -90° , to one another (Fig. 3.3). The angle between the cylinders is 60° for six cylinder engines and 90° for eight cylinder engines. V-types of engine are shorter in design compared to in-line engines. The short crankshaft provides rigidity and smooth operation at high engine speeds. In this configuration, roughly double number of cylinders can be arranged as compared to in-line engines.

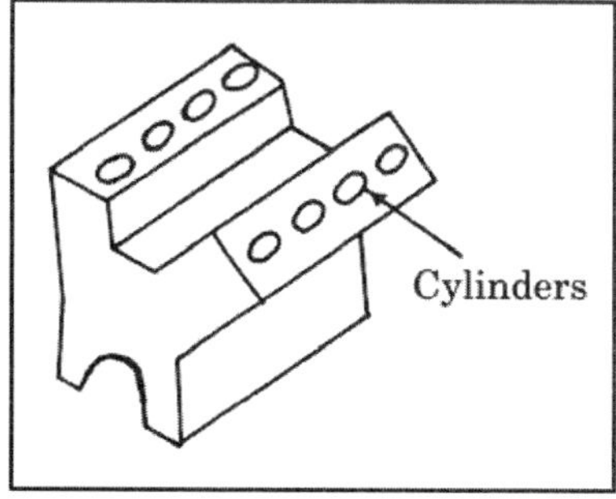

Fig. 3.3: V-4 engine

Opposed type: In this type of cylinder arrangement, the cylinders are arranged horizontally opposite to each other (Fig. 3.4). The crankshaft is shorter in design as compared to in-line cylinder configuration. It provides a good mechanical balance. Such engines are compact in design. But the length of the engine increases and hence the engine is placed in transverse direction in the tractor.

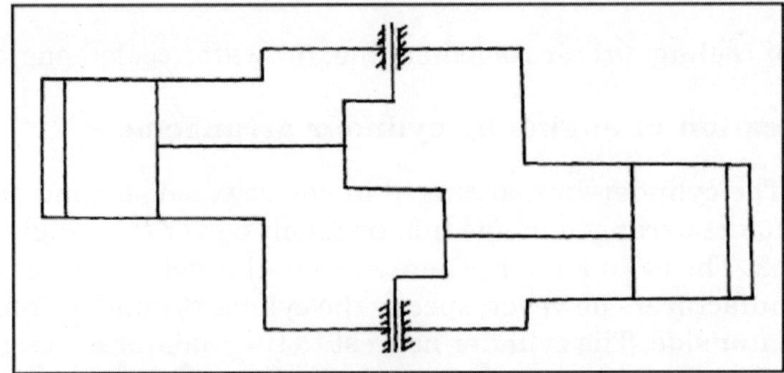

Fig.3.4: Two Cylinder Engines. (opposed type)

Radial type: In radial type the cylinders are arranged radially around the crankshafts (Fig. 3.5). This arrangement is best suited for small aircraft engines and is unsuitable for automobiles.

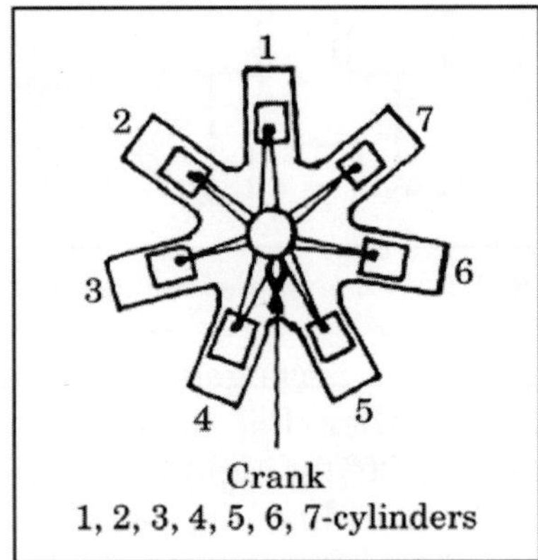

Fig. 3.5: Radial Engine

3.3.2 Classification by valve arrangements

The engines are classified into four categories according to the arrangement of inlet and exhaust valve in various positions in the cylinder head. These arrangements are termed as 'L', 'I', 'F' and 'T'. It is easy to remember the word 'LIFT' to recall the four valve arrangements. The 'I'- head design is most commonly used in tractor engine.

I-head arrangement: This arrangement is also known as over head valve system. The intake and exhaust valves are located in the cylinder head, parallel to each other in a single row. The cylinder head and block make the shape similar to the letter 'I'. In this arrangement, tappet, push rod and rocker arm are required to reverse the direction of valve movement and requires only one cam shaft to operate both the valves. In most of the engines, camshaft is fitted in the cylinder block while some engines employ overhead camshaft which runs with the help of timing chain. This type of arrangement is shown in Fig. 3.6. The I-head valve engine is more adoptable to higher compression ratio.

L-head arrangement: Both inlet and exhaust valves are placed side by side and at one side of the cylinder and operated by a single camshaft. The combustion chamber and the cylinder together are positioned in the shape of an inverted 'L' (Fig. 3.6). This design has limited use as it does not meet strict exhaust emission standards.

F-head arrangement: The inlet valve is placed in the head and the exhaust valve at one side of the cylinder block (Fig. 3.6). This engine is the combination of L-head type and I-head type. Both the valves are driven from the same camshaft. The valve in the cylinder head is operated through push rod and rocker arm while the valve fitted in the block is actuated directly by the cam shaft through tappets.

T-head arrangement: T-head engine has the inlet valve on one side and the exhaust valve on the other side of the cylinder (Fig. 3.6). Two camshafts are required one for operating inlet valve while the other for operating exhaust valve on other side.

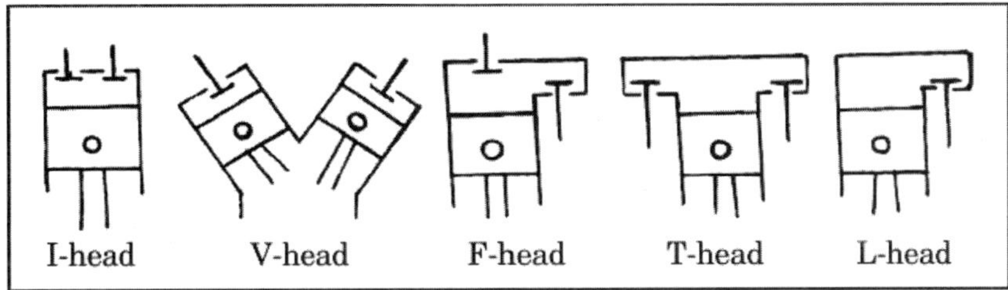

Fig. 3.6: Different types of valve arrangement

The most sought-after design is the I-head arrangement because of advantages like better cooling, better volumetric efficiency, ease in changing the shape of combustion chamber, better compression ratios and ease in maintenance.

3.4 ENGINE COMPONENTS AND THEIR FUNCTIONS

I.C. engine is basically divided into 3 parts - cylinder head, cylinder block and crankcase (Fig. 3.7). Internal combustion engine consists of a number of parts which are given below:

(1) Cylinder head, (2) Cylinder block, (3) Crankcase, (4) Cylinder, (5) Cylinder liner, (6) Piston, (7) Piston rings, (8) Connecting rod, (9) Crankshaft, (10) Camshaft, (11) Flywheel, (12) Cylinder head gasket, (13) Timing gear, (14) Intake manifold (15) Exhaust manifold.

Cylinder head: It is a detachable portion of an engine which covers the cylinder and includes valves, spark plugs, injectors (in case of diesel engine), combustion chamber, intake and exhaust manifold, passages for cooling water, oil plug etc. The cylinder head is bolted to the top of the cylinder block with the help of gasket. The cylinder head is mostly made of cast iron mixed aluminum. Some cylinder heads are made of aluminum alloy which has a higher heat conductivity than cast iron as a result of which it dissipates heat more rapidly.

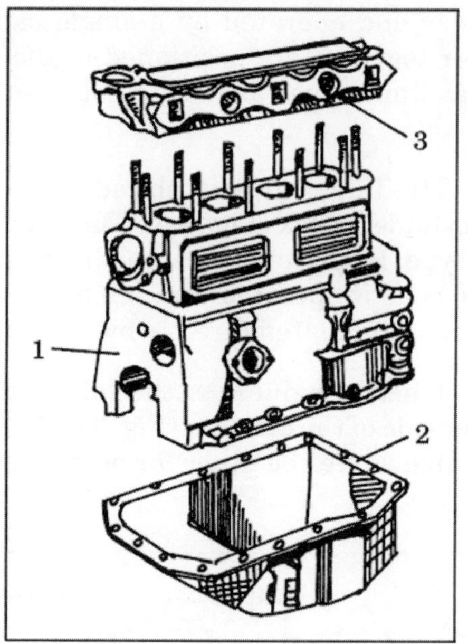

Fig. 3.7. Parts of cylinder block
1. Cylinder block, 2. Crankcase, 3. cylinder head

Cylinder block: The cylinder block is the foundation of the engine. It is a solid casting which includes the cylinder and water jackets (cooling fins in the air cooled engines). The crankcase and cylinder blocks are often cast in one piece. This is the largest and heaviest single piece of metal in the engine. When the cylinder blocks are cast in one piece, it is designated as monoblock engine. Some blocks are cast in two separate parts *i.e.*, the crankcase and block which are then bolted together. This type of block is called the split block. The lower part of the cylinder block contains supporting bearings for engine crankshaft and camshaft. The intake and exhaust manifolds are attached to the side of the cylinder block on L-head type engines. In I-head engines, the manifolds are attached to the cylinder head. The oil pan is also attached to the bottom of the block.

The other parts attached to the block include the water pump, timing gear, chain, fly wheel and clutch housing, fuel pump etc. These parts are attached to the block with sealing gasket. The spark-ignition cylinder blocks and compression-ignition cylinder blocks are very much alike. However, the diesel engine block is heavier and stronger. It must be stronger because the compression ratios and internal pressures in the combustion chambers of diesel engines are higher. The cylinder block is cast from cast iron or iron alloyed with other metals such as nickel or chromium. These added metals give the cylinder block greater resistance against wear as well as greater strength.

Crankcase: The bottom of the cylinder block plus the oil pan forms the crankcase. It encloses the crankcase. It encloses the crankshaft, camshaft and oil pan. The upper portion of the crankcase is usually integral with cylinder block. It is made of cast iron or the material same as the cylinder block.

Cylinder: Cylinder (Fig. 3.8) is a cylindrical bore made in the cylinder block. It is the basic part of the engine in which piston moves or slides inside. It is subjected to piston side thrusts and high combustion temperature. It must be constructed for resistance to wear and adequate cooling. It ·provides space in which piston operates to suck the air or air-fuel mixture. It is usually made of high grade cast iron.

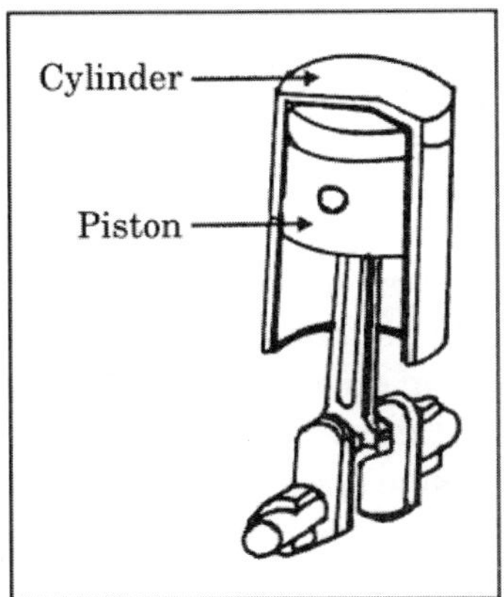

Fig. 3.8: Figure showing piston inside engine cylinder

Cylinder liner: Cylinder liner is an integral part of the cylinder. It is a cylindrical lining inserted in the cylinder (or cylindrical bore) in which the piston slides. The liner must withstand working temperature and pressure. The material used for liner is usually of nickel cast iron. The liners may be of either blockcast liner or replaceable liner. The advantage of replaceable liners over blockcast liner is the choice of obtaining a better quality of material. Chrome-hard materials can be used for liners but not for the cylinder. This makes the liners cheaper and replaceable for reboring. Hence liners are either cast into the cylinder block or installed later. In case of block-cast liner, the liners are installed in the bore and aluminum is poured around them. They then become a permanent part of the cylinder block. Liners are classified as *(a)* dry liner *(b)* wet liner (Fig. 3.9). These two kinds of liner can be installed later. Dry liners are pressed into the cylinder block. They touch the cylinder bore along their full length. Wet liners touch the cylinder block only at the top and bottom. The rest part of wet liners only touches the coolant.

Dry liners make metal to metal contact with the cylinder block casting where as wet liners come in contact with the cooling water. The water circulating inside the block first cools the block. Then the block cools the liner and the cooling of the engine is maintained. But dry liners do not come in contact with the cooling water.

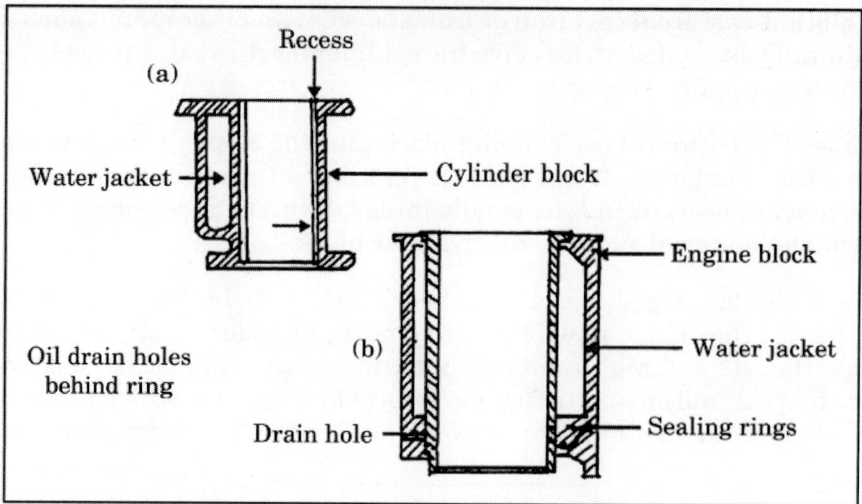

Fig. 3.9. Liners (a) dry liner (b) wet liner.

Piston: Piston is of cylindrical shape that moves up and down in the engine cylinder and its one end (top) is closed and other end (lower part) is open (Fig. 3.10). It is subjected to enormous thrust caused by burning of fuel. It is connected to the connecting rod by a piston pin. The force of the expanding gases against the closed end of the piston, forces the piston down in the cylinder. This causes the connecting rod to rotate the crankshaft. Materials used for the manufacture of a piston are cast iron due to its high compressive strength, low coefficient of expansion, resistance to high temperature. Aluminum and its alloys are preferred mainly due to its lightness. Different parts of piston are *(i)* head, *(ii)* skirt, *(iii)* ring groves, *(iv)* piston pin.

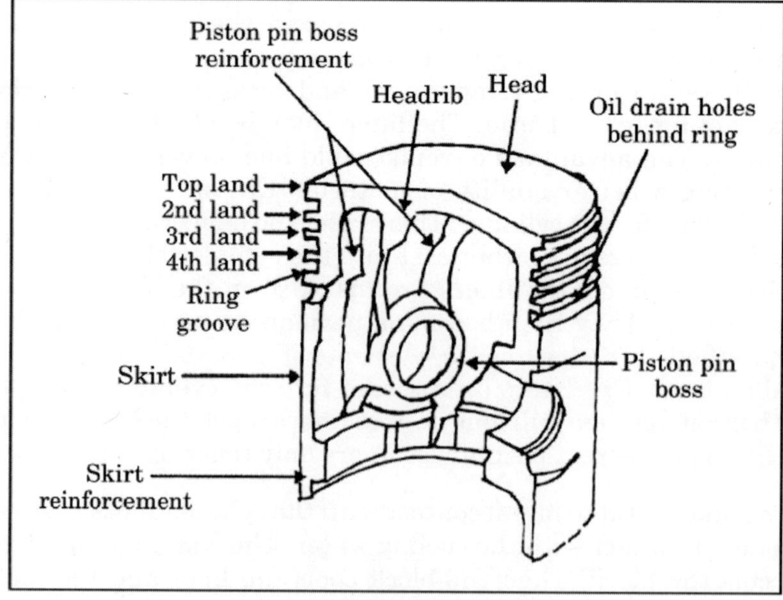

Fig. 3.10: Parts of Piston

Head of piston: It is the top of the piston. It is also called the crown of piston. It is subjected to high temperature and pressure. There are generally four different types of heads, but most commonly used is flat head whereas irregular and concave head pistons are used to give swirling (cause to move circulatory motion) action for mixing of air and fuel.

Piston skirt: Skirt (Fig. 3.11) is called the lower portion of the garment of a girl. Similarly piston skirt is the lowest part of the piston which is designed to absorb the side (lateral) movements of the piston, so that the piston becomes round in shape after getting heated and expanded. We know that metals expand when heated up, to accommodate expansion of piston, two methods are used -these are slot skirt or cam ground. In slot skirt, slot is given in piston skirt. When the piston expands due to heat, the width of slot decreases. Pistons are similarly not machined truly round but are cam ground (oval shape), mean diameter of top of piston is less than diameter at the bottom. When the piston gets heated up, the less diameter size gets space for expansion and heated piston resumes a shape of a perfect circle. The cut skirt is used to save weight and gives space for counterweights on the crankshaft.

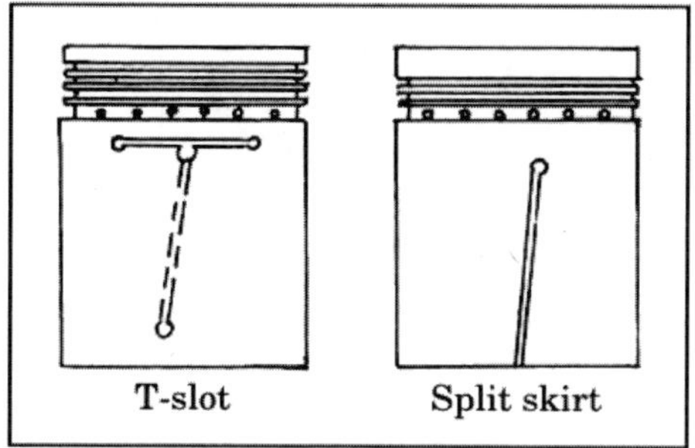

Fig. 3.11: Slots in piston skirt

Ring grooves: Grooves are provided in the upper part and lower part of the piston to fit the piston rings.

Piston pin: It links the connecting rod and the piston. This is also called the wrist pin or Gudgeon pin. There are three different ways of doing this.

(i) Pin is fixed in the piston and free in connecting rod

(ii) Pin is fixed in the connecting rod and free in the piston.

(iii) Pin is free in both piston and connecting rod.

The last arrangement is mostly used in modern engines and is called full floating design. Piston pin is generally hollow and is accurately ground and polished for tight fit in the piston bosses. It is made hollow to remove unnecessary weight, thus reducing the inertial forces developed at the end of each stroke. The pin is held within the piston bore to prevent contact with the cylinder. It is made of alloy steel.

Piston rings: They are split expansion ring, placed in the grooves of the piston. Piston rings are fitted in the grooves made in the piston. They are usually made of cast iron or pressed steel alloy. The important functions of piston rings are:

(i) They seal or have gas tight seal between piston and cylinder, so that the burnt gases are not allowed to leak or escape to the oil sump.
(ii) They control the cylinder and piston lubrication.
(iii) They must carry away the maximum heat from piston to cylinder walls.
(iv) They reduce contact area between cylinder wall and piston wall for preventing friction losses and excessive wear.

Piston rings are of two types; *(a)* Compression ring, *(b)* oil ring (Fig. 3.12).

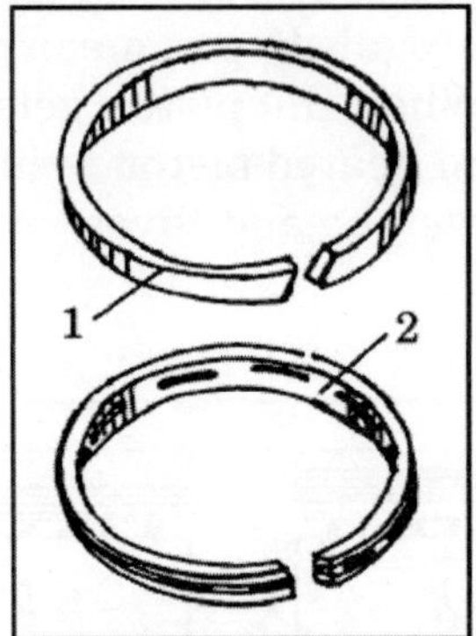

Fig. 3.12: Piston rings
1. Compression ring. 2. Oil ring.

Compression ring: Compression rings are usually plain, single piece and are always placed in the grooves nearest to the piston head. The compression rings seal in the air-fuel mixture as it is compressed. They also seal in the combustion pressure as the mixture burns. During exhaust stroke they seal the passage so that exhaust gases do not go to sump.

Oil rings: Oil rings are grooved or slotted and are located either in the lowest groove above the piston pin or in a groove above the piston skirt. They scrape excessive oil from the cylinder wall and return it to the oil pan or oil sump. They control the distribution of lubrication oil in the cylinder and the piston. Oil rings are provided with small holes through which excess oil returns back to the crankcase chamber. Ring clearance is the gap at the joint of the ring, measured when the ring is inside the cylinder. The gap is usually 1 mm per 200 mm diameter of the piston. This clearance is necessary for expansion of the ring in heated condition, without which the ring may break or buckle.

Connecting rod: Connecting rod (Fig. 3.13) is attached to the piston at one end and crankshaft on the other end. It transmits the power of combustion to the crankshaft and makes it rotate. Its small end is fitted with bronze bushing and big end is provided with bearings, split into two shells. It is usually made of drop forged steel and is made with an I-beam cross section. The upper portion where gudgeon pin is fitted is called eye, mostly called small end and the lower portion which is fitted to crankshaft pin is called big end or head of connecting rod. The centre portion which joins small end and big end is called shank.

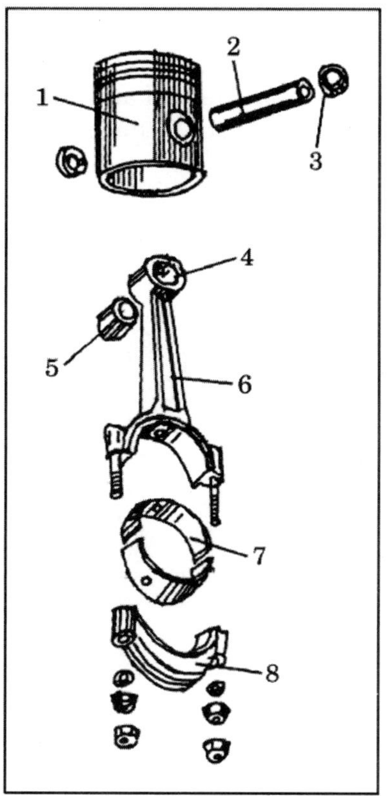

Fig . 3.13: Parts of connecting rod & piston.
1. Piston. 2. Piston pin. 3. Cirlip. 4. Small end of connecting rod , 5. Bush, 6. Connecting rod . 7. Bearing, 8. Cap

To provide connecting rod big-end bearing lubrication, oil travels the following path. The oil pump sends oil to oil lines or galleries in the cylinder block. From these oil lines, the oil moves to the main bearings, oil travels through oil passages drilled in the crankshaft. Finally the oil moves through the connecting rod bearing oil holes on to the bearing surface. The piston pin is lubricated by oil scraped from the cylinder walls. Some of this oil flows through slots or holes in the piston to lubricate the piston pin.

Connecting rod has oil passages drilled from the connecting-rod big end to the piston-pin bushing. The oil goes to the piston-pin bushing in the connecting rod by the following path. From the oil pump, oil travels through oil lines (gallery) in the block, and through oil passages in the crankshaft to the connecting rod big end bearings. The oil then moves through the oil passage in the connecting rod to the piston-pin bushing.

Crankshaft: The crankshaft (Fig. 3.14) is a strong one piece casting of heat treated alloy steel. It converts the reciprocating motion of the piston into rotary motion of the fly wheel. It is subjected to bending as well as twisting from the connecting rod thrust. It must be strong enough to withstand all these forces. Bearings are provided in the engine wherever there is rotary motion between engine parts. The part of the shaft that rotates in the bearing is called a journal connecting rod (big end) and crank shaft (also main) bearings are of split type. The part that rotates in the crankshaft bearing is called the main journal and the part which rotates is big end bearing is the crank journal. To prevent the back-and-forth movement of the crankshaft the main bearings and big end bearings are of split types. The main parts of the crankshaft are as follows;

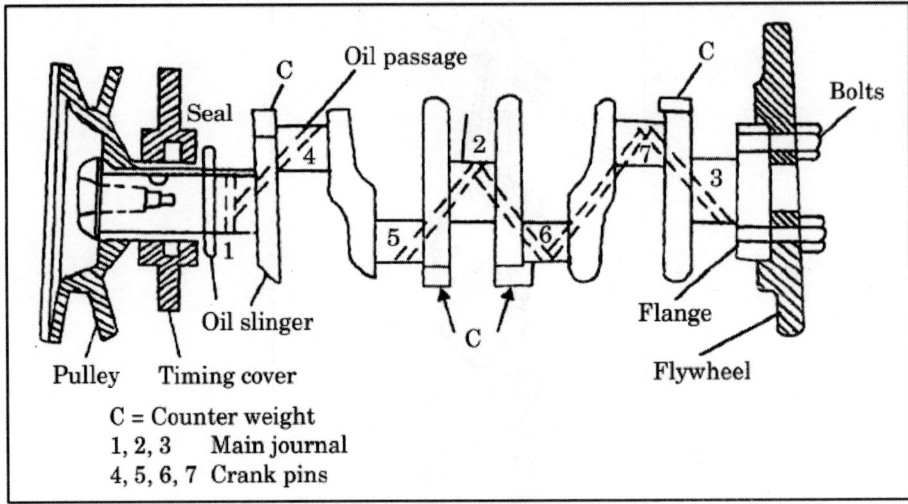

Fig. 3.14: Crankshaft

(i) **Main bearing journal:** Bearing surfaces for support on main bearings.

(ii) **Crankpins (or throw):** Bearing surfaces connected to big end of connecting rods.

(iii) **Crank arms or throws (big end pin):** It provides leverage to rotate the crankshaft. For each cylinder, there is one pin. Arm is the portion of crankshaft which joins big end pin with main journal.

(iv) **Counter-weights:** Counter weights are bolted to arm of crankshaft to balance the crankshaft while rotating.

Web is the portion of crankshaft which joins big end pin with main journal (Also it is called arm of crankshaft (Fig. 3.15). The main function of the crankshaft is to make the engine power available for useful work. The fly wheel is attached to the rear end of the crankshaft and crank gear is attached to the front end.

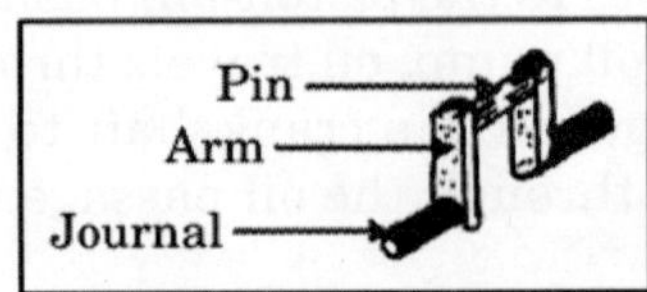

Fig. 3.15: Parts showing Pin, Arm and Journal of Crankshaft

Camshaft: Camshaft (Fig. 3.16) is a straight shaft made out of high carbon steel with cams built on it. It raises and lowers the inlet and exhaust valves at proper time. The cam shaft is generally located at the upper crankcase level and is driven from the crankshaft by means of positive drives *e.g.*, gears or chain and sprocket. It is mounted in the crankcase parallel to the crankshaft. The speed of the cam shaft is exactly half the speed of the crankshaft in 4 stroke engine. Camshaft operates the fuel pump, ignition system, valves etc.

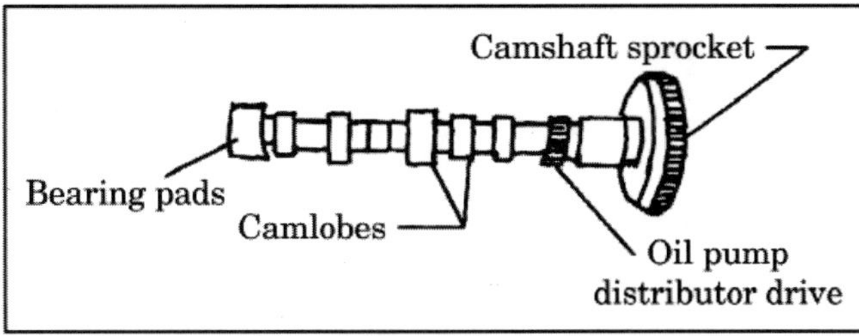

Fig. 3.16: Camshaft

Fly wheel: It is made of cast iron and is mounted on the crankshaft (Fig. 3.17). Its main functions are as follows:

(i) It stores the excess energy during power stroke and supplies back the same energy during the idle stroke, providing a uniform rotary motion by virtue of its inertia

(ii) It assists in engine balancing.

(iii) The rear surface of the fly wheel serves as one of the pressure surfaces for the clutch plate.

(iv) Sometime the flywheel serves the purpose of a pulley for transmitting power.

(v) It serves as starting aid for the engine.

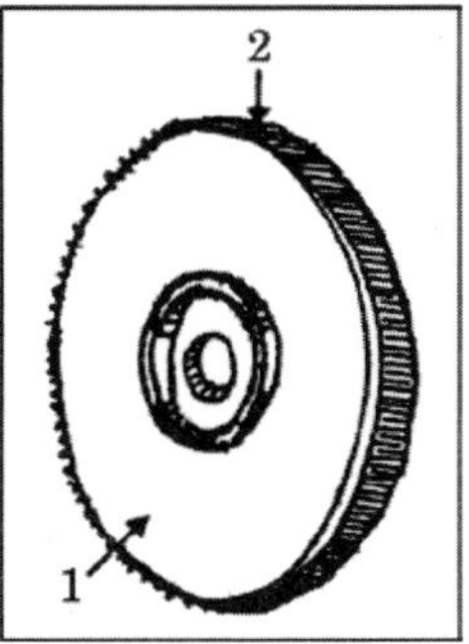

Fig. 3.17. Flywheel
1. Flywheel, 2. Ring gear.

Cylinder head Gasket: A gasket is a thin layer of soft material such as paper, cork, rubber or synthetic material. It is placed between two flat surfaces to make tight seal. Cylinder head gasket is fitted over the cylinder block under the cylinder head so that gases do not leak. Gasket seals the combustion chamber. In most cases, copper asbestos gaskets are used.

Timing gear (Fig. 3.18): Timing gear is a combination of gears, one gear of which is mounted at one end of the camshaft and the other gear on the end of the crankshaft. Camshaft gear is bigger in size than that of the crankshaft gear and it has twice as many teeth as that of the crankshaft gear. For this reason, this gear is commonly called half time gear. Timing gear controls the timing of ignition, timing of opening and closing of valves as well as fuel injection timing.

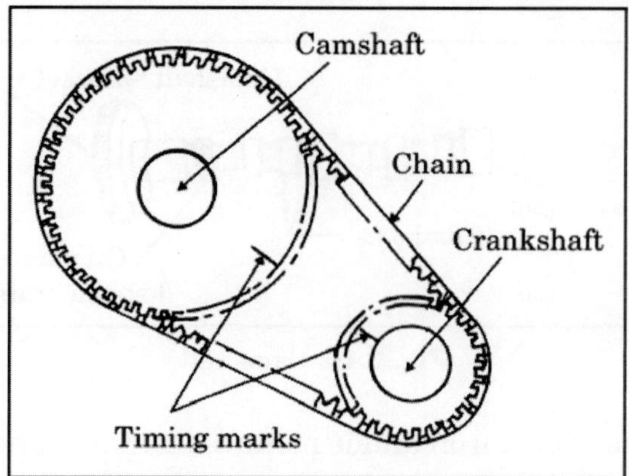

Fig. 3.18. Timing Gears.

Intake manifold: It is that part of the engine through which air or air-fuel mixture enters into the engine cylinder. It is fitted by the side of the cylinder head.

Exhaust manifold: It is that part of the engine through which exhaust gases go out of the engine cylinder. It is capable of withstanding high temperature of burnt gases. It is fitted by the side of the cylinder head (Fig. 3.19).

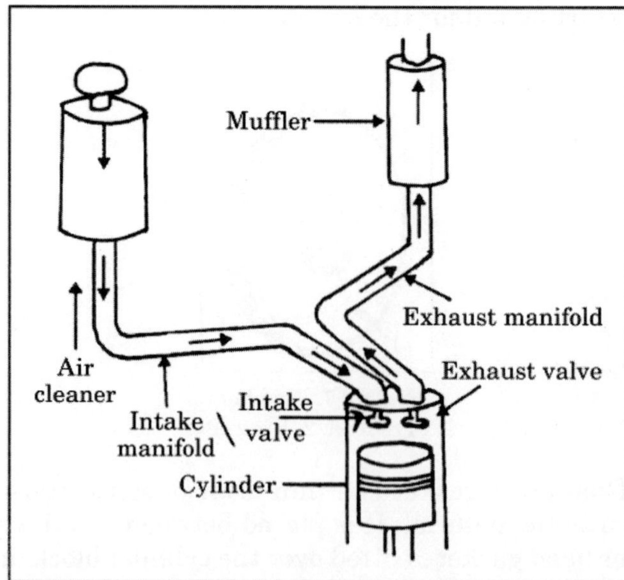

Fig. 3.19: Intake manifold and exhaust manifold.

3.5 PRINCIPLE OF OPERATION OF I.C. ENGINE

An internal combustion engine works either on 4-stroke cycle or 2-stroke cycle.

Stroke: One up to down or BDC (bottom dead centre of engine cylinder) to TDC (top dead centre) movement of piston is known as one stroke.

Cycle: The repetition of events in regular order is known as cycle.

3.5.1 4-stroke cycle engine

In 4-stroke cycle engine, the whole sequence of events *i.e.* suction, compression, power and exhaust are completed in four strokes of piston and two complete revolution of the crankshaft.

(*i*) **Suction stroke:** when we revolve the engine crankshaft through flywheel, the inlet valve opens and piston being tied up with crankshaft with the help of connecting rod travels from TDC to BDC. When piston moves towards BDC, a partial vacuum is created on the top of the piston. Due to this partial vacuum, air is drawn into the combustion chamber through open inlet valve. When piston reaches BDC, the inlet valve closes. The figure of suction stroke is shown in fig. 3.20 (a).

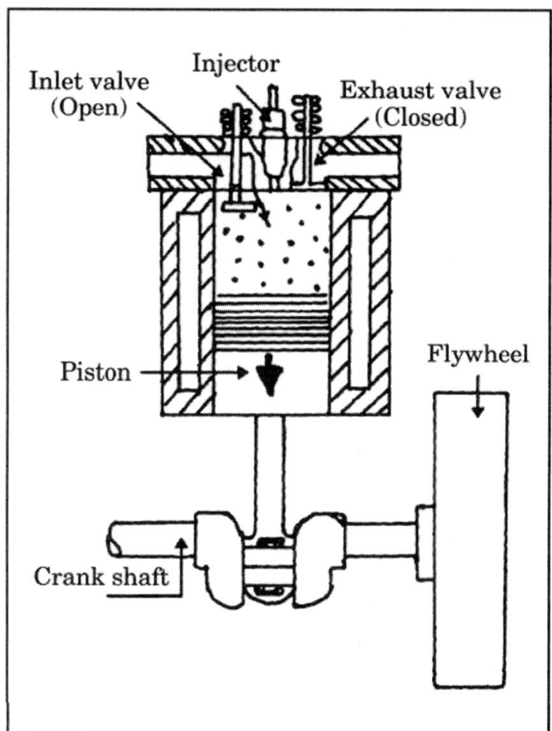

Fig. 3.20(a): Four stroke diesel engine suction stroke.

(ii) **Compression stroke (Fig. 3.20b):** After the piston has come down in suction stroke, the piston then moves upward towards TDC due to the rotary motion of-crankshaft. As soon as the piston starts going up in compression stroke, inlet valve which was kept open in suction stroke closes. Exhaust valve is also closed. The air starts compressing. By compressing the temperature and pressure of the air rises. This stroke finishes when piston reaches from BDC to TDC.

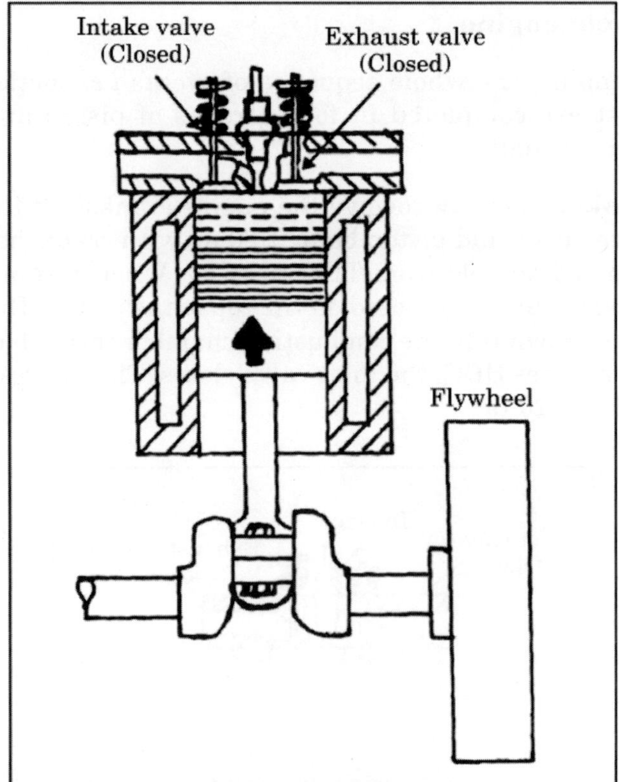

Fig. 3.20 (b): Four stroke diesel engine compression stroke.

(iii) **Power stroke (Fig. 3.20c):** In this stroke both the valves remain closed. Before piston reaches TDC in compression stroke, injector starts injecting fuel and fuel is ignited with the help of heat of compression. As soon as charge is ignited, spontaneous burning of charge takes place. Due to burning of charge, gases are produced and put pressure on the crown or upper part of piston making it to travel down with force. Since the piston is connected with crankshaft with the help of connecting rod, the crankshaft revolves. This pressure decreases as the piston approaches BDC.

(iv) **Exhaust stroke (Fig. 3.20d):** In this stroke, exhaust valve opens and piston starts moving up from BDC to TDC due to momentum or inertia of flywheel. As the piston moves up, it pushes the burnt gases out through exhaust valve.

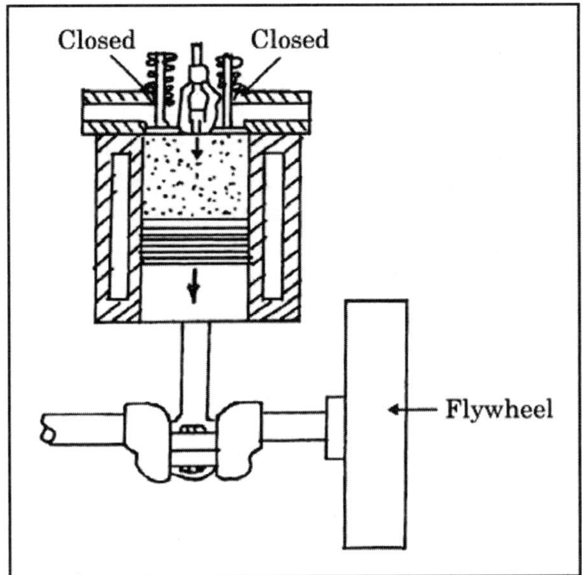

Fig. 3.20 (c): Four stroke diesel engine power stroke.

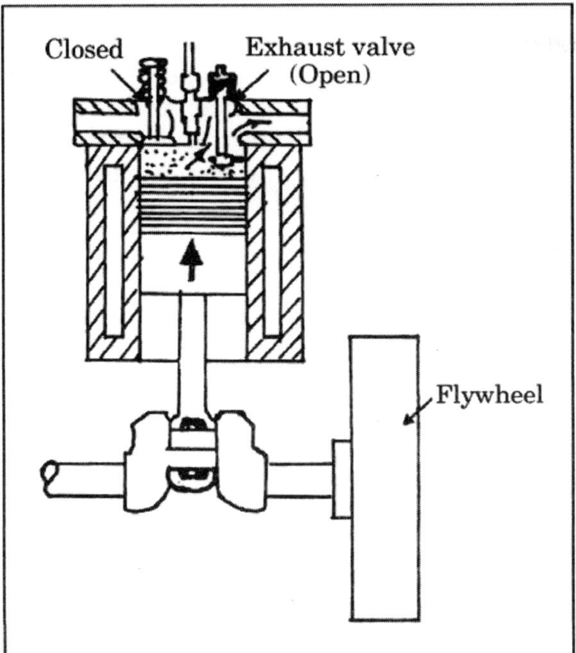

Fig. 3.20 (d): Four stroke diesel engine exhaust stroke.

In four stroke cycle engine, there is only one power stroke and rest three strokes are idle strokes. Flywheel stores the energy received in power stroke and keeps the crankshaft rotating in idle strokes.

3.5.2 Two stroke cycle engine

In a 2-stroke cycle engine, the whole sequence of events *i.e.* suction, compression, power and exhaust are completed in two strokes of piston and one complete revolution of the crankshaft. No valves are provided in this type of cycle engine but in the substitute ports are present. The crank case of the engine is air tight in which the crankshaft rotates. Both suction and compression strokes occur in one stroke of the piston and power as well as exhaust strokes occur in another stroke of the piston. To reduce the number of idle strokes as occurred in four stroke engine, two stroke engine was invented. Two-stroke cycle engine is also called Clark cycle engine named after the inventor Mr. Dugad Clark who invented it in year 1880. In this cycle engine, three ports *i.e.* inlet port (suction port), exhaust port and transfer port are provided in the cylinder liner but exhaust port and transfer port are placed almost opposite to each other (Port is the opening in the side of a structure).

First stroke (Suction + compression)

When the piston moves up from BDC to TDC in the cylinder, it covers two of the ports *i.e.* exhaust and transfer port and uncovers the inlet port. Due to this upward movement of piston, fresh charge of air-fuel mixture (in case of petrol engine) or fresh air (in case of diesel engine) enters into the crank case owing to vacuum created in it. This fresh charge is temporarily stored in this stroke and is utilized in the next stroke of piston (movement from TDC to BDC). For further upward movement of piston, the charge that has already entered into the cylinder is compressed. Just before the end of this stroke, the charge in the cylinder gets ignited as in four stroke cycle engine. Due to the burning of charge, the gases expand spontaneously putting thrust on the piston head making it to move down and to start the next stroke (Fig. 3.21(a)).

Second stroke (power + exhaust)

When the piston goes down, it partially compresses the air-fuel mixture (in case of petrol engine) or air (in case of diesel engine) previously drawn into the crankcase. For further downward movement of piston, it covers inlet port and uncovers exhaust and transfer port causing the charge to move from crankcase to the cylinder. This also allows the burnt gases to flow through the exhaust port. During downward movement of the piston, it uncovers exhaust port first causing the burnt gases to escape and resulting in the decrease of pressure. Then transfer port is uncovered (opened) for easy flow of pre compressed charge to the cylinder (combustion chamber). The piston used in this type of two stroke engine is dome type. The slope of the piston top is given in such a way that it deflects the incoming charge towards top and slope on the other side deflects the exhaust gases to get out from the exhaust port. In case, the top of the piston would have been flat, then the incoming charge will find its way out through the exhaust port, resulting in loss of power and more fuel consumption. Hence both power and exhaust stroke occur in this stroke (Fig. 3.21(b)).

The deflected piston allows the escape of exhaust gas by the charge to the combustion chamber. The process of removal of burnt or exhaust gases from the cylinder is known as scavenging.

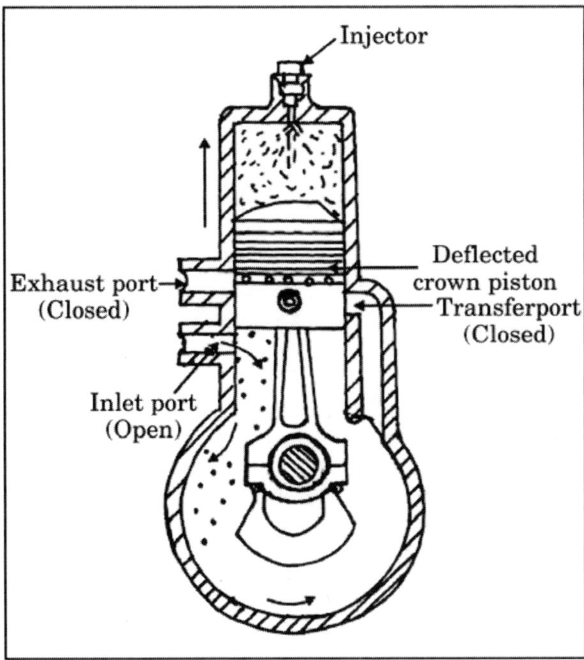

Fig. 3.21 (a): Two stroke diesel engine-upward stroke (suction & compression)

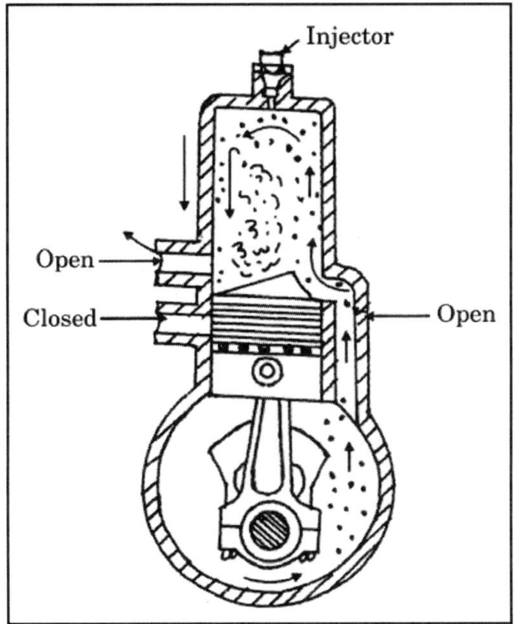

Fig. 3.21 (b): Two stroke diesel engine-Downward (Power and Exhaust)

Lubrication: In two stroke cycle of petrol engine, the charge is first kept in the crankcase and then it finds its way to the cylinder. As such, it is not possible to keep the lubricating oil in the oil sump for lubrication purpose. But the parts namely crankshaft, main bearing, piston pin, small end bearing, cylinder liner, connecting rod need to be lubricated, otherwise the engine will seize. To lubricate all such parts, certain percentage of lubricating oil is mixed with petrol in petrol engine. Mobil oil is mixed with petrol in the ratio of 1:20 *i.e.*, 1 litre of mobil is mixed with 20 litres of petrol. As such, when the charge gets into the crankcase, the lubricating oil present in the charge lubricates the different components.

Applications: Two stroke engines are generally used in mopeds, scooters, motorcycle and light vehicles.

3.6 COMPARISON BETWEEN TWO AND FOUR STROKE CYCLE ENGINE

Sl. no	*Particulars*	*2-stroke cycle engine*	*4-stroke cycle engine*
1.	No. of power stroke	One for one revolution of the crankshaft	One for every two revolutions of crankshaft
2.	Valve mechanism	Ports are provided instead of valves	Valves present
3.	Thermal efficiency	Dilution of fresh charge with exhaust gases lead to poor thermal efficiency	Escaping of gases is nearly complete, leading to better thermal efficiency
4.	Construction	Engines are normally air cooled, simple in design and easy to maintain	Water cooling system makes it complicated in design and difficult to maintain.
5.	Fuel consumption	Owing to dilution of fuel by exhaust gases (in case of petrol engine), the fuel consumption per horsepower is more	Lower fuel consumption per horse power
6.	Removal of exhaust gas	Difficult	Easy
7.	Lubrication	Fuel is usually mixed with lubricating oil	Equipped with an independent lubricating oil circuit
8.	Engine fluctuation	Engine fluctuation is less due to one power stroke for one revolution of crankshaft and therefore, lighter flywheel is provided	Engine fluctuation is more, that is why, heavier flywheel is provided
9.	Mechanical efficiency	Owing to the absence of cam, camshaft, rocker, valve mechanism etc, the mechanical efficiency is higher	Lower mechanical efficiency
10.	Volumetric efficiency	Lower volumetric efficiency due to lesser amount of air, fuel mixture enters into the cylinder	Higher volumetric efficiency
11.	Engine weight and size	For same power and stroke, engine Is lighter and occupied less space (small)	Heavier and large
12.	Air tightness of crankcase	Must be sealed for temporary storage of charge in the crankcase	Not necessary.
13.	Direction of rotation of engine	The direction of rotation can be reversed	It cannot be reversed because it has valves.

While studying the working of two stroke petrol engine, it is seen that when air fuel mixture is provided to the cylinder from transfer port, part of this is escaped through the exhaust port resulting in loss of fuel and loss of horsepower where as in 2-stroke diesel engine, it is only the air even if it escapes, we do not lose much.

3.7 COMPARISION OF DIESEL ENGINE WITH PETROL ENGINE

Sl. No.	*Particulars*	*Diesel engine (CI engine) (Compression Ignition engine)*	*Petrol engine (SI engine) Spark Ignition engine*
1.	Cycle	Diesel cycle	Otto cycle
2.	Fuel	Diesel	Petrol/Kerosene
3.	Charge	Only air is taken in during suction stroke	Mixture of air and fuel is taken in suction stroke
4.	Fuel supply	Fuel is injected into the combustion chamber through fuel injection pump and injector	Air and fuel is mixed in correct proportion (about 15:1) in carburetor and supplied to the engine
5.	Mode of Ignition	By heat of compression	By electric sparks
6.	Thermal efficiency	32-38 %	25-32 %
7.	Compression ratio	14:1 to 22:1	5:1 to 8:1
8.	Specific fuel consumption	160-200 gm/hp-hr	200 to 280 gm/hp-hour
9.	Compression pressure	35-45 kg/cm^2	6-10 kg/cm^2
10.	Temperature	About 600-800°C	About 250-300°C
11.	Engine weight per hp	High	Comparatively low
12.	Operating cost	Low	Comparatively high

Weight of diesel engine is high as diesel engine block is heavier and stronger only to absorb the high pressure created in the combustion chamber, otherwise the engine block will vibrate more. Compression ratio of an engine is a measure of how much the air or air fuel mixture is compressed in an engine cylinder. It is calculated by dividing air volume in one cylinder with piston at BDC by air volume with piston at TDC. Thermal efficiency is the ratio of energy produced in the engine to the energy in the fuel burnt to produce that much energy.

3.7.1 A comparative and constructional study between a diesel engine and petrol engine

The diesel engine (compression ignition engine) differs from the petrol engine (spark ignition) in the following respects.

1. In diesel engine, the air enters into the cylinder during suction stroke but air and fuel (petrol) mixture enters into the cylinder during suction stroke in petrol engine.

2. Diesel engine compresses only air and ignition is accomplished by the heat of compression. The petrol engine compresses a mixture of air and petrol which is ignited by an electric spark at the end of compression stroke. In diesel engine, atomized fuel is sprayed into the combustion chamber at the end of compression stroke and get ignited by the heated and compressed air.
3. The diesel engine burns fuels that vaporize less readily than petrol and runs more kilometre per litre in a vehicle than does the petrol engine because of its (diesel) higher thermal efficiency.
4. Diesel engine cylinder is fitted with a fuel injector. Petrol engine cylinder is fitted with a spark plug.
5. Diesel engine burns fuel (mostly diesel) of low volatility as compared to petrol of high volatility.
6. Diesel engine is chiefly adopted to stationary uses and tractors, buses, trucks etc. Petrol engines are used comparatively light vehicles like car, jeep scooter, motorcycle etc.

3.8 BASICS OF A HEAT ENGINE

Force: A force is an action which changes or tends to change the state of rest or of uniform motion of a body in a straight line. Its unit is Newton.

Work: Work is the moving of an object against an opposing force. It is distance times force. A force of 1 Newton acting through a distance of 1 metre in the direction of the application of force performs an amount of work equivalent to one joule. 1 Joule = 1 Newton × 1 metre.

Energy: Energy is the ability or capacity to do work. Since energy is measured by the total amount of work that the body can do, hence energy is expressed in the same unit as the unit of work.

Power: It is the rate of doing work or work done per unit time. Work can be done slowly or it can be done rapidly. The rate at which work is done is measured in terms of power.

Torque: Torque is a twisting or turning force. Torque plays the same role in rotational motion as the force in translatory motion. When force acting on a body tends to produce rotation, it is said to exert a torque. Torque is defined as the product of force times the perpendicular distance from the pivot point to the point of application of force. Unit is (N-m). Work is the distance times force but torque is the force times distance only to differentiate between them.

Horsepower: A horsepower (hp) is the power of one horse or the measure of the rate at which a horse can work. In older days when the steam engine was developed, it was felt necessary to rate the engine. Prior to introduction of the engine, horses were used for transportation purposes and it was observed that a horse can easily carry a load of 200 Ib through a distance of 165 feet in one minute. In other words, the power of horse was 165 × 200 = 33000 feet-pound / min and this load carrying capacity was taken as one horse power.

In metric system *i.e.* 1 ft = 0.3 m and 1 Ib = 0,460 kg and putting in 33000 foot-pound/min = 4500 m-kg/min. Hence 1 hp = 4500 m-kg/min = 75 m-kg/sec.

1 hp = 75 × 9.8 N - m/s ~ 740 W and 1 N. m/s = J/s = 1 W

In S.l. System, power output from an engine is measured in kilowatt (kW), which is an electrical term. It is the amount of electricity; the engine can produce if it were used to drive an electric generator.

While measuring the engine performance, dynamometer is used. Dynamometer measures engine rpm and torque, from which we are to calculate the engine power from the following formula.

$\text{Hp} = \frac{2\pi NT}{4500}$ where N = Engine rpm, T = Torque (kg-m)

Bore: Bore is the diameter of the engine cylinder (Fig. 3.22)

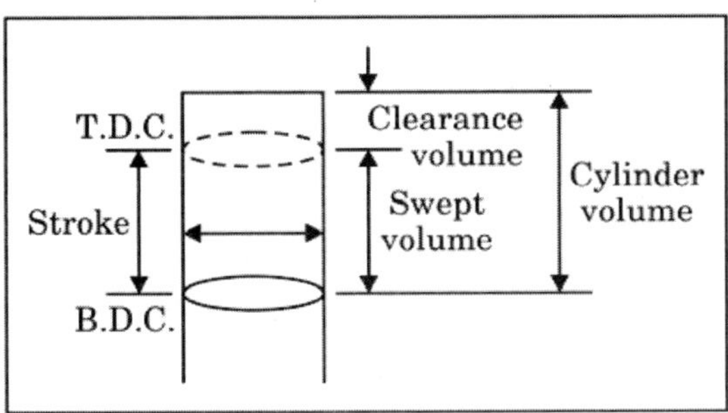

Fig. 3.22: Technical terms.

Cycle: The repetition of events in regular order is known as cycle. In an engine, the sequence repeated in regular order is Intake (Suction), Compression, Power, Exhaust.

Stroke-bore ratio: The ratio of length of stroke (L) and diameter of bore (D) of the cylinder is called stroke-bore ratio (LID). In general, this ratio varies between 1 to 1.45 and for tractor engines; this ratio is about 1 to 1.25.

Piston displacement (Swept volume): It is the volume (A × L) that the piston displaces or sweeps out as it moves in one stroke. A is the cross sectional area of piston and L is the length of stroke. Piston displacement: $\frac{\pi \times D^2 \times L}{4}$ where D is the diameter of cylinder.

Displacement volume (or simply displacement): Displacement volume of an engine is the total swept volume of all the pistons. For example, engine has eight cylinders, then displacement volume becomes equal to $8 \times \frac{\pi \times D^2 \times L}{4}$; volume is expressed in cubic centimetre (c.c.)

Compression ratio: The compression ratio of an engine is a measure of how much the air-fuel mixture is compressed in an engine cylinder. It is calculated by dividing the air volume in one cylinder with piston at BDC by the air volume with the piston at TDC.

$$\frac{\text{Volume of air or air fuel mixture in the engine cylinder when piston at BDC}}{\text{Volume of air or air fuel mixture in the engine cylinder when piston at TDC}} = \text{Compression ratio } (r)$$

The volume of air or air fuel mixture in the engine cylinder when piston is at TDC is called the clearance volume

$$r = \frac{\text{Total cylinder volume}}{\text{Clearance volume}} = \frac{\text{Swept volume + Clearance volume}}{\text{Clearance volume}}$$

'r' of 14: 1 means that during compression stroke, the air is compressed from 1/ 14 th of its original volume and 'r' of diesel engine varies from 14:1 to 22:1 and petrol engine varies from 4: 1 to 8: 1.

Volumetric efficiency: The amount of air or air-fuel mixture taken into the cylinder on the suction stroke is a measure of the engine's volumetric efficiency.

Volumetric efficiency is the ratio of the amount (either volume or weight) of air or air-fuel mixture that actually enters to the amount that could possibly enter (piston displacement).

Indicated horse power: Indicated horse power is the power that the engine develops inside the combustion chamber during power stroke and is received by the piston.

The reason we call it 'indicated power' is that we can measure it with an indicator. An indicator is an instrument which gives graphical representation of pressure inside the engine cylinder at each point of piston stroke. Indicated horse power is a term employed in research and laboratory work, but is of little practical value to the owners and operators of engines.

Indicated horse power = $\frac{PLAN}{4500}$; N = $\frac{\text{Engine 2 r.p.m}}{2}$ for 4 stroke and = Engine r.p.m for 2 stroke cycle engine. P = mean effective pressure in kg/cm^2; L = length of piston stroke in meter; A = Area of piston top or cross sectional area of piston cm^2; N = Number of Power stroke per minute. Suppose there are 8 revolution per minute, then the number of power strokes will be 8/2 = 4, for four stroke and for

2 stroke cycle engine, it will be 8. If there will be x cylinders, then total. I.H.P. = $\frac{\text{PLAN}}{4500}x$.

Indicated mean effective pressure: The pressures developed during the combustion or power stroke is at the highest level when the piston is at TDC and decreases gradually when piston moves downward and reaches at BDC. Hence indicated mean effective pressure is the average pressure in the cylinder throughout the power stroke. It is measured with the help of a special instrument called pressure indicator.

Brake mean effective pressure (BMEP): Brake mean effective pressure of an engine cannot be measured. It is merely a value used for comparing the mean pressures in the engine cylinders. It is expressed in kg/cm^2. It is calculated from the brake horse power of the engine *i.e.*, BMEP = $\frac{\text{BHP} \times 4500}{\text{LAN}}$ where N = rpm/2 (for 4 stroke) and = rpm (for 2 stroke cycle engine).

Brake horse power: It is the horsepower available on the crankshaft and is measured by a suitable dynamometer. It is the useful power available or produced to get the useful work. Engine power output is measured in terms of brake horse power (bhp). The name comes from the braking device that is used to hold engine speed down while horsepower is measured. For example when an engine is rated 5 hp, it is meant for the brake horsepower of the engine. This is the amount of usable power the engine can produce at a certain speed. The usual way to rate an engine is with a dynamometer.

Friction horse power: Friction horse power (fhp) is the power required to overcome the friction of the moving parts in the engine (parts moved in valve mechanism, piston rings, fuel pump, lubrication pump etc.).

Relation of BHP, IHP and FHP is that BHP = IHP – FHP

Brake horse power is the power delivered, IHP is the power developed in the engine and FHP is the power lost due to friction. Hence the horsepower delivered by the engine (bhp) is equal to the horsepower developed (ihp) minus the power lost due to friction.

Belt horse power: It is the power of the engine measured at the end of a suitable belt, receiving drive from the pto shaft (power take-off shaft).

Drawbar horsepower: It is the power delivered by tractor, measured at the end of the drawbar. It is that power which is available to pull loads. Drawbar is a flat bar having a number of holes and is attached to the rear axle of the tractor to pull the implements.

Power take off horsepower: It is the power of the tractor available at its PTO shaft. In general, the belt and PTO horsepower of a tractor will approximately be same and is measured by either a hydraulic or an electrical dynamometer.

Specific fuel consumption: Specific fuel consumption is expressed in terms of weight or volume of fuel supplied per horse power of engine output per unit time. Since the output of the engine is normally expressed in terms of brake horse power, hence specific fuel consumption is expressed on the basis of b.h.p. output and also sometimes called brake-specific fuel consumption.

Engine efficiency: The term efficiency compares the effort exerted and the results obtained. Engine efficiency can be compared in two ways as mechanical efficiency and thermal efficiency. Mechanical efficiency: $\frac{\text{bhp}}{\text{ihp}} \times 100$

Thermal efficiency: Thermal means "of or related to heat"

Thermal efficiency of an engine =

$$\frac{\text{Power produced in a given time}}{\text{Energy in fuel burned to produce this much power for the same time}}$$

Some of the heat produced by combustion is carried away by the engine cooling system. Some heat is lost in the exhaust gases. These are all thermal losses that reduce the thermal efficiency of the engine. They do not add to the power output of the engine. The remainder of the heat is used by the engine to develop power. Because a large quantity of heat is lost during engine operation, thermal efficiencies maybe as low as 20 per cent. They are seldom higher than 25 per cent. Now if the efficiency is determined on the basis of brake output of the engine, then it is known as brake thermal efficiency and in case it is determined on the basis of indicated horse power, then it is known as indicated thermal efficiency.

$$\text{Indicated thermal efficiency} = 100 \times \frac{\text{indicated power developed}}{\text{Heat input of fuel}}$$

$$\text{and Brake thermal efficiency} = \frac{\text{Brake horse power}}{\text{Heat input}} \times 100$$

Example 1: Calculate the brake horse power of a 4 stroke 4 cylinder I.C. engine, having the following dimensions; cylinder bore (D) = 120 mm, stroke length (L) = 150 mm, crankshaft speed (rpm) = 1000. Friction power = 8 hp; Mean effective pressure = 3 kg/cm^2. No. of cylinder = 4.

Solution:

I.H.P. = $\frac{PLAN}{4500} x$ and A = $\left(\frac{\pi}{4}\right)(12)^2$ = 113.14 cm^2 and I.H.P = 3 × 0.15 × 113.14 × $\frac{1000}{2} \times \frac{4}{4500}$ = 22.62 hp; BHP = IHP – FHP = 22.62 – 8 = 14.62 hp.

Example 2: Calculate the brake horse power of a 2 cylinder 4 stroke cycle I.C. engine of 15 × 20 cm. The mean effective pressure is 600 kPa and speed of crankshaft is 1000 revolution per minute. The mechanical efficiency is 70 %.

Solution:

D × L = 15 × 20 cm; D = 15 cm L = 20 cm 1 N= 1kg (mass) × 9.8

$$600 \text{ kPa} = 600{,}000 \text{ Pa} = 600{,}000\frac{\text{N}}{\text{m}^2} = \frac{600000}{9.8 \times 10^4}\frac{\text{kg}}{\text{cm}^2}\ 6.12 \text{ kg/cm}_2;$$

$A = \frac{\pi}{4}(15)_2 = 176.78 \text{ cm}^2$; L = 0.2 m; 1 HP $= \frac{\text{PLAN}}{4500}x = 6.12 \times 0.2 \times 176.78 \times \frac{1000}{2} \times \frac{2}{4500} = 48.08$ hp = 35.86 kW; Brake horse power = 35.86 × 0.70 = 25.10 kW.

Example 3: A 4 cylinder 4 stroke gas engine has cylinder diameter of 20 cm, store-bore ratio is 1.5, clearance volume 2500 cm^3, engine speed 300 rpm, mean effective Pressure 600 kPa and mechanical efficiences 70%, Calculate IHP, BHP, compression ratio, swept volume.

Solution:

L/D = 1.5; D = 20 cm; L = 1.5 × 20 = 30 cm; $A = \frac{\pi}{4}(20)^2 = 314.28 \text{ cm}^2$; L = 0.3 m; 600 kPa = 6.12 kg/cm^2; IHP $= \frac{\text{PLAN}}{4500}x = 6.12 \times 0.3 \times 314.28 \times \frac{300}{2} \times \frac{2}{4500} = 38.46$ hp = 28.69 kW; BHP = 28.69 × 0.70 = 20 kW;

$$\text{Compression ratio 'r'} = \frac{\text{Swept vol + Clearance vol}}{\text{Clearance vol}}$$

$$= \frac{\frac{\pi}{4} \times (20)^2 \times 30 \text{ cm}^3 + 2500 \text{ cm}^3}{2500 \text{ cm}^3} = 4.77 : 1$$

Example 4: A 2 stroke single cylinder diesel engine having 30 cm bore diameter, 45 cm stroke length has a brake wheel 2 m, mean effective pressure = 3 kg/cm^2, load on the brake in 60 kg, speed 300 rpm, oil consumed 5 kg/hr and calorific value of fuel 10500 kcal/kg. Calculate mechanical efficiency and brake thermal efficiency.

Solution:

D = 30 cm, L = 45 cm = 0.45 m; $A = \pi/4\ (30)^2 = 706.5 \text{ cm}^2$; P = 3 kg/cm^2.

$$\text{IHP} = \frac{\text{PLAN}}{4500}x = \frac{3 \times 0.45 \times 706.5 \times \frac{300}{2} \times 1}{4500} = 31.79 \text{ hp}$$

Torque = load × brake arm (radius of brake wheel)

A brake is device by means of which artificial frictional force is applied to a moving body in order to retard or stop the motion of a body. A brake dynamometer is a brake but in addition, it has a device to measure the frictional resistance.

T = 64 × 2/2 = 64 kg – m; Brake power

$$= \frac{2\pi NT}{4500} = \frac{2 \times 3.14 \times 300 \times 64}{4500} = 26.79 \text{ hp} = 20 \text{ kW}$$

$$\text{Mechanical efficiency} = \frac{26.79}{31.79} = \frac{\text{BHP}}{\text{IHP}} = 0.84 = 84\%$$

$$\text{Fuel per kW hr} = \frac{5 \text{ kg}}{20} = 0.25 \text{ kg}$$

Heat supplied per kW hour = 0.25 × 10500 = 2625 kcal; 1 w = J /s;

$$1 \text{ kW-hr} = \frac{10^3 \text{J} \times 3600\text{S}}{\text{S}} = 3600 \text{ kJ and } 1 \text{ J} = 0.238 \text{ cal.}$$

1 kWhr = 3600 × 0.238 = 857 kcal; Brake thermaf efficiency 857/2625 = 32 %

Example 5: A 4 cylinder 4 stroke engine is developing a mean effective pressure of 6.5 kg/cm². The diameter of cylinder is 10 cm, stroke length is 12.5 cm and speed of crankshaft is 1800 rpm. If the mechanical efficiency of the engine is 70%, calculate BIIP, IHP of the engine. What is the horsepower required to rotate the engine at no load at the same rpm.

Solution:

P = 6.5 kg/cm²; D = 10 cm; L = 12.5 cm; Engine rpm = 1800 rpm,

$A = \frac{\pi}{4}(10)^2 = 78.5 \text{ cm}^2$; For 4 stroke cycle, $N = \frac{\text{rpm}}{2}$; for 2 stroke cycle N = rpm.

$$\text{And IHP} = \frac{\text{PLAN}}{4500} x = \frac{6.5 \times \frac{12.5}{100} \times 78.5 \times \frac{1800}{2} \times 4}{4500}$$

$$= \frac{6.5 \times 12.5 \times 25.5 \times 78.5 \times 36}{4500} = 51.02 \text{ hp}$$

$$\text{Mechanical efficiency} = \frac{\text{BHP}}{\text{IHP}} = 0.7 = \frac{\text{BHP}}{51.02} \Rightarrow \text{BHP} = 35.71 \text{ hp}$$

FHP = IHP - BHP = 51.20 - 35.71 = 15.49 hp.

The horse power required to rotate the engine at no load in equal to FHP. Hence, horse power at no load = 16 hp = 12 kW.

Example 6: How many times, the power of an engine increase or decreases if the diameter of the piston is increased by 30 % and stroke length is reduced by 25%, all factors remaining same.

Solution:

Let the engine is a 4 stroke cycle engine

Original power = P_1 Changed Power = P_2

Original diameter = D_1 Changed diameter = $D_2 = D_1 \times 1.3$

Original stroke length = L_1 Changed stroke length = $L_2 = 0.75\ L_1$

Let the number of cylinder be 2,

$$P_1 = \frac{PL_1A_1N_1}{4500}x \text{ and } P_2 = \frac{P(0.75\ L_1)(A_2)N_1}{4500}x$$

$$A_1 = \frac{\pi}{4}(D_1)^2;\ A_2 = \frac{\pi}{4}(D_2)^2 = \frac{\pi}{4}(D_1 \times 1.3)^2;$$

$$\frac{P_1}{P_2} = \frac{PL_1\left(\frac{\pi}{4}\right)D_1^2Nx}{4500} \times \frac{4500}{P(0.75\ L_1)\frac{\pi}{2}(D_1 \times 1.3)^2\ Nx}$$

$$\frac{P_1}{P_2} = \frac{1}{0.8 \times 1.69} = \frac{1}{1.352};\ P_2 = P_1 \times 1.352$$

Hence power increases by 1.352 times.

Example 7: A 3 cylinder 4 stroke cycle engine develops 36 bhp at a speed of 3200 rpm. The specific fuel consumption (S.F.C.) is 205 gm/bhp/ hour. Calculate the amount of fuel supplied to each cylinder per injection.

Solution:

Fuel consumption per hour per cylinder

$$= \frac{\text{Horsepower} \times \text{S.F.C.}}{\text{No. of cylinder}} = \frac{36 \times 205 \text{ gm/bhp/hr}}{3} = 2460 \text{ gm/hr/cylinder}$$

For 4 stroke cycle engine no. of power strokes/min = rpm/2; No. of power stroke/ min = 3200/2 = 1600; No. of power stroke/hour = 60 × 1600; or no. of injections/hour = 60 × 1600.

Amount of fuel supplied to each cylinder per injection = = 0.02 gm........

Example 8: A high speed diesel engine operating at 2500 rpm has a single cylinder. During compression stroke, the fuel injection starts, 10° before TDC and ends at TDC. Calculate the duration in seconds the fuel injection takes place in the cylinder.

Solution:

Revolution/sec = $\frac{2500}{60} = \frac{125}{3}$; Crankshaft angle/sec. = $360 \times \frac{125}{3} = 15000^0$

Injection time for 10° rotation = $\frac{1 \times 10}{15000} = \frac{1}{1500}$ seconds.

Example 9: A diesel tractor consumes 6 litres per hour of diesel fuel of 9000 kcal/litre heat value. Calculate the power available at the drawbar if thermal efficiency is 32%, mechanical efficiency is 80% and transmission efficiency is 85%. Neglect rolling resistance and traction losses.

Solution:

Convert kcal/hr to hp. = $\frac{\text{Kcal}}{\text{hr}} = \frac{1000 \times \text{cal}}{3600 \text{ seconds}}$; But 1 Cal = 4.186 J;

$$= \frac{\text{Kcal}}{\text{hr}} = \frac{1000 \times \text{cal}}{3600 \text{ seconds}} = \frac{1000 \times 4.186}{3600}\frac{\text{J}}{\text{s}} = \frac{1000 \times 4.186}{3600}\text{W}$$; But 1 hp = 746 watt.

Hence $\frac{\text{Kcal}}{\text{hr}} = \frac{1000 \times 4.186}{3600 \times 746}$ hp = 0.0015 hp.

Fuel consumption = 6 lit/hr and calorific value = 9000 kcal/lit

So fuel consumption × calorific value = Fuel hp. = 9000 × 6 = 9000 × 6 × 0.0015 hp = 81 hp. This is the power developed by burning of fuel *i.e.* IHP

Drawbar horsepower = fuel hp × thermal efficiency (decimal) × mechanical efficiency (decimal) × transmission efficiency (decimal).

∴ Power available at the drawbar = 17.62 hp

Drawbar is a device by which the pulling power of the tractor is transmitted to the trailing implement. It consists of a cross bar with suitable holes attached to the lower hitch link and fitted at the rear part of the tractor.

Example 10: Calculate the horsepower available at the piston, engine flywheel, driving axle and drawbar of a tractor consuming fuel of 10 litres per hour and of fuel heat value 9000 kcal/litre. The thermal, mechanical and transmission losses being 60, 20 and 15 per cent respectively. Neglect rolling resistance and traction losses.

Solution:

Fuel hp	= Fuel consumption × fuel calorific value = 10 × 9000 × 0.0015 hp = 135 hp.
Thermal losses	= 60 % *i.e.* thermal efficiency = 100 - 60 = 40 %
Mechanical losses	= 20 % *i.e.*, mechanical efficiency = 80%
Transmission losses	= 15 % *i.e.*, transmission efficiency = 85 %.
Power available at piston	= fuel hp × thermal efficiency (decimal) = 135 × 0.4 = 54 hp.
Power available at the flywheel	= power at piston × mechanical efficiency (decimal) = 135 × 0.4 × 0.8 = 43.2 hp.
Power available at the driving axle	= Power available at the flywheel × transmission efficiency (decimal) = 135 × 0.4 × 0.8 × 0.85 = 36.72 hp.
Power available at the drawbar	= Power available at the driving axle - (rolling resistance and traction losses). But rolling resistance and traction losses are to be neglected. Then power available at drawbar is 36.72 hp.

Chapter 4

Thermodynamic Principle of I.C. Engine

4.1 INTRODUCTION

Thermodynamics can be defined as the science of energy. Although everybody has a feeling of what energy is, it is difficult to give a precise definition for it. Energy can be viewed as the ability to cause changes. The name thermodynamics derives from the Greek words therme (heat) and dynamis (power), which is most descriptive of the early efforts to convert heat into power. It mainly concerns with (i) the concept of energy *(ii)* the laws that govern the conversion of one form of energy into another *(iii)* the properties of the working substance or working media or working fluid to obtain the energy conversion.

The study of thermodynamics started with the analysis of heat engine processes to improve the engine efficiency, but today the scope has widened and there are important applications of thermodynamic principles outside the field of heat engines. Internal combustion engine is a heat engine and hence its efficiency is also analyzed with the thermodynamics principles. But the working medium, the gas, in case of internal combustion engine is compressed and expanded according to certain laws so as to give maximum possible cycle efficiency. These fundamental laws are called gas laws.

4.2 GAS LAWS

The vapor phase of a substance is customarily called as a gas when it is above the critical temperature. The best known equation of state for substances in gas phase is the ideal gas equation of state. The equation which relates the pressure, temperature and volume of the gas is called an equation of state. A gas which follows the gas laws at all ranges of pressure and temperature can be considered as an "Ideal gas", but no such gas exists in nature. A perfect gas or an ideal gas is defined as a gas having no forces of molecular attraction. However, real gasses tend to follow ideal gas at low pressure and high temperatures or at both. This is because the molecules are far apart at reduced pressures and elevated temperatures and the force of attraction between them tends to be small.

4.2.1 Boyle's Law

During the compression process of the gas in the cylinder, the pressure and temperature rise but the volume decreases. A rise in temperature takes place owing to the heat generated during compression. The fall in volume and rise in pressure have a definite relationship as defined by Boyle's law. It states that for a given quantity of gas, at constant pressure, volume is inversely proportional to the pressure.

Thus $V \propto \frac{1}{P}$ or PV = constant.

For different state points of the same gas, the relation can be expressed as $P_1V_1 = P_2V_2 = P_3V_3 \ldots = P_nV_n$.

4.2.2 Charles' Law

During the compression process, heat is generated and is liberated during the expansion or power stroke. Thus the temperature increases during the former process and decreases during the latter.

If the pressure of a definite mass of gas is kept constant, the volume of the gas changes by 1/273rd of its volume per degree rise of temperature above 0°C. Alternatively, if the volume of the gas is kept constant, with increase of temperature, the pressure increases by 1/273 of its value per degree rise in temperature above 0°C.

Charles' law defines the temperature-pressure-volume relationship of a gas. It states that pressure remaining constant, the volume of a fixed mass of a gas is directly proportional to its absolute temperature *i.e.* V *a* T or volume remaining constant, the pressure of a fixed mass of gas is directly proportional to its absolute temperature *i.e.* $P \propto T$.

4.2.3 Combination of Boyle's and Charles' Laws

During actual process in an engine, the pressure, volume and temperature of the gas vary simultaneously. The behavior of a gas can be studied better by combining the two laws.

By Boyle's Law $V \propto \frac{1}{P}$ (Temperature remains constant)

By Charles Law $V \propto T$ (Pressure remains constant)

$\therefore \quad V \propto \frac{T}{P}$ (Both pressure and temperature change)

or PV = CT where C is a constant of proportionality.

If M is the molecular weight of gas and N is the mole number, then, mass of gas (m) (kg) = MN and C = mR where R is the gas constant and is different for different gasses.

So, $$PV = mRT$$

The above equation is called the ideal-gas equation of state or simply the ideal gas relation. The gas that obeys this relation is called an ideal gas. Here P is in absolute pressure, T is the absolute temperature and V is the volume occupied by mass '*m*' of the gas.

Hence $$R = \frac{PV}{MP}\left(\frac{\frac{\text{Force}}{m^2}}{kg\ K}\right) = \frac{J}{kg\ K}$$

$$R = \frac{R_U}{M}$$

Where R_U is the universal gas constant and M is the molecular weight of the gas. The universal constant R_U is same for all gases. The value of R_U = 8.314 KJ/K mol. K

The ideal-gas equation of state can be written hi several different forms.

$$V = mv \rightarrow PV = mRT$$

$$mR = (MN)\ R = NR_U \rightarrow PV = NR_UT$$

Where v is the specific volume (volume/mass)

For unit mass of gas, we get PV = RT

4.3 TERMINOLOGY OF THERMODYNAMICS

System: It may be defined as a quantity of matter or a region in space chosen for study.

Surroundings: The mass to region outside the system is called the surrounding.

Boundary: The real or imaginary surface that separates the system from its surroundings is called the boundary. The boundary is the contact surface shared by both system and the surroundings. Mathematically speaking the boundary has zero thickness and thus it can neither contain any mass nor occupy any volume in space.

Universe: System + Surrounding

Isolated system: A system which can exchange neither energy nor matter with its surroundings is called an isolated system.

Closed system: A system which can exchange energy but not matter with its surroundings is called a closed system. An example of closed system is a piston and cylinder in case of I.C. engine.

Open system: A system which can exchange matter as well as energy with its surroundings is said to be an open system.

State of system: The state of the system relates to a set of properties that completely describes its condition. At a given state, all the properties of a system have fixed values, if the value of even one property changes, the state changes to a different one.

Extensive and intensive properties of a system: An extensive property of a system is that which depends upon the amount of substance and substances present in the system. The examples are mass volume and energy (internal energy). An intensive property of a system is that which is independent of the amount of the substance present in the system. Examples are temperature, pressure, density, specific heat etc.

Process: The operation by which a system changes from one state to another is called a process. Whenever a system changes from one state to another, it is accompanied by change in energy. Incase of open system, there may be change of matter as well.

Path: The series of states through which a system passes during a process is called the path of the process.

Cycle: A system is said to have undergone a cycle if it returns to its initial state at the end of the process. That is, for a cycle, the initial and final states are identical.

Isothermal process: The prefix 'iso' is often used to designate a process for which a particular property remains constant. An isothermal process is a process during which the temperature T remains constant. The change of state of a gas with respect to pressure and volume when temperature remains constant is known as isothermal change (or Boyle's law).

Isobaric process: Process with constant pressure.

Isochoric process: Process with constant volume.

Adiabatic process: A process during which there is no heat transfer is called an adiabatic process. The word adiabatic comes from the Greek word adiabatos, which means not to be passed. There are two ways a process can be adiabatic: Either the system is well insulated so that only a negligible amount of heat can pass through the boundary or both the system and the surroundings are at the same temperature and therefore there is no driving force (temperature difference) for heat transfer. An adiabatic process should not be confused with an isothermal process. Even though there is no heat transfer during an adiabatic process, the energy content and thus the temperature of a system can still be changed by other means such as work.

Let us assume that the walls of the cylinder head and piston are non conductors of heat. During compression of the gas within the cylinder, the heat produced remains inside the cylinder itself, increasing the temperature. No heat is transferred from surroundings nor is heat lost by the gas to the surroundings. The high speed of the engine does not allow heat to be either transferred from or to the surroundings. This process takes place according to the law PV^y = constant. The exponential 'y' depends on the initial pressure and nature of the gas, its value for air is 1.4. 'y' is the ratio of two specific heats of the gas *i.e.* ratio of specific heat at constant pressure and at constant volume. The adiabatic, process is also called the isentropic process. The relationships between pressure, volume and temperature in adiabatic process are (*i*) $P_1V_1^\gamma = P_2V_2^\gamma$ (*ii*) $\frac{T_1}{T_2} = \left(\frac{V_2}{V_1}\right)^{\gamma-1}$

Polytropic process: The relations of pressure and volume in polytropic process is $P_1V_1^n = P_2V_2^n$ = constant.

The value of n may be greater or less than γ in case of 'adiabatic process'.

If $n = 0$, then P = constant, the polytropic process becomes constant pressure process.

If $n = \infty$, $V = \left(\frac{C}{P}\right)^{1/n} = \frac{C}{P^{1/n}} = C$ the polytropic process becomes constant volume process.

If $n = 1$, the process becomes isothermal process

If $n = \gamma$, then $PV^\gamma = C$ which represents adiabatic process.

Working fluid: Heat engines and other cyclic devices usually involve fluid to and from which heat is transferred while undergoing a cycle. This fluid is called the working fluid.

Reversible and Irreversible Process

A process is said to be reversible if both the system and the surroundings can be restored to their original conditions. This is possible only if the net heat and net work exchange between the system and the surroundings is zero for the combined (original and reverse) process. Processes that are not reversible are called irreversible processes. The effects such as friction, heat transfer through a finite temperature difference etc, render a process irreversible.

Reversible processes actually don't occur in nature. They are merely idealizations of actual processes. Reversible processes can be approximated by actual, devices, but they can never be achieved. That is, all the processes occurring in nature are irreversible.

The reasons for reversible processes are two, first, they are easy to analyze, since a system passes through a series of equilibrium states during a reversible process; second, they serve as idealized models to which actual process can be compared.

Internally and Externally Reversible Processes

A process is called internally reversible if no irreversibility occurs within the boundaries of the system during the process. The system proceeds through a series of equilibrium states and when the process is reversed, the system passes through exactly the same equilibrium states while returning to its initial state. That is, the paths of the forward and reverse processes coincide for an internally reversible process.

A process is called externally reversible if no irreversibility occurs outside the system boundaries during the process.

Heat transfer between a reservoir and a system is an externally reversible process if the outer surface of the system is at the temperature of the reservoir.

A process is called totally reversible or simply reversible, if it involves no irreversibility within the system or its surroundings.

Reversible Cycle and Carnot Cycle

Heat engines are cyclic devices and that the working fluid of a heat engine returns to its initial state at the end of each cycle. Work is done by the working fluid during one part of the cycle and on the working fluid during another part. The difference between these two is the net work delivered by the heat engine. The efficiency of a heat-engine cycle greatly depends on how the individual processes that make up the cycle are executed. The net work, thus the cycle efficiency, can be maximized by using processes that require the least amount of work and deliver the most, that is, by using reversible processes. Hence, the most efficient cycles are reversible cycles, that is, cycles that consist entirely of reversible processes.

Reversible cycles cannot be achieved in practice because the irreversibility associated with each process cannot be eliminated. However reversible cycles provide upper limits on the performance of real cycles. Heat engines and refrigerators that work on reversible cycles serve as models to which actual heat engines and refrigerators can be compared. Reversible cycles also serve as starting points in the development of actual cycles and are modified as needed to meet certain requirements.

The best known reversible cycle is the Carnot cycle, first proposed in 1824 by French Engineer; Sadi Carnot. The theoretical heat engine that operates on the Carnot cycle is called the Carnot heat engine. The Carnot cycle is composed of four reversible processes - two isothermal and the two adiabatic. The processes are isothermal heat addition, isentropic expansion, isothermal heat rejection and isentropic compression. The Carnot cycle can be executed in a system for example a piston-cylinder device and either gas or a vapor can be utilized as the working fluid. The Carnot cycle is the most efficient cycle that can be executed between a heat source at temperature T_1 and a sink at temperature T_2 and its thermal efficiency is expressed as

$$\eta th,\ carnot = 1 - \frac{T_1}{T_2}$$

The Carnot cycle cannot be performed in practice because of the following reasons:

(i) Isothermal process can be achieved only if the piston moves very slowly to allow time for heat transfer so that the temperature remains constant. But power cycle in practice completes in a fraction of a second. Adiabatic process can be achieved only if the piston moves as fast as possible so that the heat transfer is negligible due to very short time available. In same cycle, the piston movement cannot be varied *i.e.* very slowly for part of the cycle and very fast in other part of the cycle.

(ii) It is impossible to perform a frictionless process

(iii) It is impossible to transfer heat without finite temperature difference.

Therefore it is not practical to build an engine that would operate on a cycle that closely approximates the Carnot cycle.

The real value of the Carnot cycle comes from its being a standard against which the actual or the ideal cycles can be compared. The thermal efficiency of the Carnot cycle is a function of the sink and source temperatures only, and the thermal efficiency relation for the Carnot cycle conveys an important message that is equally applicable to both ideal and actual cycles. Thermal efficiency increases with an increase in the average temperature at which heat is supplied to the system or with a decrease in the average temperature at which- heat is rejected from the system.

The source and sink temperatures that can be used in practice are not without limits. The highest temperature in the cycle is limited by the maximum temperature that the components of the heat engine can withstand. The lowest temperature is limited by the temperature of the cooling medium such as the atmospheric air.

4.4 AIR STANDARD CYCLES

A thermodynamic cycle maybe defined as a series of processes through which the working medium progress and allow the medium to return to its original state. As far as internal combustion engines are concerned, the whole process is such a complex problem that the simplifying assumptions need to be introduced. Since the complexity and the cost of engine test work is high, engine cycle analysis has become an important tool for engineers. Although I.C. engine does not operate on a thermodynamic cycles, as combustion process is not reversible and the working substance does not go through a cycle, still the concept of theoretical cycles maybe useful to show the effect of changing operating conditions on the performance characteristics, to indicate ultimate performance and to evaluate one engine relative to another. Thus, thermodynamics cycle is an artificial device employed to depict an ideal thermal cycle, in which all events are assumed to take place exactly as specified.

In gas power cycles, the working fluid remains a gas throughout the entire cycle. Petrol engine, diesel engine, gas turbine etc. are familiar examples of devices that operate on gas cycles. In all these engines, energy is provided by burning a fuel within the system boundaries. That is, they are internal combustion engines. The common working fluid in those devices is air. These devices take in either a mixture of fuel and air as in petrol engine or air and fuel separately and mix them in the combustion chamber as in diesel engine. During combustion process, the composition of working fluid changes from air and fuel to combustion products during the course of the cycle.

However, working fluid can be resembled to air due to the following reasons (i) The air is predominantly nitrogen that undergoes hardly any chemical reactions in the combustion chamber. Hence working fluid with air closely resembles air at all times *(ii)* The mass of fuel used and compared with the mass of air is very small. Therefore, the properties of mixture can be approximated to the properties of air.

Even though internal combustion engines operate on a mechanical cycle (the piston returns to its starting position at the end of each revolution), the working fluid does not undergo a complete thermodynamic cycle. It is thrown out of the engine at some point in the cycle (as exhaust gases) instead of being returned to the initial state. Working on an open cycle is the characteristic of all internal combustion engines.

The actual gas power cycles are rather complex. To reduce the analysis to a manageable level, we utilize the following approximations, commonly known as the air-standard assumptions. The simplest theoretical cycle is called the air-cycle approximation. The air-cycle approximation used for calculating conditions in internal combustion engines is called the air-standard cycle. The analysis of all air standard cycle is based upon the following assumptions:

(i) The working fluid nearly resembles air, which continuously circulates in a closed loop and always behaves as an ideal gas.

(ii) The physical constants of the gas in the cylinder are the same as those of air at moderate temperatures.

(iii) All processes that make up the cycle are internally reversible.

(iv) The combustion process is replaced by a heat-addition process from an external source.

(v) The exhaust power is replaced by a heat rejection process that restores the working fluid to its initial state.

(vi) The working medium does not undergo any chemical change during the cycle.

(vii) The specific heat capacity of gas is assumed constant throughout the cycle.

Because of many simplifying assumptions, it is clear that the air-cycle approximation does not closely represent the conditions within the actual cylinder. Although the quantities calculated from this approximation are considerably in error, the trends shown are usually correct, so that the general effect on the efficiency of such variable as inlet pressure or compression ratio may be calculated. Because of the simplicity ofthe air cycle calculation, it is often used to obtain rough answers to complex engine problems.

The most common air standard cycles used in connection with internal combustion engines are:

(i) **Otto cycle:** This cycle is also known as constant volume cycle. It is an ideal cycle for spark-ignition petrol engine.

(ii) **Diesel cycle:** This cycle is also known as constant pressure cycle. It is an ideal cycle for compression-ignition diesel engine.

(iii) **Dual cycle:** This cycle is also known as combination cycle.

4.5 IDEAL OTTO CYCLE

The Otto cycle is the ideal cycle for spark-ignition reciprocating engines. It is named after Nikolaus. A. Otto who built a successful four-stroke engine in 1876 in Germany. The cycle consists of two constant volume lines and two reversible adiabatic lines (Fig. 4.1) so as to give maximum theoretical efficiency. It is called constant pressure cycle, because when a gas is heated it expands. If the volume remains constant, the pressure rises according to Charle's Law.

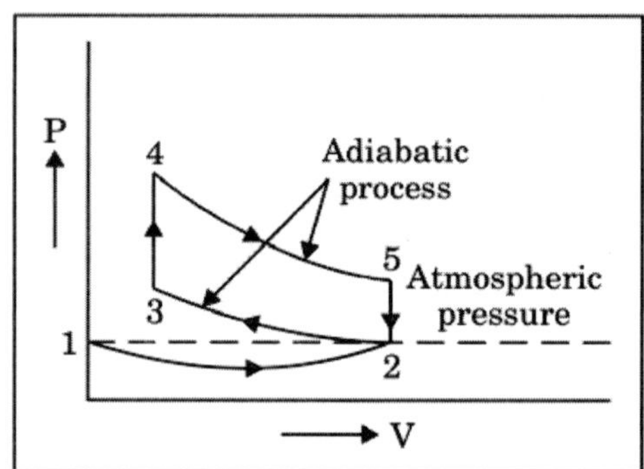

Fig. 4.1: Ideal otto cycle

Suction stroke (1-2) takes place below atmospheric pressure. Compression stroke (2-3) takes place adiabatically *i.e.* without heat exchange with the atmosphere. The piston reaches the top dead centre (T.D.C.). Heat is supplied from the hot body at constant volume (3-4). Pressure and temperature rise to P_4 and T_4, respectively. The gas expands during the working stroke (4-5) adiabatically reducing both pressure and temperature. Heat is rejected at constant volume (5-2) to the cold sink. The pressure and temperature fall back to their initial values.

The mechanical work done is proportional to area 2-3-4-5 and is equal to the difference of heat received from the hot body and heat rejected to the cold body considering 1 kg of air.

Q_1 = heat received = $C_v\,(T_4 - T_3)$

Q_2 = heat rejected = $C_v\,(T_5 - T_2)$

$$\text{Theoretical thermal efficiency} = \frac{\text{Work done}}{\text{Heat supplied}} = \frac{Q_1 - Q_2}{Q_2}$$

$$= 1 - \frac{Q_2}{Q_1} = \frac{T_5 - T_2}{T_4 - T_3}$$

For adiabatic compression and expansion,

$$= \frac{T_3}{T_2} = \left(\frac{V_2}{V_1}\right)^{\gamma-1} \text{ and } \frac{T_4}{T_5} = \left(\frac{V_5}{V_4}\right)^{\gamma-1}$$

or $$T_4 = T_5\left(\frac{V_5}{V_4}\right)^{\gamma-1}$$

or $$T_3 = T_2\left(\frac{V_2}{V_3}\right)^{\gamma-1} \text{ and } \frac{V_2}{V_3} = \frac{V_5}{V_4} = r \text{ (Compression ratio)}$$

$\therefore$ $$\eta = 1 - \frac{T_5 - T_2}{(r)^{\gamma-1}(T_5 - T_2)} = 1 - \frac{1}{(r)^{\gamma-1}}$$

Here r = 1.4 (for air)

This is the expression of air standard efficiency of the Otto cycle. This efficiency depends upon the compression ratio r, which is the ratio of total cylinder volume and clearance volume. Higher the compression ratio, higher is the thermal efficiency. But in actual practice, r, is limited due to certain operating considerations.

4.5.1 Actual Otto Cycle

Comparing the actual Otto cycle (Fig.4.2) with the theoretical one, we note the following differences.

1. The expansion and compression process don't take place exactly adiabatically.
2. Heat is added and rejected not necessarily at constant volume.
3. The mass of the gas during expansion and compression strokes varies invariably.

4. Combustion changes the chemical composition of the gas to a great extent. Thus the specific heat of the gas does not remain constant.
5. The exhaust gases do not return to their original condition and are rejected to the atmosphere. The actual Otto cycle is truly an open cycle and not a closed one.

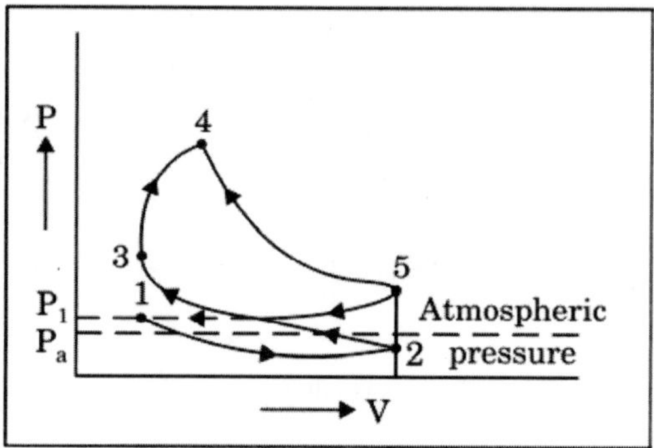

Fig.4.2: Actual otto cycle

Intake (1-2): During this process, a mixture of fuel and air is drawn into the cylinder, the amount of which depends upon *(i)* heat imparted by residual hot gases and *(ii)* amount of residual gases left after the previous exhaust stroke.

The pressure of gas during intake is at 1, which is higher than that of the atmosphere as the cylinder has certain amount of residual exhaust gas left after the previous exhaust stroke. The pressure drops from P_1 to P_a *i.e.* to that of atmosphere. As the piston travels from T.D.C. to B.D.C., the pressure of fresh gases goes down below the atmospheric pressure during the suction stroke P_2.

Compression (2-3): Fuel-air mixture is compressed adiabatically to a higher temperature and pressure, to create conditions suitable for ignition. Actually, the compression process takes place poly tropically *i.e.* according to the law $P_V^n = C$ and not adiabatically.

Combustion (3-4): The compressed air-fuel mixture is ignited by means of a spark jumping across the electrodes of the spark plug. This process actually does not take place at constant volume but begins before the T.D.C. position and ends after the T.D.C. position. The rate of combustion depends upon *(i)* compression ratio *(ii)* combustion chamber shape *(iii)* engine load *(iv)* spark timing etc.

Expansion (4-5): Some of the heat produced during combustion of fuel is converted into mechanical work. The gas at high pressure and temperature expands exerting pressure on the piston head to move it downwards. This process also does not follow the adiabatic law but follows a polytropic path. Heat exchange takes place between gas, combustion chamber and piston crown. Pressure and temperature are greatly reduced.

Exhaust (5-1): The residual gases must have minimum pressure as the higher this pressure, the greater shall be the loss in brake horsepower of the engine.

Pumping loss: The exhaust takes place at a pressure slightly higher than the atmospheric pressure due to the resistance of the exhaust value to the exhaust gases. This gives an area 1-2-1 in the form of small loop and indicates what is called the pumping loss of the engine. This area is treated as negative and hence subtracted from the area of the larger loop which is treated as positive, giving us the net work done during the cycle.

It is also to be noted in ideal p - v diagram that every corner is sharp which represents opening and closing of value instantaneously. Also the suction and exhaust take place at atmospheric pressure.

Actually, the opening and closing of valves cannot take place instantaneously but take some time by which every corner in the p-v diagram is round. Also the suction takes place at a pressure slightly lower than the atmospheric pressure due to the resistance of the inlet valve to the entering charge.

4.6 DIESEL CYCLE

The dies el cycle is the ideal cycle for CI reciprocating engines. The CI engine, first proposed by Rudolph Diesel in the 1890s, is very similar to the SI engine, differing mainly in the method of initiating combustion. In SI engines (also known as gasoline engines), the air-fuel mixture is compressed to a temperature that is below the auto ignition temperature of the fuel and the combustion process is initiated by firing a spark plug. In CI engines (also known as diesel engines), the air is compressed to a temperature that is above the auto ignition temperature of the fuel and combustion starts on contact as the fuel is injected into this hot air. Therefore, the spark plug and carburetor are replaced by a fuel injection in diesel engines.

Diesel cycle differs from Otto cycle in another aspect. The heat is added at constant pressure instead of at constant volume. Thus it comprises two adiabatic processes, one constant pressure heat addition process and another constant volume heat rejection process.

Diesel cycle is also known as the constant pressure cycle. Figures 4.3 and 4.4 show the ideal and actual diesel cycles respectively. Because with the high compression ratios of diesel engine, the maximum pressure of the cycle might be so high as to be dangerous if the constant volume cycles were used. Also the fuel injection process in diesel engine starts when the piston approaches TDC and continues during the first part of the power stroke. Therefore, combustion process in these engines takes place over a longer interval. Because of this longer duration, the combustion process in the ideal diesel cycle is approximated as a constant-pressure heat addition process. In fact, this is the only process where the Otto and diesel cycles differ. The remaining three processes are the same for both ideal cycles.

Intake: The intake of air (0-1) takes place at a pressure slightly less than atmospheric pressure.

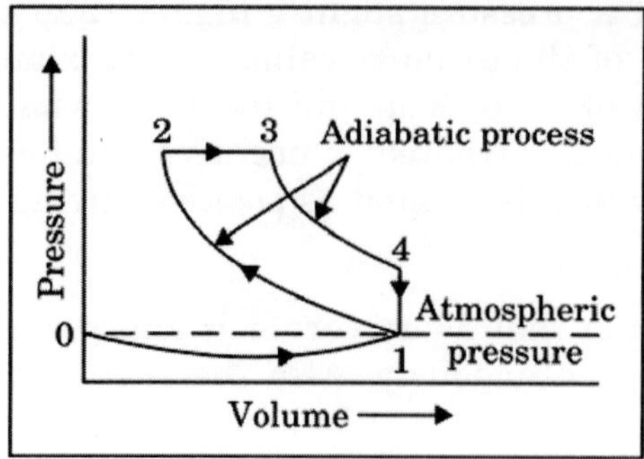

Fig. 4.3: Ideal diesel cycle

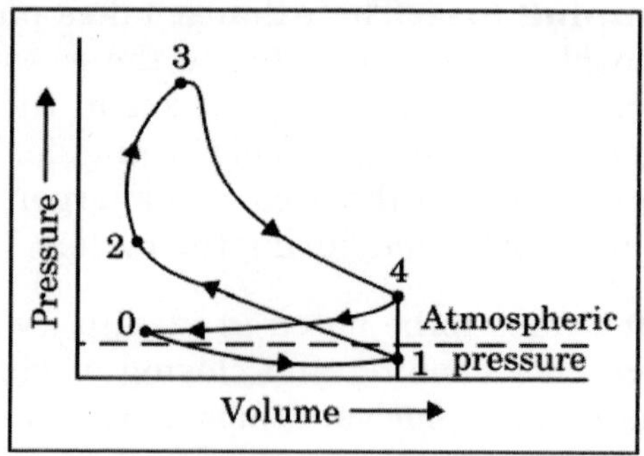

Fig. 4.4: Actual diesel cycle

Compression (1-2): Air is compressed adiabatically. But actual compression takes place poly tropically according to the law $PV^n = C$ where n ranges from 1.36 to 1.40.

Combustion (2-3): This process takes place at constant pressure. Fuel is sprayed under pressure at the end of the compression stroke. The fuel gets atomized and ignited by hot air.

Expansion (3-4): During this process, work is done by the gas on the piston. Actually, it follows a polytropic law *i.e.*, $PV^n = C$ where $n = 1.22 \rightarrow 1.25$.

Exhaust (4-0) and (4-1) in ideal cycle: Theoretically, the exhaust must take place at constant volume but actually it follows a different path. The pressure and temperature at the end of the exhaust process decrease.

Pumping loss: The area 0 – 1 – 0 represents pumping loss.

Theoretical thermal efficiency:

Heat added = $C_pP(T_3 - T_2)$; Heat lost = $C_v(T_4 - T_1)$ and

$$\text{Thermal efficiency} = \frac{C_p(T_3 - T_2) - C_v(T_4 - T_1)}{C_p(T_3 - T_2)}$$

$$\eta th = 1 - \frac{1}{\gamma}\left(\frac{T_4 - T_1}{T_3 - T_2}\right)$$

$\gamma = \frac{C_p}{C_v}$; the specific heat ratio and the cutoff ratio r_c, as the ratio of the cylinder volumes after and before the combustion process: $r_c = \frac{V_3}{V_2}$; Compression ratio $r = \frac{V_1}{V_2}$;

Applying the isentropic law to point '1' and '2' as well as '3' and '4'.

$$T_1 = T_2\left(\frac{V_2}{V_1}\right)^{\gamma-1} = \frac{T_2}{r^{\gamma-1}}$$

$$T_4 = T_3\left(\frac{V_3}{V_4}\right)^{\gamma-4} = T_3\left(\frac{V_3}{V_1}\right)^{\gamma-1} \text{ as } V_1 = V_4$$

$$= T_3\left(\frac{V_3}{rV_2}\right)^{\gamma-1} = T_3\left(\frac{r_c}{r}\right)^{\gamma-1}$$

Substituting the value of T_1 and T_4 in the thermal efficiency expression,

$$\eta th = 1 - \frac{1}{\gamma}\left[T_3\left(\frac{r_c}{r}\right)^{\gamma-1} - T_2\left(\frac{1}{r}\right)^{\gamma-1}\right]$$

$$= 1 - \frac{1}{\gamma}\left(\frac{1}{r^{\gamma-1}}\right)\left(\frac{T_3 r_c^{\gamma-1} - T_2}{T_3 - T_2}\right)$$

But $r_c = \frac{V_3}{V_2}$ and $P_2 = P_3$. Due to constant pressure $V \propto T$ so, $\frac{V_2}{T_2} = \frac{V_3}{T_3}$ or $T_3 = T_2\frac{V_3}{V_2} = T_2 r_c$

substituting the vlaue of T_3 in the above expression,

$$\eta th = 1 - \frac{1}{\gamma}\left(\frac{1}{r^{\gamma-1}}\right)\left(\frac{T_2 r_c^{\gamma} - T_2}{T_2(r_c - 1)}\right)$$

$$= 1 - \frac{1}{\gamma}\left(\frac{1}{r^{\gamma-1}}\right)\left(\frac{r_c^{\gamma} - 1}{r_c - 1}\right)$$

$$= 1 - \frac{1}{r^{\gamma-1}}\left(\frac{r_c^{\gamma} - 1}{\gamma(r_c - 1)}\right)$$

From the expression of efficiency of a diesel cycle, it is observed that its efficiency differs from the efficiency of an Otto cycle by the quantity in the bracket *i.e.* $\left(\frac{r_c^{\gamma} - 1}{\gamma(r_c - 1)}\right)$. This quantity is always greater than 1 as $r_c > 1$ and $\gamma > 1$. Therefore $\eta th_{(otto)}$ is more than ηth (diesel), when both cycles operate on the same compression ration. The efficiency of diesel cycle engine depends on compression ration as well as cutoff ratio.

4.7 IDEAL DUAL CYCLE

The approximation of the combustion process in internal combustion engine as constant-volume or a constant-pressure heat addition process is not quite realistic. Then a better (but slightly more complex) approach would be to model the combustion in both gasoline (petrol) and diesel engine as a combination of two heat transfer processes, one at constant volume and the other at constant pressure. The ideal cycle based on this concept is called the dual cycle and a p – V diagram for it is given in the figure 4.5.

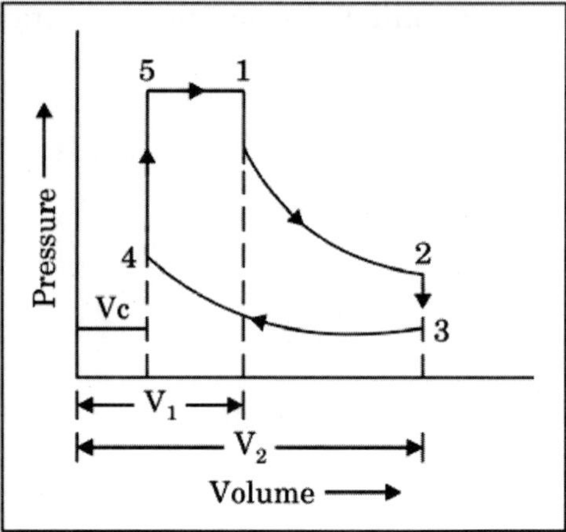

Fig. 4.5: Ideal dual combustion cycle

Air is compressed adiabatically along the path 3-4. Heat is supplied partly at constant volume (4-5) and partly at constant pressure (5-1). The expansion (1-2) is adiabatic. Heat is rejected to the cold body at constant volume (2-3).

$$\text{Thermal efficiency} = \frac{\text{Heat supplied} - \text{Heat rejected}}{\text{Heat supplied}}$$

$$\eta th = \frac{C_v(T_5 - T_4) + C_p(T_1 - T_5) - C_v(T_2 - T_3)}{C_v(T_5 - T_4) + C_p(T_1 - T_5)}$$

$$\eta th = 1 - \frac{C_v(T_2 - T_3)}{C_v(T_5 - T_4) + C_p(T_1 - T_5)}$$

$$= 1 - \frac{(T_2 - T_3)}{(T_5 - T_4) + \gamma(T_1 - T_5)}$$

where $r = \frac{V_3}{V_4};\ r_c = \frac{V_1}{V_5}$

Pressure ratio $r_p = \frac{P_5}{P_2}$ and expansion ratio $r_e = \frac{V_2}{V_1}$. Now, applying the isentropic law to points 3 and 4. $T_4 = T_3\left(\frac{V_3}{V_4}\right)^{\gamma-1} = T_3(r)^{\gamma-1}$ and applying gas laws points 4 and 5, and noting $V_4 = V_5$, $\frac{P_4}{T_4} = \frac{P_5}{T_5}$; $T_5 = T_4\frac{P_5}{T_4} = T_3(r)^{\gamma-1}\,r_p$. Similarly, applying gas laws to points 5 and 1 and noting $P_5 = P_1$, $\frac{V_5}{T_5} = \frac{V_1}{T_1}$; $T_1 = T_5\frac{V_1}{V_5}$; $T_1 = T_3(r)^{\gamma-1}\,r_p r_c$.

Applying the isentropic law to the points 1 and 2, T_2

$$= T_1\left(\frac{V_1}{V_2}\right)^{\gamma-1} = T_3(r)^{\gamma-1}\,r_p r_c\left(\frac{1}{r_e}\right)^{\gamma-1}$$

$$= T_3(r)^{\gamma-1}\,r_p r_c\left(\frac{r_c}{r}\right)^{\gamma-1} = T_3\,r_p r_c^{\gamma}.$$

Substituting the values of T_4, T_5, T_1 and T_2,

$$\eta th = 1 - \frac{\left(T_3 r_p r_c^{r} - T_3\right)}{\left\{T_3\,(r)^{\gamma-1}\,r_p - T_3\,(r)^{\gamma-1}\right\} + \gamma\left\{T_3\,(r)^{\gamma-1}\,r_p r_c - T_3\,(r)^{\gamma-1}\,r_p\right\}}$$

$$= 1 - \frac{1}{r^{\gamma-1}}\left[\frac{r_p r_c^{r} - 1}{\left(r_p - 1\right) + \gamma r_p\left(r_c - 1\right)}\right]$$

This is the required expression for air standard efficiency of Dual cycle in terms of compression ratio, pressure ratio and cut off ratio. The above expression shows that the efficiency increases with increase of compression ratio and with the increase of pressure ratio. Both the Otto and the diesel cycles can be obtained as special cases of the dual cycle. It can also be noted that for same heat input the Otto cycle gives higher efficiency as compared to diesel cycle and the efficiency of Dual cycle lies between the two.

4.8 DEVIATION OF REAL CYCLES FROM IDEAL CYCLE

Real cycles for internal combustion engines differ from the ideal cycles in many respects:

(i) The working substance is not pure air but a mixture of air and fuel gas, vapour or finally atomized liquid.

(ii) Combustion not only adds heat to the working substance but changes its chemical composition as well.

(iii) Specific heats of gases of the working substance vary considerably with temperatures.

(iv) The residual gases change the composition temperature and actual amount of the fresh charge.

(v) The theoretical constant volume or constant pressure combustion is not instantaneous.

(vi) Compression and expansion are not adiabatic and heat is exchanged in both directions between the gases and engine cylinder walls.

(vii) There is some definite time required for the combustion to occur inside the engine cylinder.

(viii) There are some heat losses due to the flow of heat from gases to cylinder walls.

Most of the factors listed above tend to decrease the thermal efficiency and power output of the actual engines. On the other hand, when taken into account, they show that the thermal efficiencies of actual engines are not so much different from the theoretically possible.

Chapter 5

Engine Balancing

5.1 INTRODUCTION

The internal combustion engines have reciprocating parts like piston and connecting rod that move once in one direction and then in the other, causing vibration while the engine is in running condition. The cycle of operations of four-stroke is completed in two revolutions of the crankshaft. With such an operating cycle, the crankshaft receives energy from the piston only during one half of its turn when the piston moves on the power stroke. During the remaining three half revolutions, the crankshaft continues to rotate by inertia and aided by the flywheel, it moves the piston on all its supplementary strokes *i.e.* exhaust, intake and compression. Therefore, the crankshaft on a single cylinder engine operating on the four-stroke principle revolves non-uniformly. It accelerates on the power stroke and decelerates on the supplementary strokes of the piston. Furthermore, the single-cylinder engine usually produces less power and creates excessive vibration. For this reason, modern tractors and automobiles are powered by multiple cylinder engines.

The nonuniformity in the rotation of crankshaft in case of single cylinder engine and uniformity in case of multi cylinder engine can be seen from the turning moment diagram of a single cylinder engine (Fig. 5.1). We see that there is a lot of variation between the torque developed during power stroke and the rest of the strokes namely exhaust, intake and compression. The mean torque is obtained by finding out the total area enclosed by the (T- 0) curve and dividing this value by 4vt or 720° *i.e.*, angular displacement of the crankshaft for completion of the cycle of operations.

A high value of maximum torque in a single cylinder engine necessitates heavier flywheel and transmission component parts. The engine performance at low speeds would be quite jerky.

Figure 5.2 depicts the turning moment diagram of a four cylinder engine. We notice that there is a peak of power stroke after every 180° of crankshaft revolution. The mean torque line is slightly below the peak torque value, it being quite different from that for a single cylinder engine. Therefore, the engine runs smoothly and requires a lighter flywheel. The flywheel has to store and release lesser amount of energy.

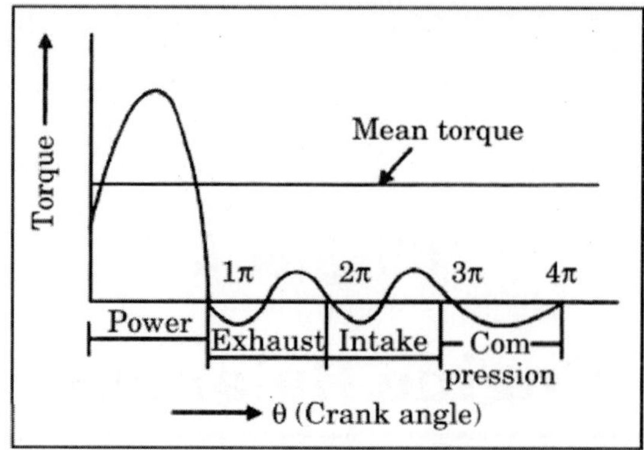

Fig. 5.1: Torque fluctuations (single cylinder engine)

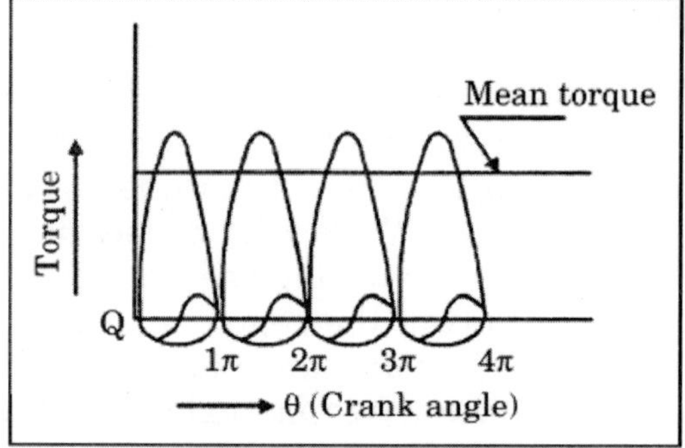

Fig. 5.2: Torque fluctuations (four cylinder engine)

The crank web weights perfectly balance the inertia forces caused due to the accelerating reciprocating parts in the case of multi cylinder engines, whereas balancing cannot be perfect in case of single cylinder engines. Hence many problems in the engine balancing can best be solved by using multi cylinder engine.

5.2 ENGINE BALANCING AND TYPES OF BALANCING

During the operation of the engine, the crank mechanism components are acted upon by various forces that change both in magnitude and in direction. Some of the forces are useful and provide for the engine to operate, while others are harmful and cause undue wear of the engine component parts.

The useful forces are those generated by the gas pressure in the cylinders. The harmful ones include the inertia forces produced by the changing motion of the reciprocating parts and friction forces resulting from the relative motion of rubbing pairs.

The reciprocating and rotary motion of the crank mechanism components give rise to the inertia forces of the reciprocating masses and the centrifugal forces of the rotating masses. These forces are transferred to the engine body and hence to the frame of the tractor or automobile. The periodic variations of the inertia forces, both in magnitude and in direction, cause the engine and the vehicle as a whole to vibrate. These vibrations weaken and the vehicle is a whole to vibrate. These vibrations weaken or loosen the screwed joints of the component parts and impose additional loads on the crankshaft bearings, thus accelerating their wear.

The balancing of an engine consists in creating such a system of forces acting in the engine under steady running conditions as will have its resultant forces and resultant moments of forces either constant in magnitude and direction or equal to zero. Counterbalancing the inertia forces is achieved by properly selecting the number of cylinders and the crank throw arrangement and by using additional moving masses *i.e.*, counterweights. These methods are usually used co jointly. The crank throws of four-cylinder engines are arranged at an angle of 1800 to one another, so that the inertia forces generated by the two extreme pistons and their associated connecting rods moving in one and the same direction are almost fully counterbalanced by the inertia forces produced by the neighboring pistons and connecting rods that move in opposite direction (Fig. 5.3a). The counterweights carried on the crankshaft webs balance the crank mechanism components associated with each cylinder.

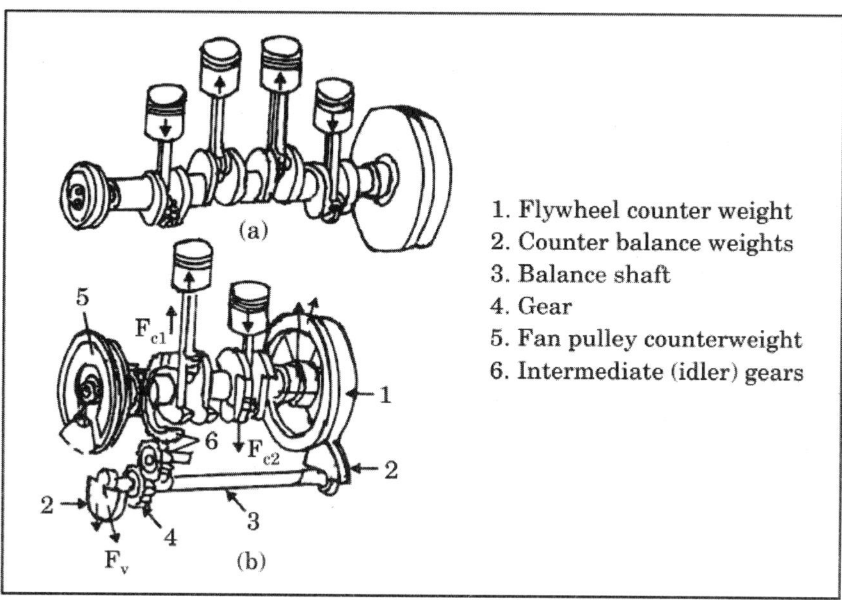

Fig. 5.3: Illustrating the balancing of engines. (a) four cylinder engine, (b) two cylinder engine.

In a two-cylinder tractor engine, the inertia forces due to the changing motion of reciprocating masses can be completely or almost completely counterbalanced by means of a special balancing mechanism (Fig. 5.3b). The mechanism consists of two counterbalance weights carried on the ends of balance shaft .The shaft is supported in plain bearing bushes and is driven from the crankshaft through the intermediary of two idler gears, so that it turns at the crankshaft speed but in the opposite direction. The fly wheel and fan pulley also carry counterweights. The inertia forces F_{c1} acting along the axis of the first cylinder are equal but opposite to the inertia forces F_{c2} acting along the axis of the second cylinder and so their sum is zero. But being equal in magnitude and opposite in direction, these forces produce a moment acting in the plane passing through the cylinder axes. This moment is counterbalanced by the resultant moment produced by counterbalance weights and counterweights. Engine balance and vibrations are most' prominent among the engine problems. Firing order is inter-related to the engine balance and vibration in multi cylinder engine.

In multi cylinder engines, smaller bore and stroke restrict the inertia forces due to the reciprocating masses of the engine, to a minimum. Thus such engines can run at high speeds without any problem. The design of multi cylinder engine is quite compact as the cylinders can be placed in several arrangements.

5.3 ENGINE BALANCING, FIRING ORDER AND FIRING INTERVAL

5.3.1 Engine Balancing

Engine balancing is related to firing order. One cylinder engine working on four stroke cycle has only one power strokes for every two revolutions of the crank shaft. Hence, it will not run smoothly and quietly, in spite of the compensating effect of a large flywheel. The operation will be very rough and to withstand it, the engine parts must be made large and heavy. Hence, single cylinder engine working on four stroke cycle is not used in tractor or automobile. Heavy vehicles or tractors adopt four, six and eight-cylinder engines. .By increasing the number of cylinders, the power strokes for each revolution of the crank shaft also increase, giving a more uniform torque and smoother operation. Above four cylinders, there is no period during which some cylinder is not delivering power and there is no time at which the fly wheel must supply all the power required to maintain the engine speed. The more cylinders in an engine, the more continuous flow of power, if the power strokes are spaced equally and the less work is to be done by fly wheel in storing and releasing the energy and the less is the vibration. A lighter fly wheel can serve the purpose in multi cylinder engine.

5.3.2 Firing Order

The sequence in which the power stroke occurs in each cylinder is called firing order. The arrangement of crank pin on the crankshaft and design of camshaft both determine the firing order.

When the cylinders are in line, the cylinder nearest to the radiator is designated as No. 1 and the one directly behind it in No. 2 and so on. The firing order is shown by the sequence of the number of cylinders in which the cylinders deliver their power strokes. For example, the firing order of a four cylinder engine will be written as l-2-4-3. This means that the firing will take place in the sequence of first, second, fourth and third cylinder respectively in a four cylinder engine.

5.3.3 Firing interval

The interval in degrees between successive power strokes in different cylinders is the firing interval. Firing interval of a two-stroke cycle engine is equal to $\frac{360°}{\text{no. of cylinder}} = \frac{\text{1 revolution of the crankshaft}}{\text{no. of cylinder}}$ = and firing interval of a four-stroke cycle engine is equal to $\frac{720°}{\text{no. of cylinder}} = \frac{\text{2 revolution of the crankshaft}}{\text{no. of cylinders}}$. The firing orders for various cylinder engines for engine balancing are described in the following sections.

5.4 POWER BALANCE

5.4.1 Single cylinder engine (Four-stroke)

In single cylinder engine, there is one power stroke in two revolutions of the crankshaft (Fig. 5.4).

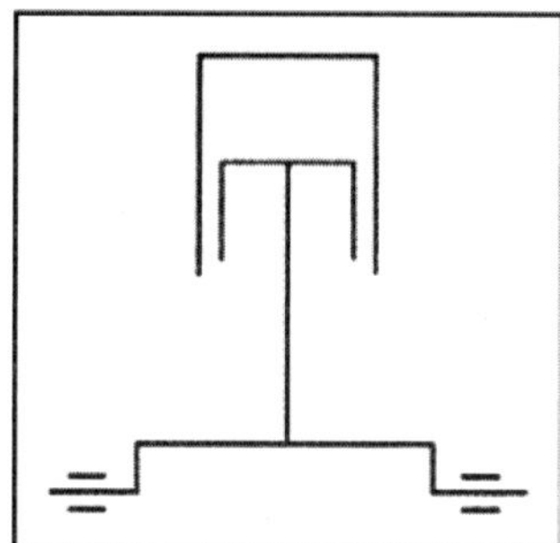

Fig. 5.4: Crank pin arrangements of single cylinder engine.

1st revolution	P	Where	P = Power stroke
	E		E = Exhaust stroke
2nd revolution	S		S = Suction stroke
	C		C = Compression stroke

Thus out of the four strokes of the piston, the power is delivered in one stroke (power stroke) and in the remaining strokes of the piston, the power is consumed in overcoming the friction resistance of the moving parts. The torque distribution during a cycle is uneven, resulting rough working and vibrations. Since there is only one piston and one connecting rod which reciprocate with no working parts to counter balance their weight, the single cylinder engine does not have mechanical balance. However the engine is balanced to some extent by using the counter weights attached to the crankshaft and also by using a fly wheel so heavy that its momentum produces a comparatively steady movement. The fluctuations high engine speed causes vibration, even in the best designs of one cylinder engines. Hence, one-cylinder engines are undesirable for use in tractors and automobiles.

5.4.2 Two-cylinder engine

It consists of two cylinders and the crank shaft is designed such that the crankpins are 180° apart (Fig. 5.5). When the engine is in operation, the two pistons move in opposite directions. Hence this construction has good mechanical balance with reference to the primary inertia forces. When piston No. 1 moves downward on power stroke, piston No. 2 moves upward on either compression or exhaust.

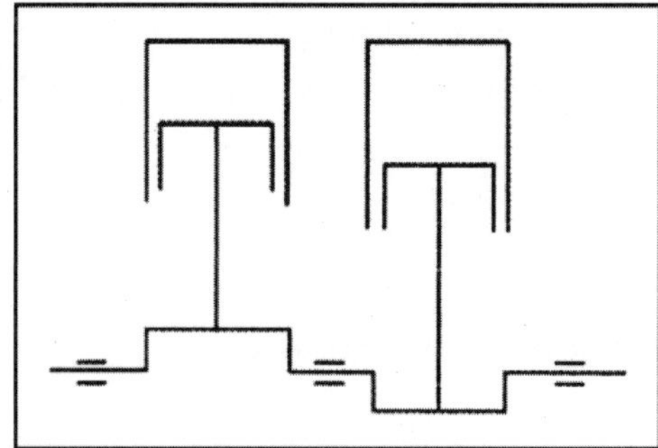

Fig. 5.5: Crank pin arrangements of two cylinder engine.

Table 5.1: Two cylinder engine- power balance sheet

	I			II	
Cylinder	1	2	Cylinder	1	2
1st revolution	P	C	1 st revolution	P	E
	E	P		E	S
2nd revolution	S	E	2nd revolution	S	C
	C	S		C	P

In Table I, the power balance is worked out with piston No. 2 moving upward on compression, in which both power strokes occur during the first revolution of the crank shaft while there are no power strokes during the second revolution. In Table II, the power balance is worked out with piston No. 2 moving upward on exhaust, in which one of power impulses is at the beginning of the first revolution and one at the end of the second revolution, producing the same result as obtained in Table I. Thus it is clear that this type of cylinder arrangement in the two-cylinder engine causes vibrations due to irregular production of power.

5.4.3 Three Cylinder Engines

In such engines, the crankshaft is designed such that the crank pins are 120°apart and the firing interval is given as

$$\text{Firing interval} = \frac{720^\circ}{3} = 240^\circ$$

The crank pins are arranged as shown in Fig. 5.6 and power balance sheet is given in Table 5.2.

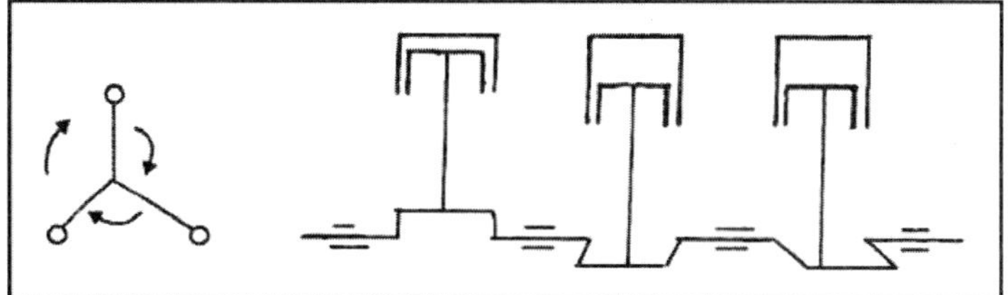

Fig. 5.6: Crank pin arrangement in a three-cylinder engine.

5.4.4 Four-cylinder Engines

Figure 5.7 shows the arrangement of cylinders and cranks. The firing interval = $\frac{720°}{3}$ = 180°, *i.e.* power strokes are executed at intervals of 180°. Piston number 1 and 4 move in the same direction together. If number one moves upward executing compression stroke, piston number 4 shall move in the same direction and execute the exhaust stroke.

Similarly, pistons 2 and 3 together move in the same direction. If piston number 2 executes the power stroke while moving downwards, then piston number 3 shall be on its intake stroke. Figure 5.7 shows crankpins 1-4 in one direction and crank pins 2-3 are placed at an angle of 180° to these. This type of placement neutralizes the effect of primary inertia forces although the secondary inertia forces are not completely balanced.

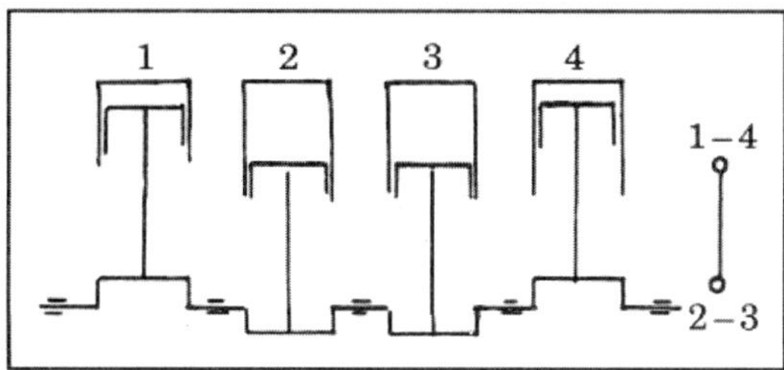

Fig. 5.7: Crank pin arrangements (four cylinder engine)

Table 5.3 depicts the four strokes executed by each piston as it completes one cycle of operations. This chart shows that there can be two possible firing orders 1-3-4-2 and 1-2-4-3.

The most suitable firing order in four cylinder engines is 1-3-4-2 as it is clear from the power flow chart (Table 5.4), that a power impulse is obtained in one of the cylinders after every 180°.

5.4.5 Six Cylinder Engines

A six cylinder engine has the firing interval $\frac{720°}{6}$ = 120° which means that six power strokes are made available after every 120° of crank shaft rotation. The crank throws No. 1 and 6, 2 and 5 and 3 and 4 are in the same radial plane. The angle between these planes is 1200.This engine gives two types of firing orders 1-5-3-6-2-4 (Fig. 5.8) and 1-4-2-6-3-5 (Fig. 5.9). The power flow diagram for six cylinder engine is presented in Table 5.5.

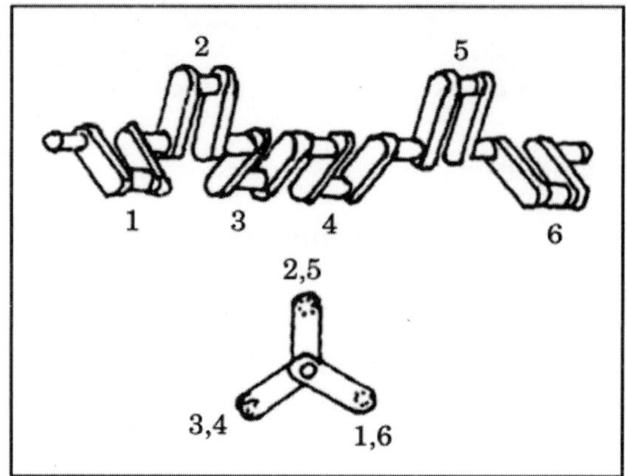

Fig. 5.8: Firing order of six cylinder engine (in-line)

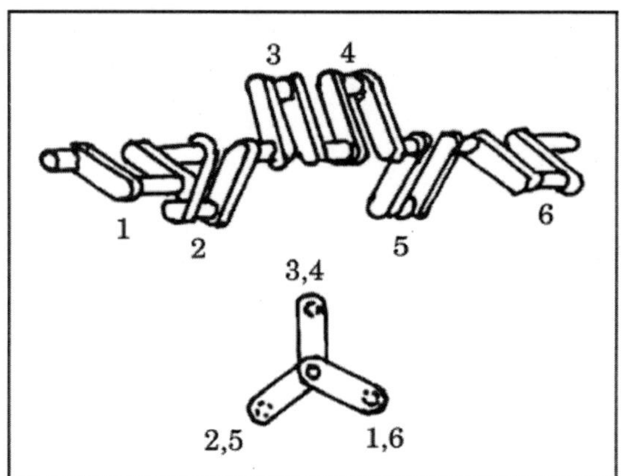

Fig. 5.9: Another type of firing order for six cylinder engine (in line)

Table 5.2: Three cylinder engines - power balance

Firing order	*Cycle of operation i.e. 2 revolutions of crank shaft*											
	One revolution of the crank shaft i.e. at 360°						*One revolution of the crank shaft i.e. at 360°*					
	One stroke or 180°			*One stroke or 180°*			*One stroke or 180°*			*One stroke or 180°*		
	60°	*60°*	*60°*	*60°*	*60°*	*60°*	*60°*	*60°*	*60°*	*60°*	*60°*	*60°*
Cylinder No.1	P	P	P	E	E	E	S	S	S	C	C	C
Cylinder No.2	S	C	C	C	P	P	P	E	E	E	S	S
Cylinder No.3	E	E	S	S	S	C	C	C	P	P	P	E

Table 5.3: Power balance chart-four cylinder engines

Cylinder	*1*	*2*	*3*	*4*	*Cylinder*	*1*	*2*	*3*	*4*
1st Rev.	P	E	C	S	1st Rev.	P	C	E	S
	E	S	P	C		E	P	S	C
2nd Rev.	S	C	E	P	2nd Rev.	S	E	C	P
	C	P	S	E		C	S	P	E

Table 5.4: Power flow diagram- four cylinder engines

Firing order	*Two crankshaft revolutions or 720°*			
	One revolution or 360°		*One revolution at 300°*	
	180°	*180°*	*180°*	*180°*
1	P	E	S	C
2	C	P	E	S
3	S	C	P	E
4	E	S	C	P

Table 5.5: Six cylinder engine-power flow diagram

Firing order	*Two revolutions or 720°*											
	One revolution or 360°						*One revolution at 360°*					
	180°			*180°*			*180°*			*180°*		
	60°	*60°*	*60°*	*60°*	*60°*	*60°*	*60°*	*60°*	*60°*	*60°*	*60°*	*60°*
1	P	P	P	E	E	E	S	S	S	C	C	C
5	C	C	P	P	P	E	E	E	S	S	S	C
3	S	C	C	C	P	P	P	E	E	E	S	S
6	S	S	S	C	C	C	P	P	P	E	E	E
2	E	E	S	S	S	C	C	C	P	P	P	E
4	P	E	E	E	S	S	S	C	C	C	P	P

5.4.6 Eight Cylinder Engines

In an eight cylinder engine, the firing interval is $\frac{720^\circ}{8} = 90^\circ$ which means that there occurs eight power strokes after every 90° of crankshaft rotation. The crank throws are arranged crosswise with an angle of 90° between them. Figure 5.10 shows the position of crank shaft and crank throws for one type of an eight-cylinder engine. The firing order is 1-6-2-5-8-3-7-4. The firing order for another type of crank shaft for an eight-cylinder engine is 1-4-7-3-8-3-7-4. The firing order for another type of crank shaft for an eight-cylinder engine is 1-4-7-3-8-5-2-6 (Fig. 5.11). The four crank throws are in horizontal plane and four in vertical plane, forming a cross in the side view. The eight cylinder engines are perfectly balanced for both primary and secondary inertia forces.

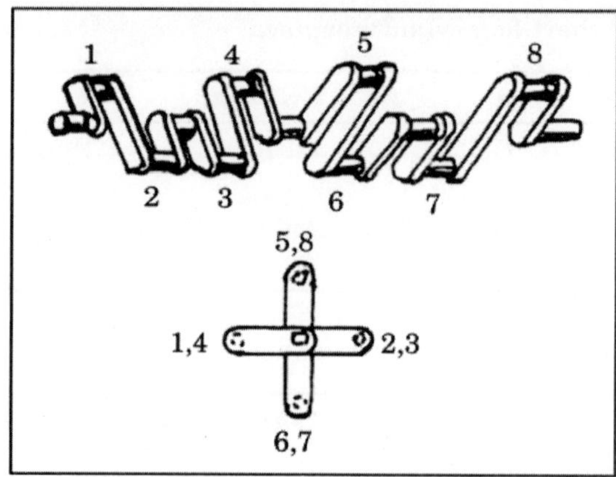

Fig. 5.10: Firing order of eight cylinder engine (in line)

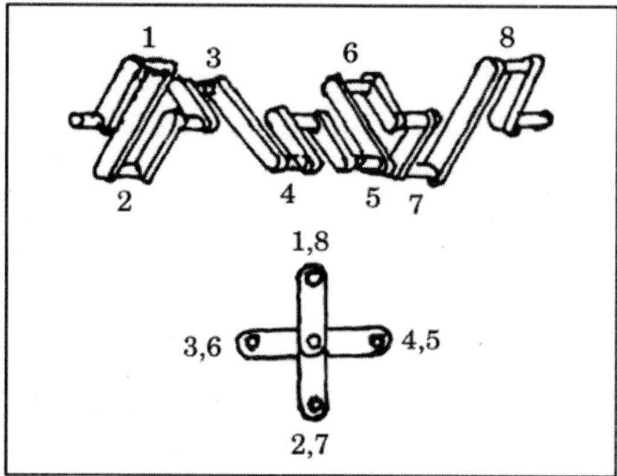

Fig. 5.11: Another type of firing order for eight cylinder engine (in line)

Knowing the firing order of an engine, one can correctly connect the ignition wires to the spark plugs or the fuel pipes to the fuel injectors and adjust the valves.

5.5 POWER OVERLAP

It has already been stated that multiple cylinder engines are smoother in operation than single cylinder engines. The more the cylinders in an engine, the more continuous will be the flow of power and smoother will be its operation.

In a four-cylinder engine, each piston in turn pushes the crankshaft through 180° rotation during its power stroke. Each power stroke begins every 180° exactly as the previous one ends. There is no overlap between the power strokes in a four-cylinder engine as shown in Fig. 5.12.

In a six-cylinder engine, each power stroke begins every 120° of crank shaft rotation and continuous for 180° (half a rotation). This produces a 60° overlap between power strokes that follow each other in the firing order of the engine as shown in Fig. 5.13. As one piston completing the last 60° of its power stroke, the next piston is operating on the first 60° of its power stroke.

Firing order	Stroke			
1	P	E	S	C
3	C	P	E	S
4	S	C	P	E
2	E	S	C	P

180°	180°	180°	180°
One revolution = 360°		One revolution = 360°	
Two crank shaft revolution = 720°			
One complete cycle			

Fig. 5.12: Power flow in a four cylinder engine (There is no power overlap)

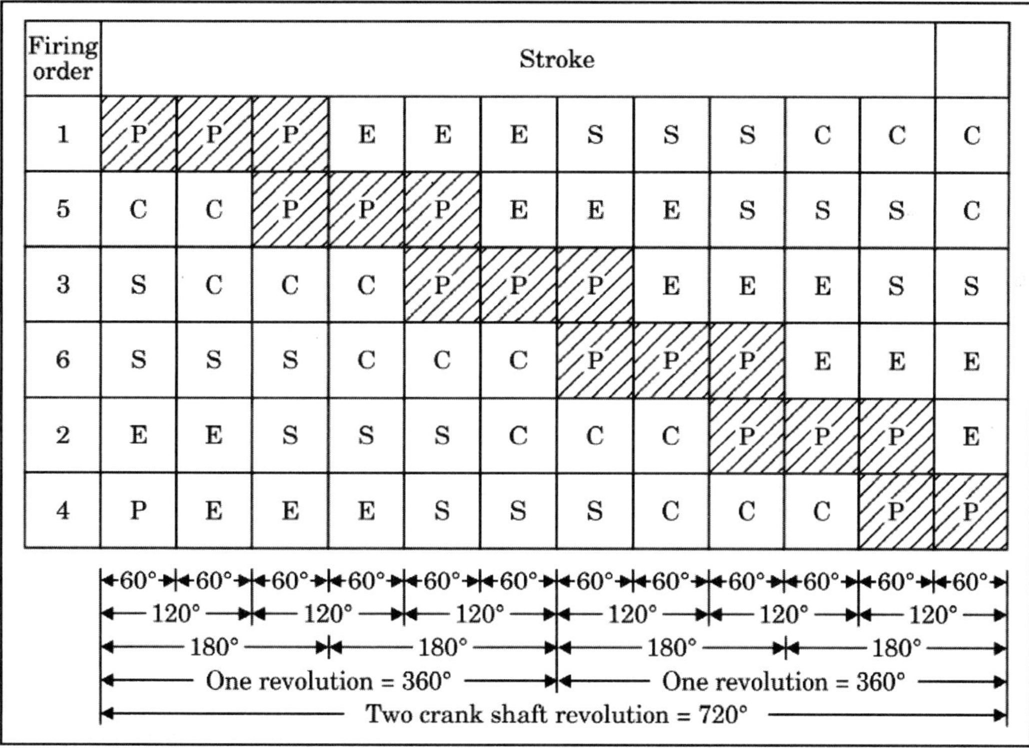

Firing order	Stroke											
1	P	P	P	E	E	E	S	S	S	C	C	C
5	C	C	P	P	P	E	E	E	S	S	S	C
3	S	C	C	C	P	P	P	E	E	E	S	S
6	S	S	S	C	C	C	P	P	P	E	E	E
2	E	E	S	S	S	C	C	C	P	P	P	E
4	P	E	E	E	S	S	S	C	C	C	P	P

60°	60°	60°	60°	60°	60°	60°	60°	60°	60°	60°	60°
120°		120°		120°		120°		120°		120°	
180°			180°			180°			180°		
One revolution = 360°						One revolution = 360°					
Two crank shaft revolution = 720°											

Fig. 5.13: Power flow in a six cylinder engine.

In an eight-cylinder engine, a power stroke begins every 90° of crank shaft $\left(\frac{720°}{9} - 90°\right)$. This produces a power overlap of 90° (180° – 90°= 90°). Similarly, a twelve-cylinder engine would have a power overlap of 120° ($\frac{720°}{12}$ – 60° and 180° – 60° = 120°). Similarly, a sixteen-cylinder engine an overlap of 135°. Obviously the more the cylinders in an engine, the more the power overlap and less the resulting vibrations produced by the engine.

Chapter 6

Valve Mechanism

6.1 A VALVE AND REQUIREMENTS OF A GOOD VALVE

A valve is a small mechanical device, used for opening and closing the passage leading to engine cylinder. Inlet valve of an internal combustion engine allows air or air-fuel mixture to go into the combustion chamber. The exhaust valve allows burnt gases to go out of the engine cylinder. Each valve is opened or closed once during each cycle. Valve operating mechanism is only present in 4 stroke cycle engine. The requirements of a good valve are as follows.

(i) Must be able to withstand high combustion pressures (30-45 kg/cm^2) and temperatures (2200°C).

(ii) Must be able to withstand erosion from corrosive combustion chemicals. Valves are chromium plated for this purpose.

(iii) Must effectively seal the cylinder to prevent the leakage of air fuel mixture and gases.

(iv) The stem of the valves must remain in perfect alignment with minimum wear.

(v) Must be of minimum weight.

(vi) Must have adequate area of contact with cylinder head or block (valve seat area)

6.2 CONSTRUCTION AND OPERATION OF VALVE MECHANISM

Valve operating mechanism in case of overhead type valve arrangement consists of several components like (i) crank gear (ii) cam gear (iii) camshaft (iv) cam follower (valve lifter or tappet) (v) Push rod (vi) rocker arm (vii) valve etc. Similarly the valve operating mechanism in case of L-head type valve (side valve) arrangement (section 2.3.2) consists of the components like (i) crank gear (ii) cam gear, (iii) cam shaft, (iv) cam follower (valve lifter or tappet) (v) valve etc. The different components of a valve are like (i) valve tip (ii) valve locks (iii) valve retainer (iv) valve spring (v) valve stem (vi) valve head (vii) valve face (viii) valve margin (Fig. 6.1).

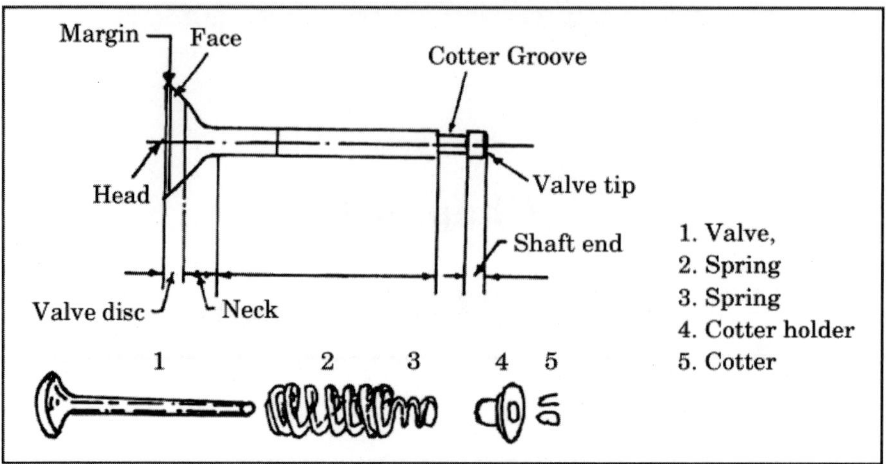

Fig. 6.1: Nomenclature of valve assembly

Every cylinder has an inlet and an exhaust valve. The valve mechanism in the L-head, T-head and F-head engines are of side valve types are shown in Fig. 6.2. The cam shaft is driven by the crankshaft. The cam shaft is connected to the crankshaft by a set of gears or by a chain. The camshaft rotates at half the engine speed. This is because we know that in a four-stroke cycle engine, the inlet valve opens once and exhaust valve also opens once in a complete cycle of two revolutions of the crankshaft. Therefore, the cam gear and the cam must operate these valves only once in the cycle and that is possible only when camshaft rotates at half the speed of crank shaft.

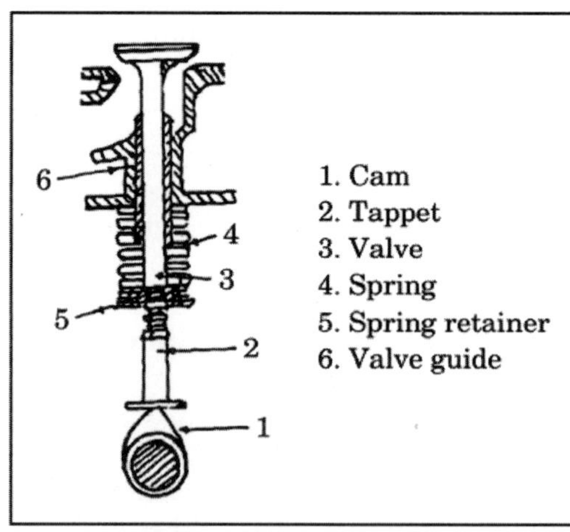

Fig. 6.2: Valve is lifted

The cam lifts the tappet and the valve. By this, valve spring is compressed and the valve opens. As the cam shaft rotates further, the tappet move down (Fig. 6.3). The valve is closed by the valve spring. The valve moves up and down in the valve-stem guide. The valve spring is held by means of a spring retainer.

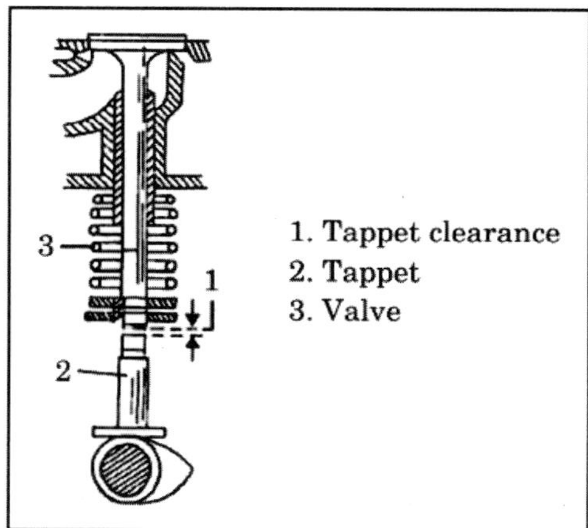

Fig. 6.3: Valve tappet clearance.

In case of overhead valve (in F and I type head design), when the camshaft rotates, the cam projection lift the tappet. This tappet operates the push rod. The push rod causes one end of the rocker arm to rotate about a shaft. The other end of the rocker arm pushes down the valve stem. Due to this action, the valve is opened. As camshaft rotates further, the tappet moves down. The valve now gets closed by the valve spring (Fig. 6.4). A tappet clearance is provided between the rocker arm and the bottom of the valve stem. This clearance can be adjusted by means of an adjusting screw.

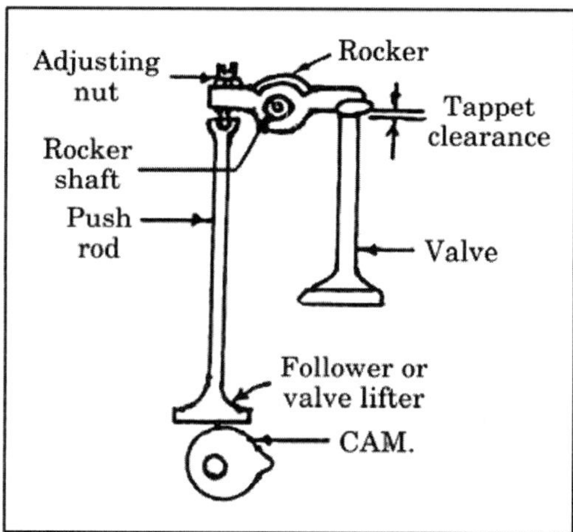

Fig. 6.4: Over head valve mechanism

6.3 VALVE MECHANISM COMPONENTS

The valve mechanism components are as follows:

Crankshaft gear: A gear fixed at the end of the crank shaft which meshes with the gear of the cam shaft is called crankshaft gear.

Cam gear: A gear fixed at the end of the cam shaft to mesh with the crank shaft gear is called cam gear.

Cam: Cam gets drive from cam shaft and pushes the cam follower (tappet).

Tappet: Tappet is also called the valve lifter. Tappet raises or lowers the valves through push rod. It receives motion from the cams mounted on the cam shaft. It opens or closes the valves at proper time. It is usually made of hardened steel.

Push rod: It is a long metallic rod, one end of which is connected to the cam follower and the other end is connected to the rocker arm. It pushes the rocker arm when the cam passes below the cam follower.

Rocker arm: It is an arm used to change the upward motion of push rod to downward motion of valve for its opening. It is a small rod, one end of which touches the end of valve stem (valve tip) and the other end touches the upper end of push rod. The rocker arm is supported at the middle by means of fulcrum.

Valve tip: It is the flat end of valve stem over which rocker arm presses.

Valve-spring attachment: This consists of valve lock, valve retainer and valve spring. One end of the valve spring presses against the cylinder head. The other end is attached to the end of the valve stem with a spring retainer and a retainer lock (valve lock). When installed, the spring is compressed with the retainer above it. Then the retainer lock is installed in the groove in valve stem. When the spring is released, the retainer presses against the locks holding it in place in the valve stem.

Valve spring: Valve is opened by the action of rocker arm but is closed by spring. The valve spring rests on steel washer on the valve guide while top of the spring is locked with valve lock at lock groove.

Valve stem: It is a round steel rod attached with the valve head.

Valve head: It is made up of specially alloy which can withstand high temperature and hammering action due to expanding gases. In most of the engines, flat head valves are used.

Valve face: It is a circular and angular seating surface machined on the valve head. Its angle is usually kept 30 or 45° to suit the similar angle of valve seat cut on cylinder head.

Valve margin: Valve margin is that part of valve stem which is in between valve head and valve face.

Valve seat: It is the place in the cylinder head where the valve head sits well. It may be made in cylinder head or cylinder block.

6.4 TAPPET/VALVE CLEARANCE

It is the gap between rocker arm and valve stem (valve tip) in case of over head type of engine. In case of side valve engine, it is the gap between the cam follower and valve stem. This gap varies from 0.006"–0.02". When clearance is more, valve will open late and close earlier. When clearance is less, the valve opens earlier and closes late. Tappet clearance may be equal for both the inlet and exhaust valves or less in the inlet and more in the exhaust valve, as the exhaust is subjected to more heat and hence will expand more.

6.4.1 Need to Adjust Tappet/Valve Clearance

Tappet/valve clearance is a must for smooth and efficient running of any engine. Proper clearance plays a major role in engine performance. A number of troubles may occur due to insufficient valve clearance. Both less and more clearance are harmful.

More clearance: The valve will open late and will close early. Hence a less amount of fresh air will enter during the suction stroke and consequently the volumetric efficiency and power developed will reduce. For exhaust valve, the exhaust gases will not get enough time to get out and hence will mix with the fresh air in the next stroke and early ignition of the mixture takes place during compression stroke, which is called pre-ignition.

6.4.2 Less Clearance

The valve opens early and closes late. For inlet valve, due to its early opening, the hot exhaust gases escape into the inlet manifold and burn the charge there which is called back firing. For exhaust valve early opening, there occurs the net loss of power in the previous stroke *i.e.*, power stroke. If there will be no clearance, the valve stem and push rod will bend.

6.5 VALVE TIMING DIAGRAM

A valve timing diagram is a diagram of crank shaft rotation (Fig. 6.5) with respect to opening and closing of inlet and exhaust valve. Hence valve timing mechanism is concerned with relative closing and opening of valves and their duration with respect to the cylinder position and the degree of crank shaft rotation.

The opening and closing points are determined by the following factors (i) the timing of crank shaft and cam shaft gears (ii) the shape of the cam (iii) the tappet or valve clearance.

Earlier it has been stated that the inlet valve opens during suction stroke and exhaust valve opens during the exhaust stroke. In practice, however, these valves open a bit earlier than the specified stroke and close after the stroke is over. Theoretically, a stroke is 180° (Piston movement from TDC to BDC or BDC to TDC), but in practice it is more than that. TDC = Top dead centre; BDC = Bottom dead centre; SVA = Suction valve open; SVC = Suction valve close; EVO = Exhaust valve open; EVC = Exhaust valve close.

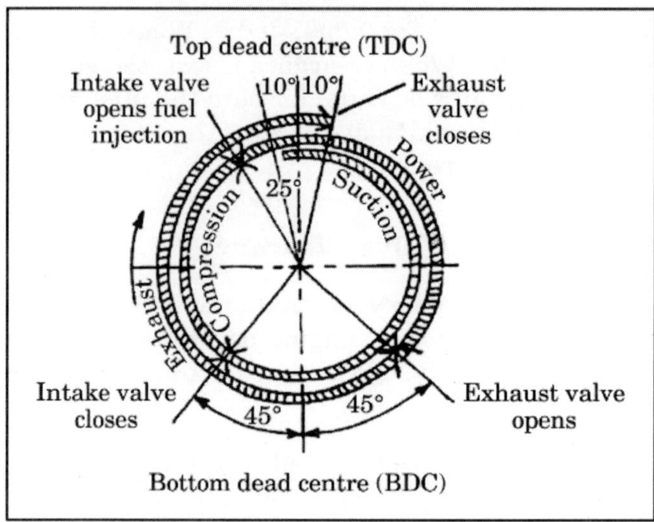

Fig. 6.5: Valve timing diagram

During suction stroke, the air is rushed into the cylinder because of the vacuum caused by the downward movement of the piston. As the engines are in high speed, the time for the air to enter is less. As a result, volumetric efficiency will be less and power developed will be less. That's why the valve is allowed to close after 35-450 from BDC.

The inlet valve is also allowed to open 5 to 10° earlier than the normal suction stroke, because the valve cannot open abruptly; the action is gradual. By allowing the valve to open earlier sufficient time is available for the valve to remain fully open. Thus sufficient air can enter into the cylinder. Hence total inlet valve opening according to the diagram is 180° + 45° + 10° = 235°.

As the piston is forced down by the expanding gases during power stroke, it is found necessary to open the exhaust valve 35 to 45° before BDC. By early opening, the burnt gases have an outlet for expansion and begin to rush out of the their own accord. This removes a greater part of the burnt gases, thereby reducing the amount of work to be done by the piston on its return stroke.

During the next upward stroke (exhaust stroke), the remaining gases are forced out of the open exhaust valve as the pressure in the cylinder exceeds than that in the exhaust manifold. Again if the exhaust valve is closed at TDC, a large portion of exhaust gases will be retained in the cylinder. The best results are obtained when the valve closes 5 to 10° after piston has started travelling down in the next suction stroke. Thus the total exhaust valve opening according to the diagram is 180° + 45° + 10° = 235°.

From the figure, it is seen that there is a overlap of 20° during which both valves are open at the same time. Most engines have valve overlap during the end of the exhaust stroke and in the beginning of the intake stroke. Then the exhaust gases are moving rapidly from the cylinder through the exhaust valve. Remaining exhaust valve opening after TDC (*i.e.*, after exhaust stroke), the gases get more time to leave. At the same time, starting to open the intake valve before TDC on exhaust stroke gives the incoming air-fuel mixture to enter into the cylinder.

Example 1: An engine is operating at 800 rpm. The inlet valve opens just at TDC and closes 40° after BDC. Calculate the time for which the inlet valve remains open.

Solution:

800 rpm = 800/60 revolution per sec. The inlet valve opens for 180° + 40° = 220° rotation of crankshaft; 1 revolution = 360°; 220° = 220/360 revolution of crankshaft; Number of revolution in 1 sec = 800/60. and for 1 revolution of crankshaft, the time required for the valve is 60/800 second. For 320/360 revolution, time taken is = 60/800 x 200/360 = 117240 second. Hence the time for which, the inlet valve remains open is 11/240 second.

6.6 VALVE MATERIAL AND VALVE COOLING

The material of the valve directly affects the valve life and performance. The operating characteristics of the engine determine the material of the valve. The inlet valve which runs at lower temperature is made of nickel chromium alloy steel. The exhaust valves which run at high temperatures are made of silicon chromium alloy steel. Valves are manufactured by drop forging method. The valve stem is chromium plated to reduce wear and erosion. The heads are nickel or aluminum plated to evade corrosive gases.

The heat flows from the head of the valve to the valve stem and then to cooling water. Some heat also flows to the cylinder head via the valve seat insert. The area of contact of the valve with seat insert must be adequate or else it will run hotter. The hottest portion of the valve is the valve head. The valve seat being near to the cooling water runs cooler. The valve length also changes due to heat. The coolest part of the valve is the valve stem which is cooled by the lubricant oil spread over the valve guides.

Chapter 7

Fuel and Combustion

Fuel is a substance consumed by the engine to produce energy. The common fuels for internal combustion engine are *(i)* Petrol or gasoline, *(ii)* Power kerosene and *(iii)* Diesel.

7.1 REQUIREMENTS OF A GOOD FUEL

A fuel must have a good vaporizing ability. It must vaporize easily from liquid to gaseous state and then evenly mix with intake air in correct proportions in the carburetor. High volatility helps in easy starting of the engine, decreases the warming up period and gives even and quiet operation at all speeds and loads. On the other hand, too much volatility results in poor economy as it vaporizes in the fuel system before being discharged from the carburetor jets.

In I.C. engine, all liquid fuel must be converted into vapour before burning. High speed diesel oil is most difficult to vapourise. Vapourising temperature of high speed diesel oil is higher than that of the petrol, hence petrol vapourizes quicker than diesel oil in the engine cylinder. This helps in easy starting of petrol engines. If all liquid fuel is not well vapourized, it would do crankcase oil dilution. The oil that vapourises quickly can be distributed well in different cylinders of the engine, hence distribution of fuel in different cylinders is better in petrol engine than that of diesel engine.

Heat or calorific value is an indicator of the power developed by the engines. It is measured by burning a unit mass of fuel in excess of air. Petrol has a high calorific value than diesel fuel *i.e.*, it liberates a large amount of heat when burnt completely.

Anti-knock quality determines the ability of a fuel to burn without causing knocking or detonation in the engine. A low octane number petrol detonate instead of burning smoothly in the combustion chamber. Similarly ignition quality refers to ease of burning the oil in the combustion chamber. Catane number determines the ignition quality of a diesel fuel. A higher cetane number fuel gets ignited as soon as it is injected into the cylinder. A lower catane number fuel results in difficult starting and knocking. Commercial diesel fuels have got cetane rating varying from 30 to 60.

Additives are added to improve the octane rating of the fuel. The fuel system serves the purpose of storage and delivery to the engine in the form of vapour. This fuel is mixed up with the proper ratio of air which forms highly explosive charge. After combustion, the waste gases are removed through the exhaust system. Hence petrol has the following characteristics.

(i) Is highly volatile *i.e.*, vaporizes quickly.
(ii) Provides better rate of combustion and flame propagation velocity.
(iii) Has higher heat content.
(iv) Has good antiknock quality *i.e.*, it burns without causing detonation
(v) Has no sulphur content and does not cause gumming effect in engine component parts owing to addition of lead while processing

Similarly diesel fuel has the following characteristics:

(i) Ready ignition and even burning throughout the combustion chamber
(ii) Good lubrication properties
(iii) Lower viscosity enables better atomization from nozzle hole
(iv) Lesser sulphur content and impurities
(v) Higher vapourising temperature than petrol.

7.2 SOURCES OF FUELS

Almost all of the fuels commonly used in farm tractors and in automobiles are derived from crude petroleum. The word petroleum is derived from two Latin words petra means rock and oleum means oil. Petroleum in a hydrocarbon.

7.2.1 Origin of petroleum

The origin of petroleum lies in plants and animals which lived on earth and in sea many millions of years ago. These organisms died and their remains became buried under the earth. Due to bacterial decomposition and under the action of earth's heat and pressure, these remains were converted to liquid and gaseous hydrocarbons, the petroleum. The gaseous hydrocarbons that were produced constituted the natural gas. Natural gas includes gases like methane, ethane, butane etc.

7.2.2 Composition of petroleum

Petroleum is an extremely complex mixture, made up of hundreds of compounds, mostly hydrocarbons. Hence petroleum is a mixture of many different hydrocarbons (compounds of carbon and hydrogen, combined approximately in the proportion of 86 per cent carbon to 14 per cent hydrogen) with some sulphur and other impurities. These hydrocarbons are grouped into four general categories.

1. Paraffms (alkanes) (C_nH_{2n+2}) [All hydrocarbons in which all carbon atoms are linked to one another by only single bonds].

2. Olefins (alkenes) (C_nH_{2n}) [Hydrocarbons containing one carbon-carbon double bond].
3. Naphthenes (C_nH_{2n}) [Cyclic or closed chain hydrocarbons in which all the carbon atoms are joined by single covalent bonds to form a ring type structure.
4. Aromatics (C_nH_{2n-6}) [Ring like structure with carbon atoms arranged in a closed chain with alternate single and double bonds]

Paraffins: The paraffin series of hydrocarbons begin with CH_4 (methane), the next higher one having one more carbon atom with two more H atoms, attached to it. The paraffin series in turn can be subdivided into normal paraffin hydrocarbons, which have a straight or open chain structure with one bond between each atom such as pentane, heptane etc. and isomers which have same number of C and H atoms and the same molecular weight but a different structure. Two or more compounds having the same molecular formula but different chemical and physical properties called isomers. Isomers of normal heptane are as follows:

```
    H  H  H  H  H                H  H  H  H  H  H  H
    |  |  |  |  |                |  |  |  |  |  |  |
 H—C—C—C—C—C—H              H—C—C—C—C—C—C—C—H
    |  |  |  |  |                |  |  |  |  |  |  |
    H  H  H  H  H                H  H  H  H  H  H  H
```

Pentane C_5H_{12} Heptane C_7H_{16}

```
   H H H
    \|/
     C  H  H  H  H  H
     |  |  |  |  |  |
  H—C—C—C—C—C—C—H
        |  |  |  |  |
        H  H  H  H  H
```

Methyl hexane

```
        H H H
         \|/
          C
          |
          C
          |
        H C H
         \|/
   H  H   C  H  H
   |  |   |  |  |
H—C—C—C—C—C—H
   |  |   |  |  |
   H  H   H  H  H
```

Ethyl pentane

Isomers of normal heptane

The paraffins are saturated, quite stable and low in gum-forming properties. Kerosene is a good example of commercial fuel made up largely of paraffins. The straight-chain hydrocarbons generally detonate badly in an engine. On the other hand, the isomers are highly knock resistant. Normal heptane (n-heptane), for example, is the low antiknock fuel (zero octane) used as a reference fuel in knock testing. Triptane, as isomer of heptane, is one of the most knock-free hydrocarbons known. These fuels have the same chemical formula but are different structurally and vary widely in burning characteristics.

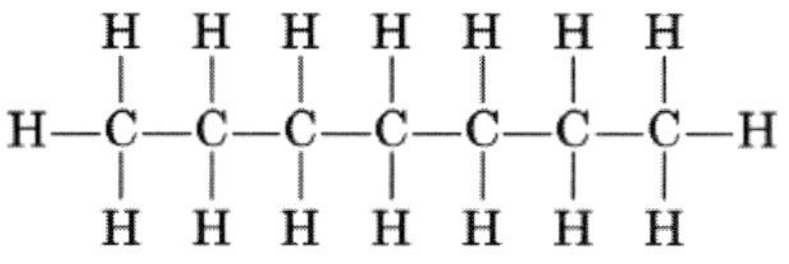

```
                                                    H           H
                                                    |           |
                                                 H—C—H     H—C—H
   H   H   H   H   H   H   H                  H    |           |    H
   |   |   |   |   |   |   |                  |    |           |    |
H—C—C—C—C—C—C—C—H                        H—C—C————————C—C—H
   |   |   |   |   |   |   |                  |    |           |    |
   H   H   H   H   H   H   H                  H    |           H    C
                                                    |
                                                 H—C—H
                                                    |
                                                    H
```

n-heptane (C_7H_{16})

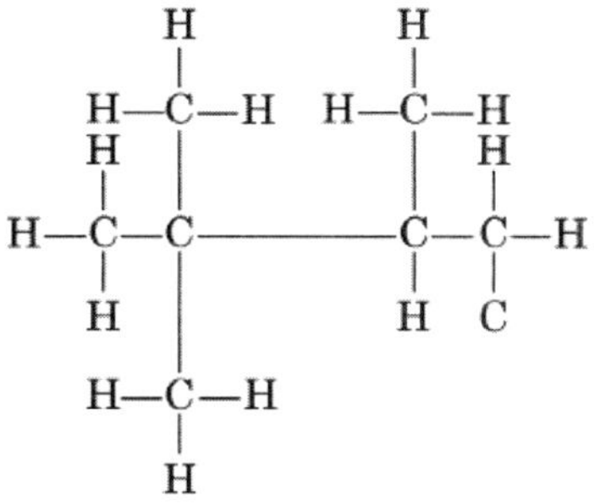

Isomer triptane

Olefins: The olefins are also straight or open chain hydrocarbons but have two atoms of hydrogen fewer per molecule as compared to paraffins. There occurs a double bond between any two carbon atoms. They are more resistant to detonation than the paraffins and are unsaturated;

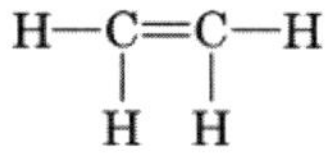

```
H—C=C—H
   |  |
   H  H
```

(Ethylene)

Naphthenes: They have the same chemical formula as the olefins, but they are saturated compounds and high antiknock values. They have molecules made up of carbon and hydrogen with a ring structure and the prefix cycle is used in their name. Cyclo propane can be illustrated as follows:

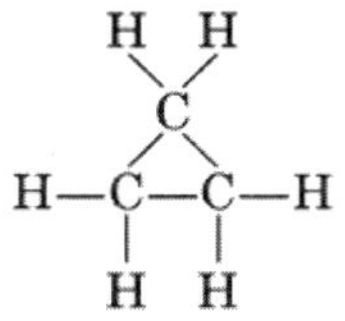

```
     H  H
      \ /
       C
      / \
  H—C—C—H
     |   |
     H   H
```

Cyclo propane (C_3H_6)

Aromatics: The aromatic series of hydrocarbons C_nH2_n-6 have a ring type structure for all or most of the carbon atoms to which are attached H or group of C and H atoms. They are unsaturated but stable chemically. The formula CgHe represents benzene having the following structure.

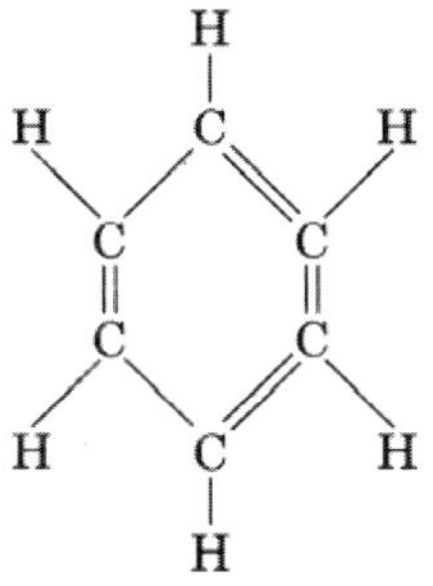

```
          H
          |
   H      C      H
    \   /   \\  /
      C        C
      ||       ||
      C        C
    /   \   //  \
   H      C      H
          |
          H
```

C_6H_6 (Benzene)

Alpha-methyl naphthalene, $C_{11} H_{16}$ is another aromatic of importance because it is used as a reference fuel in testing diesel fuel. It has high antiknock value as a fuel for SI engines but poor ignition quality for diesel fuel.

7.2.3 Petroleum refining

Petroleum as it comes from the ground is a rather viscous and highly coloured liquid. It often possesses unpleasant odour which is largely due to the presence of sulphur compounds in it. Technically it is called crude oil. Crude oil is obtained from wells drilled deep in the earth in the areas where oil-bearing layers are located. The process of purifying crude oil is called refining.

Fractional distillation: The process of dividing crude petroleum into fractions with different boiling ranges and free from undesirable impurities is called fractional distillation. To separate the hydrocarbons, the crude oil is heated and the different hydrocarbons are given off as vapours. The hydrocarbons with the lowest boiling point are given off first.

The process of refining crude oil is carried out in a tall, steel column called fractionating column (Fig. 7.1). The crude oil is pumped continuously through preheated pipes into the bottom of fractionating column. The fractionating column is used to heat the oil. The lower part of the column are very hot and the upper part has progressively lower temperature. Heat changes the oil partly into vapour and the lowest boiling fraction ascends to the top. The vapours of oil as they rise up the fractionating column become cooler and condense at the shelves at various heights. The highest boiling fraction condenses at the bottom and the lowest boiling fraction at the top. At the top of the tower, the temperature is considerably cooler. Outlets are provided in the side of the column at suitable height to withdraw a number of fractions.

Some of the refined products obtained from the fractionating column do not coincide with the commercial demand. For this purpose, various processes are used to convert some of these fractions into compounds of greater need and demand. The fuels obtained from distillation are known as straight-run fuels. The yield of gasoline from crude oil by this process is not sufficient to meet the demand for that fuel. Also many straight run fuels tend to run high in straight chain paraffins that are generally low in antiknock value. The prevailing practice is to reform the heavy fraction to light gasoline fraction by cracking to improve the anti-knock value.

Cracking: It consists of breaking down large and complex molecules into lighter and simpler compounds with lower boiling points. An illustration is the cracking of $C_{14} H_{30}$, the paraffin tetradecane, producing a lighter paraffin and an unsaturated olefins.

$$C_{14} H_{30} \rightarrow C_7 H_{16} + C_7 H_{14}$$

The two resulting molecules are in the range of gasoline. Pressure, heat and time are the factors essential to the thermal-cracking reaction. After an oil has been cracked, it must again be refined by the distillation process and separated into fuel fractions in much the same way as the crude is first fractionated.

Catalytic cracking: It makes use of a catalyst and requires somewhat lower temperature and pressures to obtain the reaction. Catalytically cracked fuels tend to contain fewer unsaturated hydrocarbons and more stable paraffin isomers with high antiknock values.

Polymerization: This involves the light gases to be converted to gasoline and heavier fuels (liquids). Polymerization, the reverse of cracking is accomplished by a catalyst, heat and pressure, principally on the olefins, propane and butane.

Hydrogenation: It is a process wherein hydrogen is added chemically under high pressure and temperature to produce more desirable compounds. It is used to convert unstable compounds to stable compounds.

Blending: Blending is a process of mixing certain refinery products to obtain a commercial product of desired quality.

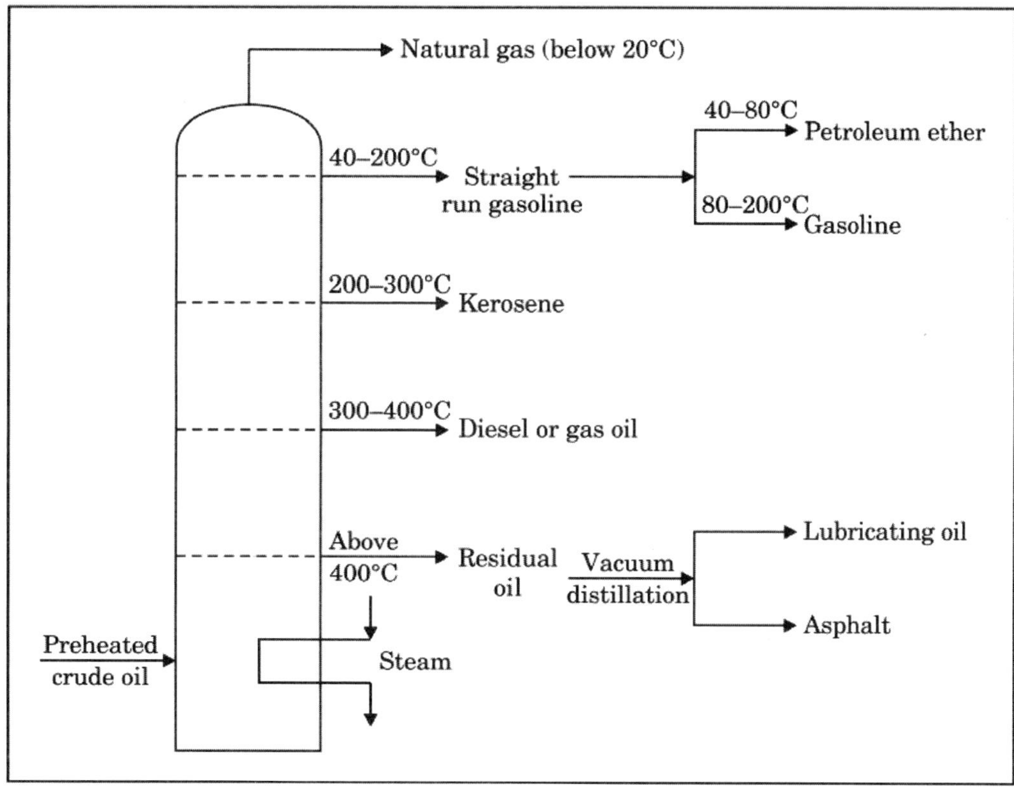

Fig. 7.1: Flow sheet of petroleum refining

Table 7.1: Composition and uses of the main petroleum fractions

Sl. no.	*Name*	*Boiling range °C)*	*Composition*	*Uses*
1.	Natural gas	Up to room temperature	C_4–C_5	Fuel gas
2.	Straight run gasoline	40–200	C_5–C_{11}	
	(a) Petroleum ether	40–80	–	Solvent
	(b) Gasoline or petrol	80–200	–	motor fuel, solvent, dry cleaning
3.	Kerosene (Power kerosene or powerine)	200–300	C_{11}–C_{16}	Illuminant, fuel for stoves for making oil gas
4.	Gas oil or diesel oil	300–400	C_{16}–C_{18}	Fuel for diesel engines, for conversion to gasoline by cracking
5.	Residual oil, Refractionated under vacuum to give	Above 400	–	–
	(a) Lubricating oil	–	C_{17}–C_{20}	Lubrication
	(b) Paraffin wax (on cooling)	–	C_{20}–C_{30}	Ointment, candles, vaseline wax,
	(c) Non-volatile residue, Aspahalt	–	–	Paints.

7.3 FUELS FOR INTERNAL COMBUSTION ENGINE

The most common fuels for I.C. engine are petrol (gasoline) powerine (Kerosene), High speed diesel oil (HSD) and light diesel oil (LDO). Most powerine operated engines are provided with two fuel tanks, a petrol tank and powerine tank. Powerine engines are started on petrol and after being warmed up, changed to powerine. Powerine being less volatile, does not vapourize well and so cannot be used for starting the engine. However, if a hot engine has stopped it can be started on powerine immediately, although such a practice should be avoided as far as possible. It is always recommended that at the end of each working day, the engine should be changed to petrol from powerine, so that it may start easily in the next day. Some of the most common fuel impurities which give a lot of troubles in operating an engine are sediment and water, higher sulphur content and increased gum forming characteristics. All of these tend to clog the carburetor. Increased sulphur content has a corrosive effect on the parts with which it comes in contact. Fuel with a high gum content causes the sticking of valves and piston rings.

Diesel engine fuels commercially sold in India are mainly of two types namely (i) HSD or high speed diesel *(ii)* LDO or light diesel oil. HSD' fuel is used for high speed engine and LDO fuel is used for slow speed cold starting engines. The important properties of different fuels for I.C. engine have been given in the following Table 7.2. The diesel fuel should be free of acid, any foreign matter, dirt and moisture, all of which can clog or injure parts of the engine or fuel injection system. It must be able to lubricate the fuel pumps and fuel injection nozzles all the time.

Table 7.2: Important properties of fuels

SI. no.	*Name of fuel degree*	*A.P.I. gravity*	*Specific gravity*	*Calorific value Kcal/kg*
1	Light diesel oil	22	0.920	10300
2	High speed diesel oil	31	0.820	10550
3	Powerine (kerosene)	40	0.827	10850
4	Petrol	63	0.730	11100

7.4 COMBUSTION

The burning of fuel in an automobile engine is a chemical reaction and is referred as combustion. During combustion, the chemical energy of fuel is converted into heat energy. Automotive fuels are made up mostly of two elements, hydrogen and carbon. So fuel is therefore called a hydrocarbon (HC). During complete combustion in the engine, these two elements unite with a third element, the gas oxygen (O). Oxygen which comes from air forms about 23 percent of it. The percentage of nitrogen is 76.9. Actually there are other minor elements present in air, but they need not be considered in combustion calculation due to their inertness. During combustion, oxygen combines with hydrogen to form water (H_2O), oxygen combines with carbon to form carbon dioxide (CO_2), heat is liberated and pressure increases in the combustion chamber. The final or desired result occurs during the power or expansion stroke when a volume change takes place and work is done.

In the engine, we do not get ideal combustion. In stead, some amount of fuel does not burn. Also some only partly burns, producing carbon monoxide (CO) instead of CO_2. The unburned fuel and partly burned fuel cause pollution of air as they exit through the tail pipe with the exhaust gases. It is therefore, necessary to introduce the required quantity of air for complete combustion of fuel. The minimum weight of air required for the complete combustion of 1 kg of liquid fuel can be estimated from the analysis of fuel.

Methane gas (CH_4), the simplest of the paraffin series, can be used to illustrate the combustion process. The calculation of the masses of the compounds is as follows.

$$CH_4 + 2O_2 + 7.6\,N_2 = CO_2 + 2H_2O + 7.6\,N_2$$

$$(12 + 4) + (2 \times 32) + (7.6 \times 28) = (12 + 32) + 2\,(\,2 + 16) + (\,7.6 \times 28)$$

$$16 + 64 + 212.8 = 44 + 36 + 212.8$$

$$292.8 = 292.8$$

This shows the balanced chemical equation and hence complete combustion will occur. From the above calculation, it can be seen that 64 kg of oxygen is required for 16 kg of fuel. Air contains 23% oxygen. Hence, amount of air required get 64 kg of oxygen $= \frac{64}{0.23} = 278$ kg of air and $\frac{\text{Air}}{\text{Fuel}} = \frac{278}{16} = \frac{17.3}{2}$.

Hence the correct air-fuel ratio becomes 17.3:1. Since nitrogen is inert and does not enter into the combustion process, it may be eliminated and air requirements may be calculated after oxygen requirement is known by means of the percentages of oxygen in the air as mentioned above.

Example 1: Find an air fuel ratio for complete combustion of fuel in a spark ignition engine using petrol.

Solution: Let the chemical composition of petrol be C_6H_{14}. Air contains 23% oxygen by weight. Atomic weight of C, O and H be respectively 12, 16 and 1.

$C_6H_{14} + O_2 \rightarrow CO_2 + H_2O$ or $2\,(C_6H_{14}) + 19\,O_2 = 12\,(CO_2) + 14\,(H_2O)$

So, $2\,(12 \times 6 + 14 \times 1) + 19\,(2 \times 16) = 12\,(12 + 32) + 14\,(2 + 16)$

or $(2 \times 86) + (19 \times 32) = (12 \times 44) + (14 \times 18)$

or $172 + 608 = 528 + 252.$

Hence the chemical equation is balanced and complete combustion will occur. Oxygen per kg of fuel = $\frac{608}{72}$ = 3.54 kg; Correct air supply per kg of fuel = $3.54 \times \frac{100}{23}$ = 15.37 kg; Hence correct air-fuel ratio = 15.37 : 1 ≃ 15 : 1

Example 2. Find an air fuel ratio for complete combustion of fuel in a diesel engine using diesel ($C_{16}H_{34}$).

Solution:

$C_{16}\,H_{34} + O_2 \rightarrow CO_2 + H_2O$ Or, $2\,(C_6\,H_{34}) + 49\,(O_2) = 32\,(CO_2) + 34(H_2O)$

Or $2(226) + 49\,(32) = 32\,(44) + 34\,(18)$

Or $452 + 1568 = 1408 + 612$

Or $2020 = 2020$

Hence equation is balanced and complete combustion will occur.

The oxygen per kg of fuel is $\frac{1568}{452}$ and Correct air supply per kg of fuel = $\frac{1568}{452} \times \frac{100}{23}$ = 15.08 and hence correct air fuel ratio = 15.08 : 1 □ 15 : 1

7.5 IMPORTANT QUALITIES OF S.I. ENGINE FUELS

The selection of a fuel depends upon the type of gasoline engine and its operating conditions. A few important properties of fuel are tested to determine the suitability of fuels for an I.C. engines. The followings are the important properties of gasoline fuel for use in gasoline engine.

Volatility: Volatility of a liquid is its ability to change into vapour. In spark ignition engines that burn gasoline, the volatility is an important characteristic. A high–volatility gasoline vapourizes very quickly. A low volatility gasoline vapourizes slowly. A good gasoline should have just the right volatility for the climate in which the gasoline is to be used. If the gasoline is too volatile, it will vapourize in the fuel system. The result will be a condition, called vapour lock. It prevents the flow of gasoline to the carburetor. Vapour lock causes the engine to stall from lack of fuel. The volatility of gasoline is seasonally adjusted by the refmers. This tends to increase the chances of vapour lock occurring under certain conditions.

On the other hand, gasoline is that do not vaporize readily may cause hard starting, slow warming up and unequal distribution of fuel to the individual cylinders. The gasoline used during winter season should have a high volatility. The same quality gasoline 1vhen used in summer period causes vapour lock. Therefore, it is essential to evaluate the volatility of fuel. The common tests are *(i)* distillation test, *(ii)* vapour pressure test.

(i) **Distillation test:** In this test, a measured amount of fuel is heated in a flask, starting from a low temperature and gradually increasing it. The amount of condensed vapour corresponding to each temperature is collected .and noted separately. This process goes on operating till the evaporation ceases. From the figure of distillation curve (Fig. 7.2) for the common fuels, 10 per cent point (10 per cent of fuel distilled) is of primary importance as a specification related to engine starting. In temperate and cold climates, the 10 per cent point should be lower in the winter than in the summer. Fifty per cent point is important as an index of engine warm-up characteristics *i.e.* the lower the 50 per cent point temperature, the faster the warm-up of the engine. The 90 per cent evaporated temperature is associated with engine acceleration, crankcase dilution and fuel economy. The presence of greater amount of fuel at the end causes poor mixture distribution in the intake manifold of an engine and this may affect engine performance during acceleration. The presence of a large portion of fuel that has not evaporated can cause crankcase dilution. The unburned portion may enter the crankcase past the piston rings and dilute the oil.

(ii) **Vapour pressure test:** Vapour pressure test is another means of evaluating the volatility of the fuel. A special instrument known as Reid vapour pressure test is commonly used to measure the vapour pressure of a fuel. The pressure is read on the Bourdon gauge at 37.8°C. The summer gasoline should have a vapour pressure of 48 to 62 kpa with the winter grade gasoline being 14 to 21 kpa higher than the summer grades.

Antiknock quality: Engine knocking or detonation or spark knock refers to violent noises heard in the engine during the process of combustion. The airfuel mixture in the cylinder of a spark ignition engine burns spontaneously in the localized areas instead of progressing from the spark. This usually causes an audible 'ping' or knock or detonation. The actual loss of power and damage to an engine because of detonation is generally not significant until the intensity becomes very severe.

However, heavy and prolonged detonation may have an adverse effect in terms of power loss and possible damage to engine.

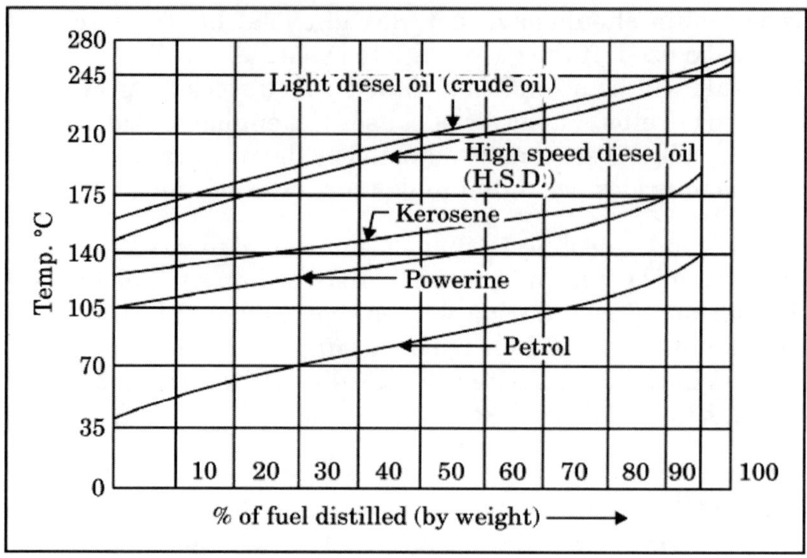

Fig. 7.2: Distillation test curves of common fuels

Detonation is an uncontrolled second explosion (after spark occurs at the spark plug) with the spontaneous combustion of the remaining compressed airfuel mixture resulting in a pinging sound. It occurs after the occurrence of spark in the spark plug and takes place during last stage of combustion. During normal combustion, the charge starts burning as soon as spark occurs at the spark plug. The flames move smoothly and evenly across the combustion chamber. Only one explosion takes place.

During abnormal combustion, the spark starts combustion in a normal way. However, before the flame can reach the far side of the combustion chamber, the last part of the charge explodes. The result is a very quick increase in pressure. This is detonation and gives a pinging sound. Detonation can ruin the engine. It may cause damage to piston, valves, gaskets and other parts. Gasoline that detonates easily is called a low-octane gasoline. A gasoline that resists detonation is called a high-octane gasoline. Detonation is caused by high compression ratio, improper cooling, carbon deposits in the combustion chamber, easily ignition etc.

Knocking can effectively be controlled by adding tetraethyl lead to petrol to raise its octane number.

Compression ratios and detonation: Over the years, the compression ratios of automobile engines have gone up. The reason is that higher compression ratios give engines more power. But a high compression ratio can cause a problem. It increases the temperature of the air-fuel mixture. The higher heat of compression may cause the remaining air-fuel mixture to explode before normal combustion is completed. This is denotation. If the air-fuel mixture explodes before the spark occurs at the spark plug, this condition is called preignition. The compression ratio must be kept at a desirable range so that the charge will not ignite prematurely from the heat of compression.

Preignition: Burning of air-fuel mixture in the combustion chamber before the piston has reached the top dead centre is called preignition. It occurs when the charge is fired too far ahead of top dead centre of the piston due to spark advance or excessive heat in the cylinder. It takes place before the occurrence of spark in the spark plug.

Octane number: It is a measure of antiknock properties of a gasoline. Percentage of Iso-octane ($C_8 H_{18}$) in a reference fuel consisting of a mixture of iso-octane and normal heptane ($C_7 H_{16}$), when it produces the same knocking effect as the fuel under test is called 'Octane number' of fuel. Iso-octane has excellent antiknock qualities and is given a rating of 100. Normal heptane would knock excessively and hence it is assigned a value of zero. Petrol of 94 octane number is defined as that which has the same qualities as that of a fuel having 94 parts of iso-octane mixed with 6 parts of normal heptane. High octane rating is necessary for high compression, high performance engines. It is generally the most important single qualify characteristic for fuel used in spark ignition engine.

Sulphur content: High sulphur content is undesirable because of formation of corrosive substances in the presence of water vapour. Hence sulphur test is done to find out the presence of corrosive sulphur compounds in the fuel. The lower the sulphur content, lesser is the amount of corrosion. The corrosive sulphur is indicated by immersing a polished strip of copper in the fuel for 3 hours at 50°C. A comparison is made with this strip and a freshly polished copper. If corrosive sulphur is present, the colour would be discoloured. For finding the total quality of sulphur, a sample of fuel is burnt in test bombs and sulphur is measured.

Gum content: A fuel should have minimum gum content. If the gum forming substances remain after the gasoline evaporates, carburetor passages may become plugged, valve and piston rings may become gummy or sticky. The hydrocarbon fuels, particularly those made by cracking and containing unsaturated unstable compounds, have a tendency to form viscous liquids or solid called gums resulting from oxidation during storage. The gum test consists of evaporating samples of fuel in a beaker. The residue is weighed in milligrams per 100 cc of fuel. The evaporation takes place in a copper dish and is continued for a longer period of time, indicating gum-forming properties that may develop upon storage and exposure to air and light. Gasolines are often blended with nonvolatile oils or additives to minimize the problem of gum formation.

Gravity test: The gravity or density of a fuel is expressed as specific gravity or as API gravity, a scale devised by American Petroleum Institute. The specific gravity of a liquid is the ratio of its weight to the weight of an equal volume of water at 15.6°C. The boiling temperature, the volatility and the heat values are somewhat related to the gravity, so that it is used as a means of estimating these values. A high API gravity fuel commonly contains more heat units per kilogram but because of its lightness, it contains a lower heat value per litre. The relationship between the API gravity scale and the specific gravity scale is expressed as follows.

$$\text{A.P.I. degree} = \frac{141.5}{\text{Specific gravity at } 15.6°\text{C}} - 131.5$$

From this relation, the API gravity of water is 10. The fuels lighter than water have lower specific gravity but higher API degree values. The instrument used for testing the specific gravity of fuel is known as hydrometer. Specific gravity of a fuel affects the spray penetration as fuel or charge is injected into the combustion chamber.

Heating values: Heat value or calorific value of a fuel is of primary importance as it is an indication of how much power the fuel provides when burnt. The heat liberated by combustion of a fuel is expressed in Kcal/kg. The fuel with the higher heating value is most valuable and gives the greater amount of work output when burnt in an engine.

The heating values maybe reported as high or low depending on the methods and conditions of the test. The heat value of a fuel is determined by burning it in a special device known as a calorimeter. When the heat values are obtained by bomb calorimeter, the unit quantity is burned with oxygen at constant volume and the water formed is cooled from a gaseous state or steam to liquid, the heat value is known as total or high or gross heat value. If the water formed in the combustion process remains as steam and the heat of vaporization (latent heat) contained therein is not accounted for it is called lower heat value. In practical application of fuel combustion in engines, the water vapour is not condensed, instead it passes out with the exhaust gases in the form of steam. Therefore, the combustion process in the engine is nearly represented by the low heating value.

It is common practice to calculate engine thermal efficiencies in terms of the high heating value. Also unless otherwise stated, printed values of heat of combustion are high values.

7.6 IMPORTANT QUALITIES OF DIESEL FUEL

Diesel engines use diesel fuel oil. The fuel oil is sprayed or injected into the engine cylinders towards the end of the compression strokes. Heat of compression ignites the fuel oil and power stroke follows.

Diesel fuel is made from crude oil by the same refining process that produces gasoline. There are two grades of diesel fuel *i.e.* Number 1 diesel and Number 2 diesel. Number 1 diesel is more volatile and also called light diesel oil. Similarly Number 2 diesel is more difficult to vaporize and also called high speed diesel oil.

Vaporising temperature of high speed diesel oil is higher than that of petrol, hence, the petrol vapourises quicker than diesel oil in the engine cylinder. Number

2 diesel is recommended for most driving condition and for high speed engine. Number 1 diesel is used where temperatures are very low. The light diesel oil (number 1) weighs heavier per litre and is the main source of fuel for slow speed cold starting engines. Diesel fuel is a light oil with the proper viscosity, volatility and cetane number for use as a fuel. The followings are the important requirements of diesel fuel for its suitability in diesel engine.

Viscosity: Viscosity refers to the ease with which a liquid flows. The lower the viscosity, the more easily the liquid flows. Diesel fuel must have a relatively low viscosity. It must flow through the fuel-system lines and spray into the engine cylinders with little resistance. If oil has high viscosity, it will not break up into fine particles. Then it will not burn rapidly enough. Engine performance will be poor. But if the viscosity is too low, the fuel oil will not lubricate the moving parts in the fuel pump and fuel injectors. Damage may occur. Number 2 diesel has the right viscosity for most driving conditions. Number 1 diesel has lower viscosity so it flows and sprays properly at low temperatures. Therefore, proper viscosity of the diesel fuel is essential for its use.

Cetane number: The ignition quality of diesel fuel is determined by a method somewhat similar to that used in determining the anti-knock quality of gasoline.

The catane number of diesel fuel refers to the ease with which the fuel ignites. With a high cetane number, the fuel ignites easily (or at a relatively low temperature). The lower the cetane number, the higher the temperature needed to ignite the fuel.

The low-catane number fuel takes a little longer time to ignite. This is called ignition lag. During this slight delay, the fuel tends to accumulate in the cylinder. Then, when ignition does occur, all the accumulated fuel ignites at once. The pressure goes up suddenly and a combustion knock results. This is similar to detonation in a spark-ignition engine. High-cetane fuel ignites as soon as it enters the cylinder. There is no accumulation of fuel. The result is a smooth pressure rise, so no combustion knock takes place.

As in case of octane number scale, the scale of cetane number represents blends of two pure hydrocarbon reference fuels. The reference fuels are cetane ($C_{16}H_{34}$) and alpha methyl naphthalene ($C_{11}H_{10}$). Cetane is a high igniting fuel and alpha-methyl naphthalene is a very slowly igniting fuel. The percentage of cetane in a mixture of cetane and alpha methyl naphthalene which has the same ignition quality as the fuel under test is called the cetane number. Thus a fuel with cetane number 60 has the same ignition quality as a mixture of 60% cetane and 40% alpha methyl naphthalene.

The relationship between the cetane number of a diesel fuel and the performance of a diesel engine should not be confused with the relationship between the octane number of gasoline and the performance of a spark ignition engine. With a spark ignition engine, raising the octane number improves potential engine performance by allowing the compression ratio to be increased, thereby increasing the power and efficiency of the engine. In diesel engine, the desirable level of cetane number is established by the requirements of good ignition quality at light loads and low temperatures.

Generally high-cetane fuels permit an engine to be started at lower air temperatures, provide faster engine warm-up without misfiring, reduce the formation of varnish and carbon deposits and eliminate diesel knock. Cetane numbers above the commercially available range may lead to incomplete combustion and exhaust smoke if the ignition delay period is too short to allow proper mixing of the fuel and air within the combustion space. The commercial diesel fuels have got cetane rating varying from 30 to 60.

Carbon residue: A diesel fuel often has a tendency to form carbon deposits in the engine. Carbon residue is an indication of the carbon forming properties of the fuel under certain engine conditions. The carbon residue can be measured in the laboratory by heating a sample of the fuel in a closed container in the absence of air. The carbon residue will remain in the container. The amount of carbon residue which is considered permissible in diesel fuel depends somewhat on the characteristics of the engine.

Ash content: Diesel engine injectors are precision made and therefore, are quite sensitive to any abrasive material in the fuel. Since the ash content is directly related to the wear of the injection system, it must be kept quite low. The maximum allowable ash content should be 0.01 per cent. The ash-forming materials in fuel contribute both to wear in the injection system and to engine deposits and hence should be kept low.

Flash point: The flashpoint is not directly related to engine performance. It is however of importance in connection with safety and protection measures, involved in fuel handling and storage. Flash point of an oil is the temperature at which the fuel must be heated to give off inflammable vapors. At that temperature, a distinct flash (sudden flame) is obtained when the flame is passed over the container.

Pour point and cloud point: Pour point is the lowest temperature at which the fuel ceases to flow. Hence it is the minimum temperature at which the movement of oil is observed last.

The cloud point of a fuel is the temperature at which paraffin wax or other solid substances begin to crystallize out or separate when the oil is chilled under prescribed conditions. The cloud point generally occurs at about 5°C above the pour point.

Diesel fuel additives: The additives in diesel fuel are for engine protection and to improve engine performance. Antioxidants, metal deactivators, corrosion inhibitors are used in diesel fuel. The ignition quality of diesel fuel can be improved by the addition of amyl nitrate or hexyl nitrate.

Chapter 8

Carburetor

8.1 CARBURETION AND A CARBURETOR

The process of conversion of liquid fuel to vapour and then mixing with fresh air for combustion is called carburetion and the device in which this process takes place is called carburetor. Hence, carburetor is a mechanical device to supply a combustible mixture of varying degrees of richness to suit engine operating conditions. It is fitted on intake manifold and below the air cleaner. The mixture must be rich (have a higher percentage of fuel) for starting, acceleration and high speed operation. The mixture must be lean (have a lesser percentage of fuel) during idling conditions and with a warm engine. Carburetor provides the varying mixture for different operating condition. It vapourizes petrol fuel and mixes the vapourised particles with air outside the combustion chamber of S.I. engine.

8.2 COMBUSTION AND VAPORIZATION OR ATOMIZATION

Fuel is a substance consumed by the engine to produce energy. The process of fuel combustion converts the chemical energy of the fuel into heat energy. Automotive fuels are made up of mostly two elements hydrogen and carbon, which have the chemical symbols H and C. It is therefore, called a hydrocarbon (HC). For combustion or complete combustion, these two elements unite with a third element, the gas oxygen (O). Oxygen is available to the fuel from the air. But oxygen forms about 23 per cent of air we breathe. Hence, required amount of air is essential for complete combustion of the fuel. In addition to air, the fuel must be converted into vaporized form for its complete combustion.

If air is only in contact with the fuel, combustion will not be complete. The best result is obtained when air is thoroughly mixed with the fuel. These conditions can be very well explained with the help of kerosene pumping stove. If we simply have kerosene in a pot and try to burn it, it will take time and even if it burns, there will be no pressure, in the flame. But as we pump the kerosene stove, oil starts coming out in smaller droplets and with a flame, the stove gives a powerful burning. The difference between the two is that in first, the fuel was in liquid form and air was only touching it while in second case, the fuel is thoroughly mixed with air in fine atomized form, causing it to burn intensely. The fuel is broken into small particles by allowing it to pass through the stove nozzle with high velocity. Therefore, it is required to vaporize or atomize the fuel before combustion. To produce quick vaporization of the fuel, it is sprayed into the flowing air and sprayed fuel is turned into fine droplets by the tearing action of high velocity air. The process of breaking the liquid into smaller droplets is called vaporization or atomization. But the liquid is not truly converted into atoms, as the name implies of atomization.

Thus the working of a carburetor is same as the working of a hand sprayer (Fig. 8.1), which is used to kill bugs and mosquitoes in the house. The oil and air come out from it in the form of fine dust. The petrol dust is formed in the carburetor on the same principle as used in the sprayer.

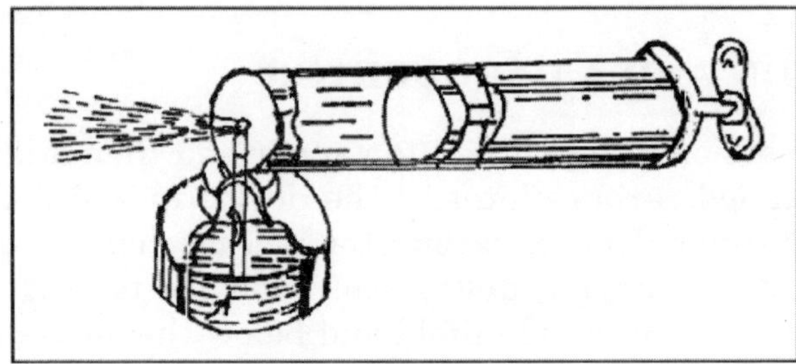

Fig. 8.1: Hand Sprayer

In I.C. engine, all the liquid fuel must be converted into vapour before burning. Diesel oil is most difficult to vaporize. Vaporizing temperature of diesel oil is higher than that of petrol, hence petrol vaporizes quicker than diesel oil in the engine cylinder. This helps in easy starting of petrol engines. If all the liquid fuel is not well vaporized, it would do crank case oil dilution. For vaporization of diesel, fuel injector is provided. Hence, fuel ignition system of diesel engine is different from that of petrol engine.

8.2.1 Function of carburetor

1. Vapourisation of fuel (petrol)
2. Mixing of air and petrol in correct proportion to get the desired results from the engine.
3. Maintaining a suitable reserve of fuel in the float chamber.

8.2.2 Requirements of a carburetor

1. Capacity of supplying correct air-fuel mixture at different engine loads and speeds.
2. Ease in starting of the engine in cold condition and hot condition.
3. Supplying required quantity of fuel at idle speed
4. Response to fast acceleration and to supply of rich mixture

8.3 AIR-FUEL RATIOS

The complete combustion of fuel (petrol) requires 15 parts by weight of air to mix with 1 part by weight of petrol. The air-fuel ratios at the various engine operating conditions are shown below.

	Air : Fuel
Starting (in cold condition)	1 : 1
Starting (normal ambient temperature)	9 : 1
Idling condition	12 : 1
Acceleration	13 : 1
Normal condition (Chemically correct air: fuel ratio)	15 : 1

Chemically correct air-fuel mixture or ratios in which the fuel is completely vapourised and then properly mixed with air to produce required power output from the engine. The air-fuel ratio more than 15: 1 is called lean mixture and less than 15: 1 is called rich mixture.

The chemical formula of petrol = C_6H_{14}

Air contains 23% oxygen by weight and $C_6H_{14} + O_2 \rightarrow CO_2 + H_2O$

After balancing, $2\ (C_6H_{14}) + 19\ (O_2) = 12\ (CO_2) + 14\ (H_2O)$

Or $2\ (12 \times 6 + 1 \times 14) + 19\ (16 \times 2) = 12\ (12 + 2 \times 16) + 14(2 \times 1 + 16)$

Or $2\ (86) + 19\ (32) = 12\ (44) + 14\ (18)$ or $780 = 780$

In complete balancing, fuel = 2 × 86 = 172 and oxygen = 19 (32) = 608. Hence oxygen per kg of fuel = 608/172 = 3.54 kg and correct air supply per kg of fuel = 3.54 × 100/23 = 15.36 kg @ 15 kg. Therefore correct air and fuel ratio = 15:1

8.4 TYPES OF CARBURETORS

There are three basic types of carburetors. These are classified according to flow of air from carburetor to the engine manifold.

1. Updraft carburetor (Fig. 8.2a)
2. Down draft carburetor (Fig. 8.2b)
3. Horizontal or natural draft carburetor (Fig. 8.2c)

In updraft carburetor, carburetor is mounted below or beside the engine manifold and mixture flows upward into the engine.

In horizontal carburetor, carburetor is fitted in line with manifold. Air moves in the horizontal direction.

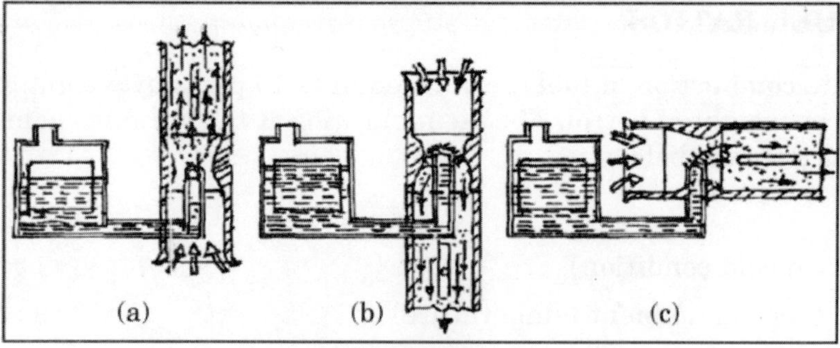

Fig. 8.2: Types of Carburettors

In downdraft carburetor, carburetor is mounted above the intake manifold of the engine, so that the air enters the upper part of the carburetor and the mixture flows downward into the manifold. Down draft carburetors are mostly used in automobiles owing to the following advantages over the other types.

(a) This gives better volumetric efficiency as the air-fuel mixture flows down assisted by gravity.

(b) The carburetor can be fixed at the top of the engine and remains easily accessible.

8.4.1 Working principle of a carburetor

The working principle of a single carburetor has been explained with the help of Fig. (8.3). As piston moves down in suction stroke, a partial vacuum is created in the cylinder and the piston sucks the air from intake manifold. A partial vacuum is any pressure less than atmospheric pressure. Atmospheric pressure pushes air or air-fuel mixture into the cylinder to fill the vacuum. But in petrol engine, carburetor is fitted in the intake manifold. As air flows towards the engine cylinder, it must pass through the carburetor. A venturi is located in the air passage through the carburetor. As air flows through the venturi, a partial vacuum is produced in the venturi. The venturi restricts the flow of air so that air pressure in the venturi is decreased due to increase of the air velocity. The pressure inside the float chamber is normal atmospheric pressure. Due to this pressure difference, the fuel comes out through main jet to the venturi and gets evaporated due to negative pressure around the tip of the main jet (in the venturi).

The high velocity air in the venturi also helps the atomization of fuel due to tearing action. The air mixes with the vaporised fuel in the venturi and moves towards the engine cylinder through intake manifold. The amount of air-fuel mixture is controlled by the throttle valve, the lowermost part in the carburetor and opens to the intake manifold. During idling condition of the engine, the engine throttle valve remains nearly closed. With the throttle valve closed, there is a high vacuum below the throttle valve from the intake manifold due to suction stroke of the piston. A small amount of air pass through the air horn. The air speed is low and very little vacuum develops in the venturi. So main jet does not discharge fuel.

Another system called idle system sends required quantity of air and fuel mixture to the intake manifold during idling condition. The idle line is opened below the throttle valve and the amount of mixture is controlled by the idling adjusting screw.

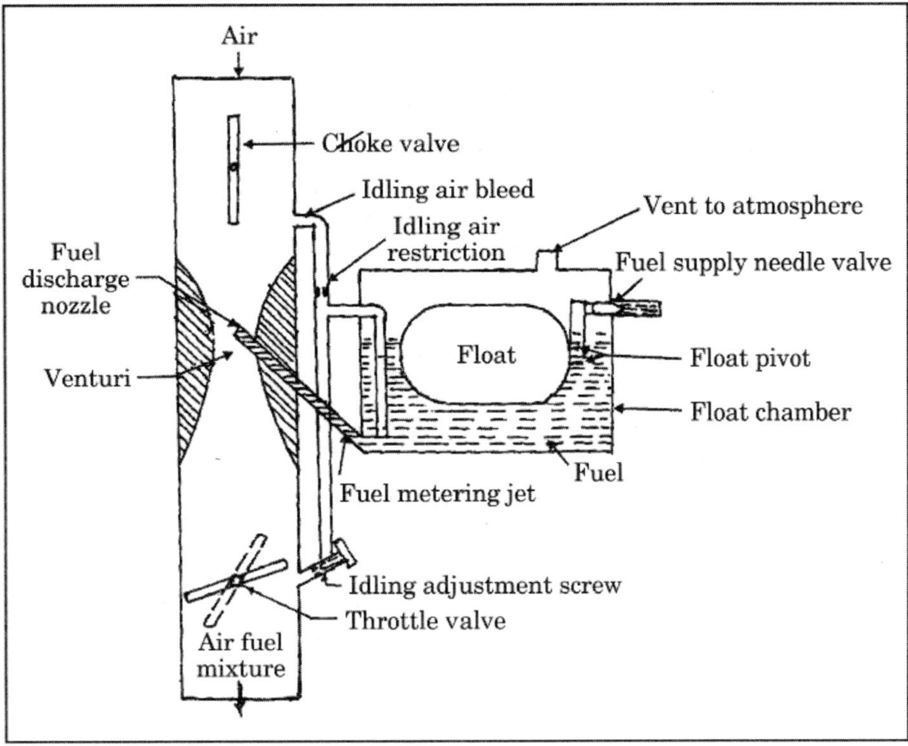

Fig. 8.3: Simple Carburetor

When the engine load increases, throttle valve opens more resulting in more supply of air-in the venturi. More supply of air in venturi results in higher pressure drop which in turn causes more supply of fuel from the float chamber into the venturi. The closing and opening of throttle valve according to load requirements is controlled by accelerator pedal, linked mechanically with it. Thus air-fuel ratio is regulated at different load conditions in the carburetor. The complete vaporization of fuel occurs in the combustion chamber during compression stroke.

8.5 COMPONENTS OF A SIMPLE CARBURETOR AND THEIR FUNCTIONS

A simple carburetor consists of the following components.

(1) Choke valve
(2) Venturi,
(3) Main jet
(4) Float chamber and float
(5) Throttle valve
(6) Idling jet
(7) Idling adjustment screw.

Choke valve: It is a valve present at the top portion of the carburetor. It is a device for restricting the air supply in the carburetor or it chokes off the air flow into the carburetor to provide rich mixture during starting or in cold condition. That's why it is named as choke valve. We have to supply rich mixture when cold engine has to be supplied. The round cylinder through which air passes into the carburetor is called the air horn.

During cranking, the air speed through the carburetor horn is very low. Vacuum from the venturi action and vacuum below the throttle valve are not sufficient to produce adequate fuel flow for starting. To produce enough fuel flow during cranking, the carburetor has the choke valve of butterfly type. It is controlled mechanically or by an automatic device. When the choke valve is closed it is almost horizontal. Only a small amount of air passes in to the carburetor. Then while cranking, a fairly high vacuum develops in the air horn. This vacuum causes the main jet to discharge a heavy stream of fuel. The quantity delivered is sufficient to produce the air fuel mixture needed for starting the engine.

As soon as the engine starts, the speed of the engine increases and engine needs more air and a slightly leaner mixture. The more air is available when the choke is made widely opened.

Venturi: It is the portion of carburetor having reduced cross sectional area. It is so designed to increase the velocity of air during flow and hence a reduced pressure inside the venturi.

Main jet: It is a narrow pipe through which fuel is supplied to the venturi from float chamber due to pressure difference. The top of the jet is always kept slightly above (about 15 mm) the level of fuel in the float chamber in order to avoid unnecessary flow of extra fuel due to vibration, jerk or bending of vehicle in turning or uphill.

Float chamber and float: The float system includes the float bowl, float and needle-valve arrangement. The float and the needle valve maintain a constant level of fuel in the float bowl. If the level is too high, then too much fuel will be fed from fuel jet. If it is too low, too little fuel will be fed. In both the cases, poor engine performance occurs. If the fuel enters the float bowl faster than it is withdrawn, the fuel level rises. This causes the float to move up and push the needle valve into the valve seat. This shuts off the fuel inlet so that no fuel can enter. Then, if the fuel level drops, the float moves down and releases the needle valve off the valve seat so that fuel inlet is opened.

Hence, fuel level is maintained constant in the float bowl. A pivot is provided in the float bowl. The float is pivoted at this point. The float moves up and down due to pivot point.

Throttle valve: Throttle valve is a butterfly type of valve between the mixing chamber in the carburetor and the inlet manifold of the engine to regulate the quantity of charge. It is directly connected to the accelerator pedal through a mechanical linkage. It is partially open at slow speed and fully open at high speed. When the driver pushes down on the accelerator pedal, the throttle valve opens widely and more air-fuel mixture enters into the cylinder to develop more power. If the driver releases the accelerator pedal, the throttle valve almost closes causing less air-fuel mixture to flow to the engine and engine power drops off.

Idling jet: It is a special type of jet which supplies fuel at idling speed or low speed of the engine. It usually consists of a passage which starts from choke valve and ends just below the throttle valve *i.e.*, on the manifold side of the butterfly. When the throttle valve is nearly closed, the sucking force in the manifold pulls the necessary idling fuel from the idling jet. The mixture of air and fuel is rich during this condition.

Idling adjustment screw: It is a screw fitted with the idling jet that decides the amount of rich air-fuel mixture should be provided during idling condition of the engine.

Air-fuel mixture is maintained due to the redesign and manufacturing of choke valve and venturi and main jet or metering tube in the carburetor. Idling adjustment, screw is situated inside bonnet in the vehicle and is adjusted by driver himself when required by opening the bonnet. Kerosene engine is similar to petrol engine and is provided with a carburetor. Air-fuel ratio is maintained in diesel engine by adjusting the nozzle to spray right quantity of fuel. Throttle valve and accelerator pedal is also attached to diesel engine but in turn attached to the governor of the diesel engine.

8.6 FUEL SUPPLY SYSTEM OF A PETROL ENGINE

The functions of fuel supply system are to supply to the engine the correct proportion of air-fuel mixture under various conditions of engine speed and load requirements. The fuel supply system of a petrol engine is shown in Fig. 8.4. It comprises of a fuel tank, fuel tubing (lines), fuel filters, mechanical fuel pump, carburetor, spark plugs etc.

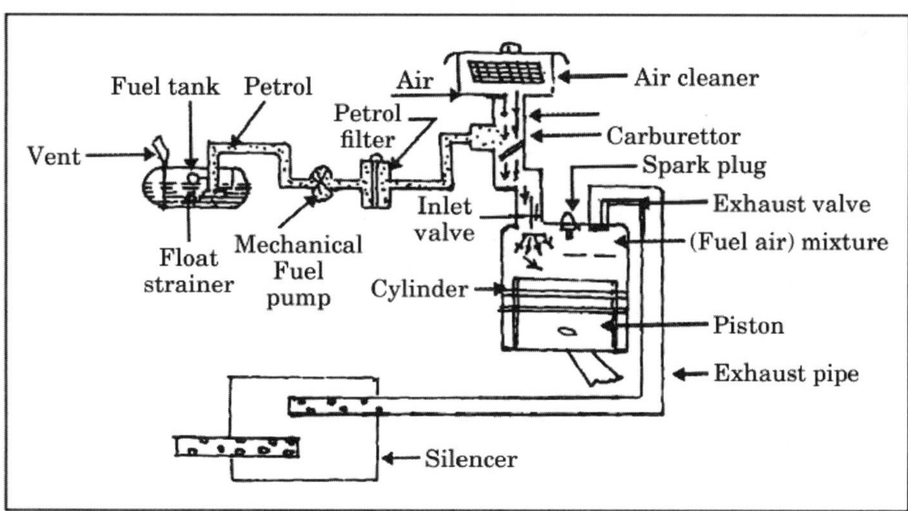

Fig. 8.4: Fuel System of a Petrol Engine

8.6.1 Tank

The fuel supply system comprises a fuel tank, mechanical or electric pump, fuel filter, carburetor and metallic flexible tubings.

The fuel tank is made up of tinned-steel sheet supported by sheet curved stiffeners. It is divided into two or three compartments, the walls of which have holes of large diameter to allow fuel to flow from one compartment to the other. The compartments help to limit the surging and frothing of fuel. The filler pipe is wide enough to have a good quantity of fuel and to allow air to escape to the atmosphere from it. A plastic float gauge with a wire operates the fuel gauge on the dash board. A drain plug located at the lowest point allows water to be drained out of the fuel tank.

8.6.2 Fuel lines

A metallic pipe of small diameter connects the fuel tank and the fuel pump. Rubber tubing is inserted at the middle of the fuel line for flexibility and ease. The ends of the flexible rubber tubing are fixed to the metallic pipe by small clips. Flexible metallic tubings with screwed connectors are used in the carburetor. The pipe connections have an union, nipple and a nut to which the metallic pipe is connected.

8.6.3 Filters

Petrol from the petrol pumps is usually contaminated with specks of dirt or water. The fine dirt enters into the small diameter passages and jets of carburetor resulting in early cylinder wear and non functioning of the carburetor.

Filters are used to trap the dirt and water thus preventing them from entering the fuel pump. These are located and fixed anywhere in the flexible pipe line before the carburetor inlet. A cylindrical fine wire brass gauge is screwed to the fuel inlet passage. The fine wire gauge is surrounded by a glass bowl for sighting any foreign matter, and for storing the fuel. A clamping nut is located at the bottom of this glass bowl and is used to empty it. The fine wire brass gauge and the bowl require cleaning at fixed intervals.

8.6.4 Mechanical fuel pump

Construction

The mechanical type of fuel pump comprises

(i) Eccentric
(ii) Pump lever
(iii) Return spring
(iv) Diaphragm with spring
(v) Diaphragm pull rod
(vi) Spring loaded inlet and outlet valves
(vii) Pump body
(viii) Upper transparent dome with a filter

Working

The eccentric which revolves on the camshaft actuates a pump lever as shown in Fig. 8.5. The inner end of this lever pushes a diaphragm pull rod to actuate the diaphragm. When the diaphragm moves down against its spring pressure, a vacuum is created above the diaphragm due to sudden increase of volume, inlet valve opens allowing fuel to flow into the pumping chamber. The outlet valve remains closed.

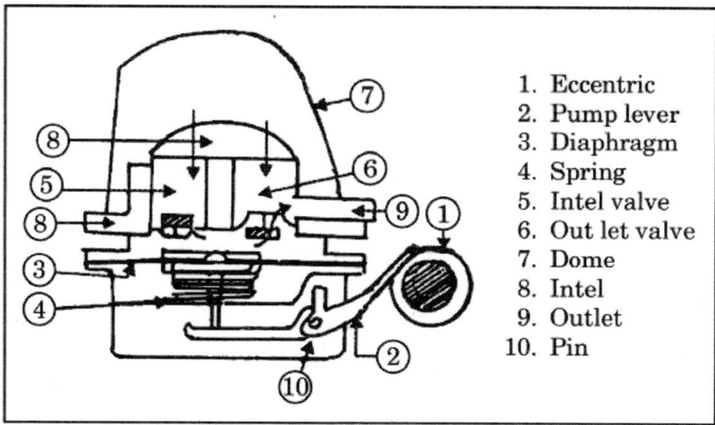

Fig. 8.5: Mechanical pump

During the upward movement of the diaphragm, the diaphragm spring applies pressure on the fuel causing the outlet valve to open and inlet valve to close. The fuel is now delivered from the pumping chamber to the carburetor through a pipe.

When the float chamber of the carburetor is full, the needle float valve sticks firmly to its seating and thus stops further fuel supply. The fuel in the pumping chamber of the pump holds down the diaphragm to its lowest position against spring pressure.

The pump lever keeps on oscillating but does not move the diaphragm. This is so because the pump lever is made in two parts, the two are pivoted together at the centre. Even if the first part rotates along with the camshaft, the second part does not. It should be noted that the fuel delivery pressure to the carburetor depends upon the magnitude of force exerted by the diaphragm spring. This force, in turn, depends upon the design parameters of the spring.

Mechanical pumps operate only when the engine runs. These are extremely reliable but the fuel can develop a vapor lock owing to engine heat.

Chapter 9

Fuel Supply System of a Diesel Engine

9.1 INTRODUCTION

In compression ignition engines (diesel engine), air is alone taken into the engine cylinder during the suction stroke. On the compression stroke, the air is highly compressed to a pressure of about 35 kg/cm^2 absolute and temperature of air rises to about 600°C. Towards the end of the compression stroke, diesel fuel is sprayed at a pressure of 100-200 kg/cm^2 into the combustion chamber. The high temperature of air automatically ignites the sprayed and atomized fuel in the combustion chamber causing the power stroke to occur. The fuel supply system in the diesel engine creates high pressurized fuel to be sprayed in the combustion chamber.

Hence the fuel supply system of a diesel engine must have the following component parts for supplying high pressure fuel to the injector.

1. A fuel tank to store the fuel oil.
2. A fuel feed pump to supply fuel to the fuel injection pump from the fuel tank.
3. A fuel injection pump to deliver fuel under pressure to the injector.
4. A hand priming pump to feed fuel from the tank to the injection pump for starting the engine and to bleed air from the system.
5. Injectors to inject fuel into the combustion chamber in an atomized (broken into smaller droplets) state.
6. A governor to regulate the input fuel supply to the injector in accordance with the engine load
7. Fuel lines with a leak-off pipe connection
8. Filters for filtration of fuel.

9.2 REQUIREMENTS OF FUEL INJECTION SYSTEM

1. Fuel supply should be in accordance with the various load and speed requirements of the engine.

2. Correct ignition timing and even burning of fuel
3. Constant supply of fuel oil
4. Proper atomization of fuel
5. Fuel must have lubrication qualities to lubricate the different parts of fuel supply system (pumps, valves, injectors etc.)
6. There must be minimum foreign materials in fuel supply line to reduce wear and tear of fuel supply parts.
7. The injection pressure must be higher than the air pressure in the combustion chamber.

9.3 PROPERTIES OF FUEL

The properties of the fuel have considerable influence on the performance and reliability of a diesel engine. The main properties of the fuel are discussed below.

Ignition: The ability of a diesel fuel to ignite by itself or self-ignite under the conditions existing in an engine cylinder is called fuel ignition quality. A fuel which will self-ignite at a low temperature is considered to be of good ignition quality. Such fuel performs better in an engine. It enables easy starting, produces less smoke and reduces fuel knock. The ignition quality is expressed by an index called cetane number.

The cetane number of a diesel fuel is a measure of its ignition characteristics. The higher the cetane number, the shorter the ignition delay period, hence, the smoother, the engine operation. The cetane number is defined as the percentage by volume of cetane in a mixture cetane (which is given the cetane value 100) and alpha-methyl napthalene (whose value is 0) that has the same ignition quality as the fuel being tested.

Volatility: The readiness with which a liquid changes to a vapour is known as volatility of the liquid. The vapourizing temperature of high speed diesel oil is higher than that of petrol. If all the fuel is not vaporized, it will go to the crankcase oil. The rapid vapourizing of fuel helps to produce more power.

Carbon residue: The carbon of a fuel indicates its tendency to form carbon deposits on the engine parts. After all the volatile matter in a sample of fuel has been evaporated, what remains is the carbon residue. This carbon residue should not exceed 0.22 to 0.25 % of the fuel sample by weight.

Sulphur content: The presence of sulphur in the diesel fuel shortens the ignition delay period and thus reduces engine roughness and knock but excessive sulphur content in the fuel may cause deposits and corrosion. This deposit increases the wear of the cylinder, piston and piston rings. The corrosion is caused when sulphur produced during combustion combines with the water of combustion to form strong acids which attack the engine components. Therefore, the limit for sulphur content should be less than 1 % by weight.

9.4 METHODS OF FUEL INJECTION

There are two methods of injecting fuel into the combustion chamber in the case of diesel engines.

1. Air injection (Fig. 9.1)

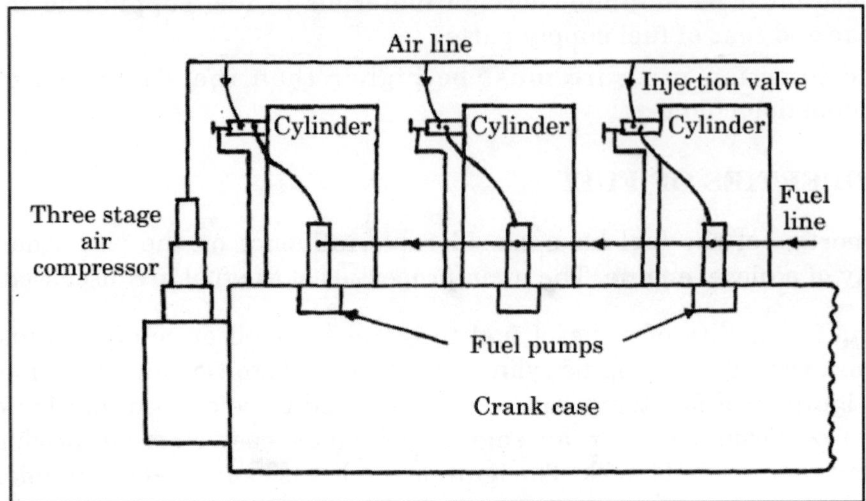

Fig. 9.1: Air Blast Injection.

2. Solid or airless injection.

In the air injection system, the fuel is injected through the injector with the help of compressed air at a pressure of about 55-80 kg/cm^2 i.e., much higher than the compression pressure of air in the combustion chamber. Fuel is pumped to the injector from the tank at a pressure of about 35 kg/cm^2, where from a separate tube air at a high pressure (55-80 kg/cm^2) is supplied by a compressor, which forces the fuel into the combustion chamber. This system is adopted in only very large stationary four-stroke cycle engines and is becoming obsolete because of the necessity of an air compressor, heavy and bulky air storage tank and other parts.

The solid injection system incorporates a fuel injection pump, which meters out correct quantity of fuel, at high pressure and at precise timings to the injectors. The injectors, in turn, atomize the fuel and spray it at high velocity to get burned when coming into contact with hot compressed air. Fuel is first pumped from the main supply tank to an auxiliary chamber by means of feed pump. High pressure pump receives the fuel from the auxiliary chamber and forces the fuel to the injector in the cylinder head.

The direct injection system is further subdivided as *(i)* common rail system *(ii)* individual pump system or in line jerk pump or timed pumped system *(iii)* distributory system.

Common rail system (Fig. 9.2): This system utilizes one pump only. Fuel is sucked from the fuel tank and is delivered to the combustion chamber by a pump. The pump delivers the fuel under high pressure to the common rail, from where the fuel is supplied to every injector. One injector is fitted for each cylinder.

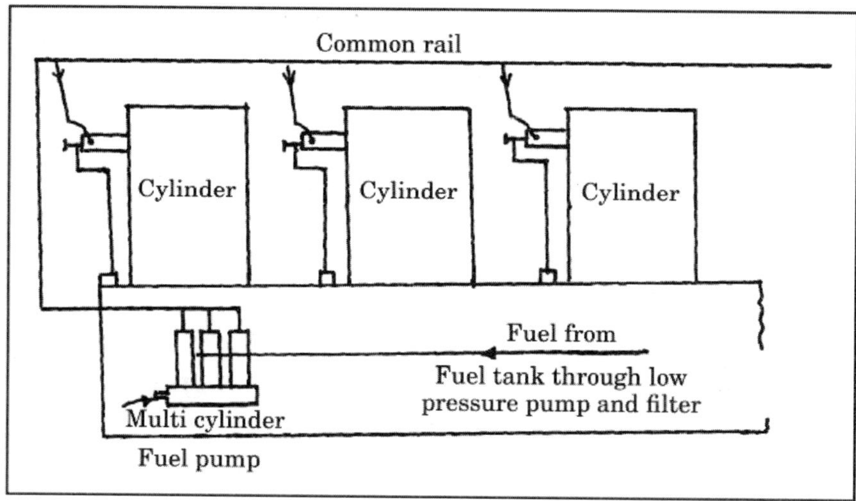

Fig. 9.2: Common Rail Injection

Individual pump system (Fig. 9.3): There is a separate pump for -each cylinder. The fuel tank supplies fuel to different pumps from where it goes to the respective cylinders. This system is most widely used in all diesel engines.

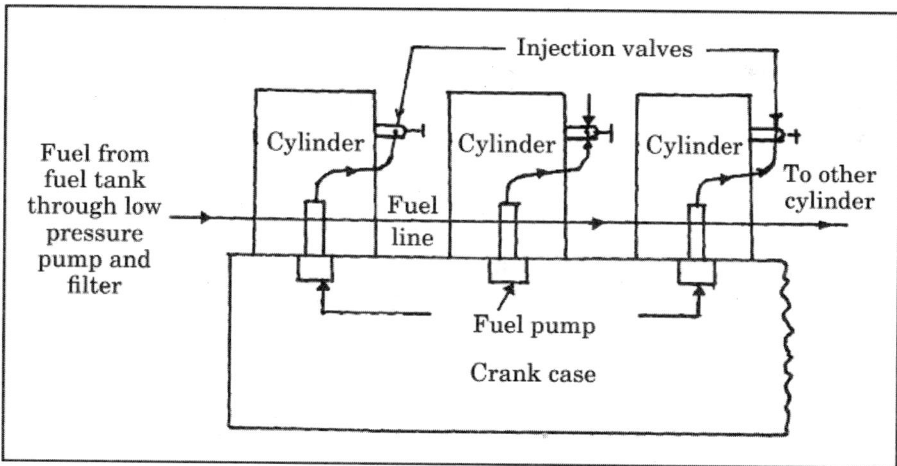

Fig. 9.3: Individual Pump Injection

Distributory system: This system consists of a distributory unit along with the pump. The pump supplies fuel to the distributory unit from where the fuel is distributed to every cylinder separately.

9.5 MAJOR PARTS OF FUEL SYSTEM

Fuel system of diesel engine (Fig. 9.4) consists of the following components. (1) Fuel tank (2) sediment bowl (3) fuel feed pump or fuel lift pump or fuel transfer pump (4) primary filter (5) secondary filter (6) low pressure line or pipe (1.5 to 2.5 kg/cm^2) (7) Fuel injection pump (8) High pressure pipe (100-200 kg/cm^2) (9) fuel injector (10) overflow valve (11) hand priming device (12) combustion chamber.

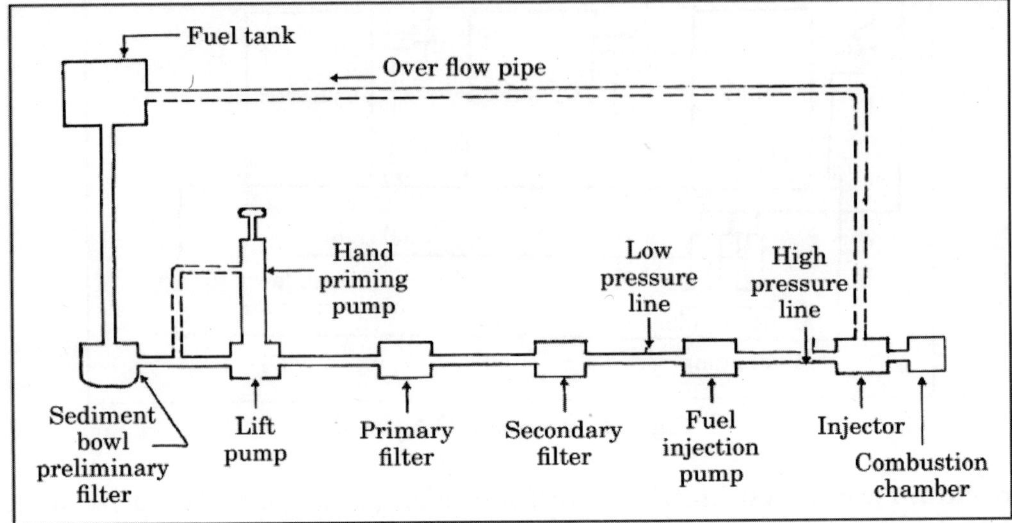

Fig. 9.4: Flow Diagram of fuel in Diesel Engine

Working: Fuel is drawn from the fuel tank by the feed pump. The feed pump is located on the front of the injection pump and is drawn by an eccentric fixed on the camshaft. Eccentric is the circular projections around the shaft of the camshaft. The fuel is forced to the injection pump through fuel filters (primary and secondary filters). The injection pump, in turn delivers fuel to the injectors in correct quantity, at high pressure and at precise timings. Fuel is injected to the combustion chamber through injection nozzles. The extra fuel which does not take part in the combustion is returned to the fuel tank through over flow pipe (or leak-off pipes). The injector injects the fuel at high pressure (100-200 kg/cm^2), 30-40 degrees before TDC of piston in a spray form *i.e.* in the atomized state. The fuel burns when coming in contact with hot compressed air which is at a pressure of about 35 kg/cm^2.Air is expelled from the fuel supply system by operating hand priming device or priming pump before starting the engine. Diesel fuel feed system is shown in Fig. 9.5

9.5.1 Fuel feed pump (Fig. 9.6)

The fuel feed pump draws the diesel from the diesel tank and supplies it to the fuel injection pump (Fig. 9.8). It consists of a barrel, plunger, plunger return spring, spindle, roller tappet, suction and delivery valves. The fuel pump is driven by the eccentric of the camshaft of the fuel injection pump. When the eccentric is at its minimum throw, the plunger moves inward by the action of the plunger return spring. During this time, the diesel from the diesel tank is drawn into the suction chamber. The suction valve is then opened. The delivery valve is closed. When the cam is further rotated, the eccentric pushes the roller tapper outward. Due to this action, the plunger is pushed forward. When this happens, the plunger return spring is compressed.

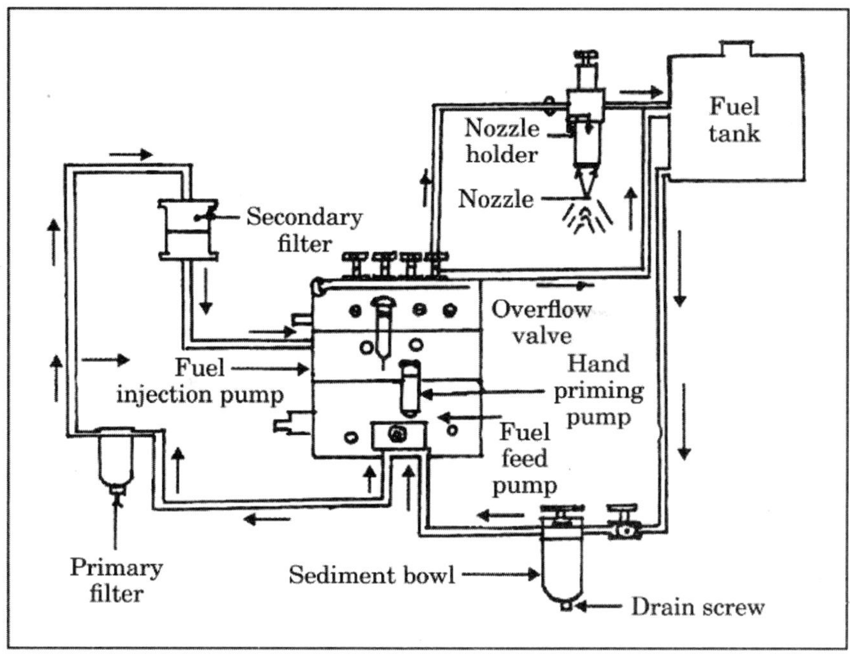

Fig. 9.5: Diesel Fuel System in Line Pump

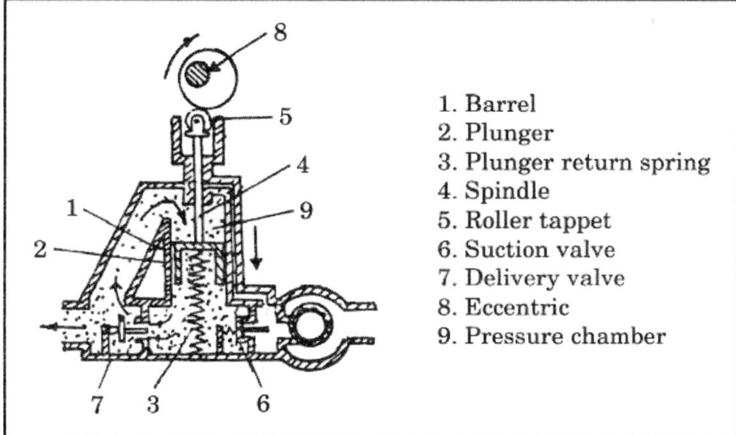

Fig. 9.6: Feed Pump

During this time the suction valve is closed. The fuel from the suction chamber is pushed out through the delivery valve. The fuel then enters and occupies the upper portion of the plunger. It is retained in that particular pressure chamber.

Now when the plunger again moves inward, the diesel oil is compressed and directly discharged to the diesel injection pump. This operation is repeated.

Suppose the fuel injection pump does not require the fuel from the fuel feed pump. In that case, the diesel oil builds up pressure in the pressure chamber of the feed pump. The plunger is thus pushed against the action of the return spring. By this operation, the roller tappet just about keeps away from the cam, although touching it. This is not operated by the cam of the camshaft. Therefore, no further stroke takes place in the fuel feed pump. Once the fuel injection pump requires this oil, the pressure is immediately relieved. The plunger moves inward and the operation is once again started.

9.5.2 Hand priming pump (Fig. 9.7)

The hand priming pump is located on the top of the suction valve of the fuel feed system (Fig. 9.5). The hand primer can be operated for bleeding the fuel feed system. When the engine is not running, the diesel injection pump can be supplied some fuel from the hand primer. The hand primer consists of a barrel, a plunger and a spindle. By operating the spindle, a sufficient amount of the fuel can be supplied to the filter and diesel injection pump. This helps in actually removing the air from the fuel feed system. It also helps in checking the operation of the diesel injection pump. After operating this pump, the hand primer should be properly screwed in place. Figure (9.5) shows the various parts of a fuel feed pump along with a hand primer.

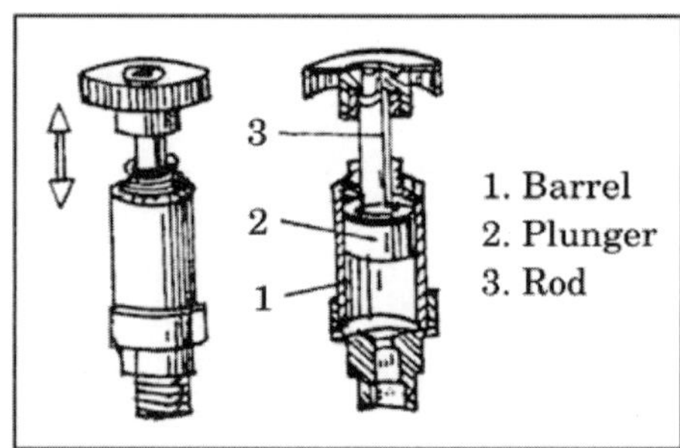

Fig. 9.7: Hand Priming Pump

9.6 FUEL INJECTION PUMP

The fuel injection pump serves the purpose of delivering highly pressurized, accurately timed and right quantities of fuel to the cylinder in accordance with top of the barrel, a one way or non-return delivery valve is fixed. The plunger is lifted by cam lobe of the cam shaft and returns to its original position with the help of a spring when cam lobe passes off. The fuel injection (F.I.) pump is known as the heart of the diesel fuel supply system, as it controls speed, load and fuel economy of the vehicle. F.I. pump unit (Figure 9.8) consists of the following essential components.

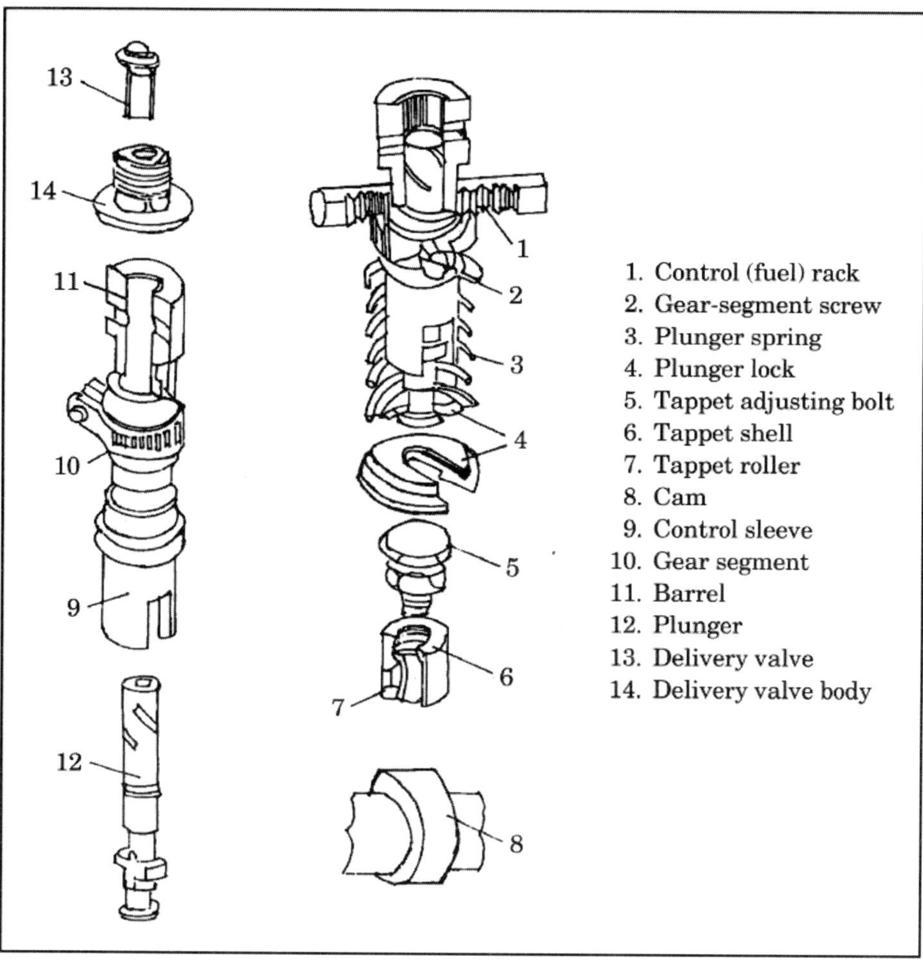

Fig. 9.8: Pumping Element of a Fuel-injection Pump

(1) Cam
(2) Tappet roller
(3) Tappet
(4) Cut washer (plunger lock)
(5) Plunger retaining spring (retainer)
(6) Plunger
(7) Barrel
(8) Control sleeve
(9) Control rod
(10) Delivery valve seat
(11) Delivery valve spring
(12) Delivery valve
(13) Delivery valve holder
(14) Pump body

The plunger reciprocates inside the barrel. The clearance between the plunger and barrel is very small and of about 0.001 to 0.002 mm. The top upper part of the plunger has a helical groove and a vertical groove. The barrel at its upper part has two opposite side openings located at different heights. The upper opening, the inlet port serves to fill the barrel space above the plunger with fuel. The lower opening, the spill port is used to bypass the fuel to the inlet side of the pump or oil gallery in the pump housing. The lower part of the plunger is fitted with a circular recess, called control sleeve. The sleeve has toothed quadrant. The teeth gear of the control rod is in mesh with the gears of the toothed quadrant. The control rod is attached to the governor and governor is attached to the accelerator pedal with the help of mechanical linkage. The plunger stroke remains constant but the effective stroke is variable with the help of the helical groove in the plunger. The effective stroke of the pump plunger is from the point at which the suction port (inlet port) and spill port are closed by the plunger to the point at which it is again opened to the spill port by the helical grove.

The output of an engine is controlled by either increasing or decreasing the amount of fuel by turning the plunger. The rotation of plunger changes the position of its helical groove with respect to the barrel port (spill port) .The movement of plunger is governed by the control rod of the pump. This control rod is connected to the governor by a linkage.

The amount of fuel injected is varied by changing the effective length of the plunger stroke. When less engine power is needed, the effective length of the plunger strokes is short. Relatively small amounts of fuel are delivered. When more power is needed, the effective length of the plunger strokes is increased. More fuel is delivered. The actual length of plunger travel is always the same. However, the quantity of fuel injected is changed by turning the pump plungers.

Control of fuel supply by plunger

The injected fuel quantity is varied by changing the plunger's effective stroke. To do so, the control rod (rack) turns the pump plunger in the barrel with the help of control sleeve, so that helical groove (helix), which runs diagonally around the plunger circumference, can open the spill port sooner or later and in doing so change the end of delivery point and with it the injected fuel quantity. In case of maximum delivery, helix opening with the spill port cannot take place until the maximum effective stroke has been reached. With the partial delivery, the helix opening with the spill port occurs sooner depending upon the exact position of the pump plunger. At zero-delivery position, the plunger position is so changed that helix faces the spill port immediately. No fuel is delivered. This is the position to which the plunger is rotated when the engine is switched off.

Effective stroke: It is the length of plunger's movement from the point at which the ports are closed by the plunger to the point at which the helix coincides with the spill port. By rotating the plunger according to the speed and load of the engine, the effective stroke is shortened and lengthened. But the total stroke of the plunger from BDC to TDC remains same.

The successive positions of plunger during delivery stroke have been illustrated as follows

1. In position, when the plunger is at BDC, the pressure is reduced in the barrel, causing the diesel fuel under low pressure to enter into it from the oil gallery through the inlet port cut in the barrel.

2. In position, when the plunger starts going up from BDC towards TDC, it closes the both inlet and spill ports. As such, the fuel gets trapped in the barrel above the plunger and when the plunger moves further up, this trapped oil gets pushed (pumped) out to the injector.
3. In position, when the plunger further goes up, the helix (helical groove) comes in coincidence with the spill port resulting in the escaping of fuel to the oil gallery instead of going through the delivery valve.
4. In half load position, the position of plunger is so moved that the helix cut in the plunger faces the spill port when it has traveled half of its maximum stroke.
5. In idle speed position, the position of plunger is so moved that the helix faces the spill port after the plunger is lifted a little bit and as such very little fuel can be pumped to make the engine run at idle speed.
6. In stop position, the position of plunger is so changed that the long slot of the helix faces the spill port of barrel, resulting in the no trap of fuel in the barrel. In this case, the fuel instead of being pumped up finds its way back to the gallery through the spill port. This position is known as stop position and the engine stops working as no fuel is being pumped to the injector. The position of plunger at different load conditions have been shown in Fig. 9.9.

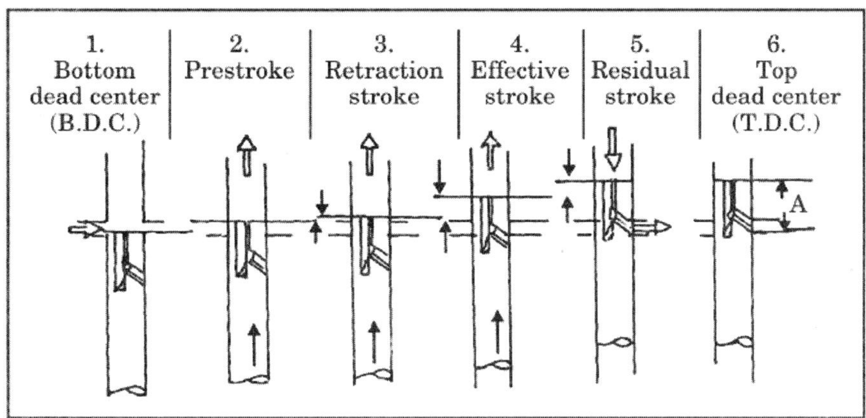

Fig. 9.9: Plunger Stroke Phase

The delivery of the fuel-injection pump is increased by moving the control rack forward. The control-rack movement is controlled by a governor mounted on the rear wall of the pump housing. When control rack pulled out (full injection) and pushed in (no injection) from the no injection position to the full injection position (full rack movement), the contour of helix advances from the closing of the ports and beginning of the injection.

9.7 FUEL INJECTOR

Fuel injector (Fig. 9.10) is a device by means of which, the fuel is injected into the combustion chamber. This is also known as the atomizer or sprayer. From the fuel injection pump, the fuel passes through the delivery valve and high pressure pipe to the injector. At the end of the injector is a nozzle through which the fuel is sprayed in fine atomized form into the cylinder.

Hence the function of the fuel injector is to deliver finely atomized fuel under high pressure to the combustion chamber of the engine. The main component of the injector is the nozzle comprising nozzle body and nozzle needle valve.

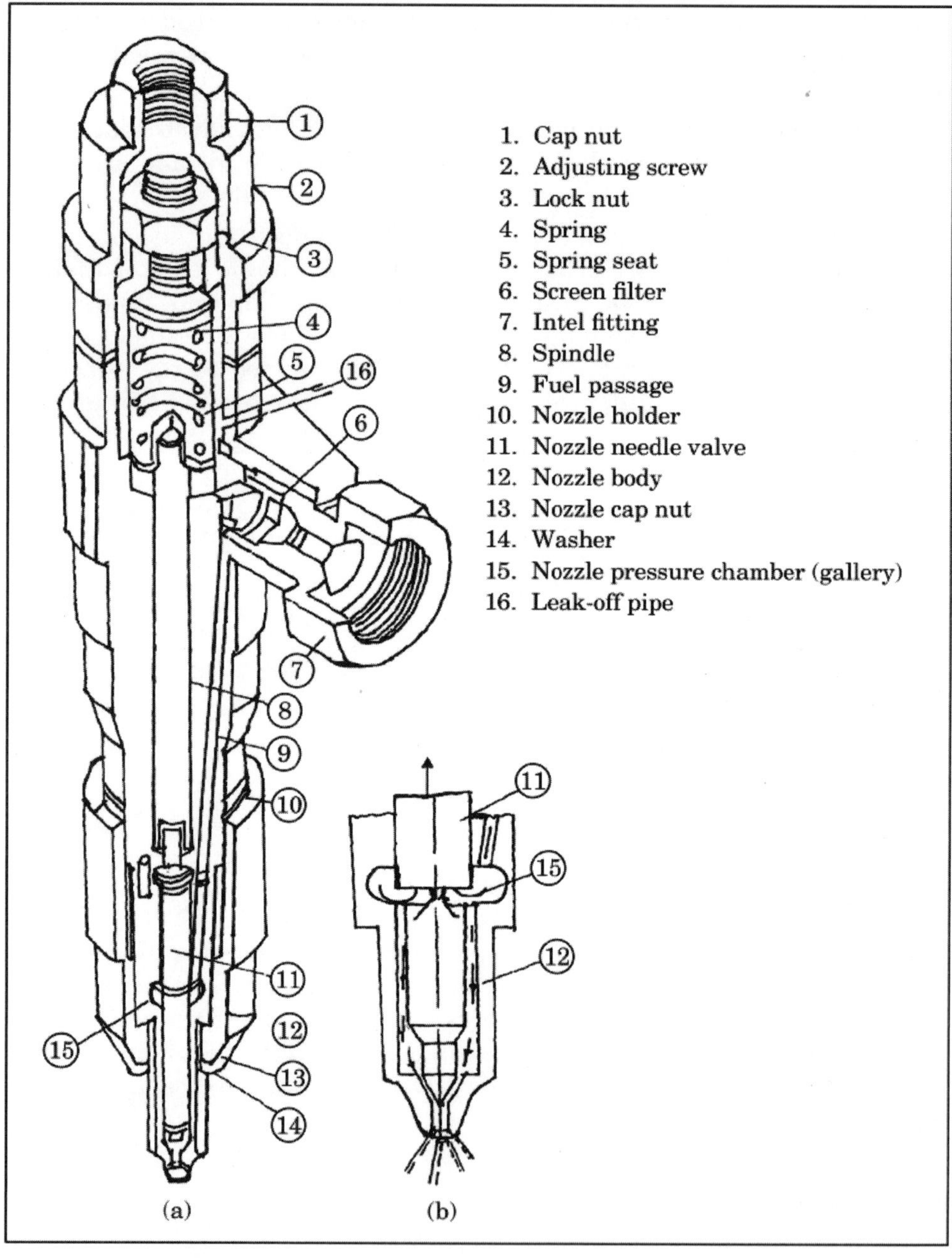

Fig. 9.10: Injector (a) Construction (b) Illustrating the Operation of the Injector

The nozzle is a type of jet through which the fuel is correctly distributed in the combustion chamber. The function of the nozzle is to convert the pressure energy of the fuel into the kinetic energy by passing through the orifice of the nozzle outlet.

***Operation*:** The nozzle is operated by fuel pressure. Fuel from the FI pump passes to the nozzle holder through the delivery pipe, which holds the injection nozzle. Then the fuel from the annular groove and with pressure comes to the pressure chamber where pressure is sufficiently built up to raise the nozzle needle which is held (kept) on its seat by means of a spring. When the valve is lifted, the fuel is injected at high pressure through the small discharge area of spray holes into the combustion chamber. The valve lift during injection is between 0.2 and 0.25 mm.

When the delivery pressure falls off, the spring returns the nozzle valve to its seat. The pressure at which the nozzle needle valve opens and injection commences is determined by the compression of the valve spring. This can be adjusted manually by adjusting screw. A very small portion of the fuel escapes from the upper part of the nozzle valve by overflow pipe (leak-off pipe) ensuring effective lubrication.

The main parts of an injector are (1) end cap (cap nut), (2) adjusting screw (3) lock nut (4) spring (5) spring seat (6) nozzle valve rod (Nozzle spindle) (7) Nozzle holder (8) Nozzle cap (9) Nozzle body (10) Pressure chamber (11) Nozzle needle valve (12) Nozzle hole (13) Leak -off pipe.

9.7.1 Types of fuel injection nozzles (Fig. 9.11)

The various types of fuel injection nozzles are briefly discussed below.

Single hole: A single hole nozzle has one hole at its centre. The fuel is sprayed through this nozzle. The spray of the oil is vertically downward.

Multi-hole: These nozzles have many holes arranged around the nozzle in a circle. These holes may vary according to the engine. Their numbers as well as their sizes and hole angles may all vary.

Pintle: There is a pin or pintle at the tip of the nozzles valve. Therefore, this nozzle valve is called the pintle nozzle. The open and closed positions of the pintle nozzle are shown. When the pintle is lifted, the fuel is sprayed uniformly. The advantage of this nozzle is that it prevents the carbon from depositing in the hole.

Delay: Figure (9.11d) gives the various stages of the delay nozzle. A requirement of the combustion chamber is that the fuel is supplied to it gradually.

When the pintle nozzle is lifted, initially a small quantity of the fuel is sprayed. When the nozzle is lifted to the maximum, the quantity of the fuel sprayed is also maximum. In this manner, the supply of the fuel is very gradual. This is the advantage of the delay nozzle. This design is preferred in certain pre-combustion chamber type engines.

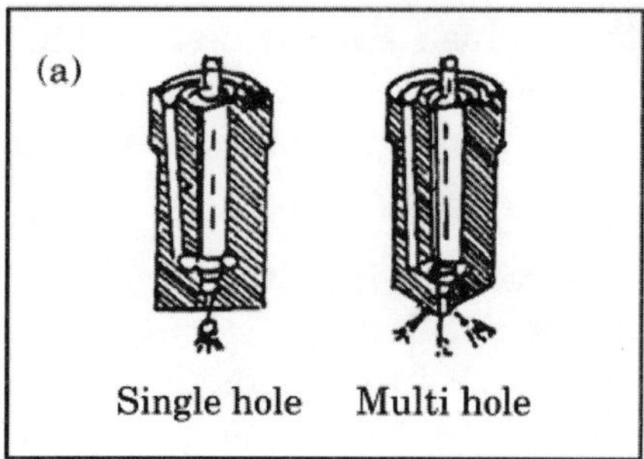

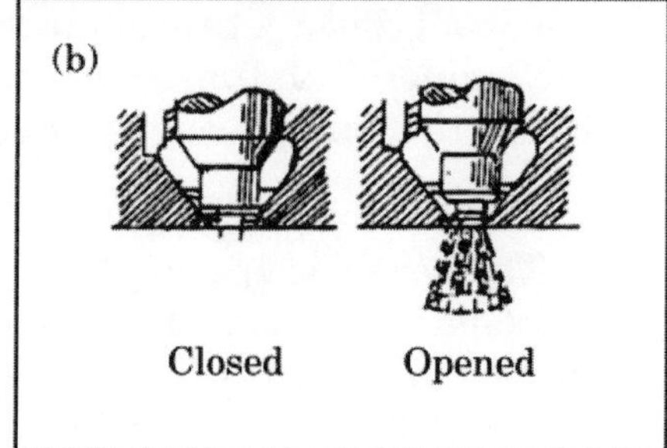

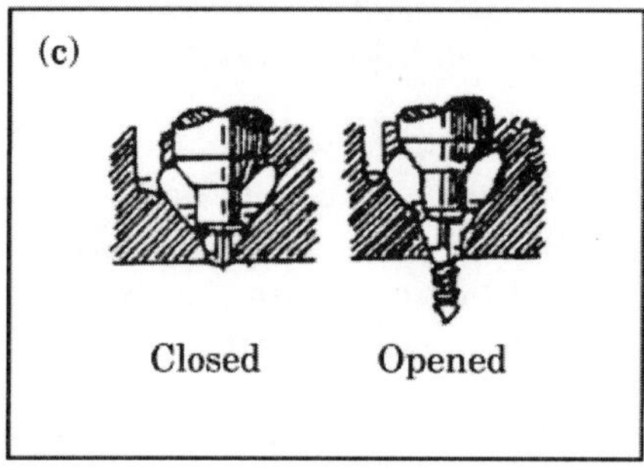

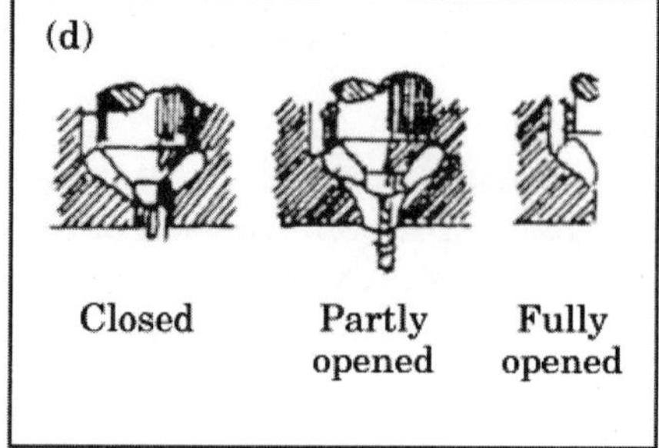

Fig. 9.11: Types of Nozzles (a) Types of Nozzles. (b) Pintle Nozzle. (c) Pintle Nozzle. (d) Delay Nozzle

9.7.2 Selection of nozzle

The choice of a nozzle is determined mainly by the combustion process and the design of the combustion chamber. The following data are important for the ultimate choice of a nozzle: power and speed of the engine, fuel consumption, duration of injection, injection pressure and spray angle. Without reliable knowledge of the shape of the combustion chamber it is not possible to accurately determine in advance the type of nozzle required.

Hole type nozzles are used in direct injection type engines, but pintle type nozzles are used in the pre-combustion chamber, turbulence chamber and air-cell engines. For ready reference, the specifications of mico nozzles available on indigenous tractors are given in the following Table.

Size	*Pintle nozzle*		*Hole Nozzle*		
Body length, mm	27	40	32.5	43	53.5
Height of collar mm	–	10	8	10	25
Diameter of collar, mm	17	22	17	22	17
Stem diameter, mm	14	18	14	18	–
Spray angle, degrees	(0 to 35°)		hole angle up to 180°		
Opening pressure, Kg/cm^2	80 to 125		150 to 250		

Chapter 10

Engine Governing System

10.1 PURPOSE OF GOVERNOR

The governor is a mechanical device used to maintain the speed of the engine constant regardless of the changes in the load on the engine. Tractors and other stationary engines run at varying load. If there is no device for controlling their speeds, speed variations will occur frequently by the changes of the load. If the load becomes less, the speed will go up and if there is heavy load, the speed will come down, provided fuel supply is not controlled.

Consider a chaff cutter, operated by an oil engine cutting forage crops *i.e.*, working on full load. When one lot is finished and during the feeding of the next bundle, the machine requires minimum power, but the constant supply is for the maximum load. At this stage, *i.e.*, in the idle condition, the speed may become so high that the machine parts may fly away. Hence in order to maintain a steady speed at varying loads and to protect the engine parts from high speeds, governor is used.

Generally, the load on the engine of a tractor varies depending on the lay of the land, soil characteristics, types of field work etc. As a result, the speed of travel, speed of belt pulley and rpm of PTO shaft are affected. Under these circumstances, it becomes difficult for the operator to regulate always the throttle lever by accelerator pedal to meet the temporary changes in the engine load. A governor automatically regulates the engine speed on varying load condition and thus the operator is relieved of the duty of constantly regulating the throttle lever to suit different load conditions. Therefore, the purpose of the governor is to provide automatic control of the idling and maximum speed of the engine. The accelerator under driver's foot controls the intermediate speeds of the engine. The fuel pump fitted with a governor has been shown in Fig. 10.1.

10.2 METHODS OF GOVERNING

The governing of speed of the engine according to the load is done by one of the following methods:

1. The fuel supplied to the engine is completely cut-off during few cycles of the engine. This is known as Hit and Miss Governing. This is generally used for gas engine.
2. The fuel supplied per cycle of the engine is varied according to the load on the engine. This is known as "quality governing". The Air: Fuel (A:F) ratio is changed according to the load on the engine. Rich mixture is supplied at high loads and lean mixture is supplied at low loads. This is used for diesel engine.
3. The quantity of air-fuel mixture supplied is varied according to the load on the engine. The A:F ratio of the mixture supplied to the engine at all loads remains nearly constant, therefore, it is known as "Quantity governing". This is used for petrol engine.

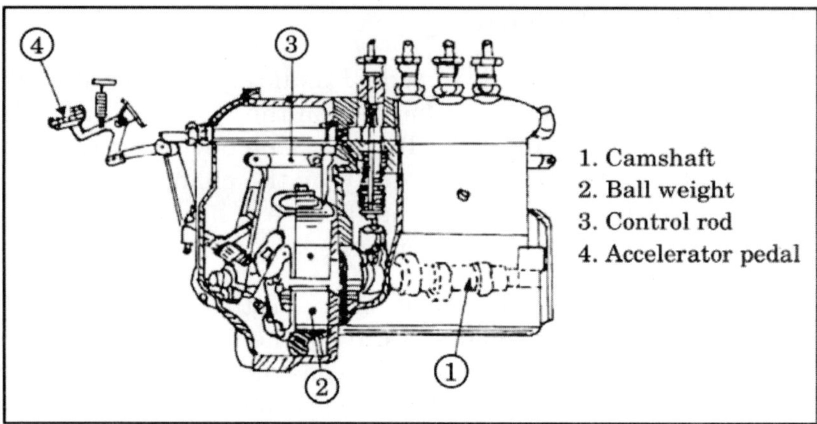

Fig. 10.1: Fuel Pump Fitted with a Governor.

10.3 TYPES OF GOVERNORS

The governors, according to the principle of operation, may be classified into two groups.

(a) **Pneumatic governors:** They operate by the depression created by the air flow in the inlet manifold.

(b) **Mechanical governors:** These work on the principle of centrifugal force due to rotation of two fly weights. Such governors are further divided into two.

  (i) Constant speed governors: These are mounted on engines which are required to run at constant speed, for example stationary engine used for an electricity generation plant.

  (ii) Variable speed governors: These keep the speed of the engine constant with changing loads within prescribed limits.

Tractor engines are fitted with variable speed governors in view of the following advantages.

(i) Easy and convenient operation of the tractor and implements.
(ii) Longer service life of tractor and implement on account of operating at slow speed.
(iii) Reduced fuel consumption at partial load
(iv) High rate of field work due to reduced time required to change from gear to gear while operating the tractor with implements.

10.4 CLASSIFICATION OF GOVERNING SYSTEM

According to the- method of governing, the governing system is basically classified as

1. Hit and miss system and 2. Throttle system

***Hit and miss system*:** Hit and miss means fire and misfire respectively. That means the system controls the engine speed by providing the engine conditions of misfiring. Misfiring means no power generation in the power stroke. Therefore, in this system, the frequency of explosions or power strokes of engine are regulated. When the engine sped exceeds the rated speed, the exhaust valve gets open, which prevents fresh air-fuel mixture or fuel to enter into the engine cylinder during suction stroke.

The main principle in this method is to keep all explosions alike at maximum intensity but to vary the number of explosions per unit time, depending upon the power requirement. A detent or a latch is provided on the exhaust valve push rod which prevents the valve from closing when the speed is too high. As a result, no charge is drawn into the cylinder on the suction stroke and no power is developed in the engine cylinder due to complete absence of air (in diesel engine) or air + fuel (in petrol engine) and as a result, the speed falls. Thus it is found that explosions are missed intermittently. That means, the governor pushes the exhaust valve to remain open till the speed comes to normal. By controlling the frequency of explosions, the engine speed is controlled by varying load.

Its use is limited to only single cylinder old farm engines, because of its disadvantages like (a) very heavy flywheel is needed to keep the engine running during missed strokes (or idle strokes) (6) it may cool down during these missed stroke,

***Throttle system*:** This system, consists in controlling the amount of fuel mixture or fuel during suction stroke and thereby changing the explosion intensity in the cylinder. Here the number of explosions are not reduced, only intensity of explosion is changed. There is uniform firing in the cylinder throughout the period of operation. Throttle, as explained in carburetion system, is a butterfly valve situated in the air inlet tube, which controls the amount of mixture entering the engine. The governor of this system operates the throttle, thereby maintaining a uniform engine speed at varying loads. If the speed decreases due to heavy load, the governor will open the throttle fully, so that maximum mixture may go in and the engine may produce a power sufficient enough to maintain the previous speed at the increased load. Similarly as the speed goes up by removing the load, the governor pushes the butterfly valve to nearly closed position, resulting in minimum mixture entering the engine, producing lesser power and thus the speed comes down to its normal position. This is the most effective system and is universally used in tractors, trucks, motors etc.

Throttle system of governor can be divided into two types; (a) Centrifugal governor (&) Pneumatic governor.

10.5 CENTRIFUGAL GOVERNOR

The main principle of a centrifugal governor lies in the fact that when a weight rotates about a point, it tends to fly outward due to centrifugal force. If the weight is constrained by a linkage, it can be made to operate a control lever.

The centrifugal governor is common on tractor and stationary engines. It consists of spring loaded centrifugal weights, a sliding collar and connecting rod linked either to control the throttle valve in the throttle governing system or to actuate the exhaust valve in the Hit and Miss system. The centrifugal weights maybe either mounted on the engine crankshaft or on another shaft driven by the crankshaft. The sliding collar moves along the axis of the shaft depending upon the speed of centrifugal weights. Since the weights are mounted on the shaft rotated by engine, its speed is same as the engine speed. The weights are fixed to the shaft with the help of spring placed in them. The spring fitted in the weights counterbalances the centrifugal force produced by the fast moving of the shaft.

Centrifugal governor (Fig. 10.2) thus consists of (i) spring loaded centrifugal weights *(ii)* sliding collar *(iii)* spring *(iv)* throttle valve *(v)* connecting rod. In case of petrol engine, at low loads, the engine speed tends to go high, the centrifugal weights tend to fly outward against spring tension, thereby closing the throttle valve and reducing the entry of charge inside the engine. Similarly at high loads, the speed tends to become slower and the centrifugal weights come closer due to spring tension. The contraction of the weight causes the throttle to open fully, allowing more fuel to come in the engine cylinder to increase the speed of the engine. Hence engine speed is controlled at low loads (or maximum speed) or high load (or minimum speed). The accelerator pedal is also properly linked with the throttle valve of the carburetor. Without depending upon the operation of the governor, the driver can accelerate or decelerate the engine speed by pressing or releasing the accelerator pedal. He can also operate at intermediate speeds based on the road condition and speed requirements. The governor only controls the idling and maximum speeds. The other intermediate speeds are controlled manually by the driver.

In case of diesel engine, the speed is controlled by the quantity of fuel injected due to the position of control rod in the injection pump. The control rod is used to rotate the plunger, thus altering the effective portion of its stroke and varying the amount of fuel injected. The movement of the fly weights is transmitted to the control rod *via* connecting linkage. The rod is moved in the direction of STOP as the engine speed increases *i.e.* the speed is limited because the injection pump delivers less fuel. When the engine speed decreases, the control rod is moved in the direction of FULL LOAD *i.e.*, the fuel injection pump increases the fuel delivery rate and the engine speed increases. The governor thus maintains automatically all speeds steady, including idling and maximum speed. The intermediate speeds are controlled manually by the driver.

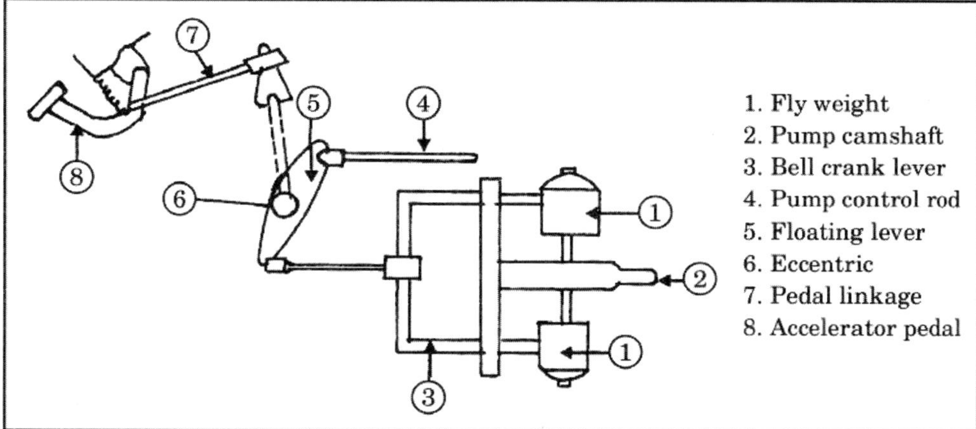

Fig. 10.2: Mechanical Governor.

10.6 PNEUMATIC GOVERNOR

This is a simple device used on both carburetor and diesel engines to control the quantity of fuel - air mixture or fuel depending upon the requirements of the engine at different loads. The pneumatic governor functions due to the vacuum created by the suction in the air inlet pipe of the engine. The venturi effect in the carburetor is that when air is made to pass through restricted passage, its speed increases and at that place, partial vacuum is created. Taking advantages of this principle, pneumatic governors are designed.

Pneumatic governor (Fig. 10.3) consists of two parts. One part is the venturi air flow control unit on the intake manifold of the engine and the other part is the diaphragm unit on the fuel injection pump in the diesel engine and on the connecting lever to throttle valve in the petrol engine.

There are two chambers in the diaphragm unit (a) Atmospheric chamber, which is connected to the atmosphere (6) Vacuum chamber, which is connected to the venturi unit. A leather or rubberized diaphragm separates these compartments. A breather is used in the atmospheric chamber so that clean air can enter into it. The venturi tube connects the air tight vacuum chamber and venturi of the intake manifold.

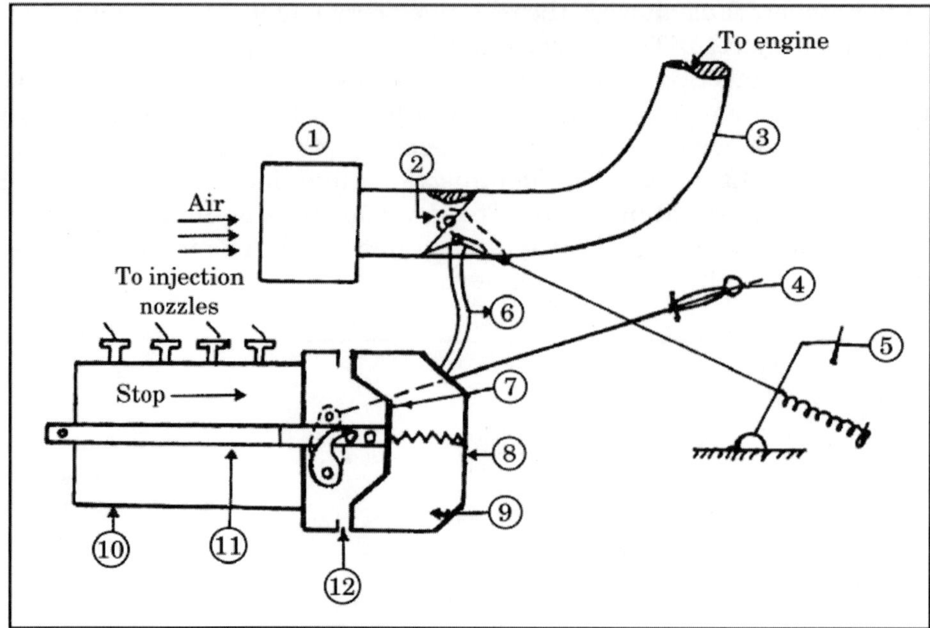

Fig. 10.3: Pneumatic Governor

The different components of pneumatic governor are (i) air filter *(ii)* venturi valve *(iii)* intake pipe *(iv)* start and stop knob (*v*) venturi tube *(vi)* diaphragm *(uii)* governing spring *(viii)* vacuum compartment *(ix)* atmospheric compartment *(x)* fuel injection pump assembly *(xi)* control rod *(xii)* accelerator pedal.

***Venturi control unit*:** It is mounted in the engine intake manifold. The air filter is attached to the venturi control unit. The throat of the body is machined to form a venturi. A butterfly valve and the connection for the vacuum line are located at the narrowest point. The butterfly valve is connected to the accelerator pedal through a control lever and the linkage.

***Diaphragm unit*:** Diaphragm unit consists of two chambers *i.e.* vacuum chamber and atmospheric chamber as explained earlier. Diaphragm is linked to control rod with the help of a pin and is kept in full load position with the help of diaphragm spring .As soon as engine starts, there is suction of the air through the air inlet pipe. A partial vacuum is thus created at the venturi. This is transmitted to the diaphragm spring side. The difference of pressure on both sides of the diaphragm causes it to move. When the diaphragm moves, the control rod also moves. As the speed of the engine increases, the vacuum in the venturi increases, the diaphragm is pulled out, control rod is moved to the stop position and fuel supply is reduced. This condition is also achieved by releasing the accelerator pedal making the butterfly valve to remain partially closed and high vacuum is created. When the speed of engine decreases, vacuum in the venturi decreases, diaphragm is pushed in and control rod is moved to FULL position and fuel supply is increased .Similar condition is also achieved by pressing the accelerator pedal, making the throttle valve to open fully, causing the decrease of vacuum in the venturi throat. To stop the engine, a stop knob is placed in the atmospheric chamber which when pulled pushes the diaphragm along with control rod to stop position.

Hence the movement of diaphragm actuates the control rod of fuel injection pump or the throttle lever of the carburetor to increase or decrease the fuel supply, thus controlling the engine speed.

10.7 MAXIMUM SPEED ADJUSTMENT

The maximum speed is governed by the diameter of the venture throat. The speed is reduced by the adjustment of the control valve movement. This is done by adjusting a screw in the manifold.

10.8 IDLE SPEED ADJUSTMENT

To reduce the idling speed, there is a set-screw in the system. By adjusting the set-screw, the required amount of venturi is opened. This either increases or reduces the idle speed. If the screw is turned in slightly, the idling speed is increased. The idling speed is reduced by slightly loosening the screw.

10.9 MAINTENANCE OF PNEUMATIC GOVERNOR

The operation of the pneumatic, governor can only be effective if the system is air tight *i.e.* vacuum chamber is air tight. To ensure this, the pipe system and diaphragm should be checked for leakage. In stationary engine, generally centrifugal governor is used where as in tractor and automobile, pneumatic governor is used.

Chapter 11

Engine Lubrication System

11.1 PURPOSE OF LUBRICATION

When one metal piece slides over another metal piece, it gives rise to the following effects, *(i)* friction is caused *(ii)* the surface is broken down *(iii)* heat is produced. To prevent these effects, a thin film of oil is applied between the moving parts. This is called lubrication. I.C. engine has many moving parts. Due to continuous movement of two metallic surfaces over each other, there occurs the wearing of the moving parts, generation of heat and loss of power in the engine. Hence, lubrication of moving parts is essential to prevent all these harmful effects.

When one solid surface is made to slide over another, the motion is resisted by friction between the surfaces and heat is generated to an extent depending on the speed and the load. Extreme example is the primitive method of producing fire by rubbing together two pieces of wood. Friction can be reduced, movement can be facilitated and overheating prevented by inserting or interposing a fluid film between the two surfaces. The fluid is called the lubricant and its application provides lubrication. Lubrication of Internal Combustion (I.C.) engine is concerned with metal-metal contacts and mineral oil as lubricants.

11.2 FUNCTIONS OF LUBRICATING OIL IN THE ENGINE

The lubricating oil performs several jobs in the engine:

(i) **Reducing wear:** The primary function of lubrication is to reduce friction and wear between two rubbing surfaces. The oil lubricates moving parts to minimize wear. Clearances between moving parts (for example the bearing and rotating journal) are filled with oil. The parts moving on the layer of oil are freed from excessive wear.

(ii) **Reducing power losses:** Parts moving on layers of oil require a minimum of power to make them move. The oil minimizes power losses in the engine.

(iii) **Cooling agent:** The heat, generated by piston, cylinder, bearing etc., is removed by lubrication to a great extent. The oil after picking up heat drops down into the oil pan and gives up the heat. Therefore, the oil serves as a cooling agent.

(iv) **Load cushioning effect:** The clearances between bearing and rotating journals are filled with oil. When heavy loads are suddenly imposed on the bearing, the oil helps to cushion the load. This reduces the wearing of the bearing (also between piston and cylinder).

(v) **Cleaning agent:** The oil acts as a cleaning agent. As it circulates through the engine, the oil picks up particles of metal, carbon, dirt etc and carries them back down into the oil pan. Larger particles fall to the bottom of the oil pan and smaller particles are filtered out by the oil filter.

(vi) **Sealing agent:** The lubricant enters into the gap between the cylinder liner, piston and piston rings. Thus it prevents leakage of gases from the engine cylinder. It forms a good seal between the piston rings and cylinder wall.

11.3 LUBRICATION THEORY

There lies two theories regarding the application of lubricants on a surface. (a) Fluid film theory (b) Boundary layer theory

Fluid film theory: According to this theory, the lubricant is supposed to act like mass of globules, rolling in between two surfaces. Due to their rolling effect, friction between the rubbing surfaces is reduced.

Boundary layer theory: According to this theory, the lubricant is soaked in "rubbing surfaces and forms oily surface over it. Thus the sliding surfaces are kept apart from each other, thereby reducing friction.

11.4 CLASSIFICATION OF LUBRICANTS

Lubricants are classified according to their source of availability and physical property (Fluidity). Depending on the source of supply, they are classified into the following.

(i) Animal oil (Fish fat), animal oils are obtained after treatment and purification of animal fats. These decompose on being heated and so are not fit to be used in automobiles.

(ii) Vegetable oils (linseed, castor oil etc.) Vegetable oils also decompose soon and give out deposits after sometime.

(iii) Mineral oils (obtained by refining crude petroleum and are generally used in all types of engines). Petroleum lubricants are less expensive and suitable for I.C. engines.

According to fluidity, they are divided into the following types.

(i) Fluid lubricants (which include oils)

(ii) Semi-fluid lubricants (which include grease and heavy oils)

(iii) Solid lubricants (which include graphite, mica etc.)

11.5 PROPERTIES OF LUBRICATING OIL

The properties that an engine lubricating oil should have, include the following.

***Viscosity*:** Viscosity is a measure of how easily oil flows. A low viscosity oil is thin and flows easily. A high viscosity oil is thicker. Therefore, it flows more slowly. Engine oil should have the proper viscosity so that it flows easily to all moving engine parts. But the oil must not be too thin. Low viscosity reduces the ability of the oil to stay in place between moving engine parts. If the oil is too thin (has a very low viscosity), it will be forced out from between the moving parts. Then rapid wear occurs. If oil is too thick (has high viscosity), it flows very slowly to engine parts, especially when the oil and engine are cold. This causes rapid wear of the engine parts.

Hence proper engine oil with proper viscosity is needed for the lubrication of the engine. To take care of this, viscosity-index improvers are added to engine oil, so the viscosity stays more nearly the same, hot or cold conditions.

***Viscosity Index*:** This is a measure of how much the viscosity of an oil changes with temperature. We know that an increase in temperature reduces the viscosity of oils and *vice versa*. A number between 0 and 100 represents the rate at which an oil thins out with the rise of temperature. A high index number shows a small change in viscosity and a low index value shows a large change. Paraffinic base oil is given a viscosity index (V.I.) of zero and naphthenic base oil has a V.I. of 100.

Viscosity number: There are several grades of lubricating oil. They are rated for winter and for other than winter. Winter-grade oils are supplied in three grades, SAE 5 W, SAE lOW and SAE 20W. The SAE stands for the Society of Automotive Engineers, which developed the grading system. The "W" stands for winter. For other than winter use, the grades are SAE 20, SAE 30, SAE-40 and SAE 50. The higher the number, the higher the viscosity (thickness) of the oil.

***Multiple viscosity oil*:** Many engine oils have a viscosity-index improver added that allows the viscosity of the oil to remain relatively the same, hot or cold. For example, a multiple-viscosity oil may be graded SAE 10W-30. This means that the oil is the same as SAE 10W when cold and SAE 30 when hot. This oil works satisfactorily in engines running in a wide range of outside temperatures. Manufacturers recommend the use of multiple-viscosity oil in many automotive engines.

Pour point: The lowest temperature at which oil can be poured is termed as pour point. It is this property which enables the oil to flow through oil galleries. The addition of an additive enables the oil to flow at even low temperature.

***Carbon formation*:** The oil is heated due to the high temperature of the cylinder valves and piston. With only high temperature, it would begin to break down and form carbon. The additives do not allow the carbon to form.

***Oil oxidation*:** When the oil is heated and churned, it combines with oxygen. Oxidized oil forms tar and varnish. The formation of this is prevented by adding additives to the oil. The additives also make the oil slightly alkaline. This neutralizes the acidity in the oil. Thus corrosion in reduced.

***Foaming resistance*:** As oil is churned up in the crankcase by the rotation of the crankshaft, the oil tends to foam or aerate. This can reduce the lubricating effectiveness of the oil because it has air in it. Also, the foaming can cause the oil to overflow and pass up through the crankcase ventilating system to the intake manifold and air cleaner. To prevent foaming, anti-foaming additives are mixed with the oil.

***Detergents*:** These additives act somewhat like ordinary soap. They loosen and detach particles of carbon and dirt from engine parts and carry them down to the oil pan as the oil circulates.

11.6 PARTS: TO BE LUBRICATED

Parts to be lubricated in the engine by lubrication system are

(i) Crankshaft bearing (main bearing)
(ii) Big end bearing
(iii) Small end bearing of connecting rod.
(iv) Bushes of gudgeon pin
(v) Inner wall of cylinder
(vi) Piston rings
(vii) Valve operating mechanism
(viii) Timing gear
(ix) Camshaft bearing (Tappet, valve lifter, push rod, rocker arm)

The various points of lubrication where ball bearings are used and need lubrication are

(i) Wheel grease bearing
(ii) Universal splined shaft
(iii) Leaf spring
(iv) Pedal shaft
(v) Clutch shaft
(vi) Brake cable
(vii) Fan shaft
(viii) PTO shaft
(ix) Differential
(x) Axle bearing.

SAE 30 is for accelerator, clutch linkage; SAE 90 for steering box gears, gear box and differential; SAE 1703 J is used as brake fluids.

11.7 TYPES OF LUBRICATION SYSTEM

Three types of lubrication systems are in use. These are *(i)* splash system (Figs. 11.1 and 11.2) *(ii)* Pressure feed (Fig. 11.3); wet sump lubrication (Fig. 11.4) and dry sump lubrication (Fig. 11.5) *(iii)* combined splash and Pressure feed (Semi-pressure system) (Fig. 11.6). Splash system is used in single cylinder engine. The disadvantages of splash system are:

(i) There is no uniformity in lubrication

(ii) Maintenance of proper level of oil is necessary for effective lubrications.

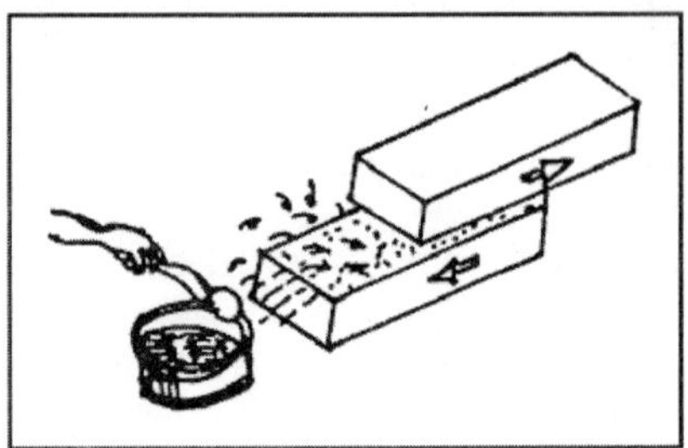

Fig. 11.1: Splashing Oil between Two Metal Blocks.

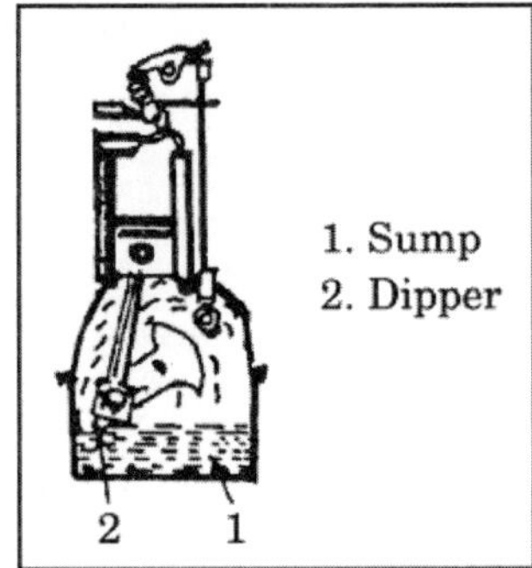

Fig. 11.2: Splash Lubricating System.

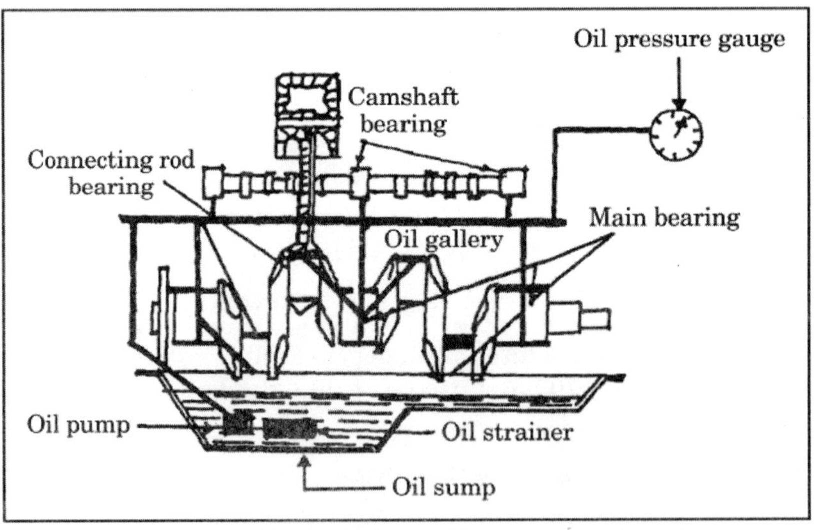

Fig. 11.3: Full Pressure Lubrication System

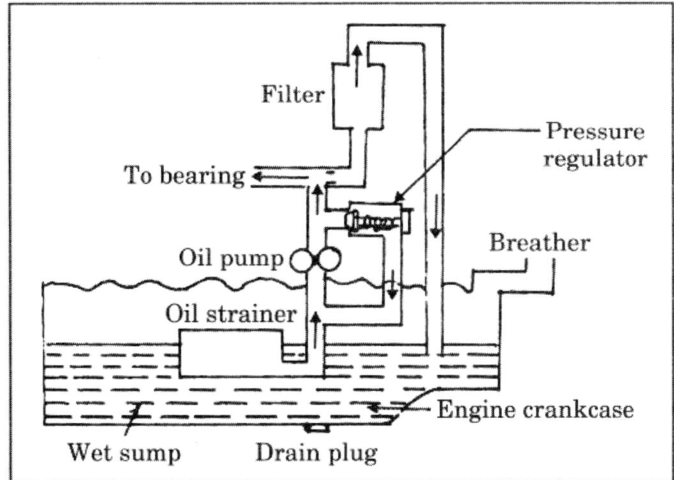

Fig. 11.4: Wet Sump Lubrication.

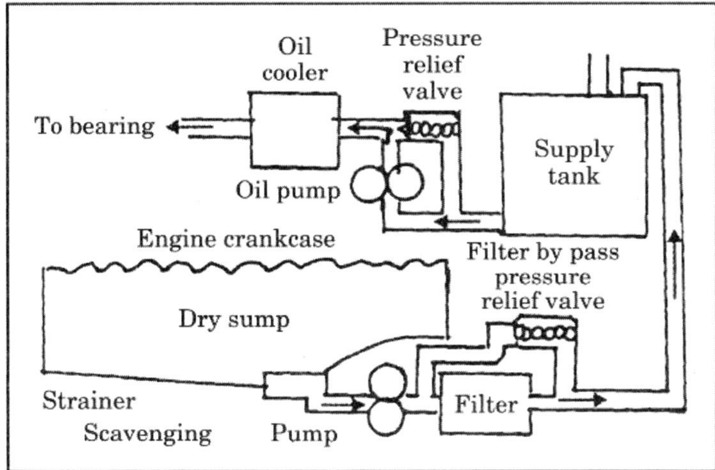

Fig. 11.5: Dry Sump Lubrication.

In wet sump lubrication, there is one pump which is used to deliver oil under pressure to the oil galley. In dry sump lubrication, there are two sump and two pumps (Scavenge pump, pressure pump). Different parts of the lubricating system are: (i) oil pan *(ii)* oil floating screen (strainer) *(iii)* oil pump *(iv)* pressure relief valve (*v*) oil filter *(vi)* oil gallery, *(vii)* pressure gauge *(viii)* dipstick *(ix)* drain plug or drain valve. Relief valve is provided to control the quantity of oil circulation and to maintain correct pressure in the lubricating system. The pressure of oil *i.e.* the lubrication system of tractor engine is around 3 kg/cm^2. Oil pressure gauge is used to indicate the oil pressure in the oil lines. It serves to warn the operator of any irregularity in the system. Oil cooler; the oil which returns from various engine parts and then falls into the sump is hot. This hot oil should be cooled. If overheated, the oil loses its lubricating proper ties. The oil cooler maybe an aluminum casing with fins in which the oil filter is housed. Oil is cooled by natural current of air.

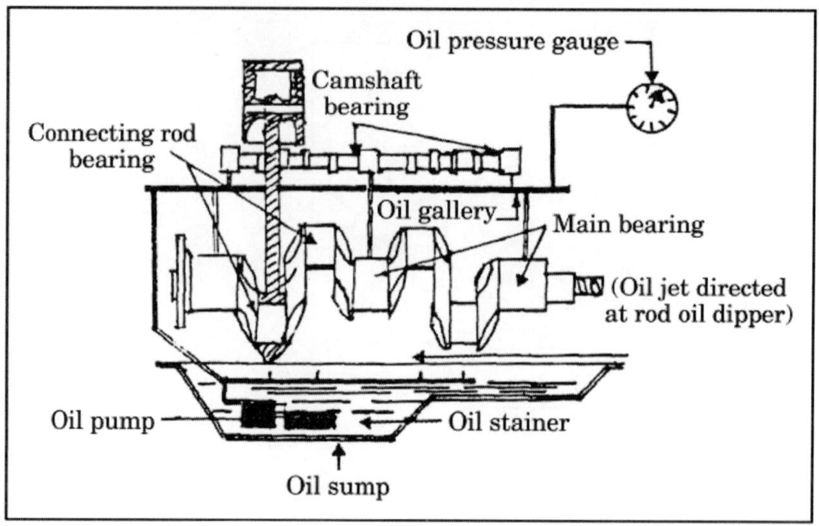

Fig. 11.6: Combined Splash and Pressure System.

11.8 DIFFERENT PARTS OF COMBINED SPLASH AND PRESSURE LUBRICATION SYSTEM

The engines studied in this section use a combined lubrication system whereby the engine components that are loaded most heavily are lubricated by forced oil circulation, while other, less loaded parts, are lubricated by splash and by gravity.

Lubricating by forced oil circulation are the crankshaft main bearings and connecting rod big-end bearings, the valve mechanism, the camshaft bushings, and timing gear bushings. The engine lubrication system includes oil pan, oil pump, oil filter, oil cooler, oil ducts and lines, oil pressure gauge and oil filter. The oil level in the crankcase is checked with dipstick when the engine stands still.

The forced oil circulation circuit in most automotive engines is the same. Figure 11.7 illustrates schematically the operation of the engine lubrication system. With the engine running, the crankcase oil is drawn by the gear type oil pump which delivers it under pressure to the oil filter. The clean oil leaving the filter is cooled in the oil cooler; hence it passes into main oil gallery. From the main oil gallery the oil flows through ducts in the cylinder block to the crankshaft main bearings and the camshaft bearing journals.

Flowing through the angular oil passages drilled in the crankshaft, the oil enters crankpin oil cavities, where it is additionally cleaned, and emerges on the crankpin surfaces to lubricate the connecting rod bearings. From the first crankshaft main bearing the oil is ducted to the spindle of idler gear and the fuel pump drive gear bushing.

Through a cross-drilling in one of the cam shaft bearing journals the oil is fed intermittently into hollow rocker-arm shaft by a duct extending upwards through the cylinder block and cylinder head and an external supply pipe. The oil is distributed to each rocker bearing bushing through radial drillings in the rocker-arm shaft. The oil escaping from the rocker bearings runs down the push rods to lubricate the valve lifters and cams.

The cylinder walls, piston skirts and pins, and timing gears are lubricated by splash. With the crankshaft rotating rapidly, the oil escaping from the crankshaft bearings and that draining from the valve mechanism is flung off in fine droplets from the big end bearings to form an oil mist. The oil droplets setting onto the surfaces of the cylinders, pistons, and cams lubricate them and run down into the oil pan, where the oil starts again on its circuit. The piston pin is lubricated with the oil droplets that get into the hole in the connecting rod small end. In engines, having oil passages drilled through their connecting rods, the piston pins are lubricated by forced oil circulation.

The operation of the engine lubrication system is monitored by watching pressure gauge that reads the oil pressure in the main oil gallery. Some engines are additionally equipped with an oil temperature gauge and an oil pressure indicator light. The light is off when the oil pressure is normal. But if the pressure drops too low, the light comes on.

11.8.1 Wet Sump and Dry Sump Lubrication

In a wet sump lubrication system (Fig. 11.4), the oil is always contained in the sump, which is drawn by the pump through a strainer. But in case of dry sump lubrication system (Fig. 11.5), the oil from sump is carried to a separate storage tank outside the engine cylinder block. The oil from sump is pumped by means of a sump pump through filters to the storage tank. Oil from storage tank is pumped to the engine cylinder through oil cooler. Oil pressure may vary from 3 to 8 kg/cm^2. Pressure lubrication system is generally adopted in case of dry sump lubrication engines. Dry sump lubrication system is generally adopted for high capacity engines. Hence the system in which the lubricating oil is stored in the oil sump is called wet sump system, like the pressure system. But the system in which the lubricating oil is not kept in the oil sump is known as dry sump system. The dry sump lubrication system has two pumps the scavenge pump and the pressure pump. The former is located at the lowest point of an oil pump and it pumps out oil in a tank. The pressure pump takes oil from the tank and feeds it to the crankshaft main and big end bearings after proper filtration through the main filter. This pump is driven by a gearing on the camshaft. The rest of the oil circuit is the same as the wet sump system.

11.8.2 Advantages and disadvantages of wet/dry lubrication

The following are the advantages of dry sump system over the conventional wet sump system.

1. A separate tank has fins around it in order to constantly cool the oil by atmospheric air.

2. Impurities can be drained out on settling down from this tank, by a drain plug. Comparatively pure oil flows into the system.
3. Oil galleries remain clean and oil pressure improves.
4. Lesser formation of varnish and sludge.

The only disadvantage of such a system is that it has high initial cost and is expensive to maintain.

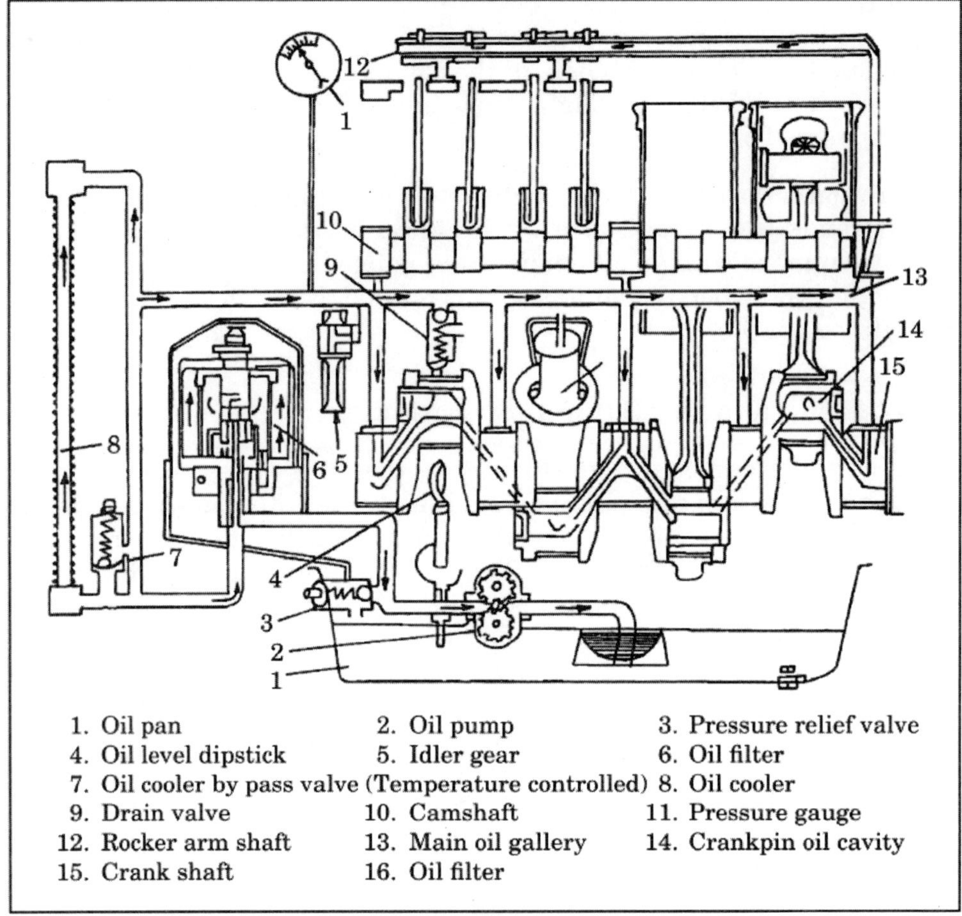

Fig. 11.7: Schematic Diagram of an Engine Lubrication System

11.9. CRANKCASE VENTILATION (BREATHER CAP)

Crank case ventilation is employed to reduce dilution and to prevent a pressure build up in the crankcase. A constant stream of air is passed through the crank case, particularly oil pan which carries with it most of the harmful vapours to the atmosphere. Dilution of lubricating oil in the oil pan occurs by the combustion product that may sometime escape past the piston rings to the crankcase.

Air must circulate through the crank case when the engine is running. This removes water that appears in the crankcase when the engine is cold. Also it removes blowby gases (burnt gases in the combustion chamber) from the crankcase. Unless the water and blowby gases are removed from the crank case, sludge and acid will from. Sludge is a thick and black substance that sometimes forms in the crank case when water is mixed with the oil. Water comes from the combustion product and also water is formed when engine is stopped or allowed to cool. Sludge can clog the oil lines and starve the lubricating system. Acids corrode metal parts. These effects can damage the engine.

11.9.1 Requirement of crank case ventilation

We know that when engine gets heated up, air in empty space in the sump also gets heated up and expands. There is no place for its expansion; it has to expel out otherwise pressure may build up and it may leak out with oil from the oil rings to the combustion chamber. Hence there is the requirement of allowing the heated air to get out from the crank case. Hence breather is provided in the engine. Similarly, after stopping the engine when the temperature goes down, the air in the crank case gets condensed and water droplets are formed. In order to prevent this, outside air is circulated.

The crank case breather filter is mostly made of oil wetted screen. It is fitted at the top of a pipe connecting the crankcase to the outside atmosphere. Sometimes, the oil filler cap is provided with a breather filter.

Chapter 12

Engine Cooling System

12.1 INTRODUCTION

Fuel is burnt inside the cylinder of an Internal Combustion (I.C.) engine to produce power. The temperature of the burning gases during power stroke in the engine cylinder reaches up to about 1500 to 2000°C. This range of temperature is above the melting point of the material of the cylinder body and cylinder head (Platinum, a metal which has one of the highest melting points, melts at 1750°C, iron at 1530°C and aluminum at 657°C). The cylinder and cylinder head are usually made of cast iron and piston in most cases are made of aluminum alloy. If the engine is not cooled, the film of lubricating oil between the rubbing components of the engine is burnt resulting in carbon deposits and undue wearing of the components. The movement of piston in the cylinder is seized due to its expansion, distortion of engine components occur due to the thermal stresses. Hence, the main function of the engine cooling system is to maintain correct engine operating temperature and to dissipate surplus heat resulting from the combustion of the fuel in the cylinder. The cooling system of an Internal Combustion engine maintains the engine temperature of about 71°C to 82°C for petrol engine and 85° to 90°C for diesel engine.

It is estimated that about 30 % of the heat produced in the engine is removed by cooling system, 5 % of the heat is carried away by lubricating oil and heat loss by radiation, about 35 % is passed to the atmosphere *via* exhaust system and only about 30 % is used to produce useful power.

An excessive heat removal from the engine (engine over cooling) reduces the engine power and increases the consumption of fuel, because of poor air-fuel mixing conditions and increased friction losses due to poor lubricating properties of oil at low temperatures. Excessively low operating temperatures cause incomplete burning of the heavier fuel fractions, resulting in heavy carbon deposits accumulating on the combustion chamber walls, pistons and valve heads, causing the seizure of the piston rings and valves. Thus, the overcooling of the engine is as undesirable as its overheating. Overheating is also as harmful as overcooling.

12.2 NECESSITY OF COOLING SYSTEM

Due to excessive temperature rise in the engine, the following bad effects occur and there is the necessity of the cooling system in I.C. engine.

(i) Cylinder and piston may expand to such an extent that the piston seizes in the cylinder and engine is stopped running.

(ii) Lubricating quality of the oil inside the engine moving components is destroyed i.e. the oil becomes too thin, causing wear and tear of the rubbing surfaces due to high temperature.

(ii) Due to high temperature, preignition of fuel mixture takes place, resulting in engine knocking as well as loss of power.

(iv) Due to high temperature, the engine components are distorted because of thermal stresses.

(v) Due to preignition, engine working cycle disturbs, causing decrease of compression ratio and volumetric efficiency of the engine.

12.3 REQUIREMENTS OF EFFICIENT COOLING SYSTEM

The important requirements of an efficient cooling system are:

1. It must be capable of removing only about 30 % of the heat generated in the combustion chamber. Too much removal of heat lowers the thermal efficiency of the engine.
2. It should remove heat at a fast rate when the engine is hot. When the engine is cold, the system should transfer heat at a slower rate so that engine temperature is reached soon.

12.4 TYPES OF COOLING SYSTEM

A system which controls the engine temperature is known as a cooling system. There are two types of cooling system, *(i)* Air cooling system (Fig. 12.1) *(ii)* Water cooling system (Fig. 12.2).

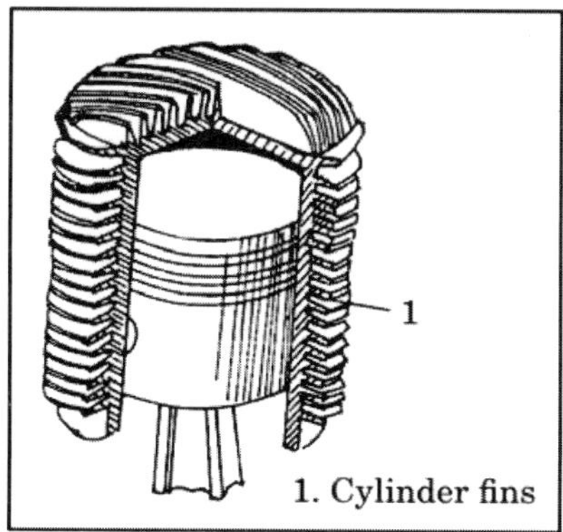

Fig. 12.1: Air Cooling.

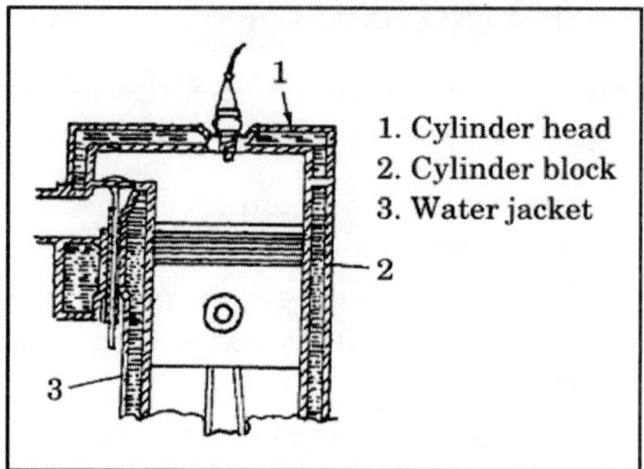

Fig. 12.2: Water Cooling.

12.4.1 Air cooling

In air cooling system, heat transfer from gases to the cylinder walls occurs by conduction and then convection and radiation to the air. Air cooling depends upon the temperature difference between engine and atmosphere, total radiating surface area and mass of cooling air supplied. Fins on the outer surface of the cylinder and head increase the radiating surface area.

The amount of heat carried off by the air cooling system depends upon the following factors.

(i) The total area of the fin surface (heat transfer increases due to large exposed surface).

(ii) The velocity and amount of the cooling air.

(iii) The temperature difference between the fins and the cooling air.

Air cooling is mostly used in motor cycle, scooter, small cars and small air craft engines where the forward motion of the machine gives good velocity to cool the engine. In this method of cooling, a strong blast of air is forced all around the cylinder head which is provided with fins to increase the surface area. The fly wheels of air cooled engines are generally equipped with fan blades to create the air blast. Normally the air cooled engines run hotter than water cooled engines and heavier (greater viscous) lubricating oil is recommended.

12.4.2 Water Cooling

In this method of cooling, the cylinder wall and head are provided with jackets through which water circulates.-The heat is transferred from high temperature gas in cylinder to cylinder wall by conduction and then to the cooling liquid by convection. The liquid becomes heated through the jacket and is itself cooled by means of air blast through radiator fin surface. The heat from liquid in turn is transmitted to air. For heat extraction purposes, water and air behave differently as it requires about 50 volumes of air to remove the same quantity of heat as 1 volume of water. Furthermore, it is difficult to bring air in intimate contact with the metal surface as compared to water as far as heat dissipation is concerned. The heat dissipation from a metal surface depends upon

(i) The conductivity of the metal

(ii) The temperature difference between the surface and cooling medium.

(iii) The area of cooling surface.

Advantages of air-cooled engine

1. The design of air cooled engine is simple
2. It is cheaper to manufacture
3. Operating temperatures are reached soon due to lesser warm-up time required as compared to water cooled engines
4. Absence of radiators, water jacket and pumps minimizes the operating costs and maintenance.
5. Air-cooled engines can operate at higher temperatures than water cooled engines.
6. This system of cooling is particularly advantageous where there is scarcity of water.
7. No risk of damage from water freezing for engine component such as cracking of cylinder jacket or radiator water tubes.

Disadvantages of air-cooled engine

1. Such a system is suitable only for low horsepower engines.
2. The fan absorbs a large amount of power.
3. Air cooled engine produces more sound, as there is no water jacket to damp down sound.
4. In a multi-cylinder engine, the cylinders are not cooled equally.

Advantages of water cooling system

1. Water has higher specific heat than air *i.e.*, it can carry more heat per kg of mass. Specific heat is the quantity of heat required to raise the temperature of unit mass by 1°C. This quantity is higher in water than air.
2. Lesser amount of water is required to transfer a given amount of heat as compared to air.
3. Water, as a cooling medium is available free of cost, unlimited quantity, low viscosity and high boiling point.
4. Uniform temperatures are maintained within the engine, reducing the risk of distortion due to temperature stresses.
5. Chances of engine overheating are greatly reduced.
6. Water acts as a noise or sound dampener.

Disadvantages of the system (Water-cooling)

1. Longer warm-up time in cold condition is required.

2. The total engine size becomes bulkier and complicated
3. On freezing into ice, water expands causing bursting of cylinder blocks
4. Water causes scale formation in jackets and corrosion of metals.
5. Greater maintenance is required *i.e.*, topping up with water daily, cleaning of water jacket frequently.

12.5 TYPES OF WATER COOLING SYSTEM

Water cooling system is mostly used in automobiles as well as in tractor, Water cooling system is of basically four types (i) Direct or non-return system (ii) Hopper or open jacket system (iii) Thermosyphon or natural circulation system (iv) Pump/ forced circulation system.

12.5.1 Direct or non-return system

Such a system is utilized in low hp engine pumps used for irrigation purposes. This is suitable for large installation and where plenty of water is available. The water from a storage tank is directly supplied to the engine cylinder. The water after circulation in the engine becomes hot and is simply discharged and riot reused.

12.5.2 Hopper or open jacket system

It is a simple device and is used in single cylinder stationary engines. It consists of a hopper or a jacket containing water which surrounds the engine cylinder. So long as the hopper contains water, the engine continues to operate satisfactorily. As soon as water starts boiling, it is replaced by cold water. The hopper is large enough to run for several hours without refilling and is open to the top. Due to this, it cannot be used in moving engines. It is required to change the water daily for which a drain is provided in the bottom of the hopper. This system is not common in present days. The hopper cooling system is shown in Fig. 12.3.

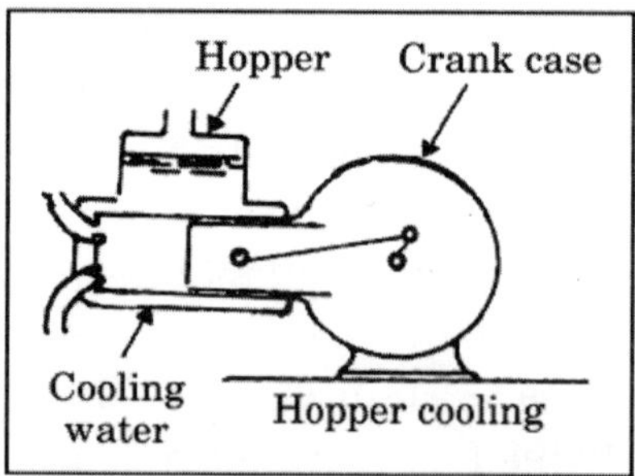

Fig. 12.3: Hopper Cooling System

12.5.3 Thermo-syphon water cooling system

Thermo-syphon system (Fig. 12.4) consists of a radiator, water jacket, fan and hose connections. The radiator must be at a higher level than the cylinder. The water enters into the jacket through gravity. The radiator is connected to the engine block by means of two pipes, *i.e.*, upper hose pipe and lower hose pipe. This system works on the principle that hot water being lighter rises up and cold water being heavier goes down. As the water circulates in the jacket, it becomes hot and rises to the top of the radiator. Cold water from radiator takes the place of rising hot water and in this way, a circulation of water is set up in the system. The hot water while flowing in the radiator tube is cooled down by the air blast from the fan. The fan is rotated by the crankshaft. The upper tank of the radiator must be filled to the upper level mark for satisfactory flow of water. The passages of the system are made bigger so as to make the flow unrestricted.

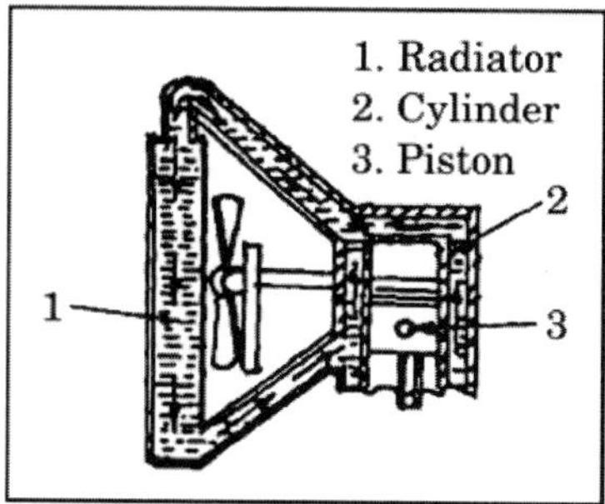

Fig.12.4: Thermosyphon System

If water level in the upper tank of the radiator falls below the level, the circulation of water ceases. Naturally as soon as the circulation ceases, boiling of water begins and further loss of water in the form of steam occurs. The rising high temperature damages the engine and the engine needs to be switched off and allowed to cool. This system is suitable for small and stationary engines.

Disadvantages

1. Rate of circulation is very slow
2. Circulation commences only when there is a marked difference in temperature.
3. Circulation stops when the level of water falls below the top of the delivery pipe.
4. The radiator being at a higher level than cylinder, the engine becomes bulkier. Hence this method is suitable for small engines.

For the above reasons, this system becomes obsolete and is no more in use. This system can be used on small engines where the loads are light and only a relatively simple cooling system is necessary.

12.5.4 Pump or force circulation water cooling system

This system (Fig. 12.5) is similar to the thermo-syphon system except that in thermo-syphon system, water circulates by the law of convection, while in force circulation, a pump is provided in the bottom which forces the water to circulate between the radiator and water jacket. After circulation, water comes back to the radiator where it loses its heat by the process of radiation. The pump is of centrifugal type and is driven by a reinforced rubber V-belt getting its drive from the crankshaft. The radiator helps in the heat exchange from hot water to the cool air being blown over it by the fan. This system is employed in cars, trucks, tractors etc.

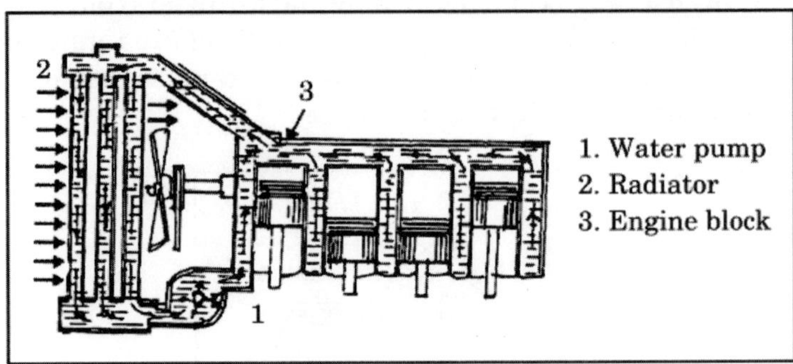

Fig. 12.5: Water Circulated by Pump.

12.6 COMPONENTS AND WORKING OF FORCED CIRCULATION WATER COOLING SYSTEM

The components of forced circulation water cooling system are as follows:

(1) Water pump
(2) Fan
(3) Radiator
(4) Hose pipes
(5) Upper tank and lower tank
(6) Water jacket
(7) Thermostat valve
(8) Radiator pressure cap
(9) Overflow pipe
(10) Temperature gauge.

Water pump: Water pump (Fig. 12.6) is a centrifugal type pump. It has a casing and an impeller, mounted in a shaft. The casing is usually made of cast iron. It is mounted at the front end of the cylinder block between the block and the radiator. It is driven by a V-belt, getting its power from the crankshaft. When the impeller rotates, the water between the impeller blades is thrown outward by centrifugal force and thus water goes to the cylinder under pressure. The bottom of the radiator is connected to the suction side of the pump. Water pumps are often designed with a by-pass opening which permits the circulation of water within the cylinder block and cylinder head only when the engine is cold and thermostat valve is closed. The rate of circulation of water by the pump should not be less than 0.5 litre/BHP/min.

***Fan*:** Heat is largely dissipated by a radiator by convection rather than radiation. If a radiator is to dissipate all the surplus heat of the engine by radiation, then the size of the radiator would be enormous. In case of automobiles, when moving at high speeds, a large quantity of air naturally passes through the radiator (usually mounted on the front) but for slow speed operation at high loads or for stationary engines, in order that the radiator may be able to dispose of the requisite quantity of heat, an engine driven fan is mounted directly behind the radiator.

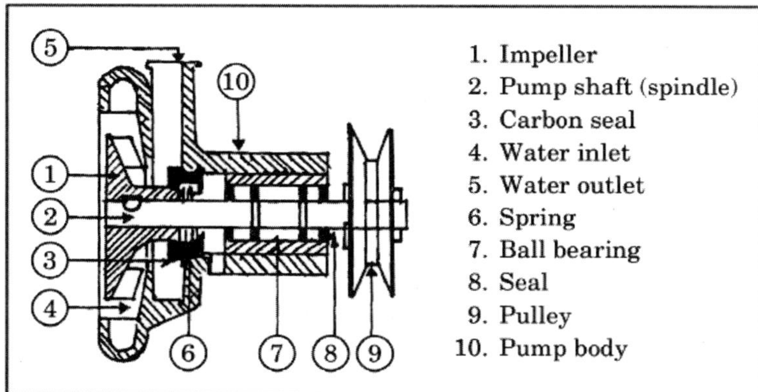

Fig. 12.6: Water Pump (Centrifugal Type).

The fan is usually mounted on the water pump shaft. It is driven by the same belt that drives the pump and dynamo (Fig, 12.7). The purpose of the fan is to provide strong draft of air through the radiator to improve engine cooling.

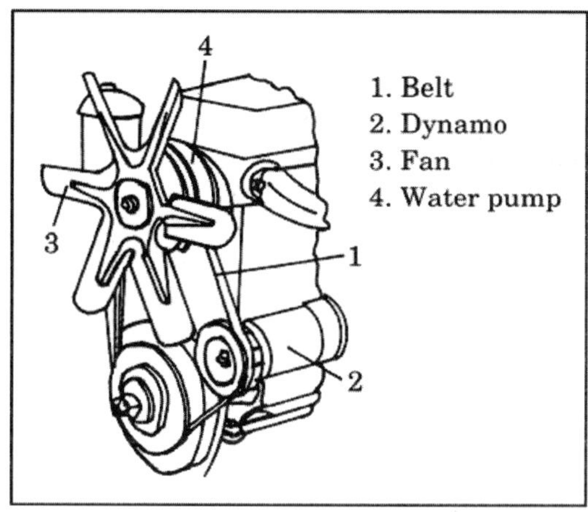

Fig. 12.7: Use of Fan Belt.

***Radiator*:** The radiator (Fig. 12.8) is a heat exchanger that removes heat from the liquid passing through it. It is mostly fitted in the front of the engine. The radiator holds a large volume of water in close contact with a large volume of air so that heat will be transferred from the liquid to the air. The radiator core is divided into two separate compartments. Water passes through one and air passes through the other. Several types of radiator cores have been used. The most common type is the tube-and-fin type of radiator.

A tube-and-fin radiator consists of a series of tubes extending from the top to the bottom. The tubes run from inlet tank to the outlet tank. Fins are placed around the outside of the tubes to improve heat transfer. Air passes between the fins. As the air passes by, it absorbs heat from the fins which in turn absorbs heat from the water. The parts of the radiator are (*i*) upper tank *(ii)* lower tank *(ii)* air fins *(iv)* water tubes (*v*) radiator pressure cap (*vi*) over flow pipe or leak off pipe or filler cap.

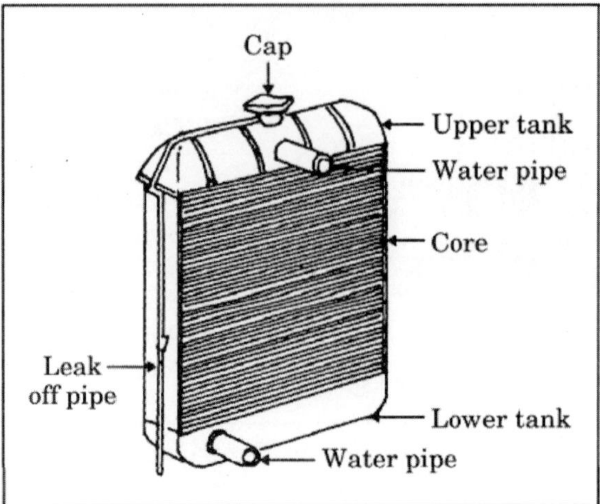

Fig. 12.8: Main Parts of Radiator

In every radiator, there are two small tanks. One at the top and the other at the bottom. Hot water from the engine flows into the upper tank. The lower tank receives cold water after it has passed through the tubes. The heat contained in the hot water is conducted to the fins provided around the tubes. Figure 12.9 shows the honey-comb arrangement of radiator tubes. An overflow pipe connected to the upper tank, permits excess water or steam to escape to the outside atmosphere.

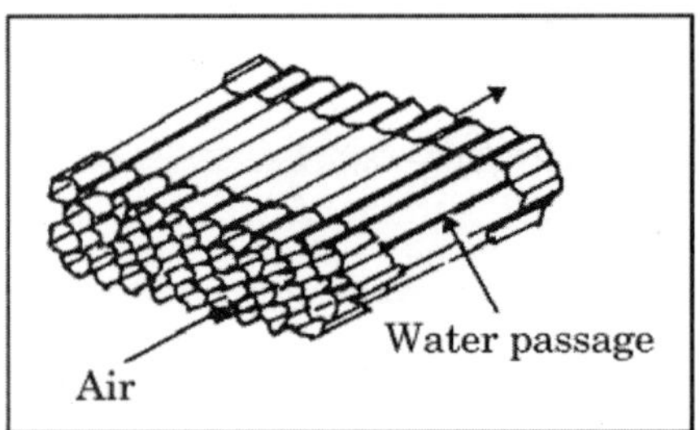

Fig. 12.9: Honey Comb Arrangement of Radiator Tubes

Radiator types:

Tube and Fin type radiator

The great majority of radiators now used on automobiles have cores of continuous fins and tubes (Fig. 12.10). The staggering of tubes in this case increases the rate of transfer as compared to inline depth wise arrangement. Usually dense or pimples are pressed in the plates to increase the rate of heat transfer. In this case the number of soldered joints is limited and therefore, it is more robust. Sometimes tubes and corrugated fin plates are used as shown in Fig. 12.11.

In this case vertical cooling flattened tubes are connected by corrugated conducting fins inserted between adjacent lines of tubes. In this way smaller front area of radiator is obtained, than with the plain fin construction.

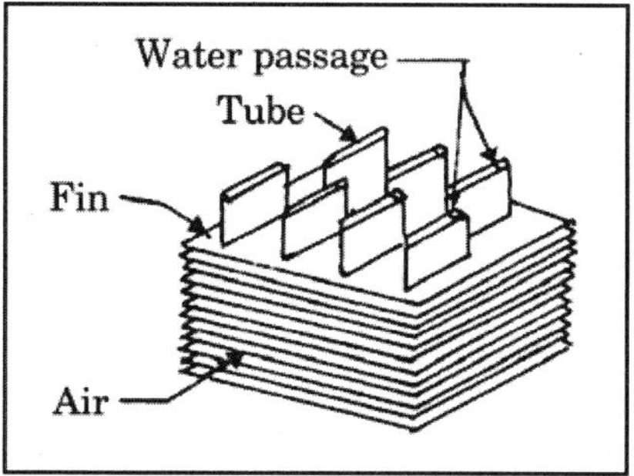

Fig. 12.10: Tube and Fin Radiator

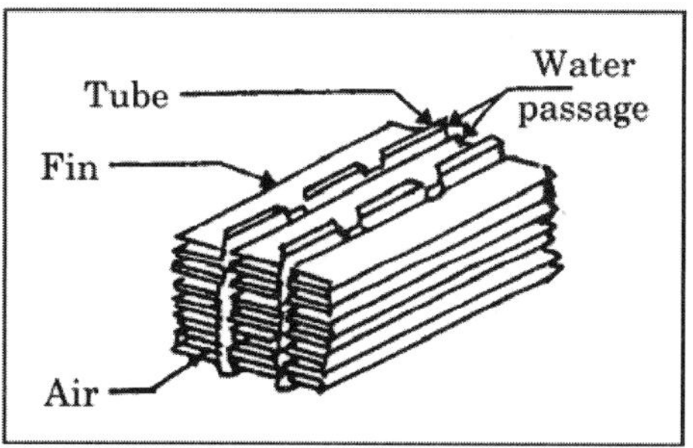

Fig. 12.11: Tube and Fin Type Radiator

Cross flow radiator

In this case large number of horizontal coolant passages are provided with their horizontal air passages. The outer opening of the coolant passages are connected respectively, to left and right hand vertical water passage. The usual overflow tank with its filler cap and overflow pipe is located at the top or behind the top of the radiator.

Round tube radiator

In the case of round tubes, water flows are provided. The diameter of the tubes may be 6 to 12 mm with crinkled fins arranged as discs, or spirally, along the length of the tubes. The latter fit into upper and lower radiator water tanks.

In general, for all radiators tubes made of brass or copper, with fins of same material, soldered in position in order to ensure good contact, are used .Sometimes rubber bushings are also used to secure the tubes to the upper and lower tanks so that replacement is simple. Radiators using aluminum tubes and fins have also been developed.

Hose pipes: There are two bent pipes called hose pipes provided in the radiator. One is provided at the top to carry hot water to the radiator upper tank, from where the water drops in the tube to the bottom tank. From bottom tank, another hose pipe (lower hose pipe) is connected to the inlet side of the water pump for circulation of cold water in the jacket surrounding the cylinder and cylinder head of the engine.

***Upper tank and lower tank*:** There are two tanks provided in the radiator one at the top and the other at the bottom. Upper tank is usually made of copper sheet. It has a mouth used for filling the radiator on which radiator cap is fixed, Also one rubber pipe or nylon pipe is fixed, called the leak-off or overflow pipe for removal of steam or excess pressure from the radiator. The lower tank is also made of same material as upper tank. The hot water after being poured into upper tank comes to the lower tank through radiator tube and collects there for being pumped to the cylinder block,

Water jacket: Water jackets are casted around the engine cylinder *i.e.* around cylinder block and head so that water can circulate freely around the cylinder as well as around the valve opening.

***Overflow pipe or leak-off pipe*:** It is a small narrow rubber or nylon or metallic tube provided at the upper tank of the radiator to remove steam or excess pressure from the cooling system to the outside environment.

***Temperature gauge*:** Temperature gauge is provided in the control panel to know the prevailing temperature inside the cooling system. It indicates the temperature of water after circulating in the cylinder block. The driver is warned if the temperature of water in the cooling system goes too high. An abnormal heat rise is a warning of abnormal conditions in the engine. The indicator warns the driver to stop the engine before serious damage is done.

***Thermostat valve*:** It is a kind of check valve which opens and closes with the effect of temperature. It is fitted between the water outlet of the engine and top of the radiator. During warm-up period, the thermostat is closed and the water pump circulates the water only throughout the cylinder block and cylinder head. When the normal operating temperature is reached, the thermostat valve opens and allows hot water to flow towards the radiator.

There is a by-pass valve between the thermostat and the pump. When the pump rotates, the water is circulated to the various parts of the cylinder. By this arrangement, the water takes away the heat from the cylinder block. But if the temperature of the water does not reach to a certain level, say 70-75°C, the thermostat is completely closed. This means that the water is not able to go to the radiator. The same water is circulated along the by-pass valve. The by-pass valve is a valve which opens only in one direction, *i.e.*, one way valve. Therefore, when water presses the valve down, the water naturally comes back to the pump. The pump recirculates the same water.

Thermostat valve starts opening at 70 to 75°C and fully open at 82°C. When temperature of water increases to 70-82°C, the thermostat valve opens. The circulating water instead of going back to the pump through by-pass valve passes through the thermostat valve to the radiator for cooling. The by-pass valve now closes the passage to the cylinder block. The hot water from radiator comes to the bottom tank through radiator tube and gets cooled by the air blast. The closing and opening of thermostat valve is shown in Fig. 12.12. The thermostat valve is of three types; (a) Bellow type (b) Bimetallic type (c) pallet type.

***Bellow type valve*:** It is made of a flexible corrugated metal bag (bellow of thin brass or aluminum) closed at both ends and filled with liquid like alcohol or ether. The liquid expands with the increase of temperature and raises the valve off its seat. This permits water to circulate between the engine and the radiator. When the bellows is heated, the liquid vaporises, creating enough pressure to expand the bellows. When the unit is cooled, the gas condenses. The pressure reduces and the bellows collapse to close the valve.

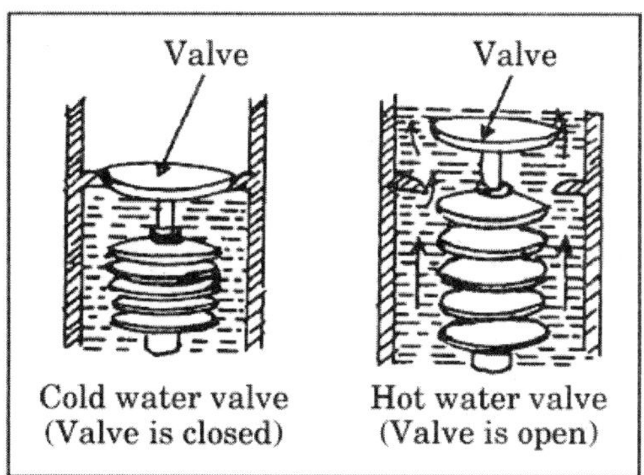

Fig. 12.12: Thermostat

Bimetallic type: This consists of a bimetallic strip. The unequal expansion of two metallic strips causes the valve to open and allows the water to flow to the radiator.

***Pallet type*:** A copper impregnated wax pallet expands when heated and contracts when cooled. The pallet is connected to the valve through a piston, such that on expansion of the pallet, it opens the valve. A coil spring closes the valve when the pallet contracts.

Radiator pressure cap: The cooling system of the engine are sealed and pressurized by a radiator pressure cap. There are two advantages of sealing and pressuring the cooling system. First the increased pressure raises the boiling point of the liquid. This increases the efficiency of the cooling system. Secondly, sealing the cooling system reduces the water losses from evaporation. There are two valves in the cap, one bigger (pressure relief valve) and one smaller (vacuum valve). They open in opposite direction. Basically radiator pressure cap (Fig. 12.13) consists of vacuum valve, vacuum valve spring, lower sealing gasket, pressure relief valve, pressure relief valve spring, upper sealing gasket. The pressure relief valve opens while pressure inside the radiator increases and vacuum valve opens when engine cools down and temperature and pressure in the cooling system decreases. The pressure relief valve is set between 0.75 - 1.0 kg/cm^2. At 1 kg/cm^2, the boiling point of water is raised by 7.7°C. But high pressure rise in the cooling system requires special strengthening of hose pipes and radiator.

We know that ordinarily water boils at 100°C and in case we boil the water at higher altitude, *i.e.*, on a mountain, the water boils at 90° C to 94°C and in case we boil the water in a deep well, it will boil at 110°C. This is because the atmospheric pressure is more in deep well. Making use of this phenomenon, the pressure is increased in cooling system.

As the pressure goes up, the boiling point goes up. Therefore, water can move at a temperature higher than 100°C without boiling. The higher, the water temperature, the greater the difference between it and the outside air temperature. This difference in temperature is what causes the cooling system to work. The hotter the water, the faster the heat moves from the radiator to the cool passing air. This means that the pressurized and sealed cooling system can take heat away from the engine faster. Therefore, the cooling system works more efficiently when the water is under higher pressure.

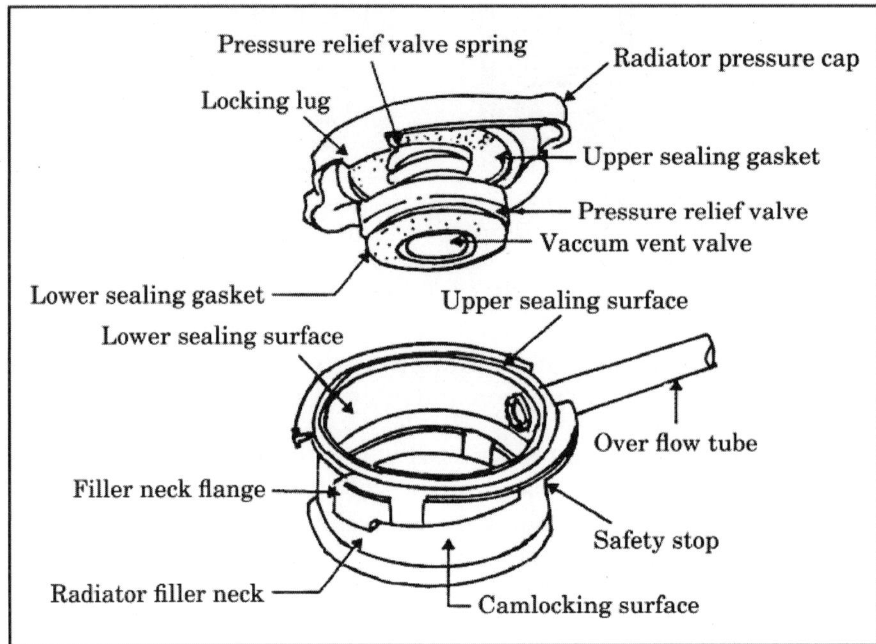

Fig. 12.13: Radiator Pressure Cap Removed from the Radiator Filler Neck.

However, the cooling system cannot be pressurized too much. If the pressure in the system gets too high, it can damage the radiator and hose pipes. To prevent this, radiator cap has a pressure-relief valve. When the pressure gets too high, it raises the valve so that excess pressure can escape into the atmosphere by overflow pipe. When the pressure decreases, the pressure relief valve comes to the original position due to spring tension.

When the engine cools, the water level in the radiator goes down and a vacuum is created. The vacuum valve protects the system from developing a vacuum that may collapse the radiator. Cold water takes less space than the hot water. Due to high atmospheric pressure on the valve and less pressure from inside, the valve opens downward and atmospheric air enters into the cooling system through overflow pipe to fill the vacuum. After equalizing the pressure, the valve closes. The working of radiator pressure cap is shown in Fig. 12.14.

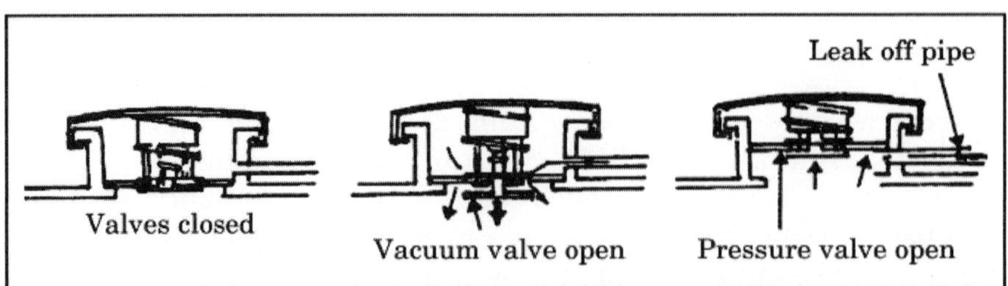

Fig. 12.14: Pressurized Cooling System Showing Working of Pressurized Cooling System Cap.

Chapter 13

Intake and Exhaust System

13.1 AIR CLEARNER, ITS NECESSITY AND UTILITY

Intake and exhaust system of an automobile and tractor consists of the following : (1) pre-cleaner (2) air cleaner (3) intake manifold, (4) exhaust manifold (5) exhaust pipe (6) exhaust muffler. The general layout of intake and exhaust system of an automobile/tractor is shown in Fig. 13.1. The intake and exhaust system deals with the inflow of fresh air and the outflow of used and burnt gases respectively in the engine.

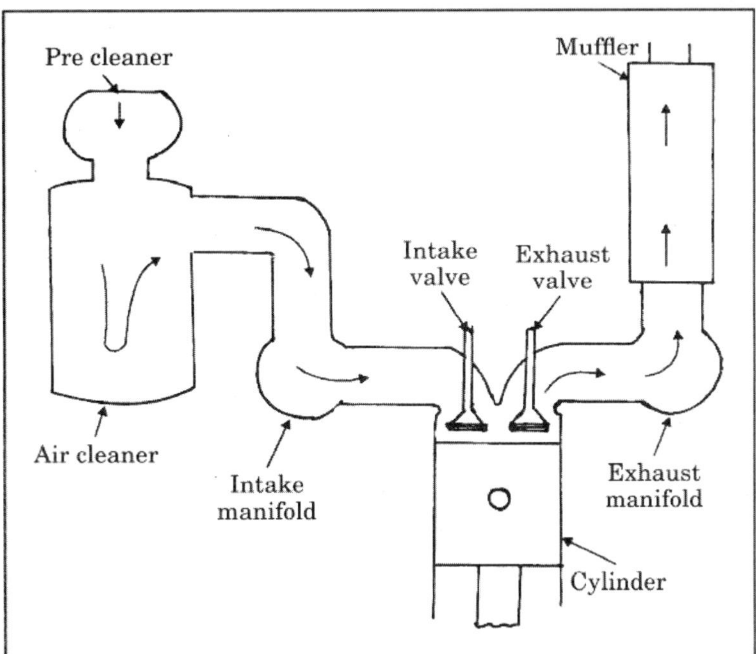

Fig. 13.1: Layout of Intake and Exhaust System.

Intake system: This system allows fresh air to enter into the engine. Its important parts are *(i)* pre-cleaner *(ii)* cleaner *(iii)* supercharger (auxiliary unit) *(iv)* intake manifold *(v)* intake valve.

Necessity of air cleaner: The air surrounding an operating tractor, contains, a large amount of dust. For example, when operating tillage machines in dry weather, the dust content of the ambient air comes to 2.5 gm/m^3, reaching 6 gin/m^3 in desert condition. In one hour of operation, medium-power tractor engine inhales around 200 m^3 of air. If the air is not cleaned, several kilograms of dust enter into the engine. Hard dust particles enhance the wear of the cylinder, piston and other rubbing components of the engine. Therefore it is not permissible to operate a tractor or automobile without cleaning the air entering into the cylinder. The operating efficiency, good performance and durability of an engine depend mainly upon its air cleaner. The dust in the intake air also deteriorates valve guide, valve seat. Sometimes, the air leaks during compression in the chamber and mixes with oil. This dusty oil affects the crankshaft and camshaft bearings and engine needs early overhauling. As such dust has to be trapped from the air before it enters into the cylinder with the help of air cleaner, otherwise dust makes a viscous paste when mixed with lubricating oil and makes the moving parts wear faster.

Utility of air cleaner

1. It traps the dust from the incoming air and does not allow it to go to the cylinder.
2. It also reduces the hissing sound produced by fast moving of air to intake pipe (due to zig zag, not direct movement of air).
3. In petrol engines, the air cleaner does not allow the flame caused due to back-firing to come out of the intake system or outside engine.

Back firing occurs if air-fuel mixture is ignited in the cylinder before the intake valve closes (due to faulty setting of ignition system or early ignition or pre-ignition).

13.2 TYPES OF AIR CLEANER

Air cleaners are of three types, (a) dry type air cleaner (b) wet type or oil bath type air cleaner (c) oil wetted mesh type.

Dry type air cleaner: The filtering element in this case is of dry type *i.e.*, paper, multi-wire netting of nylon hair, felt (matted fabric). For small vehicles, such as scooter and cars, filter paper is used in -corrugated form to give more cleaning area. Due to larger surface area of the filter element, the air speed becomes relatively low and dirts are deposited on the small pores of the filter. The dry filter is cleaned by blowing the compressed air (pressurized air) in it at each service.

The dry air cleaner is easy to service but it is costlier to maintain than the oil bath because of its frequent replacement. Also in dry air cleaner, there may be the chances of entry of dust particles into the cylinder.

Pre-cleaner is of dry type air cleaner. The general construction of paper type of dry type air cleaner is shown in Fig. 13.2.

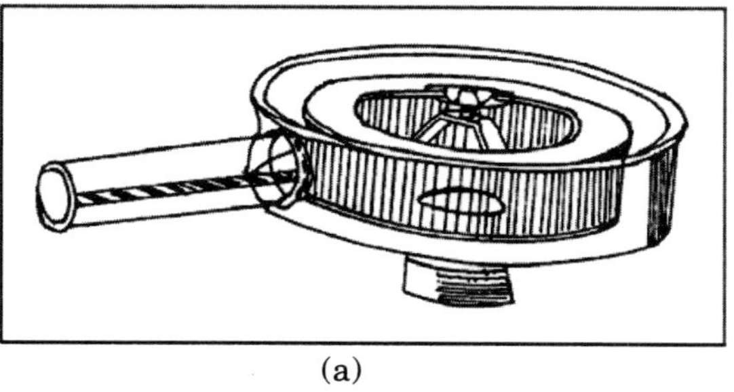

(a)

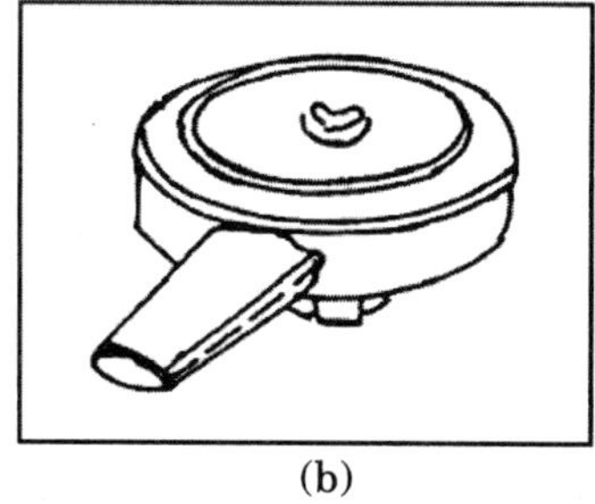

(b)

Fig. 13.2: (a) Dry Type Air Cleaner with Paper Element, (b) Air Cleaner Assembly Dry Type.

Wet type or Oil bath type: The main difference between a dry and oil bath cleaner is that in the latter case, oil is used for cleaning air. This type of air cleaner is generally used in all automobiles and tractor. The air with dust after passing through pre-cleaner (dry type air cleaner), moves with high velocity and strikes the oil present in the oil cup at the bottom. As the air touches the oil and undergoes a sudden reversal in its direction of flow (making a sharp 180° turn), some of the dust and foreign particles are clinged into the oil surface and move down into the oil cup. This accumulation or clinging of dust particles with oil occurs by the virtue of their greater inertia or reluctance to change the direction of motion. Due to mixing of gushing (high velocity) air with oil, some amount of oil also moves along with the air. To stop the oil coming to the manifold, wire gauge filter is placed just above the oil cup. The wire gauge element stops the oil to go to the manifold and it remains moist with oil. The moist wire gauge also traps the remaining dust which are escaped from the oil cup. As such only clean air goes to the intake manifold. The oil bath type of air cleaner is shown in Fig. 13.3.

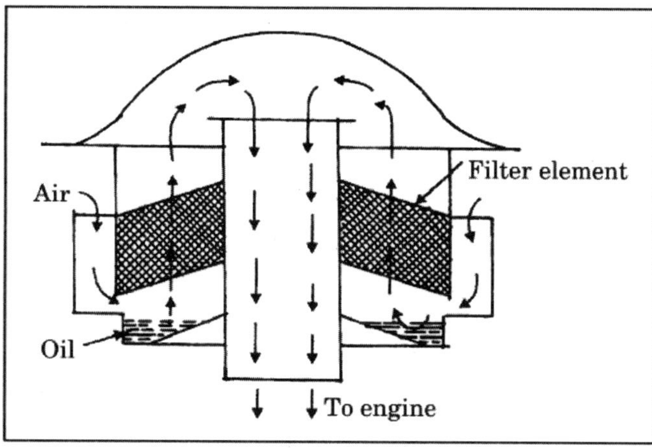

Fig. 13.3: Oil Bath Type Air Cleaner.

Oil wetted mesh type: This is an air cleaner of wire mesh type being wetted by the oil (Fig. 13.4). Being moist or viscous, the dust in the air adheres to the viscous surface of filter element, leaving dust free or clean air to manifold.

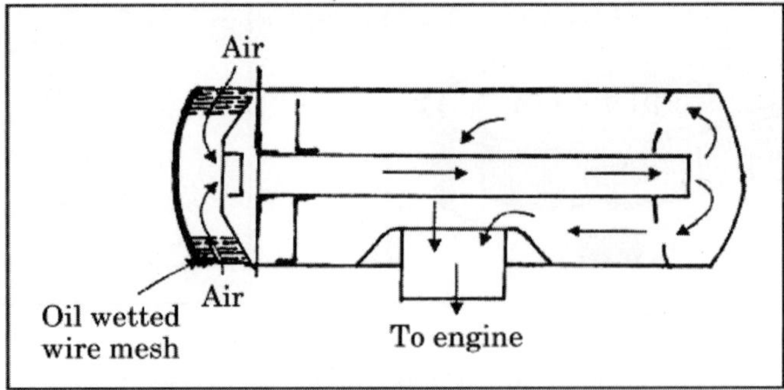

Fig. 13.4: Oil Wetted Type Air Cleaner.

13.3 PRE CLEANER

Pre cleaners are used for all such machines which work in very dusty conditions such as tractor, heavy earth-moving machines etc. These are fitted in the upper portion of the main cleaner. When the engine is running, the air is drawn through the pre-cleaner to the inlet tube of the main cleaner. Hence coarse dust particles are removed from the air stream in the pre-cleaner and reducing much of the load on the main cleaner.

Most of the pre-cleaners are of cyclone (circulating air) type or function on the centrifugal principle (Fig. 13.5). On the outer surface, *i.e.*, on intake passage, a fine wire' mesh screen in fitted, which traps the heavier dust particles from entering into the pre-cleaner. Once the air passes through the screen, it is passed through the vanes and baffles, resulting in the rotary motion of the air. The heavier dust particles are thrown out due to centrifugal force and drop downward into the dust cap. Then the pre-cleaned air passes in the central tube to the main cleaner. In some of the pre-cleaners, glass bowl is attached for dust collector, so that dust can be easily seen, when full and requires immediate cleaning.

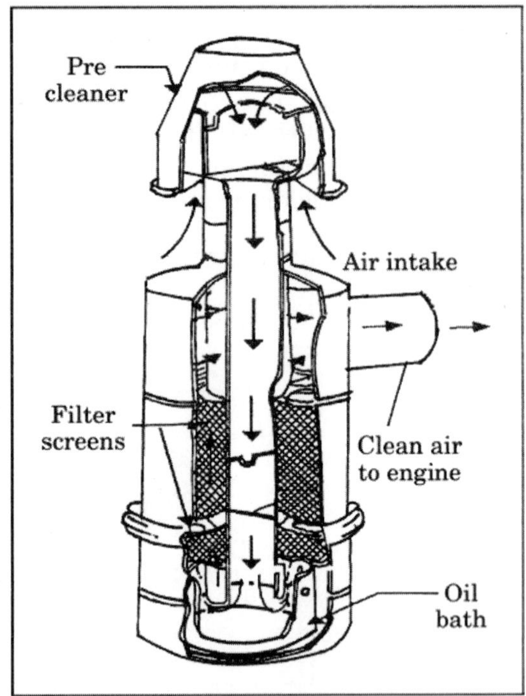

Fig. 13.5: Precleaner along with Bath Typs Cleaner.

13.3.1 Super chargers

A supercharger is a device for increasing the air pressure into the engine so that most fuel can be burnt and engine output can be increased. The pressure inside the manifold of a supercharger will be greater than the atmospheric pressure.

Supercharged air is provided either by positive displacement rotary blowers or by centrifugal blowers. This may be driven by the engine itself or exhaust gas turbine. This is an auxiliary unit and only high horsepower range engines are provided with superchargers.

13.4 INTAKE AND EXHAUST MANIFOLD

13.4.1 Intake manifold

Manifold (Fig. 13.6) is a pipe or casting with several openings through which gas or liquid is gathered or distributed. Intake manifold is a passage through which air or air fuel mixture flows from air cleaner or carburetor to the intake valve or ports in the cylinder head. It is made up of cast iron or aluminum alloy. It is also bolted from separate castings into a single unit. The manifold flange (collar) is connected to cylinder head by means of asbestos-copper gaskets, studs and nuts.

Fig. 13.6: Intake Manifold.

13.4.2 Intake valve

It is a valve which permits air or air fuel mixture to enter into the cylinder. Intake manifold is connected with the intake valve.

13.4.3 Exhaust system

The exhaust system collects exhaust gases from the engine and expels them out. The system consists of *(i)* exhaust valve, *(ii)* exhaust manifold, *(iii)* exhaust pipe, *(iv)* exhaust muffler, (silencer) *(v)* tail pipe.

13.4.4 Exhaust valve

It is a valve which permits the burnt or used gases to escape from the engine cylinder.

13.4.5 Exhaust manifold

Exhaust manifold (Fig. 13.7) is a pipe or casting with several passages (openings) attached to the exhaust valves of various cylinders. It collects exhaust gases from all the exhaust valves and conducts them one end to a central exhaust passage.

The central passage is connected to exhaust pipe and then muffler. It is usually made of cast iron.

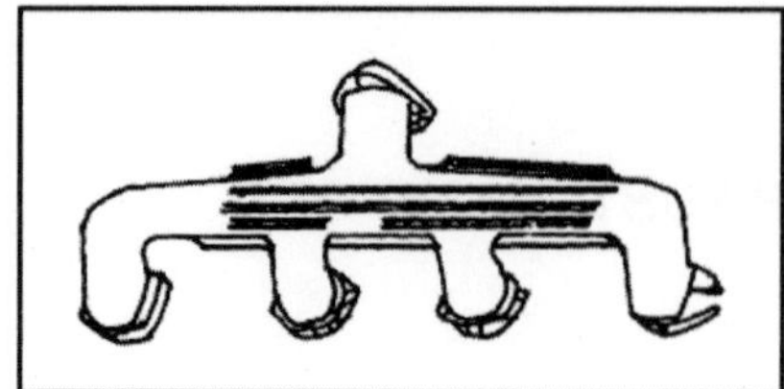

Fig. 13.7: Exhaust Manifold.

13.4.6 Exhaust pipe

Exhaust gases are let out from the exhaust manifold by a round curved pipe (exhaust pipe) whose one end is fixed with the manifold with suitable flange and the other end is clamped with exhaust muffler.

13.4.7 Exhaust muffler

The main purpose of exhaust muffler as shown in Fig. 13.1 is as follows,

(i) To reduce the temperature of exhaust gases
(ii) To damp down the sound of exhaust gases
(iii) To trap the flame of hot gases or unburnt gases
(iv) To reduce the speed of out-going gases.

The exhaust gases must be discharged into the atmosphere with minimum restriction. The restriction in flow causes back pressure.

13.4.8 Back pressure

The restriction to flow of exhaust gases causes back pressure on the piston head during exhaust stroke. The exhaust gases are not released at atmospheric pressure. These are released when their pressure is higher than the atmospheric pressure. Due to the restriction in the flow of exhaust gases in exhaust manifold and muffler, the movement of out flowing exhaust gases is hampered causing backpressure in the piston head. The back pressure results in *(i)* higher b.h.p. loss *(ii)* dilution of fresh charge with exhaust gases and *(iii)* increased fuel consumption. The back pressure can be reduced by improving the design of exhaust system (exhaust manifold, muffler etc.), so that exhaust gases will be escaped without any restriction in their path.

13.5 TYPES OF MUFFLER

Mufflers are of three types *i.e.* (i) Straight flow type (ii) Reverse flow type and (iii) Baffle type

Straight flow type: It has a single perforated tube throughout the length of the muffler (Fig. 13.8). That is why the exhaust gases are made to pass through central pipe with perforations. When the exhaust gases pass through the pipe, sound is damped and speed is reduced. When the exhaust gases get more space for expanding, velocity is reduced and loose their temperature. Fibre glass wool is spread in between the outer metal wall and the perforated tube to absorb the sound.

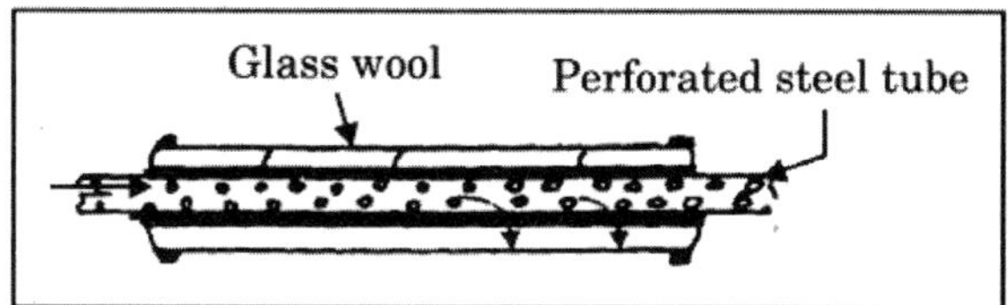

Fig. 13.8: Straight Flow Silencer.

Reverse flow type: The exhaust gases are made to pass through a greater length of perforated tube (Fig. 13.9). The tubes are interconnected not in same line but in reverse direction, resulting in movement of exhaust gases to and fro in the muffler compartment. Due to movement of gases in greater length, sound is decreased but back pressure is increased.

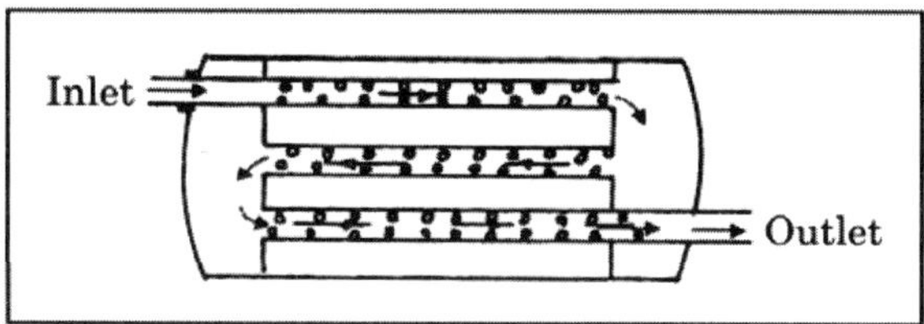

Fig. 13.9: Reverse Flow Type.

Baffle type: The exhaust gas passes through a series of baffles which causes maximum restriction and hence back pressure (Fig. 13.10). The noise reduces because the length of travel of exhaust gases increases considerably.

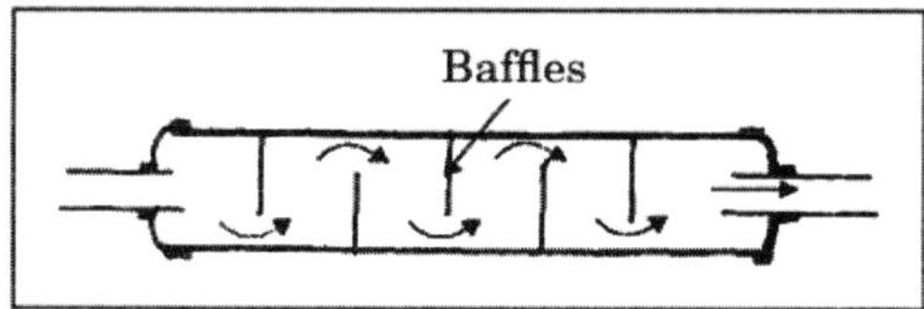

Fig. 13.10: Baffle Type.

Tail pipe: Tail pipe is fixed at one end with exhaust muffler with a suitable clamp while the other end is hanged with a flexible strap at the rear end of the chassis. So that exhaust gases are allowed to discharge at the back of the vehicle.

13.6 RAIN TRAP OR RAIN CAP

Where exhaust pipe is fitted in horizontal direction, there is no risk of rain water entering into the engine cylinder, but where muffler is fitted in vertical direction, some measures to be taken to prevent rain water to enter into the cylinder.

The rain cap is permanently installed on the exhaust pipe. The weight fixed on the cap makes the cap sit on the exhaust pipe when the engine is not working. When the engine is started, the rain cap gets lifted up due to pressure of exhaust gases.

Chapter 14

Clutch

14.1 INTRODUCTION

Clutch is a device, used to connect and disconnect the engine power with or from the transmission gears. It is located in between the engine flywheel and the gearbox.

14.2 NECESSITY OF CLUTCH

The clutch in any automobile or tractor is essential due to the following reasons.

(i) Engine power is disconnected from gear box to start the engine, warm it up and to run at high speed to develop enough power to move the tractor from rest.

(ii) Engine power to gearbox is disconnected for easy shifting of gear, so that damage to the gear teeth can be avoided.

(iii) When the belt pulley of the tractor works in the field, it needs to be stopped without stopping the engine. This is done by the clutch.

14.3 ESSENTIAL FEATURES OF A GOOD CLUTCH

(i) It should be easily disengaged and engaged without jerks.

(ii) It should have good ability of taking load without dragging and chattering.

(iii) It should have higher capacity to transmit maximum power without slipping.

(iv) Friction surface should be highly resistant to heat effect, high coefficient of friction.

(v) The operation or control by hand lever or pedal lever should be easy.

(vi) Clutch assembly should be well balanced to avoid production of vibration at high speed.

14.4 PRINCIPLE OF OPERATION

The clutch works on the principle of friction in which when one stationary surface is brought into contact with a rotating surface, the stationary surface also starts rotating. Let us take the simplest case: A driven plate mounted on a transmission shaft is enclosed in between two driving plates. When the driving plates are away, they will rotate together, but the driven plate will remain stationary. Now if by some means, the two driving plates are brought nearer, the driven plate will be pressed in-between the two rotating plates and will be bound to rotate. The basic clutch action is shown in Fig. 14.1.

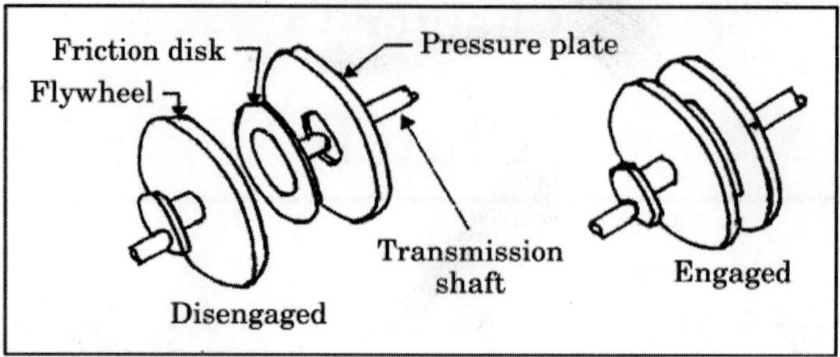

Fig. 14.1: Basic Clutch Action.

14.5 CLASSIFICATION OF CLUTCH

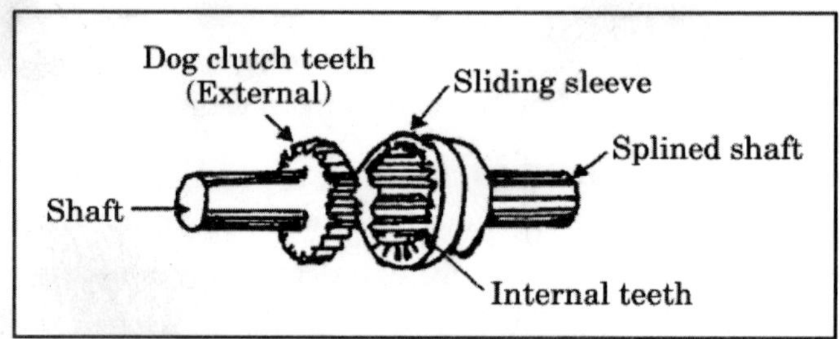

Fig. 14.2: Dog and Spline Ctutch.

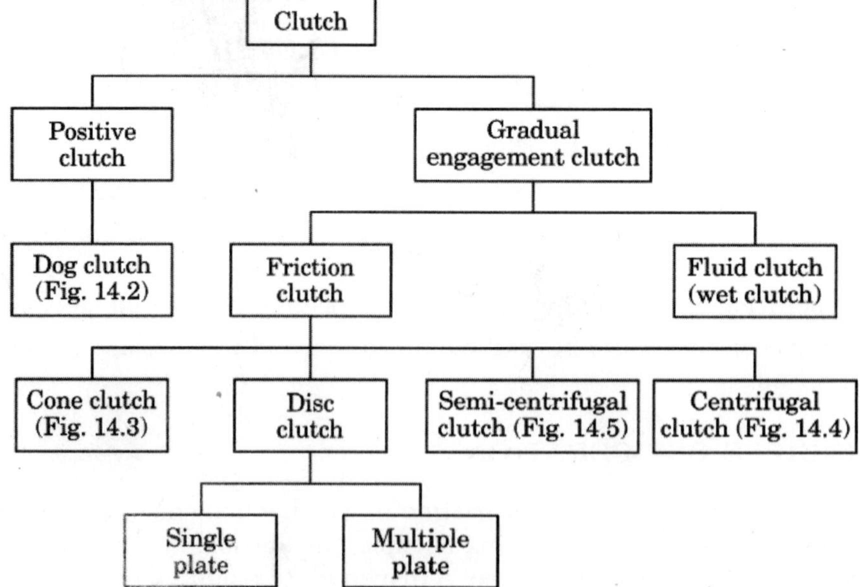

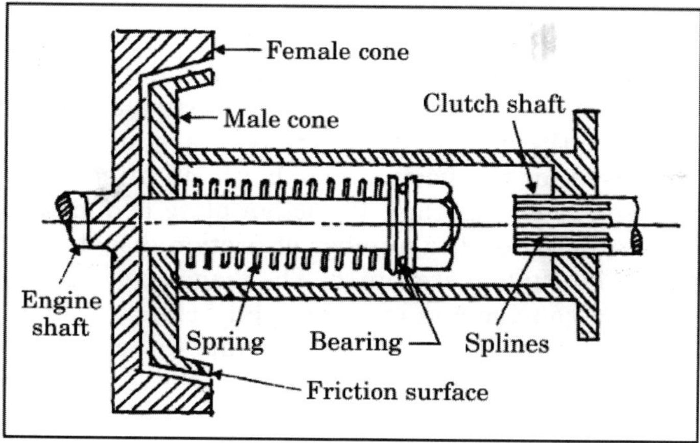

Fig. 14.3: Cone Clutch.

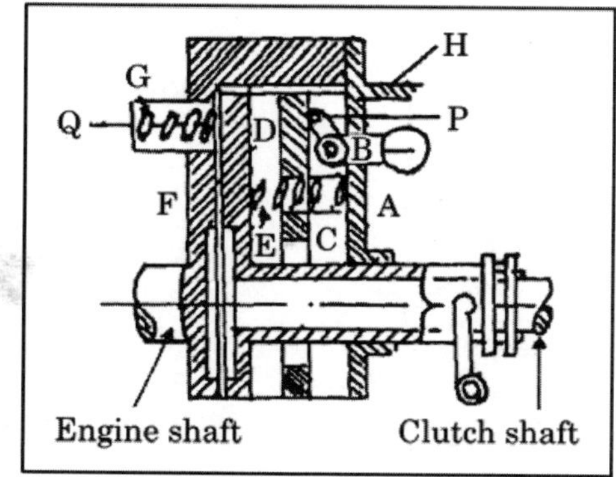

Fig. 14.4: Centrifugal Clutch.

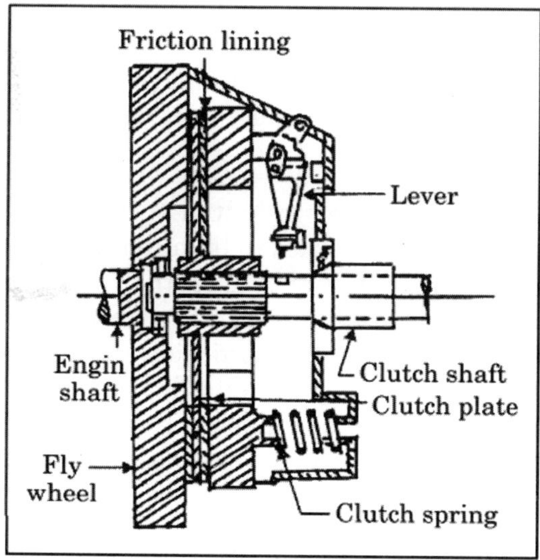

Fig. 14.5: Semi centrifugal Clutch.

14.6 FRICTION CLUTCH

Friction clutch produces gripping action, by utilizing the frictional force between two surfaces. These surfaces are pressed together to transmit power. The driven plate is gripped firmly to the driving plate. Transmission of power depends upon the kind of material used for the friction members and intensity of the force (or axial pressure) pressing them together. If friction plate is of disc type, then it is called disc clutch. Depending on the number of friction disc, the disc clutch is of single plate clutch and multiple plate clutch. The diagram of single plate clutch is shown in Fig. 14.6. The components of single plate clutch system are (i) clutch plate (friction plate or friction disc) (ii) pressure plate (iii) clutch cover (iv) clutch release fork (v) release bearing (throwout bearing) (vi) clutch release levers (release fingers) (vii) pressure spring, (viii) eye bolt (ix) pin (x) anti-rattle spring (release lever spring), (xi} clutch shaft (xii) linkage.

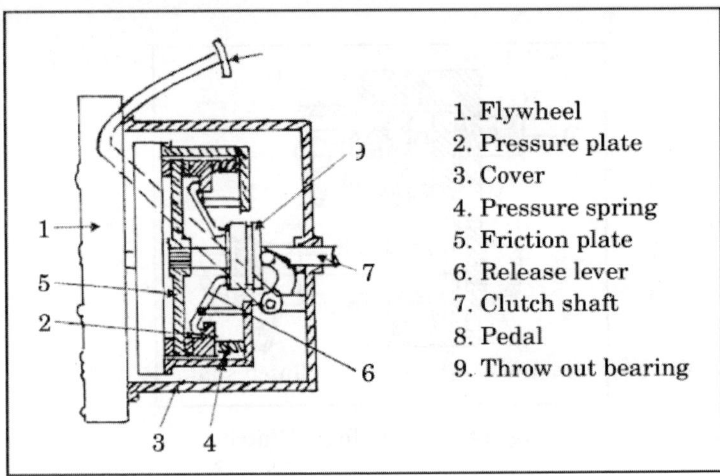

Fig. 14.6: Single Plate Clutch.

Driving plates: Driving plates are fly wheel and pressure plate assembly. They are bolted with each other to rotate as a single unit when fly wheel rotates. Clutch plate is the driven plate.

Working principle: The clutch cover is bolted to the flywheel along with the pressure plate. The clutch plate has splines in its hub that match the splines on the input shaft. These splines consist of two sets of teeth. The internal teeth in the hub of the clutch plate match the external teeth in the transmission input shaft. When the clutch plate is driven, it turns the transmission input shaft. The front end of the shaft is supported by a ball bearing (called pilot bearings) carried in a bore in the fly wheel end face and the other end passes through the release bearing. In between the pressure plate and the clutch cover, pressure springs are placed all around the circumference. Three fingers, also known as release levers, are centrally fulcrumed to the clutch cover with their outer ends connected to the pressure plate. These release levers are fulcrumed (pivoted) to the clutch cover with the help of eye bolts. A release bearing slightly away from the free ends (inner ends) of the release levers moves to and fro on the input shaft with the help of a fork which is connected to the clutch pedal through the linkage.

When we do not press the clutch pedal, the clutch plate remains pressed between the pressure plate and fly wheel facing, with the result, the drive is transmitted to the gear box. But as soon as we press the clutch pedal, the clutch release fork presses the release bearing towards the fly wheel. This release bearing, when being pressed, presses the clutch release fingers. As the release levers are pivoted to the pressure plate, the pressure plate is lifted or pushed backward against the tension of pressure springs due to pressing of the inner side of the release levers. The pressure plate no longer exerts any pressure on the frictions or clutch plate, making it to become free and stop revolving with the flywheel. This is known as the disengaging of clutch. As soon as we remove (release) our foot from the clutch pedal, the pressure plate moves ahead to press the clutch plate due to the pressure spring, causing the clutch plate and fly wheel to rotate together as one unit. Thus the.power of the engine goes to the gear box for onward transmission to rear wheels. The condition of disengagement remains till the clutch pedal remains depressed. In this way, connection and disconnection of clutch between flywheel and power transmission system are maintained as per the requirements of the vehicle or tractor.

14.6.1 Friction plate (Fig. 14.6a)

When the pressure plate presses the friction plate, it rotates the clutch shaft. This develops torsional vibration. In order to dampen this vibration, torsion springs are provided in the friction plate. There is also a waved cushion spring. On both its sides, the friction materials are fitted by means of small rivets.

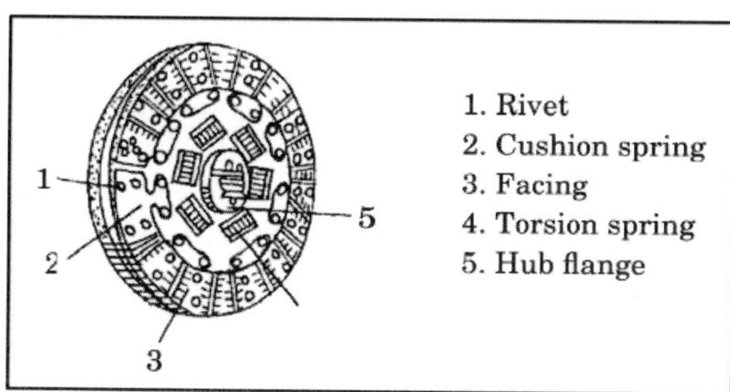

Fig. 14.6a: Friction Plate

When the friction plate is pressed by the pressure plate and flywheel, the cushion springs get compressed. Thus their thickness is reduced from 1.53 mm to 1.27 mm. This cushioning helps the clutch to engage smoothly without any chattering. There is also a hub flange provided with splines. This is fitted to the clutch shaft.

The friction material is made of asbestos fibre. This is either woven or in a moulded form. This woven facing is riveted to the friction plate on both sides. Sometimes this facing is made of threads of brass or copper woven along with long asbestos fibre and cotton.

14.6.2 Pressure plate assembly

The pressure plate assembly (Fig. 14.7) consists of a cast iron pressure plate (13) having depressions to house thrust springs (12). These springs lie between the pressure plate and the clutch cover (10). The cover is, in turn, bolted to the flywheel and rotates with it. The release levers (15) are pivoted by release lever pins (4), and these pass through the adjustable eye bolts (3). The inner ends of those levers must be in one plate; while the outer ends are linked with the pressure plate by struts (7). The levers are prevented from vibrating by anti-rattle spring (6). Adjusting nuts (8) pass through the cover and are used to adjust the height of the release levers.

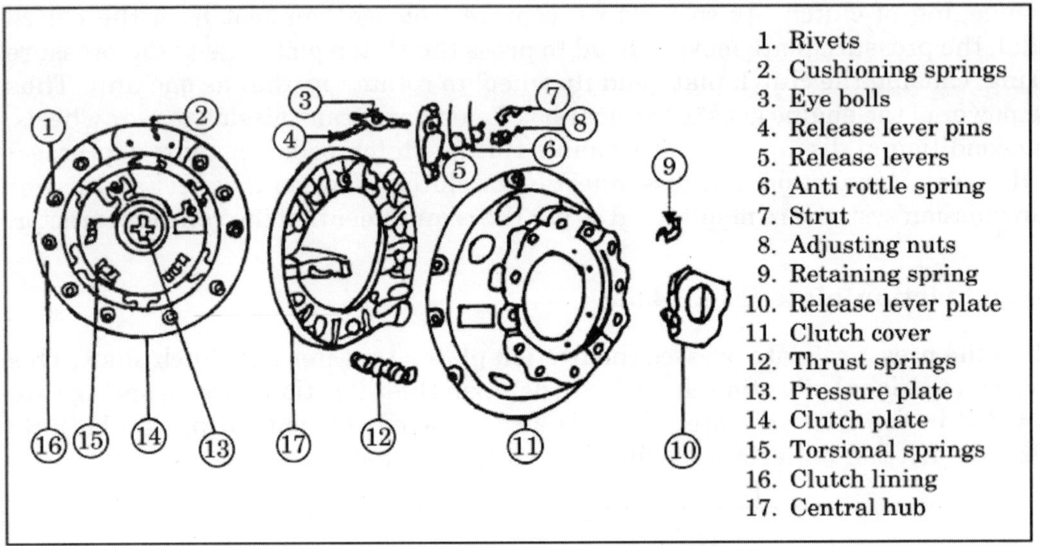

Fig. 14.7: Pressure Plate Assembly.

The inner ends of the release levers are depressed by a release bearing having a graphite block. A clearance of 3 mm is kept between the ends of the levers and the bearing.

14.7 MULTI PLATE CLUTCH

A multi-plate clutch is provided with more than one friction plate. In fact, in this clutch there are two pressure plates and two friction plates as shown in Fig. 14.8. These pressure plates are linked to the clutch cover by means of studs. The clutch cover is fitted to the fly-wheel. The first friction plate is between the first and second pressure plate. The second friction plate is between the second pressure plate and flywheel. The link mechanism is the same as the one used in the single plate clutch. The two friction plates are connected to the clutch shaft by means of a spline arrangement.

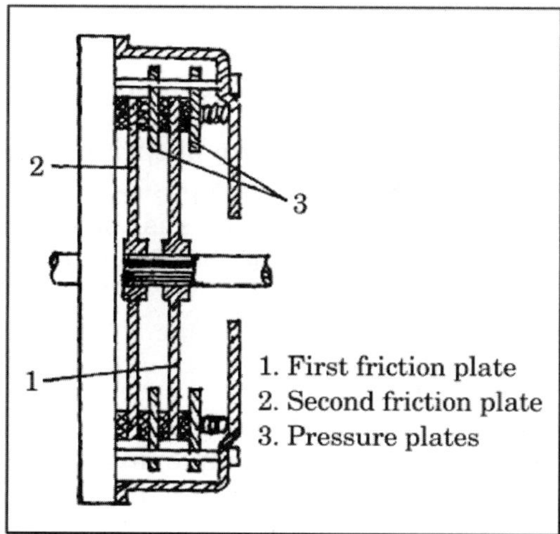

Fig. 14.8: Multi Plate Clutch.

While the flywheel is rotating, the pressure plates rotate and press against the friction plates. This causes the friction plates to also rotate..The clutch shaft is then rotated. When the pedal is pressed, the flywheel continues to rotate. The friction plates are then released. This happens because they are not fully pressed by the pressure plates. The friction plates are thus free of rotation. The clutch shaft also stops rotating.

Diaphragm clutch

This clutch has a diaphragm spring in place of coil springs. It has no need for any release lever. This is an advantage because the spring itself acts as a series of levers.

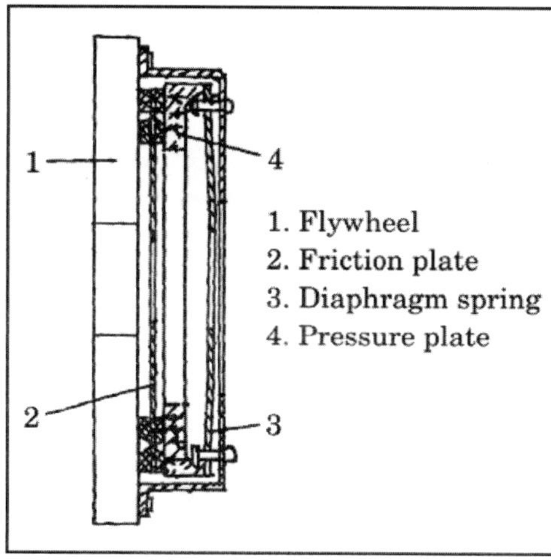

Fig. 14.8a: Diaphragm Spring Clutch (Engaged)

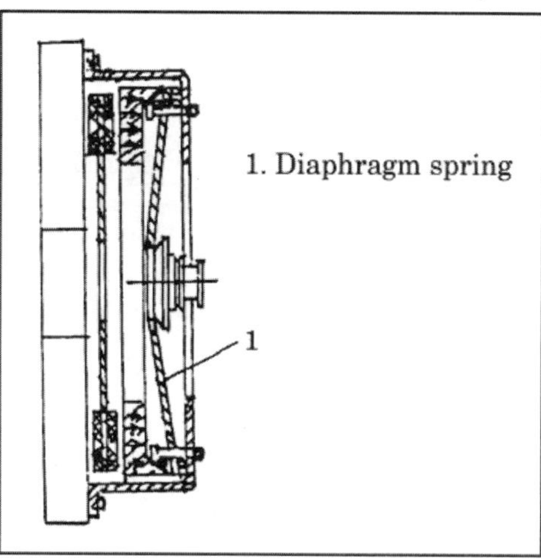

Fig. 14.8b: Diaphragm Spring Clutch (Disengaged)

Fig.14.8a shows the diaphragm spring, friction plate, collar and fork of the diaphragm spring clutch. It also shows the engaged position. Here the friction plate is pressed by the pressure plate against the flywheel.

Figure 14.8b shows the flywheel, diaphragm spring, pressure plate and friction plate. It shows the friction plate in a released position. It is free to rotate when the pressure of the pressure plate is removed. This is the disengaged position. In this position the friction plate does not rotate the clutch shaft. Here the diaphragm dishes inward.

14.8 FLUID FLYWHEEL CLUTCH

Fluid flywheels are used as clutches in employing automatic transmissions. It comprises (i) turbine (driven member) *(ii)* pump (driving member) *(iii)* oil seal *(iv)* clutch shaft and *(v)* flywheel (Fig. 14.9).

The turbine and the pumps have a number of vanes and the two are placed facing each other. The pump is driven by the engine crankshaft, and the turbine drives the clutch shaft. These two units are placed in a housing filled with oil and are separated from each other by a small clearance. The driven and driving members are not connected mechanically or rigidly. The impeller and turbine runner together form a casing completely filled with oil that is of mineral lubricating oil.

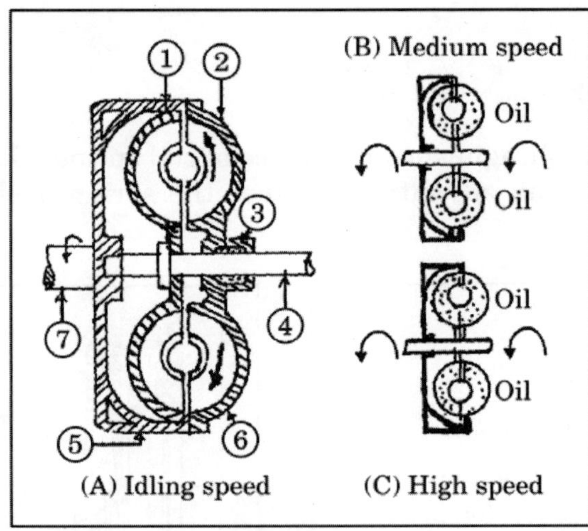

Fig. 14.9: Fluid Coupling or Fluid Flywheel Clutch.

14.8.1 Working

At engine idling speeds, the oil does not experience sufficient centrifugal force to move the turbine and hence the vehicle does not move. As the driver depresses the accelerator pedal, oil moves into the turbine and rotates the clutch shaft due to the centrifugal force. Thus, the vehicle is set in motion at slow speed. However, there is a good amount of slip. The hydraulic oil travels back to the pump after releasing some energy to the turbine.

At high engine speeds, centrifugal force is large enough to cause high rotational speeds to the turbine in order to transmit full power. The slip is reduced to two per cent; slip indicates the difference between the output and input shafts.

The fluid flywheel has the following advantages over conventional clutches.

(i) Lesser maintenance costs;
(ii) Torsional vibrations are damped
(iii) No adjustment is required
(iv) Clutch pedal is eliminated altogether
(v) The transmission shocks are absorbed by the fluid.

14.8.2 Principle of fluid coupling

Fluid flywheel is a coupling run by oil and is fitted after the engine, before the gearbox. To understand the working of fluid flywheel, let us study the following examples.

Place two electric fans in front of each other as shown in Fig. 14.10. Connect the electric supply to one fan. Due to air current of the fan Worked by electricity, the fan blades of the other fan will also start moving. The speed of the second fan will be less than that of the first fan as some of the air current will escape sidewise and will not come in contact with the blades of the second fan.

Now in case we fix up both the fan blades in a box and instead of air, we fill up the box with oil, when we switch on the first fan, its blades will start moving. With this, oil will also start moving. This moving oil will make the blades of the second fan also move at much faster rate as compared to the one illustrated in Fig. 14.10.

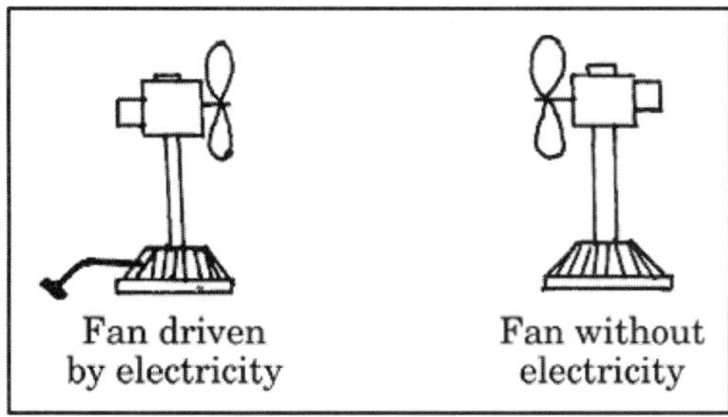

Fig. 14.10: Figure showing principle of fluid coupling

To give positive coupling, the blades are to be redesigned. Instead of ordinary air blades, we make cups on these blades as shown in Fig. 14.11A. and instead of working it with electricity, we work it with hand. Now let us fill oil in the cups of this fan, move the blades as shown in Fig. 14.11B. You will see that when' the blades move the oil in cups will spill out.

Now if we place another fan with identical cups made in it on top of this and move the first fan, the oil in the cup would revolve in the upper and lower cups as shown in Fig. 14.11C. With the arrangement, speed of the second fan will be practically the same as that of the first fan.

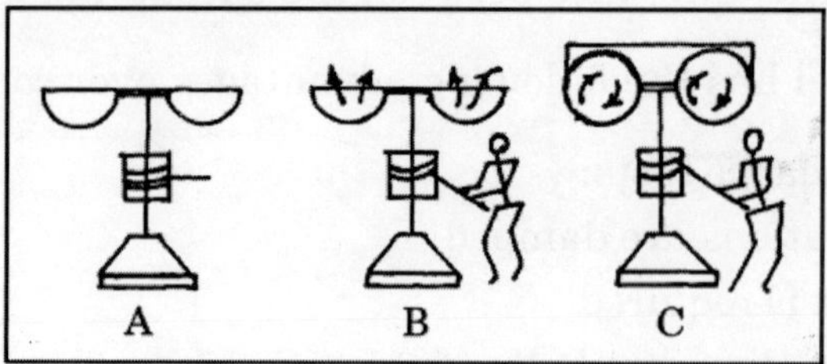

Fig. 14.11: Figure showing principle of fluid coupling

The fluid flywheel works exactly on the same principle and it has two casings- one is known as runner and the other casing. Casing is fixed with the flywheel with the help of bolts where as runner is kept free in the casing and is coupled to gearbox main shaft or drive shaft whereas other end is held in bearing fitted in hub of fly wheel. On drive shaft a bellow seal is fitted which does not allow oil to leak off.

Chapter 15

Gear Box

15.1 INTRODUCTION

The units required to transmit engine power from the engine to rear drive wheels come under transmission system. The word transmission here means a system which provides the variation of engine torque and speed into the torque and speed required by the wheels of the tractor or automobile for each task. The transmission system also provides a means of allowing the tractor or automobile to be reversed or backed when required. The unit used for the above purpose is the gear box, which is located in between the clutch and differential. The gear box is an assembly of sets of gear and shaft enclosed in a metal box.

The gear used in the gear box has teeth that mesh to transmit force and motion from one gear to another. The simplest gear is the spur gear, which is like a wheel with teeth. The gears used in the transmissions are helical gears. In these gears, the teeth are set at an angle to the gear center line. The teeth have a wiping action which improves their contact and lubrication.

15.2 PURPOSE OF TRANSMISSION

The purpose of the transmission is to provide high torque at the time of starting, hill climbing, accelerating and pulling a load. When a vehicle (or tractor) is starting from rest, hill climbing, accelerating and meeting other resistances, high torque (twisting force, or turning force or tractive effort) is required at the driving wheels. Because we know that more power (torque) is required to move the stationary machine or tractor, but as soon as the tractor sets in motion, less torque is required. Similarly, we have noticed that when we start pushing a stationary tractor, more force is required and as soon as the tractor starts rolling, less force is needed. To move the stationary tractor, engines do not have enough torque (*i.e.*, torque of crankshaft). Hence a device must be provided to permit the engine crankshaft to revolve at a relatively high speed, while wheels turn at slower speeds. Because in gear system, speed reduction means torque increase. The torque multiplication or increase in the drive wheels is possible with the help of gear box. The vehicle speed is also changed with the help of the transmission keeping the engine speed same with certain limit.

The tractors are also used for many kinds of work. Sometimes, it is required to pull a heavy load, for which we need that the rear wheels should run slowly, while engine at its rated speed. It is also required to drive a tractor at high speed for pulling lighter loads on smooth roads. In order to meet the different requirements, several sets of gears are provided in the gear box. So with the engine speed remaining the same, the running speed of the tractor or automobile can be changed with the help of gear box.

15.3 Principle of Gearing

Any combination of gear wheels by means of which motion is transmitted from one shaft to another shaft is called a gear train. Consider a single train in which the driving gear A has 10 teeth and driven gear B has 20 teeth . The gear A is rotating with a speed of 100 rpm. A is the driving gear and B is the driven gear, then speed ratio or gear ratio

$$= \frac{\text{Speed of driving gear (A)}}{\text{Speed of driven gear (B)}} = \frac{\text{No. of teeth of driven gear}}{\text{No. of teeth of driving gear}}$$

$$\text{i.e. gear ratio} = \frac{N_A}{N_B} = \frac{T_B}{T_A} = 2 : 1;$$

$$\text{But torque ratio} = \frac{N_B}{N_A} = \frac{T_A}{T_B} = 1 : 2;$$

Hence, when the smaller gear is driving the larger gear, the gear ratio is 2 : 1. However, the torque ratio is 1: 2. The larger gear turns at half the speed of the smaller gear. As a result, the larger gear will have twice the torque of the smaller gear. Power $2\pi\ N_A T_A - 2\pi\ N_B T_B$. Putting the values of N_A. T_A and T_B;

$$N_B = \frac{N_A T_A}{T_B} = \frac{100 \times 10}{20} = 50 \text{ rpm.}$$

The direction of rotation of gear B will be opposite to that of gear A. However, if an additional gear is introduced, the direction of the final gear will be the same, as that of the driving gear (Fig. 15.1).

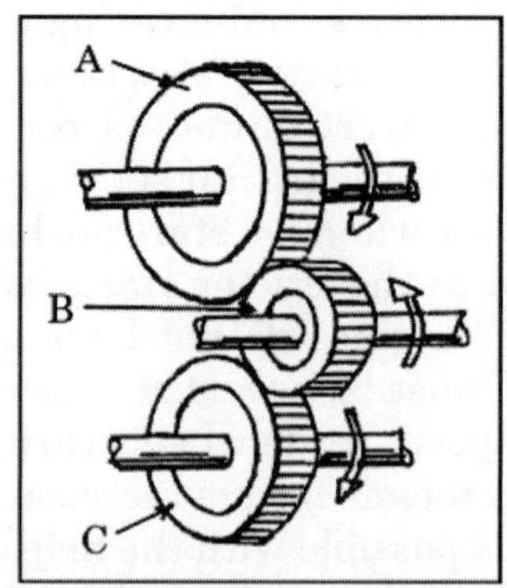

Fig. 15.1: System of Three Gears.

15.4 TORQUE AND SPEED OF ENGINE AND DRIVE WHEELS

Torque is twisting or turning force. It is measured in (kg-m) or Newton-meter (N-m). Any shaft or gear that is being turned has torque applied to it. The engine pistons and connecting rods push on the cranks on the crankshaft. This applies torque to the crankshaft and causes it to turn. The crankshaft applies torque to the gears in the transmission so that gears turn. This turning force or torque, is carried through the power train to the driving wheels so that they turn.

We know that the horsepower delivered to the rear wheels depends upon the turning effort and speed of rotation. Mathematically, it is represented as

$$\text{Horsepower} = \frac{2\pi \times \text{N(rpm)} \times \text{T(torque)(kg}-m)}{4500}$$

If the engine horsepower is constant, from the above relation, it is obvious that for higher torque at wheels, low speed is required and *vice versa*. Therefore, gearbox is incorporated in between the engine and rear wheels which is a solution for the variable torque and speed.

To understand the necessity of a gear box, let us consider a tractor in which the engine power is delivered to the rear wheels, directly through a set of bevel gears at right angles. Such a tractor will run at constant speed, which of course, will be very high and a fixed tractive effort will be available at the rear wheels. This situation is not desirable, because, depending upon the magnitude of the load being pulled and soil characteristics, the tractive effort required on the rear wheels varies. Therefore gearbox is provided for speed reduction and torque multiplication to achieve desired tractive efforts.

15.5 TYPES OF TRANSMISSION IN TRACTOR

The transmissions are classified into several types according to the following characteristic features.

(a) Number of shafts – two, three and four shaft types

(b) Manner in which gears are meshed – sliding and constant-mesh type

(c) Shifting method – manually shifted, hydraulically shifted and automatic (self-shifting) types.

(d) Number of available forward speed ratio – two speed, three sped, four speed etc.

The speed ratios provided by tractor transmissions may be arbitrarily divided into three groups., main, transport and low.

Main speed ratios: These provide for tractor land speeds required by the majority of field tasks. These speeds range from 5 to 15 kmph (1.4 to 4.2 m/s).

Transport speed ratio: These are used in carrying goods by tractor-trailer combination. In wheeled tractors, these speed ratios provide for land speeds in the range 4.2 to 9.5 m/s and in crawler tractors, 4.2 m/s.

Low speed ratios: These provide for tractor land speeds required by some field tasks, such as transplanting and root-crop harvesting. These speeds are in the range 0.16-0.4 m/s.

Automobile transmissions have only one reverse speed where as tractor transmissions may have several reverse speeds, since tractor runs in reverse direction for various operations.

15.6 TYPES OF GEAR BOX

The Gears in the tractor are of selective type and are generally of constant mesh, sliding mesh and synchromesh types.

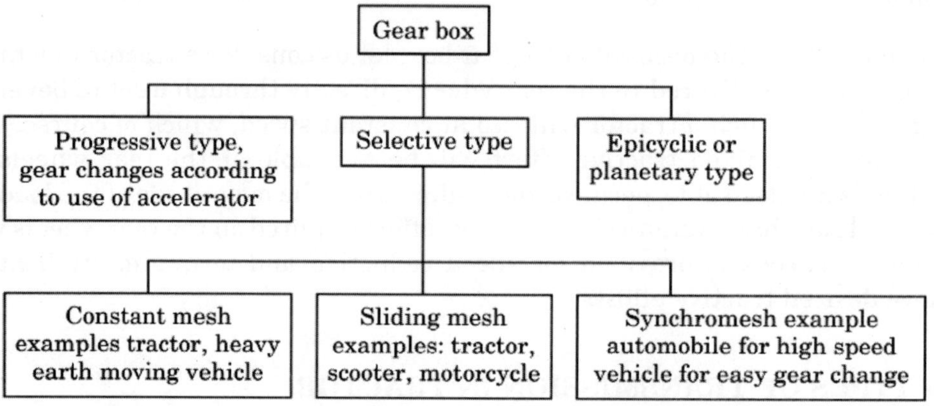

15.6.1 Gear box construction

The gear box (Fig. 15.2) consists of gear housing, clutch shaft (input shaft), main shaft (output shaft), countershaft (lay shaft), reverse idler gear shaft and gear shifting lever. The gear housing is made of cast iron. It is rigid in construction and serves the purpose of automobile or tractor frame. The clutch shaft and main shaft are arranged coaxially. The countershaft is parallel to the main shaft. The clutch gear is rigidly fixed to the clutch shaft. It always remains connected to the drive gear of the countershaft. The clutch gear is rigidly fixed to the clutch shaft. It always remains connected to the drive gear of the countershaft. The other gears on the countershaft are rigidly fixed to it. Hence all the gears in the countershaft run in the same speed as that of its drive gear. The drive gear of the countershaft gets drive from the clutch gear. Gears are generally made of alloy steel. Steel shafts on which the gears are mounted are hardened to reduce wear and trouble free service. There are no spline arrangements in clutch shaft, counter shaft and reverse idler shaft. But main shaft is splined over which the gears slide freely in horizontal direction to mesh with a suitable gear on the countershaft in sliding mesh type of gear. The reverse idler gear mounted on reverse idler gear shaft always remains connected to the reverse gear of the countershaft. The reverse idler gear is used to reverse the direction of rotation of the main shaft.

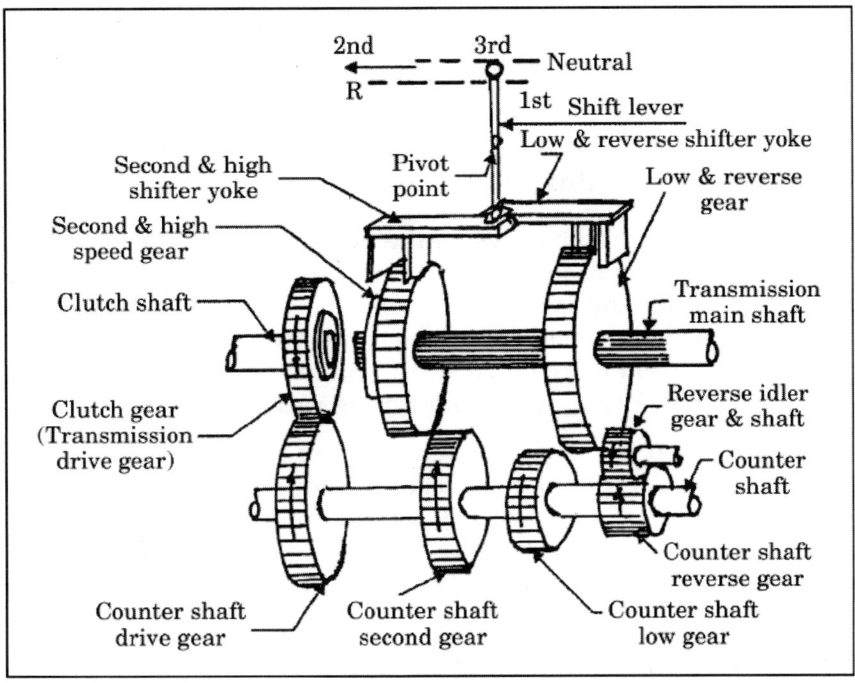

Fig. 15.2: Sliding Mesh Gear Box (Gears in Neutral).

The front end of clutch shaft is supported by a ball bearing (pilot bearing) installed in the central bore in the flywheel end face and the rear end is supported by a ball bearing in the front wall of the gear box housing. The front end of the main shaft is supported by a roller bearing housed in a bore in the end face of clutch shaft and its other end is supported by a bearing mounted in the rear wall of the gear box housing. The counter shaft and reverse idler shaft are also supported on ball bearings placed in the same gear box casing. SAE-90 grade of oil is filled in the casing up to the desired level for lubrication purpose. Gear shift rails and forks are used to shift gears from one position to another. The gears in the main shaft are not splined to the splined main shaft and they rotate freely over it in constant mesh type of gear box.

When the engine is running and clutch is engaged, the clutch shaft gear drives the countershaft gear. The countershaft rotates opposite in direction of the clutch shaft. In the neutral position, only the clutch shaft gear is connected to the countershaft gear. Other gears are free and hence the transmission main shaft is not turning. The vehicle is stationary.

Gear: A gear is a toothed wheel which when meshed with another toothed wheel (the pinion), transmits rotation from one shaft to another. The primary function of a gear is to transfer power from one shaft to another. It occupies less space compared to chain drive and belt drives.

Pinion: Smaller of two meshed gears.

Spur gear: Parallel axis gears with straight teeth, teeth parallel to the axis.

Helical gear: Parallel axis gear with teeth cut at an angle to the axis.

Bevel gear: Non parallel and coplanar gears with shafts at right angle to each other.

15.6.2 Sliding Mesh Gear Box

In sliding mesh gear system, the gears on the main shaft slide along it. The gears which drive the splined main shaft have internal splines which match with the main shaft. By operating the gear shift lever, the gears on the main shaft are allowed to slide on it and are meshed with the respective gears on the countershaft. Thus power transmits from clutch gear to the countershaft gear and to the main shaft gear for producing output speed and torque ratios.

Gears in neutral: When the engine is running and clutch is engaged, the clutch shaft gear drives the countershaft gear. The counter shaft rotates opposite in the direction of the clutch shaft. In neutral position, only the clutch shaft gear is connected to the countershaft gear. The gears on the main shaft are free and hence the transmission main shaft is not turning.The vehicle is stationary (Fig. 15.2).

First or low speed gear: By operating the gear shift lever, the larger gear (low gear) on the main shaft is moved along the shaft to mesh with the first gear of the countershaft. The main shaft turns in the same direction as the clutch shaft. Since the smaller counter shaft gear is engaged with the larger main shaft gear, a gear reduction or gear ratio of approximately 3:1 is obtained (Fig. 15.3). That is the clutch shaft turns three times for each revolution of the main shaft. Hence speed of drive wheel is reduced. The ratio between engine speed and wheel speed = 3:1.

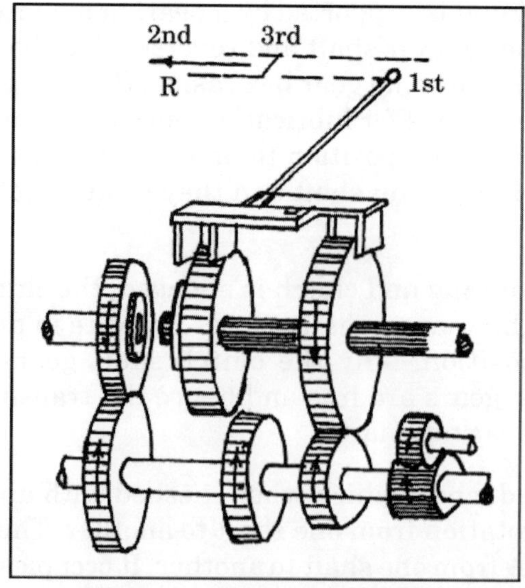

Fig. 15.3: First Speed Gear.

Second Speed gear: By operating the gearshift lever, the larger gear of the main shaft is demeshed from the first gear of the countershaft and then the smaller gear (second gear) of the main shaft is meshed with the second gear of the counter shaft. The main shaft turns in the same direction as the clutch shaft. A gear reduction of approximately 2:1 is obtained (Fig. 15.4) .That is, the clutch shaft turns two times for each revolution of the main shaft.

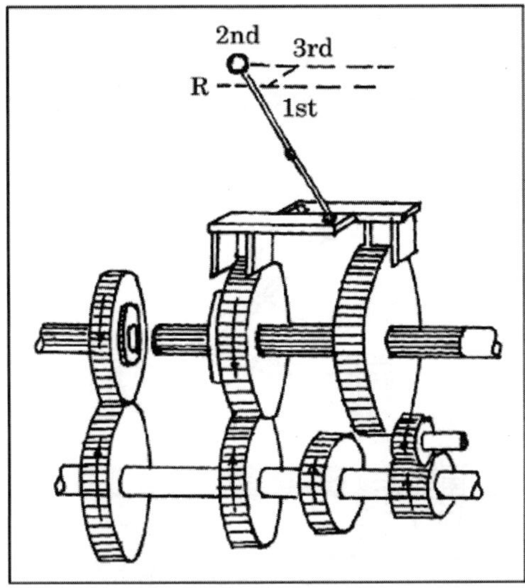

Fig. 15.4: Second Speed Gear.

Third or top or high speed gear: By operating the gear shift lever the second gear of the main shaft and countershaft are demeshed and then the second or top gear of the main shaft is forced axially against the clutch shaft gear. External teeth on the clutch shaft gear mesh with the internal teeth in the second and top gear. The main shaft turns with the clutch shaft and a gear ratio of 1:1 (Fig. 15.5).

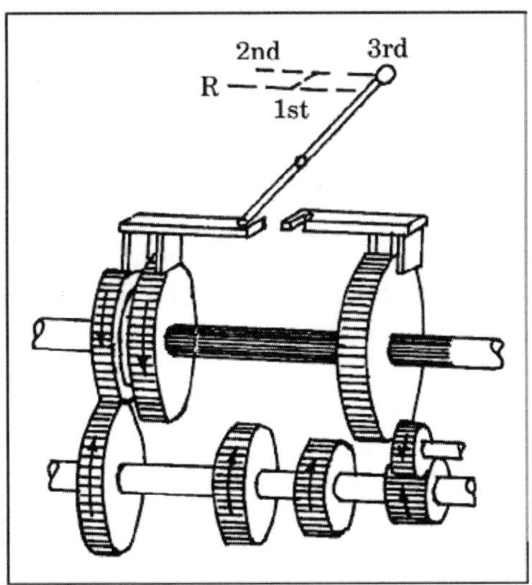

Fig. 15.5: Third Speed Gear.

Reverse gear: By operating the gear shift lever, the larger gear of the main shaft is meshed with the reverse idler gear. The reverse idler gear is always in mesh with the countershaft reverse gear. Interposing the idler gear between the countershaft reverse gear and main shaft bigger gear., the main shaft turns in the direction opposite to that of the clutch shaft. This reverses the rotation of the wheel so that the vehicle backs (Fig. 15.6).The reverse gear ratio is lower than the first gear ratio. The gear ratio in case of reverse gear arrangement is independent of the intermediate (reverse gear). That's why the reverse gear is called as the idler gear. The gear ratio depends on the driven and driver gears. While reversing the vehicle, brake is applied and the vehicle is stopped and then reverse gear arrangement is made.

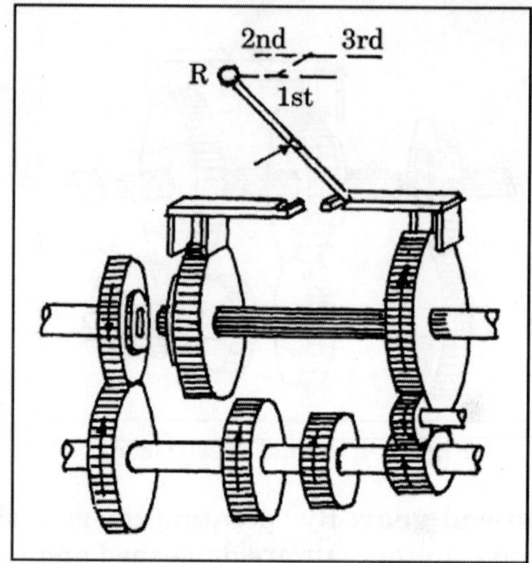

Fig. 15.6: Reverse Gear.

15.6.3 Constant mesh gear box

In constant mesh gear box, all the gears of the main shaft are in constant mesh with the corresponding gears of the countershaft. The main difference between the sliding and constant mesh gear box is that in the former, gears actually slide along the main shaft and in the latter, the counter shaft and main shaft gears always remain in mesh. The dog clutch or coupling device or splined sleeve or collar in the constant mesh gear type lock the gears in the main shaft as and when required .Two dog clutches are provided on the main shaft - one between the clutch gear and the second gear and the other between the first gear and the reverse gear. The main shaft is splined, but the gears on it are not splined with the shaft .The coupling device or dog clutch which have internal splines is mounted on the splines of the main shaft. Thus the main shaft gets drive through the driven gear and coupling unit. The gears on the main shaft are mounted on the bushing and are free to rotate along with counter shaft gear without affecting main shaft when in neutral.

The sliding mesh gear box gives a high mechanical efficiency but has certain disadvantages, like gear noise, being rough in operation and requires skill and judgement in the change of gear. The constant mesh gear box employs helical gears for quieter operation. The gears on the sliding mesh type are spur gear.The teeth of these gears are usually noisy. Hence constant mesh gear box is used to overcome these short comings.

The sliding dog clutch is slided by the selector forks to the left or right to obtain the required gear ratio.

When the left hand dog clutch is made to slide to the left by means of the gear shift lever, it meshes with the clutch gear and the top speed gear is obtained when the left hand dog clutch meshes with the second gear, the second speed gear is obtained. Similarly by sliding the right hand dog clutch to the left and right, the first speed gear and reverse gear are obtained respectively (Fig. 15.7).

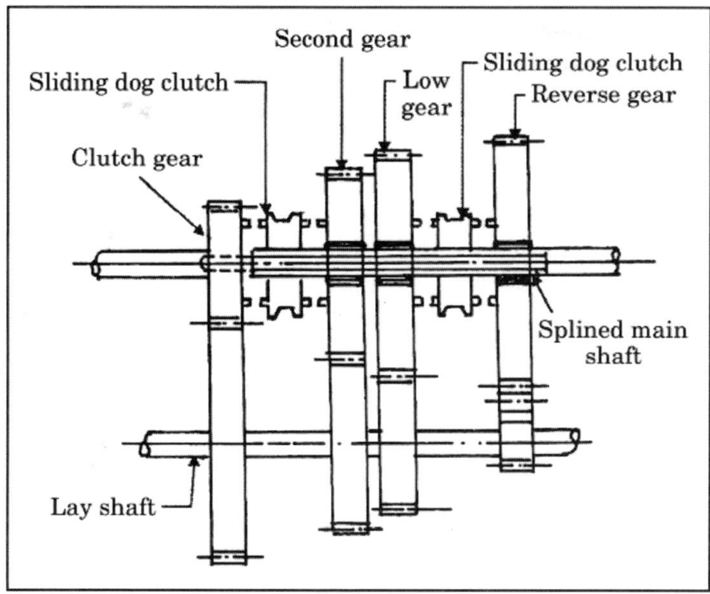

Fig. 15.7: Constant Mesh Gear Box.

In this type of *gear* box, because, all the gears are in constant mesh, they are safe from being damaged and unpleasant meshing sound does not occur while engaging and disengaging them.

15.6.4 Synchromesh gear box

The synchromesh gear box is basically a constant mesh gear box with an extra device, called a synchronizer to equalize the speed of meshing gears before they engage. With the help of synchromesh device, two gears to be engaged are first brought into frictional contact which equalizes their speed after which they are engaged smoothly. In synchromesh gear box, synchromesh units are used in place of dog clutches in constant mesh gear box.

Synchro means nearly equal *i.e.*, the speed of the main shaft gear and the dog clutch must be nearly same before their engagements. Although a constant mesh gear box have improvements over the sliding type, yet the gears which are connected to the dog clutch, produces noise and jerk. To avoid this, the synchromesh gear box is used. The synchronising device is used for smooth meshing. The synchromesh gear box is most commonly used in high speed automobiles and is very rarely used on tractor transmission.

We know that in running condition, when we press the clutch and put the gear box in neutral position, still the gears are revolving. All the gears donot revolve at the same speed and when we have to engage 2 gears running at different speeds by the sleeve or dog clutch, there is going to be some sound. This sound canbe reduced by a careful and trained driver, but in case the driver is not careful, there canbe lot of sound when gears are changed. To stop this, synchromesh units are used by which the shifting of gear is very smooth, *i.e.*, without jerk or sound and the life of gear teeth is also more as compared to sleeve or dog clutch type of arrangement.The synchromesh unit is usually fitted for top and 3rd speed, rest of the gear shifting is by the sleeve or dog clutch arrangement.

Synchronizing unit (Fig. 15.8) is an assembly of hub and sleeve. The hub is splined to the shaft and the synchronizer sleeve is mounted on the hub. When the gear lever is moved, the synchronizer cone meets with a similar cone on the selected gear to be meshed. Due to friction, the rotating gear is made to rotate at the same speed as the synchromesh unit. Further movement of shift fork or gear lever, forces the sliding sleeve towards the selected gear. The internal teeth on the sliding sleeve match the external teeth on the gear. With both the gear and the main shaft rotating at the same speed, the sleeve slides over the teeth on the gear without clashing. Now the gear is locked to the main shaft through the sliding sleeve and the shift is completed. The complete procedure of synchronization has been shown in Fig. 15.9.

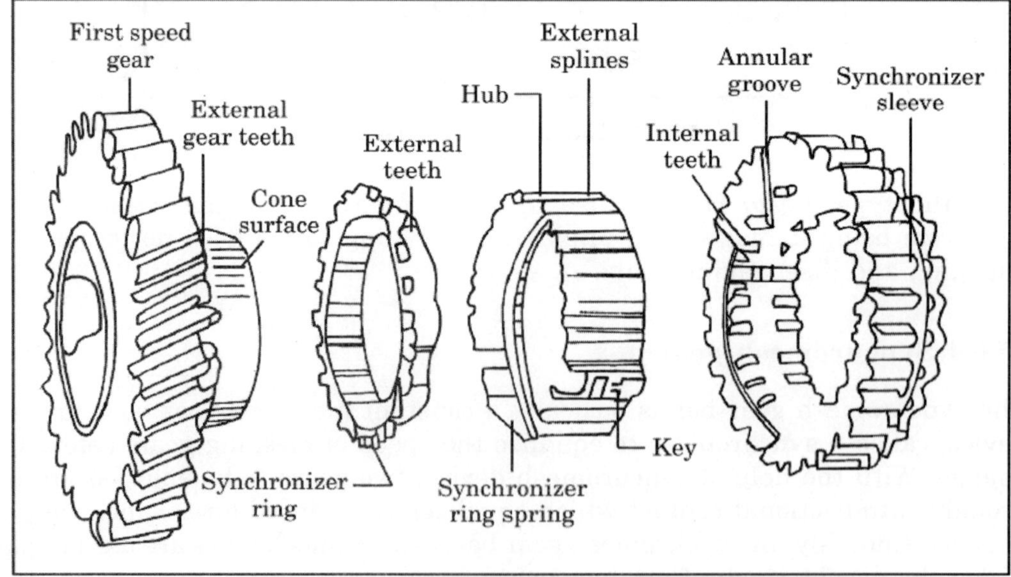

Fig. 15.8: A Disassembled Synchronizer.

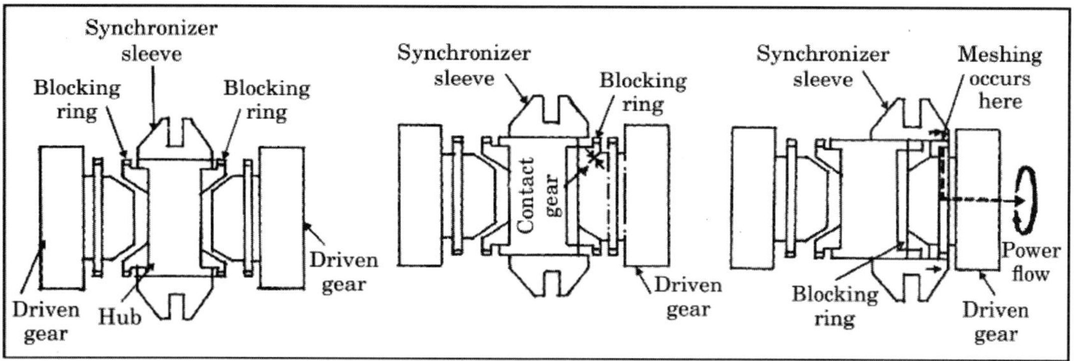

Fig. 15.9: Operation of Synchronizer Unit

First gear: In first gear, the first-second synchronizer sleeve is moved to the right. Its internal splines engage the external splines of the first speed gear.

Second gear: In second gear, the first-second synthconizer sleeve is moved to the left. Its internal splines engage the external splines of the second speed gear.

Third gear: In third gear, the third-fourth synchronizer sleeve is moved to the right. Its internal splines engage the external splines of the third speed gear.

Fourth gear: In fourth gear, the third -fourth synchronizer sleeve is moved to the left. Its internal splines engage the external splines of the clutch gear.

Reverse gear: In reverse gear, both synchronizer sleeves are in neutral.The reverse gear, which is a sliding gear splined to the transmission main shaft, is moved to the left, so that it engages the reverse idler gear. Now the reverse idler gear causes the main shaft to turn in the reverse direction and the vehicle moves backward. The advantages of using the synchromesh device are as follows.

(i) Gear changing is much simplified
(ii) Reduction in gear wear
(iii) Allows the usage of helical gears which run quietly.

The disadvantages are as follows;

(i) Complicated design
(ii) Reduction of synchronizing effect due to wear in the synchronizing cones
(iii) A quick change of gears results in noise or crashing.

15.6.5 Planetary gear system (Figure 15.10)

The planetary gear set is used in automatic transmission. A planetary gear system consists of three types of gears.

1. An outer ring gear having inside teeth. It is sometimes called internal gear.

2. Three planet pinions held on pinion shafts in a cage or carrier.
3. A sun gear at the centre of the three planet pinions.

The planet pinions mesh with the ring gear internally and with the sun gear externally. The planet gear system gets its name from the fact that the pinion revolve around the sun gear and rotate at the same time, just as the planets in the solar system rotate and revolve around the sun.

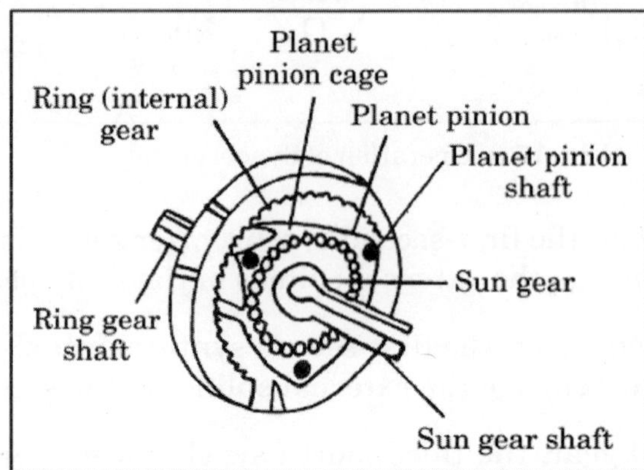

Fig. 15.10: Parts of a Planetary Gear System,

The working of the planetary system is as follows; the ring gear is attached to the output shaft. The three planet pinions are assembled into a cage that is splined to the transmission main shaft. The sun gear has an arrangement where by it may be permitted to turn, or it may be locked in a stationary position. When it is locked, the ring gear and hence the output shaft is forced to turn faster than the transmission main shaft. That is, the output shaft overdrives the transmission main shaft.

A planetary gear system can be used for various functions by holding one of the three members (ring gears, sun gear and planet pinion cage) (Fig. 15.11) stationary and turning another member. The various combinations are as follows.

(i) **Speed increase:** Hold the sun gear stationary and turn the planet pinion cage, the ring gear will rotate faster than the planet pinion cage. The ratio between the planet pinion cage and the ring gear depends upon the sizes of the different gears.

(ii) **Speed increase:** Another combination to get increased speed is to hold the ring gear stationary and turn the planet pinion cage. The sun gear will rotate faster than the cage.

(iii) **Speed reduction:** Hold the sun gear stationary and turn the ring gear. The planet gear case will turn more slowly than the ring gear.

(iv) **Speed reduction:** Hold the ring gear stationary and turn the sun gear. The planet pinion cage will rotate at a speed less than the sun gear speed.

(v) **Reverse:** Hold the planet pinion cage stationary and turn the ring gear. In this case, the planet pinions act as idlers and they cause the sun gear to turn in the reverse direction to the ring gear rotation. Thus, the system functions as reverse rotation system, with the sun gear turning faster than the ring gear.

(vi) **Reverse:** Hold the planet pinion cage stationary and turn the sun gear. The ring gear will turn in a reverse direction, but slower than the sun gear.

(vii) **Direct drive:** If any two of the three members are locked together, then the entire planetary gear system is locked out, and the input shaft and output shaft must turn at the same speeds. On the other hand, if no member is held stationary and no two members are locked together, then the system will not transmit power at all. The input shaft may turn, but the output shaft does not.

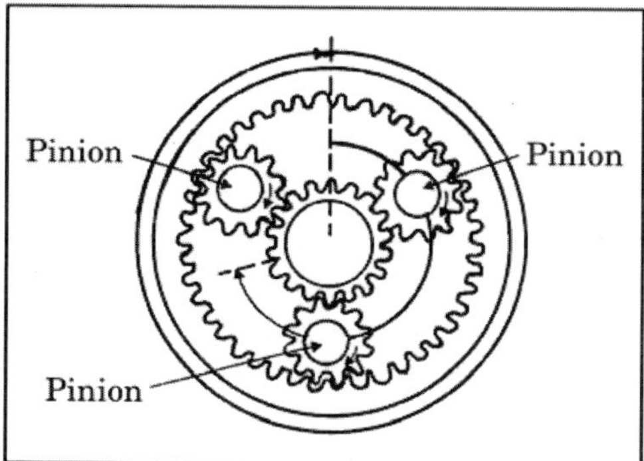

Fig. 15.11: Relative Sizes of Gears in a Planetary Gear System.

The epicyclic or sun and planet type gear box uses no sliding dog clutches, or gears to engage. Hence no noise occurs while engaging the gears and no wastage of time occurs while waiting for synchronization. Hence the advantages of epicyclic gear box are as follows :

(i) The gears are always in constant mesh and hence dog clutches or sliding gears are not used.

(ii) It provides a more compact unit operating about a common central axis.

(iii) A wide variation in gear ratios in forwad and reverse can be obtained.

(iv) Engagement of gears is smooth and noiseless.

(v) The gear and gear housing are comparatively smaller in overall dimension.

(vi) Instead of having the load or only one pair of gears, it is distributed over several gear wheels.

(vii) External contracting band brakes of relatively small dimensions are used for changing the gears.

Chapter 16

Differential and Final Drive

16.1 INTRODUCTION

Power from the engine is transferred to the drive wheels through the clutch, gearbox, differential and axle shafts. Differential unit is a special arrangement of gears to permit one of the rear wheels of the tractor or automobile to rotate slower or faster than the others. It is named differential as it allows different speeds to the rear drive wheels. If the two rear wheels are rigidly fixed to a rear axle, the inner wheel will slip causing rapid tyre wear and the vehicle becomes difficult to control during turning. Therefore, there must be some devices to provide relative movement to the two rear wheels when the vehicle/tractor is taking a turn. The device, differential avoids these problems. It allows the outer wheel to turn faster than the inner rear wheel during a turn.

Functions of differential

1. To transmit the power from transmission shaft at right angle to the axle shaft to move the drive wheels.
2. To differentiate the speed of two rear wheels while turning *i.e.* the outer wheel to run faster than the inner wheel.
3. To allow both the rear drive wheels to get equal power and speed during straight travel.
4. To reduce the speed of rotation.

16.2 COMPONENTS OF DIFFERENTIAL UNIT AND THEIR ARRANGEMENT (FIG. 16.1)

1. Drive bevel pinion
2. Crown wheel (ring gear)
3. Differential cage (differential casing)
4. Differential pinion (2 nos.) (star pinions or planet pinions)
5. Differential side gear (2 nos.) (Sun gears)

6. Differential pinion shaft
7. Rear axle (half shafts)
8. Drive wheels.

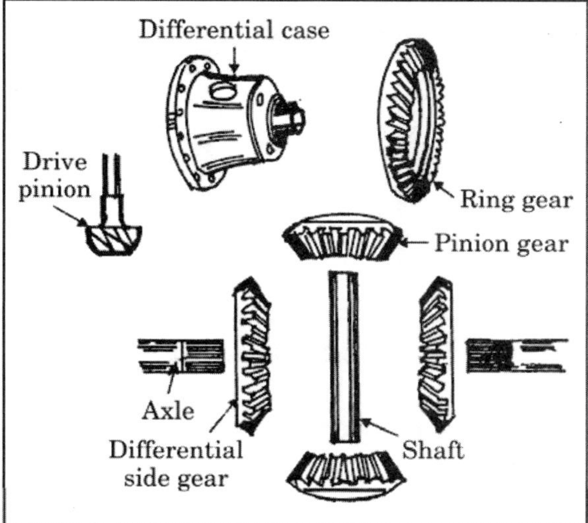

Fig. 16.1: Basic Parts of a Differential.

Arrangement of components

The drive pinion on the shaft transmits engine power to the crown wheel (Fig. 16.2). Both crown wheel and drive pinion make bevel gear arrangement with curved teeth. The ratio of teeth on the crown wheel and the driven pinion is usually 4:1 to reduce speed. The drive pinion is constantly meshed with the ring gear. The differential casing is rigidly attached with the crown wheel and rotates like one unit. The differential cage is assembled on the left axle. The differential cage has bearings that permit it to rotate independently of two axles. The cage supports two planet pinions (called differential pinions) on a shaft (differential pinion shaft). The shaft is of single unit and is rigidly fixed to the cage. The two differential pinions are free to rotate on the shaft. These two differential pinions are constantly meshed with two side gears (sun gears) (differential bevel gears) which are attached to the inner ends of the half shafts. Each half shaft is attached

Each half shaft is attached to each rear wheel. Thus instead of crown wheel being keyed directly to a solid shaft between the rear wheels, the drive is taken back from the indirect route through differential casing, differential pinion and half shaft of the rear wheel. The internal spline of the each sun gear is meshed with the spline of the respective half shaft. All gears in the differential unit are'bevel gear. Bevel gear is used when power is to be transmitted at right angle. The *axis* of two shafts are at right angle and can intersect each other. The differential side gears are splined on to the inner ends of the *axle* shaft. The other ends of the shafts are attached to the driving wheel hubs by means of flanges.

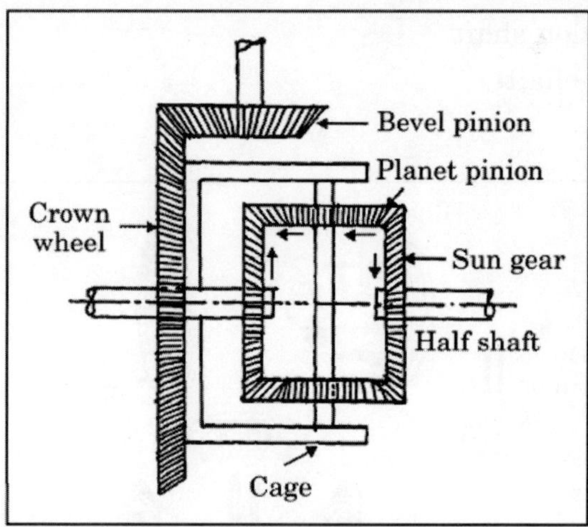

Fig. 16.2: Differential.

16.3 OPERATION OF DIFFERENTIAL

When the vehicle is on a straight road (Fig. 16.3a), the ring gear, differential cage, differential pinions, two differential side gears, all turn as one unit. The two differential pinions do not rotate on the pinion shaft. This is because, they exert equal force on the two differential side gears (sun gears). As a result, the side gears turn at the same speed as the ring gear, which causes both drive wheels to turn at the same speed also. The crown gear gets drive from drive bevel pinions. With the rotation of crown gear, the differential cage also rotates. When the differential cage rotates, the differential pinions and their shaft move around is a circle with the differential cage. Because the two sun gears are meshed with the pinions, the sun gears must rotate, causing the axle shafts and their respective driving wheels to turn. With equal resistance applied to each wheel, the differential pinion do not rotate. They apply equal torque to the side gears (sun gears) and therefore both driving wheels to rotate at one and the same speed.

However, when the vehicle begins to move around a curve or to make a turn (Fig, 16.3(*b*)), the differential pinion rotate on the pinion shaft. This permits the outer wheel to turn faster than the inner wheel. When the vehicle makes a turn, the resistance to the drive wheels is unequal. With unequal resistance, the differential pinions rotate on their shaft as well as drive (move) round the differential cage. Supposing that one of the axle shafts (or of respective drive wheel) is prevented from rotating, the differential pinions would have to walk around the stationary sun gear, causing the other sun gear to rotate at twice its normal speed. When the pinions are carried round by the cage, they drive the half shafts in the same direction but when they are rotated on their own shaft (pinion shaft), they drive the half shafts in opposite direction *i.e.*, rotation of differential pinion adds motion to one shaft and subtract motion from the other shaft. This effect causes the outer wheel to run faster than the inner wheel.

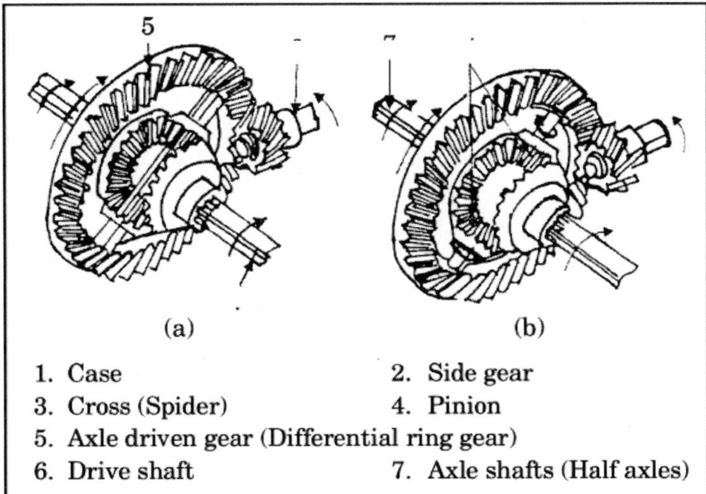

Fig. 16.3: Differential (a) illustrating the operation of the differential on a Straightway. (b) Illustrating the operation of the differential when rounding a Turn.

Also the action in a typical turn is shown in Fig. 16.4. The differential case speed is considered to be 100 per cent. The rotating action of the pinion gears carries 90 per cent of this speed to the slower rotating inner wheel. It sends 110 per cent of the speed to the faster - rotating outer wheel.

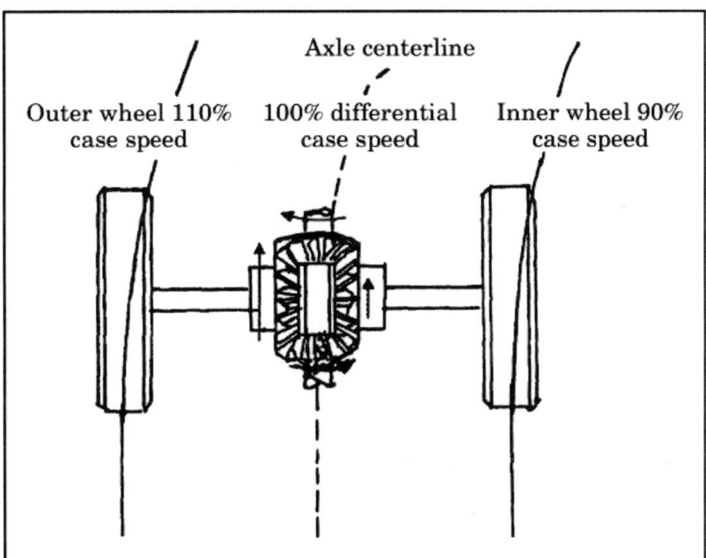

Fig. 16.4: Differential Action of Turns.

16.4 DIFFERENTIAL LOCK

Differential lock is a device or a coupling sleeve to join both the half axles of the vehicle so that even if one wheel is under more resistance, it comes out from mud or soft ground or loose soil *by* equalizing the speed among two drive wheels. Whenever one wheel offers less resistance, it turns faster causing a loss of traction. If one wheel gets in mud or loose soil, the wheel on the solid ground will not be driven while the other wheel spins around due to the differential action. To overcome this problem, all vehicles are provided with a locking system known as differential lock. With the help of differential lock, both the wheels move with the same speed and traction.

16.5 FINAL DRIVE

The final drive is the last stage in transferring power from engine to the drive wheels. It brings down further the speed of drive wheel by mounting it near the rear-drive wheels of the tractor. Hence, it is a gear reduction unit in the power trains between the differential and the drive wheels. The tractor rear wheels are not directly attached to the half shafts but the drive is taken through a pair of spur gears. Each half shaft terminates in a small gear which finally meshes with a large gear. The large gear is mounted on the shaft carrying the tractor rear wheel. Final drive provides higher torque with less speed on driving wheels. The device for final speed reduction, suitable for rear drive wheels is known as final drive mechanism.

Chapter 17

Steering System and Front Axle

17.1 INTRODUCTION

The steering system governs the angular movement of front wheels of the automobile or tractor so that it can be turned left or right without exerting much effort to the operator. The main function of a steering system is to convert the rotary motion of the steering wheel into angular displacement of the front wheels. While travelling straight, both the front wheels must be parallel with each other. During a turn, each wheel must move on an arc and these arcs should have a common centre. Tyre wear and slip is reduced to a minimum with such a condition. Steering is effected by moving the axes of rotation of the front wheels with respect to the chassis frame. To satisfy this condition, the inner wheel must turn through a greater angle than the outer one . If not, tyre wear is greatly increased. The system mainly includes the steering wheel, which the driver controls, the steering gear that changes the rotary motion of the wheel into straight line motion of the linkage with a mechanical advantage and the steering linkages.

17.2 FUNCTIONS OF STEERING SYSTEM

The functions of a good steering system are as follows:

(i) To turn the vehicle at the will of the driver

(ii) To control or minimize the wear and tear of tyres

(iii) To convert the rotary motion of the steering wheel into the angular displacement of the front wheels.

(iv) To multiply the effort of the driver for easy operation.

(v) To provide directional stability and rolling action of the wheels on the road surface

(vi) To absorb road shocks and to prevent them from reaching the driver.

17.3 REQUIREMENTS OF A STEERING SYSTEM

A smooth performance of steering system calls for the following requirements:

(i) The front wheels should roll without lateral skid while making turns.

(ii) There should be proper proportion between the angles turned by the front wheels.

(iii) The system should be light to operate *i.e.* should not call for excessive effort for steering the machine.

(iv) The wheels must automatically come to the straight ahead position after making the bend when going straight, the wheels must maintain the neutral position.

(v) The angular oscillation of the wheels must be minimum.

17.4 COMPONENTS AND WORKING OF STEERING SYSTEM

The different component of the steering system are *(i)* steering wheel (ii) steering shaft (steering column) *(iii)* pitman arm (drop arm), *(iv)* drag link *(v)* steering arm *(vi)* tie rod, *(vii)* king pin, *(viii)* stub axle.

When the steering wheel is turned, motion is transmitted to the steering box by the steering shaft rotating in a hollow steering column. When wheel is rotated, the cross shaft in steering box (gear box) oscillates. The cross shaft is connected to the drop (pitman) arm. The angular movement of the pitman arm is further transmitted to the steering arm through the drag link and tie rods. Steering arms are keyed to the respective kingpins which are the integral part of the stub axle on which wheels are mounted. The movement of the steering arm causes the angular movement of the front wheel. If driver wants to turn the vehicle to the left, he turns the steering wheel to the left and if he wants to turn the vehicle to the right, he turns steering wheel to the right, otherwise the steering wheel is in its middle position and the vehicle is going in a straight line. The simplified diagram of steering system is shown in Fig. 17.1.

Steering wheel: Made of polyurethane or hard plastic, the steering wheel consists of a circular rim with a hub at the centre. The rim is slightly elliptical in cross section for maintaining strength and to give a good hand grip.

Steering column: This is a circular column which houses the steering shaft in bush ball or roller bearing. The column itself is supported to the vehicle by brackets.

Steering shaft: It is made of good quality steel. One end is fixed in steering wheel with the help of splines or key and is kept tight by nut. The other end is secured firmly with the steering box. This shaft transmits steering torque and absorbs forces which bring about changes in its length.

Drop arm: This is forged out of good quality steel. One side of it has splines which matches with the splines of cross shaft (sector shaft or pitman arm shaft or pitman shaft) and held on sector shaft by a nut. The other end has a taper hole in which ball end is held tight with the help of nut.

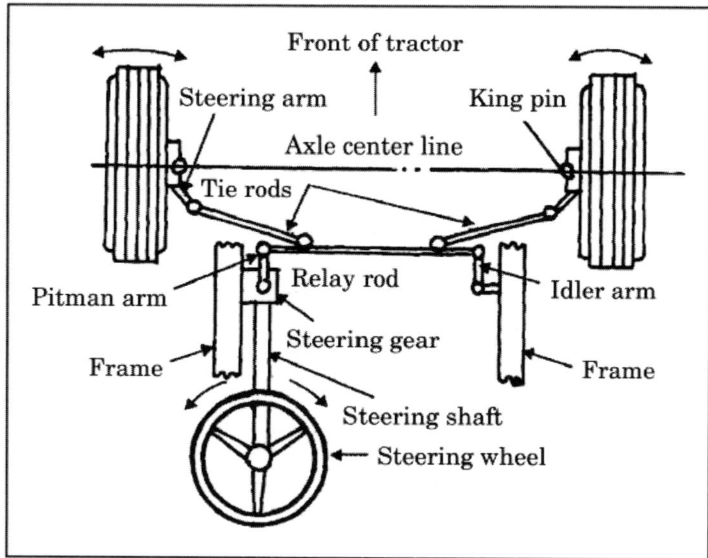

Fig. 17.1: A simplified Pitman Arm Steering System.

Steering box: The function of the steering box is to convert the rotary motion into to and fro motion of drop arm. The drag link tied up with drop arm can be pushed or pulled resulting into moving stub axle to right or left as desired by the driver. The steering gears are enclosed in this box and hence this is also called as the steering gear box.

Steering ratio (steering gear ratio)

If the steering wheel is connected directly to the steering linkage, it would require a great effort to move the front wheels. Therefore to assist the driver, a reduction system is used having a movement ratio between 10 : 1 to 22 : 1. The steering ratio is the angular ratio between the rotation of the steering wheel and the steering arm. If by turning the steering wheel through 360°, the angular movement of the steering arms is 36°, then the steering ratio would be 360°: 36° (10:1). The steering ratio is generally low for small, light vehicles and high for heavy vehicles. The low steering gear ratio produce fast steering while high ratios produce slow steering. Slow steering means the steering wheel has to be turned many degrees to produce a small steering effect. Low steering ratios is also called fast or quick steering which requires much less steering wheel movement to produce the desired steering effect.

Turning radius: Turning radius is the radius of the circle on which the outside front wheel moves when the front wheels are turned to their extreme outer position. The turning radius is usually proportional to the wheel base of the vehicle. This radius usually varies from 5 to 7.5 m for buses, trucks etc.

Steering geometry: While a vehicle is moving along a curve, all its wheels should roll truly without any lateral slip. For this, the axles of all four wheels should intersect at one point. This point will be the centre about which the vehicle will be turning at that instant. This arrangement is termed as the steering geometry of the wheels.

Referring to Fig. 17.2, the rear wheels rotate along two circles The centers of both these circles are at (0). The front wheels (1) and (2) have different axles. They rotate along two other circles with the same centre point. For the correct functioning of any steering system, the centers of the wheels of the rear axle and of wheels (1) and (2) should coincide.

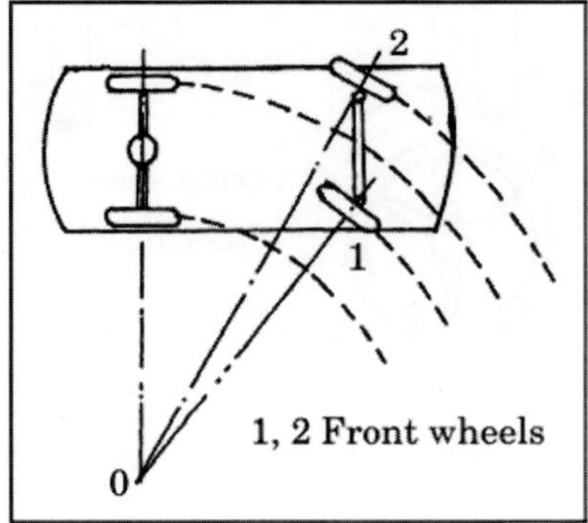

Fig. 17.2: Steering Geometry.

17.5 TYPES OF STEERING GEAR BOX

The steering box enables the driver to exert a large force on the road wheel with minimum effort applied at the steering wheel. It also changes the rotary motion of the steering wheel into lateral movement of the tie rod. The various types of gears normally available on different make of tractors are

1. Worm and sector steering gear
2. Worm and ball bearing nut steering gear
3. Worm and roller steering gear
4. Cam and lever steering gear

Worm and sector steering gear

Essentially the steering gearbox makes connection between the steering shaft and the steering linkage. This type of gearbox consists of a steering tube fitted with a worm at one end. When the steering wheel is rotated, the worm meshes with what is called a sector. The worm thus rotates the sector by a certain angle. The sector mounted on the cross shaft rotates it which is connected to the drop arm. Its motion in transmitted to the wheel through the linkage (Fig. 17.3).

Worm and ball bearing nut steering gear

In this steering box, when the steering wheel is rotated, the steering column rotates. The worm is an integral part of the steering column. Therefore, when the steering column is rotated, the worm also rotates. The worm is in mesh with a nut arrangement. Thus when the worm rotates, the nut is able to move. This movement takes place along the axis of the column either up and down. This enables the cross shaft to rotate in an arc. This in turn helps the drop arm to move also in an arc. This arm transmits the motion to the linkage (Fig. 17.4).

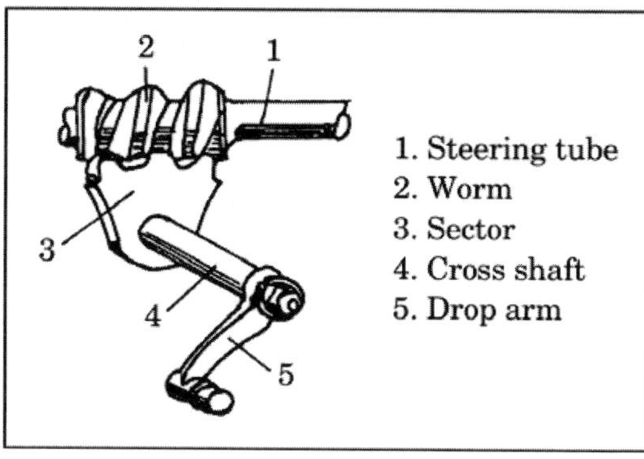

Fig. 17.3: Worm and Sector Type.

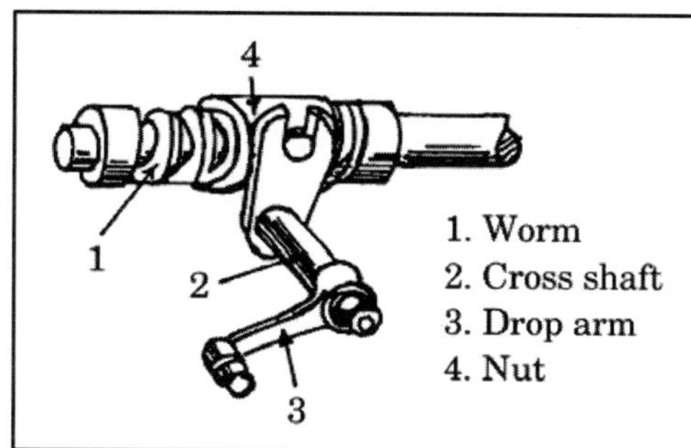

Fig. 17.4: Worm and Nut.

Worm and roller steering gear: In this steering box, the worm is at the end of the steering column. The diameter of the worm is more at its end. Its diameter is gradually reduced at the centre. A roller is in mesh with this worm. The roller is connected to the cross shaft and cross shaft is connected to the drop arm. When the steering wheel is rotated, the column also rotates. Also when the worm of the shaft is rotated, the roller rotates in an arc, which is turn moves the linkage (Fig. 17.5).

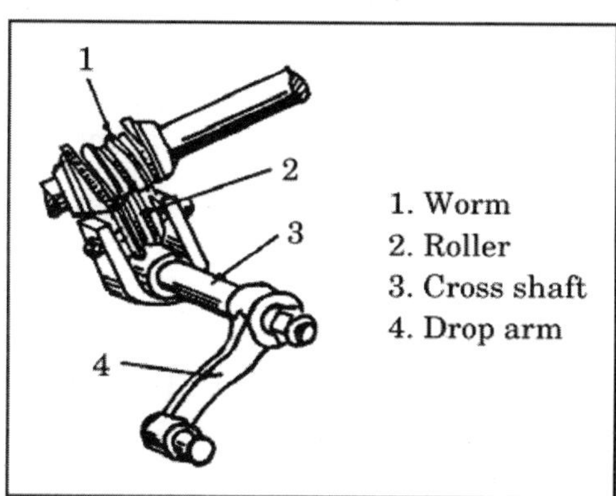

Fig. 17.5: Worm and Roller

Cam and lever steering gear: The International B-275 tractor is fitted with this type of steering system. In this system, the worm is shaped in the form of a cylindrical cam. The cam groove is tapered, being narrower at the bottom. The cam is supported on ball bearings which take up the end thrust. The lever is connected on the grooves of the cam. The lever is then fitted to the cross shaft and drop arm.

When the steering wheel is rotated, the steering column rotates. The cam moves the lever to and fro, causing the cross shaft to also rotate. This in turn causes the drop arm to rotate in an arc. This is how the cam and lever steering gear functions (Fig. 17.6).

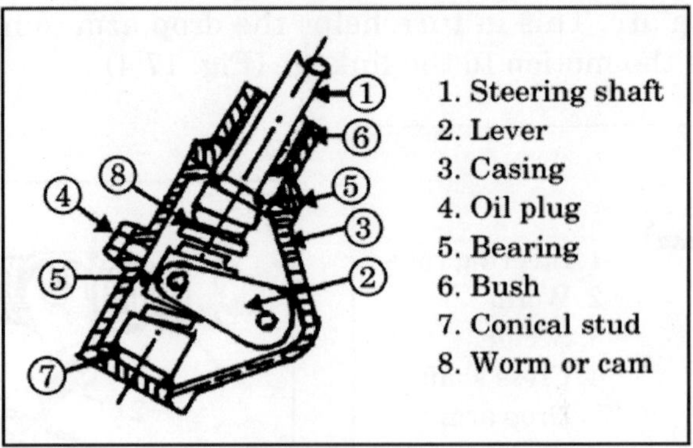

Fig. 17.6: Cam and Lever.

17.6 WHEEL ALIGNMENT

Wheel alignment refers to the positioning of the front wheels and steering mechanism that gives the vehicle directional stability, promotes ease of steering and reduces tyre wear to a minimum. The front wheel alignment depends upon the following geometrical characteristics:

(i) Camber
(ii) Castor
(iii) King pin inclination
(iv) Toe-in and
(v) Toe-out during turns.

Camber: The angle between the centre line of the tyre and the vertical line when viewed from the front of the vehicle is known as camber (Fig. 17.7).

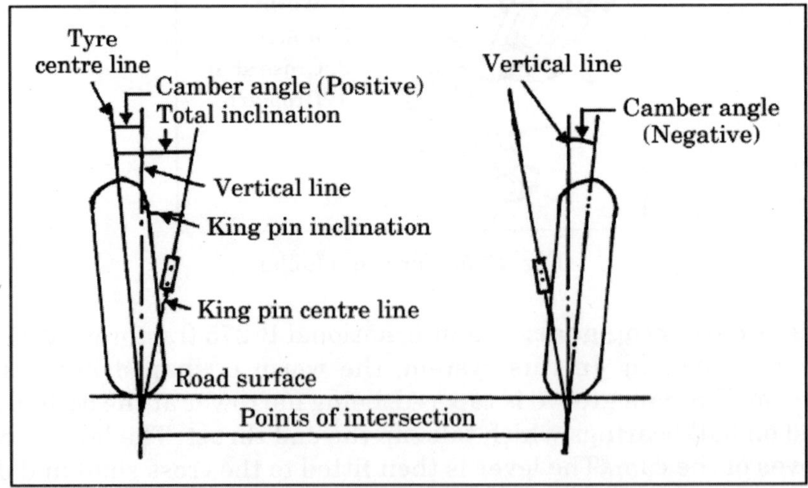

Fig. 17.7: Camber Angle (Positive and Negative) and King Pin Inclination.

When the angle is outward, so that the wheels are farther apart at the top than at the bottom, the camber is positive. When the angle is inward, so that the wheels are closer together at the top than at the bottom, the camber is negative. The camber angle is kept equal on both the front wheels. Camber should not exceed 2°. The effects of wheel camber are summarized below.

(i) Bending stress in the king pin and stub axle are reduced
(ii) Steering effort is greatly reduced
(iii) Shock loads are not transmitted to the steering wheel at high vehicle sped.

King-pin inclination: The angle between the central line of the king pin when viewed from the front of the vehicle and vertical line is called the king-pin inclination (Fig. 17.7). It generally ranges from 4 to 8°. It must be equal on both the sides. If it is greater on one side than the other, the vehicle will tend to pull to the side having the greater angle. Also if the angle is too large, the steering will become exceedingly difficult. The advantages of king-pin inclination are:

(i) It helps provide steering stability by returning the wheels to a straight ahead position after the vehicle has turned. This is called returnability.
(ii) It reduces steering effort, particularly when the car is stationary
(iii) It helps in reducing the wear on the tyre.

Included angle: The combined and king-pin inclination is called the included angle. The angle is important because it determines the point of intersection of the wheel and king-pin centre lines. This in turn determines whether the wheel will tend to toe-out or toe-in. If the point of intersections is above the ground, the wheel tends to toe-in. If it is below the ground, the wheel tends to toe-out. If it is at the ground, the wheel keeps its straight position without any tendency to toe-in or toe-out. In this position, the steering is called centre point steering (Fig. 17.7 and 17.8).

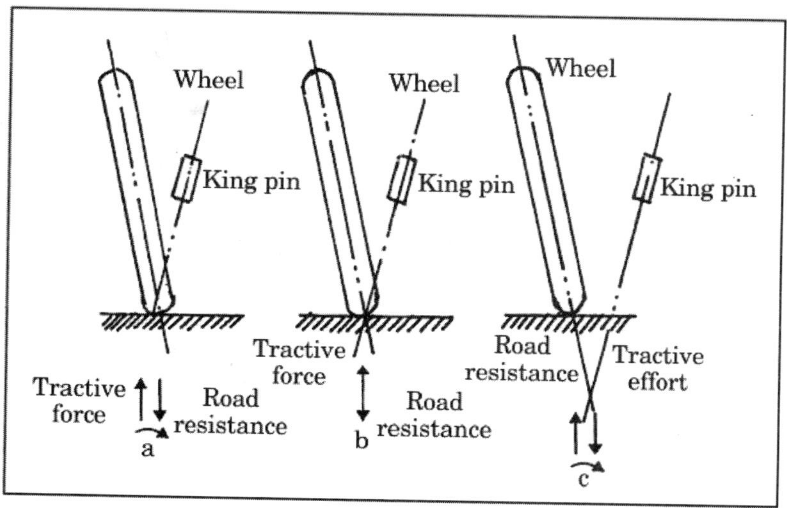

Fig. 17.8: Effect of Included Angle.

Caster: In addition to being tilted inward toward the centre of the vehicle, the king pin axis may also be tilted forward or backward from the vertical line. This tilt is known as caster. Thus, the angle between the vertical line and the king-pin centre line in the plane of wheel (when viewed from the side) is called the caster angle.

The angle is positive when the king-pin axis tilts backward and negative when the king-pin axis tilts forward (Fig. 17.9). The caster measured in degrees is kept between 2° and 7°. There are three reasons why caster is used.

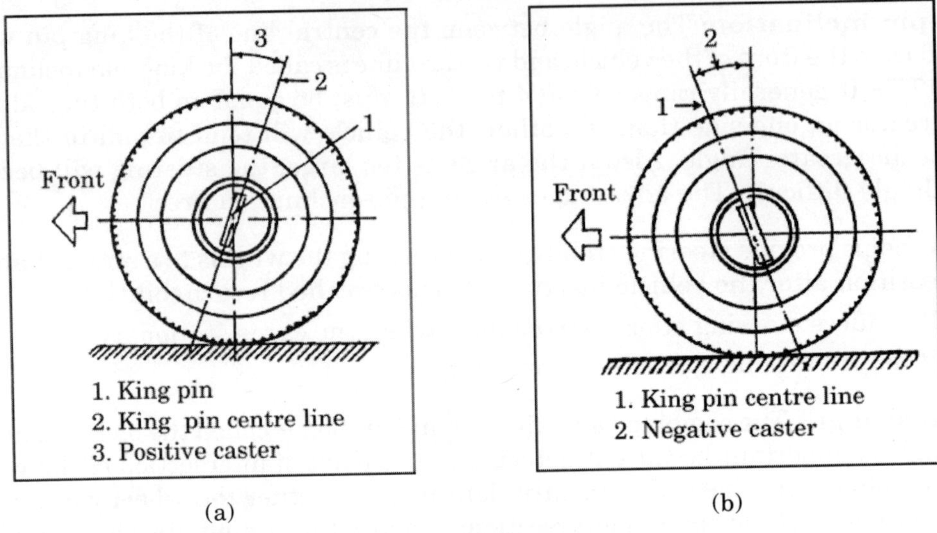

Fig. 17.9. (a) Positive Caster. (b) Negative Caster

(i) To maintain directional stability and control
(ii) To increase steering returnability
(iii) To reduce steering effort.

Toe-in and Toe-out: Toe is the amount in inches or millimetres by which the front wheels point inward or outward. Its purpose is to ensure parallel rolling of the front wheels, to stabilize steering, to prevent side slipping and excessive wear of the tyres. With toe-in, the front wheels attempt to roll inward instead of straight ahead. This is because the tyres are slightly closer together at the front than at the rear.

The distance at the front end between the front wheels is less than the distance between them at the rear, is called toe in (Fig. 17.10a), When the distance between the front wheels at the front end is greater than the distance between them at rear, then the vehicle position is called toe-out position (Fig. 17.10b).

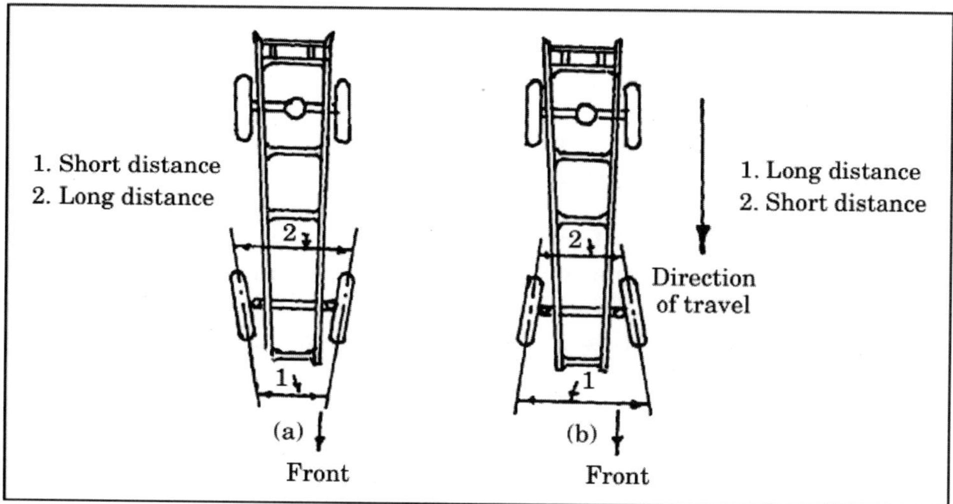

Fig. 17.10. TOE-IN and TOE-OUT

17.7 STEERING MECHANISMS

Fundamental equation for correct steering

When the vehicle takes a turn, the outer wheels move faster than the inner wheels. The four wheels must roll on the road so that there is a line contact between the road surface and tyres. This is essential to prevent tyre wear. The rolling motion of the wheels on the road surface is possible only if these describe concentric circles on the road at an instantaneous centre, when the vehicle is taking a turn. In order for turning the vehicle to the left or right, its two front wheels are mounted on short axles, known as stub axles, pivoted to the axle beam. The pivots are known as king pins. The axis of these axles, when produced meet at an instantaneous centre which lies on the common axis of the rear wheels. The axis of the inner wheel makes a larger turning angle *0* than the angle θ made by the axis of outer wheel.

Let a = CD, wheel track ; b = AB, distance between the points of front axles: l = AE, wheel base; I = Common instantaneous centre of all four wheels.

Draw IP, perpendicular from I to AB produced meeting at *P*,

Then, $b = \text{AP} - \text{BP} = l \cot \phi - l \cot \theta = l (\cot \phi - \cot \theta)$ or $(\cot \phi - \cot \theta) = \frac{b}{l}$.

The above equation is the fundamental equation for correct steering. If this equation is satisfied, there will not be any lateral slip of the wheels when the vehicle is taking a turn. The mechanisms used for automatically adjusting the values of ϕ and θ for correct steering are known as steering gear mechanism.

Types of steering mechanism

There are two types of steering gear mechanisms:

1. Davis steering gear mechanism 2. Ackermann steering gear mechanism. The main difference between the two steering gear mechanisms is that the Davis steering gear has sliding pairs, whereas the Ackermann steering gear has only turning pairs. The sliding pair has more friction than the turning pair, therefore, the Davis steering gear will wear out easier and become inaccurate after certain time. Hence Davis steering mechanism is no more used in vehicles. The Ackermann steering gear is much simpler than Davis and consists of a four-bar chain having turning pairs only.

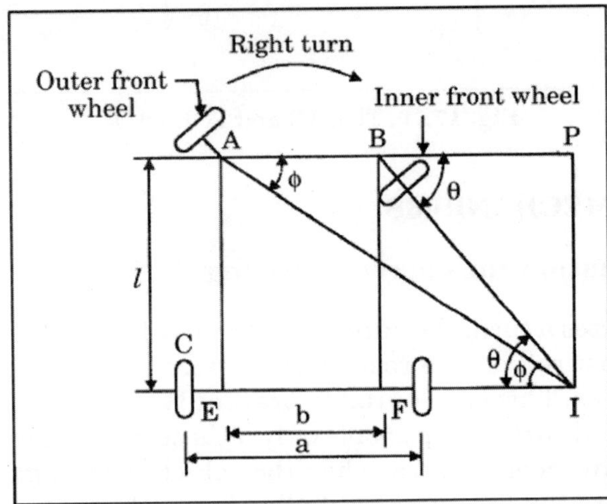

Fig. 17.11: Steering Gear.

In case of Ackermann steering gear mechanism, the correct steering can be obtained of the relation $\cot\phi - \cot\theta = \frac{b}{l}$; as derived earlier.

It will be seen that for a given configuration, the difference between cot ϕ and cot θ for each set of corresponding values of θ and ϕ is equal to the ratio of distance between king pins of front wheel (b) to the distance of point of intersection of front wheel axes from AB (l). The advantages of such a mechanism are: (i) lesser wear of tyres (ii) lower friction in pairs (iii) simplicity and durability of pin joints.

17.8 FRONT AXLE

The front axle (Fig. 17.12) is used to carry the weight of the front part of the vehicle *as* well as to facilitate steering and absorb shocks due to road surface variations. It must be rigid and robust in construction. It is made of I-section steel, pivoted at the centre. The pivot pin is housed on bushings on a support bolted to the front of the engine. As the track width of a tractor is adjustable, the front axle has provisions to be extended on both sides. The I-section type front axle has three-piece axle. Two side pieces are bolted to the middle piece with the help of two bolts through a set of holes to get desired track width. Both the outer ends of the axle have provisions for holding king pin. These king pins are supported in the axle housing with the help of metallic or nylon bushes. A thrust bearing is normally placed on the lower side of the king pin.

Fig. 17.12: Front Axle.

Chapter 18

Wheels and Tyres

18.1 INTRODUCTION

The power developed by the engine has been transferred through the clutch, gear box and differential to the rear axles. Wheels are connected to the rear axle. As the rear axle turns, wheels also turn and the vehicle moves on the road. The ultimate purpose of the power developed by the engine is to turn the wheels so that vehicle moves on the road. Modern wheeled tractors use pneumatic-tyred disc wheels. As a result of the driving wheel tyres gripping the soil, the rotary motion of the wheels is transformed into the translational motion of the tractor or vehicle.

According to the purpose, wheels are classified as driving and driven steerable types. Trucks and the general purpose wheeled tractors have all their wheels of one and the same size. Row-crop tractors have their rear wheels larger than the front wheels. The rear wheels carry the major portion (up to 70 %) of the load due to the weight of the tractor, which provides for better traction. The front wheels are loaded lighter and this makes them easier to turn and provide good directional steering stability. In this chapter, we will discuss about the wheels, tyres and tubes.

18.2 WHEEL ASSEMBLY

The wheel assembly is generally thought to consist of hub, disc or spokes, rim, tyre and tube. Wheels are as important part of a vehicle as the other parts. All the parts being perfectly in working order, the vehicle cannot move on the road without the wheels. The wheels not only support the weight of the vehicle, but also protect it from the road shocks. The rear wheel moves the vehicle and the front wheels steer it. All the four wheels must resist the braking stresses and withstand side thrust. To perform the above functions, the wheels must be (i) strong enough to withstand the weight of the vehicle; (ii) flexible to absorb the road shocks; (iii) able to grip the road surface (iv) perfectly balanced dynamically and statically (v) light enough to enable the tyre to follow the contour of the road.

18.3 TYPE OF WHEELS

The wheels are usually of disc type and wire type.

18.3.1 Disc wheel

This type of wheel consists of a steel rim and a pressed steel disc. The rim is a rolled section, sometimes riveted but usually welded to the flange of the disc. The disc performs the function of spokes. The wheel assembly is bolted to the brake drum. The hub cap or cover is usually held in position by spring clips attached to the disc. The disc is often perforated with slots near the rim, which act as fan to blow air on the brakes. A hole in the rim, serves to accommodate tube valve. Figure 18.1 shows a sectional view of a disc wheel. This type of wheel is simple, cheap and robust in construction. It is most commonly used in heavy motor vehicles - Cars, buses, trucks, tractors etc.

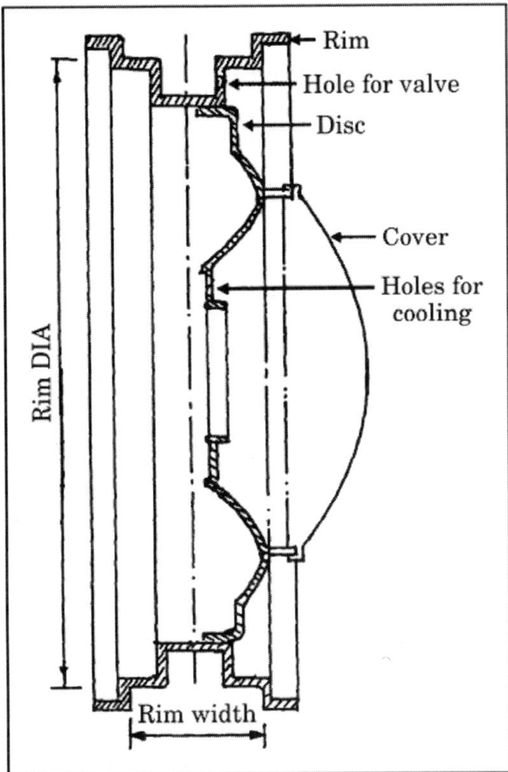

Fig. 18.1: Disc Wheel.

18.3.2 Wire wheel

This type of wheel consists of a separate hub connected to the rim with a number of wire of spokes. The headed inner ends of the spokes fit in the hub holes and the threaded outer ends fit in the rim holes, where tubular nuts are screwed through the rim holes to tighten the spokes. All the spokes must be of correct length and at correct tension to hold the rim, centrally around the hub. The spokes do not stick straight out as radii from the hub, but alternate spokes are screwed to slope forwards and backwards towards the rim. This arrangement of spokes serve special purpose of the wheel. The forward sloping spokes absorb braking torque and the rear ward sloping spokes convey driving torque. Figure 18.2 shows a section and arrangement of spokes of a wire wheel.

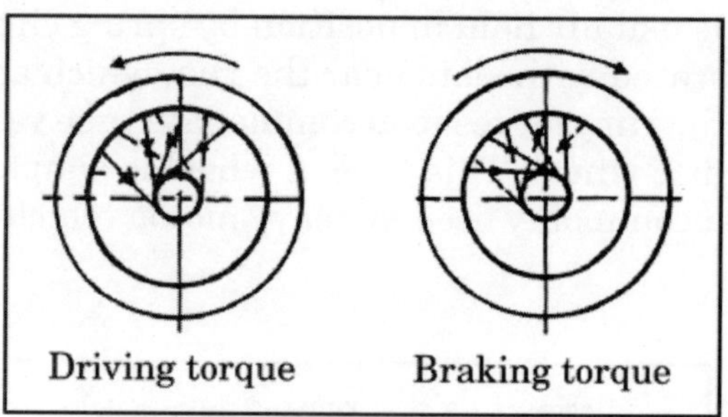

Fig. 18.2: Arrangement of Spokes.

The holes in the hub are arranged in inner and outer rows so that one set of spokes slope towards to the rim from the outer row of the hub and the other set slopes outwards to the rim from the inner row of the hub. These sideways inclinations of the spokes hold the wheel upright against cornering loads and side thrusts. A rubber chaffing band is fitted in the wall of the rim to keep tube touching the spoke nuts. The tension of the spokes may be tested, when the wheel is free of load, by tapping them, which should produce an equal ring from each spoke. The wire wheels allow free circulation of air around the brake drum. They are lighter in weight than other types of wheels and used in light motor vehicles-racing cars, scooters, motor cycles etc.

18.4 CONSTRUCTION OF WHEELS OF A ROW-CROP TRACTOR

Figure 18.3 shows the wheel of a row-crop tractor. The driving and driven or steerable wheels have been shown in the figure. Each wheel comprises hub, disc with rim and tyre with inner tube. The rim is welded to the disc and the disc is bolted to the hub. The driving wheel tyres are of low pressure type and have heavy tread bars for better traction. The driving wheel hub is keyed to axle shaft and is fixed in place by means of bolted-on insert with worm whose threads mesh with the rack teeth cut in the half axle. By turning the worm, one can change the position of the wheel on the axle shaft to obtain the desired track width.

The hub of the front steerable wheel is supported by two taper roller bearings mounted on the steering knuckle spindle and fixed in place by adjusting nut which is also used for adjusting the preload on the bearings. The bearings are lubricated with grease put in the hub. In service, grease is injected into the hub through a lubricating nipple. The front wheel pneumatic tyre consists of a casing and an inner tube. For better wheel grip, the tyre is provided with grousers.

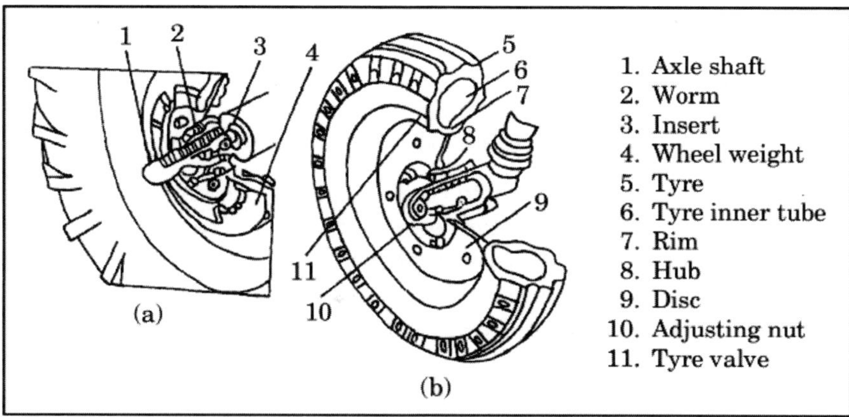

Fig. 18.3: Row Crop Tractor Wheels.

18.5 CONSTRUCTION OF WHEELS OF A GENERAL-PURPOSE TRACTOR

Figure 18.4 shows the wheel of a general-purpose tractor. General-purpose tractors use interchangeable single wheels with low-pressure tyres. Each wheel includes tyre casing, tyre inner tube and disc that is mounted on the eight wheel studs' of the wheel-hub reduction gear. The track width is more when the wheels are installed with their tyre valves facing inwards and less when they are installed with the valves facing outwards. To change over from the wide to the narrow track or vice versa, the wheels are switched from one side of the tractor to the other. The wheel tyres are provided with a ground grip tread pattern.

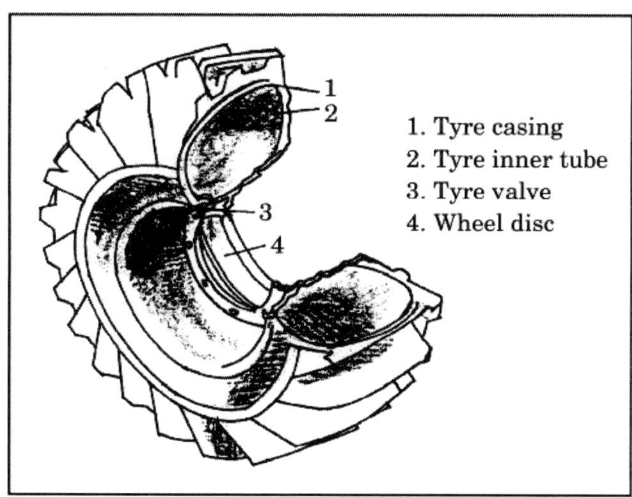

Fig. 18.4: General Purpose Tractor Wheel.

18.6 PURPOSE OF TYRES

The tyre is mounted on the wheel rim. The tyres are the final contact point between the road and the vehicle. They take all the load of the vehicle. They are flexible.

They absorb most of the shocks when the tractor is moving on an undulated soil surface and on rough roads.

The surface of the tyres has certain patterns. These enable the tyres to grip the road and provide good traction. Hence, the functions of tyres are as follows:

(i) It carries the load of the vehicle (ii) It absorbs the small road shocks (iii) It also damps down the vibration to some extent, (iv) It transmits power from the engine and transmission to the ground with which the tractor moves, (v) It also absorbs the retarding force when tractor is braked and keeps on skidding on application of brakes till the tractor comes to stop (vi) The treads made on the tyres grip the field (soil) for better traction. That is why, different types of treads are recommended for different soil conditions.

18.7 TYPES OF TYRES

Pneumatic tyres are of two types, (i) Conventional tubed tyre and, *(ii)* Tubeless tyre.

18.7.1 Tube tyre

It is a traditional tyre. It encloses a tube in which air is forced to a high pressure as a cushioning medium. The outer portion of the tyre which rolls on the road is made of synthetic rubber and is called tread. Figure 18.5 shows the cross-sectional view of such a tyre. The conventional tubed tyre basically consists of (i) a casing or carcass, *(ii)* tread. The carcass is made of four to six tyres of fabric; the layers are called plies. Each layer is a sheet wound of cotton or nylon impregnated with rubber. Greater strength is obtained if each layer or sheet is placed at an angle to the adjacent layer. Cords of rayon or nylon offer greater resistance to the heat produced due to flexing of tyres. The carcass or casing, is the basic structure of the tyre which bears the shock, loads, side thrusts and the vehicle's weight. The number of layers of cord depends upon tyre usage. Motor car tyres have 4 to 6 plies. Heavy duty bus or truck tyre have up to 22 plies where as tractors and earth-movers have up to 30 plies.

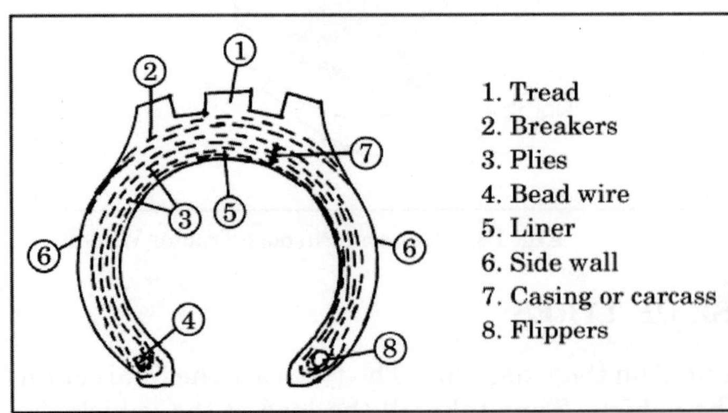

Fig. 18.5: Conventional Tyre.

Tread: The portion of the tyre that comes in contact with the road surface is known as the tread. The primary function of a tread is to improve the tyre's ability to transmit driving and braking torques. The tread is bonded directly on the casing. The materials used are natural or synthetic rubber, the former helps the tyre to remain cool during running.

Tread design: The tread pattern helps the tyre in many ways such as positive traction, firm grip, cooler running; smoother rolling, even wear; steady on wet roads, better grip on loose soil

There are zigzag patterns at the centre which throw out the water when running on wet roads. Tractors have wide and deep grooves on the shoulder of the tread to throw away the loose upper soil and stones and finally to get a firm grip on a firmer surface. Different tyre tread pattern are shown in Fig. 18.6. Pattern (a) provides good midways adhesion and good grip , pattern (b) for little sideways adhesion but good fore and aft grip, pattern (c) for good sideways, fore and aft grip but irregular wear on hard roads, pattern (d) for good wear resistance and steering characteristics and pattern (e) for rough and loose surface, giving maximum grip and sideways stability.

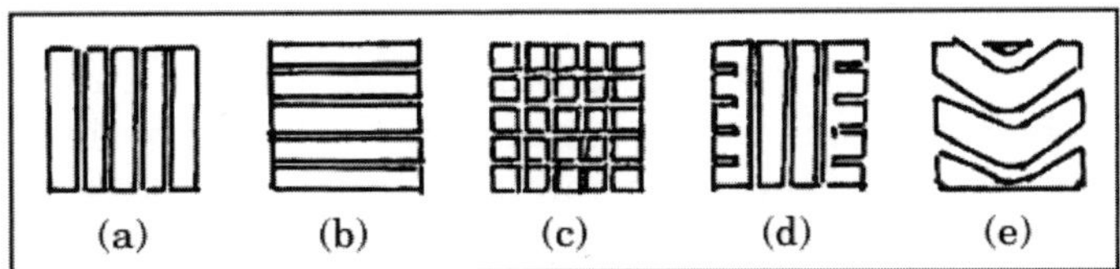

Fig. 18.6: Tyre Tread Patterns.

Breakers: The top two plies of the tyre are known as breakers. These are widely spaced to help in spreading the shocks from the road.

Sidewalls: These protect the casing but are subjected to maximum flexing action, creating a large amount of heat build up. Sidewalls are made of a rubber compound.

Beads: A number of coils of bronze-plated steel wire which retain this tyre on the rim are called beads. The plies of cord are fastened to them and therefore, serve as a metal foundation of the tyre.

Flippers: A number of layers of additional cords are wound around bead wires for extra strength and to avoid stress concentration. The flippers are wound upto a small distance on the sidewalls.

18.7.2 Tubeless tyre

Tubeless tyre does not enclose the tube . The air under pressure is filled in the tyre itself. The inner construction of this tyre is almost the same as that of the tube tyre. The material and the design of the carcass and the tread remain similar to the tubed type of tyre. The inside of the casing is lined with a soft rubber lining which forms an air-tight seal with the rim. A non-return valve is fitted to the rim through which the air is forced inside the tyre. Figure 18.7 shows the sectional view of a tubeless tyre.

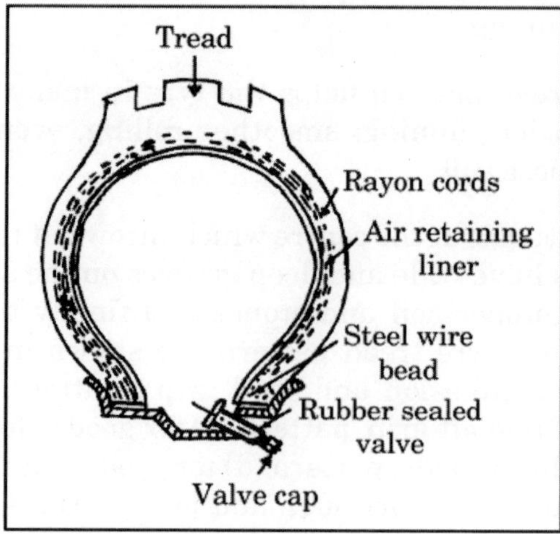

Fig. 18.7: Tubeless Tyre.

The tubeless tyres are lighter and run cooler than a tube tyre. The main advantage of a tubeless tyre is that it retains air for a long period even after being punctured by nails, provided the nail remains in the tyre. But the tube tyre releases the air almost immediately after being punctured. Also any hole in the tubeless tyre can be repaired simply by rubber plugging. Ordinary punctures can be repaired without removing the tyre from the wheel.

The tubeless and tube tyres are both called pneumatic tyre in which air is used as a cushioning medium. In almost all tractors, tube tyre is used. The tubeless tyre's use is mostly limited to the children's tricycles. Tubeless tyre is mounted on the wheel rim so that air is retained between the rim and the tyre.

18.8 TYRE SIZE AND PLY RATING

Tyre size is marked on the sidewall of the tyre. Every tyre is marked with its size. Let us see what is the meaning of this marking. If a tyre is marking with 8.25 × 20 × 10 PR (ply rating), then it means that the first number *i.e.*, 8.25 inch represents the sectional width (the width or thickness of tyre from shoulder to shoulder when tyre is inflated to the correct pressure. The second number *i.e.*, 20 inch represents the diameter of the rim and the third number represents ply rating 10 PR means the tyre consists of 10 plies. The ply rating gives an idea of tyre strength factor used to identify a particular size of tyre with its load carrying capacity.

A tyre of any specific size can be available in more than one ply rating. Hence, for any specific requirement, the ply rating must be regarded as an integral part of the tyre specification. The dimensions of a tyre have been shown in Fig. 18.8.

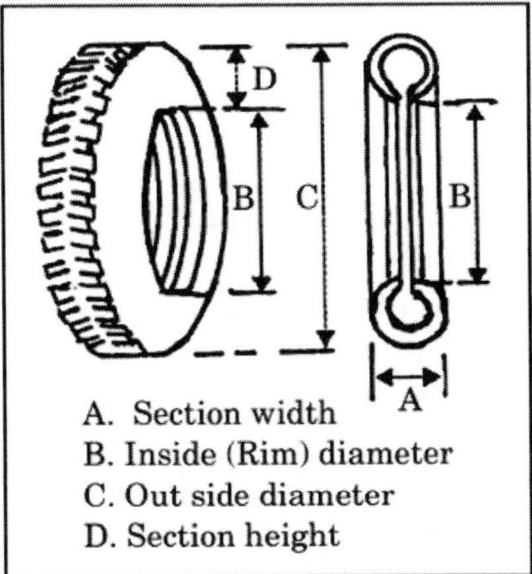

Fig, 18.8: Tyre Dimensions

18.8.1 Aspect ratio

The ratio between the tyre section height (distance from tread to rim) and the tyre section width (when the inflated to correct pressure) is called the aspect ratio. This is expressed in percentage. Now-a-days tyres are being made with lower aspect ratios. The advantages of lower aspect ratio are better performance at high speeds, lesser wear, lower slip angle, better load carrying capacity etc.

18.9 TYRE MATERIALS

Natural rubber was used to impregnate the carcass or casing cotton cords. This process continued for many years. Later, natural rubber was replaced by styrene butadiene rubber (SBR). SBR is used to impregnate the nylon cords resulting in a tread which gives better road grip on wet roads, better abrasion resistance and smoother ride. Polybutadiene (PB) mixed with SBR and other additives like carbon black, oil and sulphur imparts anti-wear, anti-skid and anti-heat properties.

The cords can also be of glass fiber which gives extra strength and elastic properties. But a special process is required to give it a satisfactory bond with synthetic rubber.

18.10 WHEEL BALLASTING

To improve traction, wheel weights are used on the driving wheels. This is known as ballasting and can be achieved by any of the following methods.

(i) Bolting cast iron weights to the wheel

(ii) Use of liquid in the inner tube.

In addition to improved traction, ballasting helps in (i) reducing the slippage (ii) increasing the drawbar pull (iii) slower tread wear (iv) less fuel consumption. For road operations and light work, such as cultivating, harrowing, planting etc., ballasting is not required.

Adding weights to wheel

Weights are attached to the rim by extension bolts. This method of ballasting has an advantage over water ballasting in that these can be easily removed when not required. The maximum amount of weight a tyre can carry depends upon the tyre size, ply rating and maximum pressure. Hence, it is always recommended to follow the operational manual to decide the weights.

Front wheels are also sometimes counter balanced with weights to increase the steerability when heavy load are superimposed on the drawbar or heavy equipment is to be mounted on the rear end of the tractor.

Use Of liquid in inner tube: The inner tubes of the driving wheel tyres are filled $^3/4$ th full with water. Except in areas where water-freezing possibilities exist, clean water is used for this purpose. In areas where water freezing can take place, a 25 % solution of calcium chloride, whose freezing point is around – 30°C is used instead of water. After filling the inner tubes with water or solution, the tyres are inflated with air to their normal inflation pressure. When the tractor is used for transport purposes, the wheel weights should be removed and the liquid is drained from the rear-wheel tyres.

18.11 FACTORS AFFECTING TYRE LIFE

The main factors are

(i) Inflation pressure
(ii) Wheel alignment
(iii) Driving manners; these include sudden acceleration, high speed, sudden braking, driving on bad roads etc.
(iv) Tyre maintenance; this includes tyre rotation at fixed intervals (in km) and checking of tyre and wheel balance.

18.11.1 Inflation pressure

Inflation pressure depends upon tyre size, tyre type, speed and load. The inflation pressures are recommended by the vehicle manufacturer. It is always recommended to maintain the correct tyre pressure. Under inflation causes the following defects:

(i) Uneven tread wear, more wear at the tyre sides
(ii) Lack of directional stability
(iii) Increased rolling resistance leading to increased fuel consumption
(iv) Excessive flexing of side walls causing excessive buildup.

Over inflation causes the following defects:

(i) Reduced tread contact area with the road surface. This results in rapid wear in the tread at the centre.

(ii) Reduced tyre grip

(iii) Reduced impact resistance

(iv) Increased vibrations resulting in uncomfortable ride.

(v) Increased stresses may cause tread separation and cracks in the side walls.

Different stages of tyre inflation with regards to tread contact with the road have been shown in Fig. 18.9.

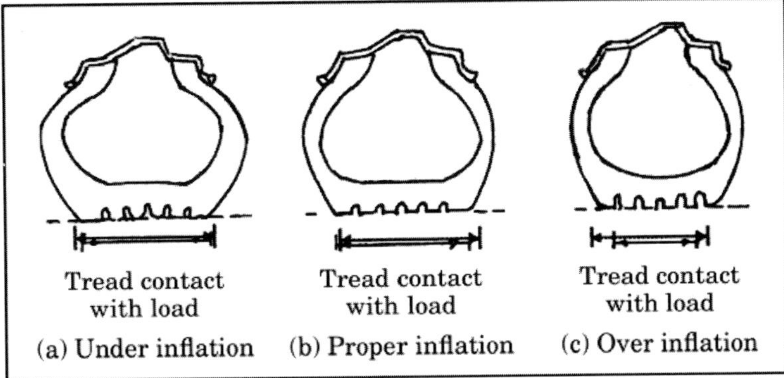

Fig. 18.9: Stages of Tyre Inflation.

18.11.2 Wheel alignment

This refers to proper steering geometry i.e., castor and camber angle, king pin inclination angle and toe-in and toe-out distances. If proper angles and distance are not maintained, the wheels skid along the road instead of rolling freely. Improper comber angle in the front wheels causes the tread to wear on one side. Improper castor causes tyre wobbling.

18.11.3 The manner of driving

This refers to driving at high speeds, constantly braking and accelerating the tractor. The tread wear increases rapidly due to increased temperatures. Tyre slip is increased while taking bends. Road irregularities produce high bumps causing scuffs in tyres.

18.11.4 Proper maintenance of tyres

Weekly checking to ensure uniform tyre wear, tyre change after every 500 km interval in case of tractor and removal of any foreign materials such as stones are recommended. Tyre pressure and size of some tractors have been given in Table 18.1

Table 18.1: Tyre size and pressure of some tractors

Sl. no.	*Size* *Tractor front tyre*	*Ply rating*	*Tyre pressure* *Pound/ inch²*	*Kg/cm²*
1.	5.50 × 16 (Swaraj)	4	32	2.25
2.	5.50 × 16 (Zeter 3011}	6	52	3.6
3.	6.00 × 16 (Zetor 4011)	4	32	2.25
4.	6.00 × 19 (275 international)	6	48	3.4
5.	6.50 × 20 (Byelarus)	6	44	3.1
Power tiller (max. speed 16 kmph)				
1.	6.00 × 12	4	24	1.7
Tractor rear tyre (max. speed 32 kmph)				
1.	11.2 × 28 (Swaraj)	4	16	1.1
2.	12.4 × 28 (Zetor 3011)	4	14	1.0
3.	13.6 × 28 (Zetor 4011)	6	20	1.4
4.	14 × 28 (275, International)	6	18	1.25
5.	8 × 32 (Byelarus)	6	32	2.25

Chapter 19

Brakes

The tractor has been started, accelerated and is running on the road. It is now to be stopped. Stopping of the vehicle or tractor is as necessary as its starting. Once the vehicle is started, it must be stopped somewhere. Brake action may be defined as that force which stops any motion. Brakes are applied to slow down or stop its motion.

19.1 PRINCIPLE OF BRAKING

A moving vehicle possesses kinetic energy which is converted into heat energy on the application of brakes. This heat is transferred to the surrounding air. In the simplest form, a brake comprises a stationary brake shoe with a friction lining on it and a brake drum. The wheel is fixed to the rotating brake drum. The driver applies force on the brake pedal which gets amplified and pushes the stationary shoe to make contact with the brake drum and stops its rotation due to frictional resistance. The heat generated due to braking action is proportional to the force which brings the shoe in contact with the drum.

19.2 REQUIREMENTS OF BRAKE

Brakes, in general, are required to slow, stop or hold the tractor and convert the kinetic energy of motion into heat, and then to dissipate this heat. The requirements of a brake in the vehicle are as follows :

(i) Application of brake should bring the vehicle or tractor to a relatively quick stop on any type of road - wet, dry, even, uneven, uphill or downhill.

(ii) The braking components must require minimum maintenance.

(iii) The braking pedal efforts required to produce maximum deceleration should be negligible and should not vary with the condition of road.

(iv) The braking components should allow minimum time between application of pedal effort and actual braking effect on the drums.

(v) The braking action should not involve any noise or drift the vehicle away from its desired path.

(vi) Provisions for quick heat dissipation must be incorporated.

(vii) A secondary braking component must be incorporated if primary braking component fails.

19.3 CLASSIFICATION OF BRAKE

Broadly, brakes are classified as, *(i)* Drum brakes and *(ii)* Disc brakes. The operating systems for such brakes in present tractors available in our country are restricted to *(i)* Mechanical type *(ii)* hydraulic type. The drum brakes under mechanical type are of two types (a) internal-expanding shoe brakes (b) external contracting shoe brakes.

19.4 PURPOSE OF MECHANICAL BRAKE SYSTEM

Suppose tractor is moving and it is to be stopped for some reason. For this, brakes should be applied to all four wheels. This is achieved by the mechanical brake system. There are cams and brake shoes on the four wheels. Cams are connected to the brake pedal by means of levers and flexible cables. When the brake pedal is pressed, the force is transmitted to the cams by means of flexible cables. The cables operate the cams and in turn brake shoes are operated. All the wheels are thus braked. In this system, levers, cables and cams are provided for the purpose of braking. It is called the mechanical braking system as no fluid is used.

19.5 CONSTRUCTION ARID OPERATION OF MECHANICAL BRAKE SYSTEM (INTERNAL EXPANDING SHOE BRAKE)

MF-1035, Ford, Escort 335 types of tractor use mechanical internal expanding shoe brake. Figure 19.1 shows a fixed brake plate attached to the axle housing. A fulcrum is shown in this fixed plate at the bottom. Two brake shoes are fixed to it. These shoes are lined on the outer side with asbestos or fiber material. A revolving cam in fixed to the top of the brake plate. When the cam rotates, these two brake shoes expand. A spring connects both the brake shoes and brings them closer.

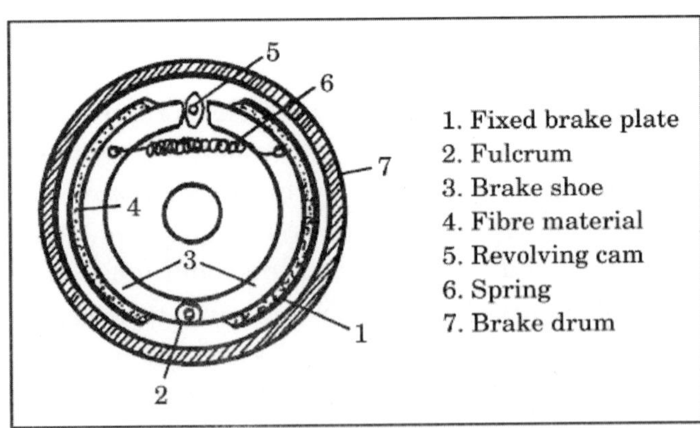

Fig. 19.1: Operation of Brake Shoes.

The cam is shown linked by means of a camshaft and a lever as shown in Fig. 19.2. The lever is operated with a rod by means of a pedal. When the pedal is pressed, the cam rotates by a slight amount because of the links. This pushes the ends of the brake shoes outward. These brake shoes press against the inner portion of the brake drum. This stops the brake drum. Thus the rotating wheel is fully stopped.

When the brake pedal is released, the spring brings both the shoes closer. The pressure on the inner portion of the brake drum is removed. The wheel is thus relieved of the grip of the brake shoes. This is how a mechanical brake operates.

In Fig. 19.2, there is one fulcrum for each brake shoe. The mechanical brake is linked to all the four wheels of the vehicle by means of proper links. When the pedal is operated, the cams on all four wheels are simultaneously rotated. Now the brake shoes on all the four wheels are also in operations. These simultaneously grip the brake drums in all four wheels. .When the brake pedal is released, the springs in these brake shoes bring them closer. The grip on the inner portion of the brake drums is thus relieved. The wheels are now free to rotate.

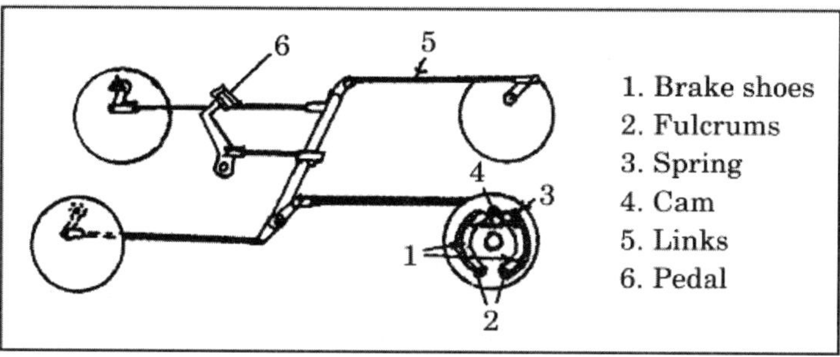

Fig. 19.2: Mechanical Brake.

19.6 CONSTRUCTION AND OPERATION OF EXTERNAL CONTRACTING SHOE BRAKE

This type of brake is normally available on crawler tractors. The hand brake used in automobile vehicle is also of same type. It is simple in construction and easy in operation. The drum mounted on the drive axle is directly surrounded by the brake band. Figure 19.3. shows the gearbox, hand brake, universal joint, propeller shaft, differential and wheel. The hand brake is shown separately in Fig. 19.4. The brake band is linked to the hand brake by means of a lever. When the hand brake is applied, the brake band first tightens its grip on the drum. This stops the motion of the drum. The vehicle is thus stopped. When the hand lever is released, the brake band loses its grip on the drum and the vehicle is now ready for starting. The hand brake is also applied while parking a vehicle. This prevents it from moving forward or backward.

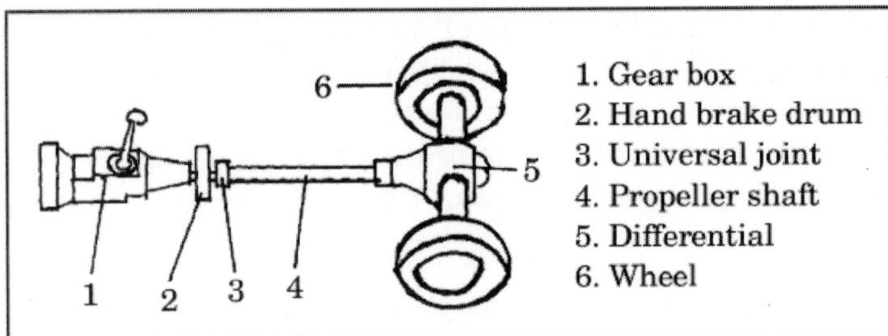

Fig. 19.3: Hand Brake.

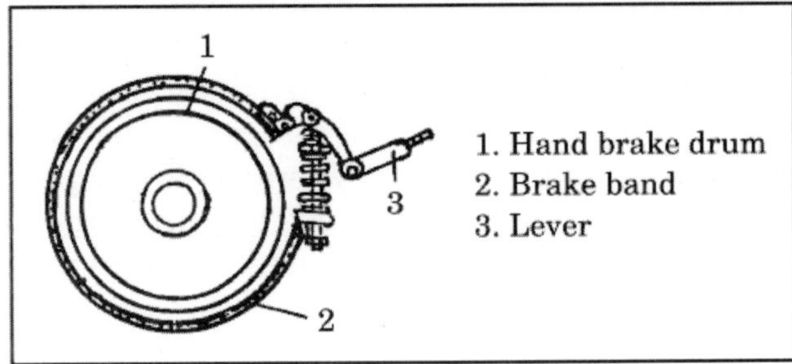

Fig. 19.4: Hand Brake.

19.7 HYDRAULIC BRAKES

19.7.1 Purpose of hydraulic brake

The hydraulic brake has the following purposes over the mechanical brakes.

(i) It gives higher efficiency than any other system.
(ii) It applies equal pressure on all the four wheels.
(iii) It consists of minimum component parts making maintenance easy.
(iv) It allows smooth and easy application of brakes.
(v) Brake shoe thrust can be altered by changing the piston area of the wheel cylinder.

19.7.2 Principle

A hydraulic fluid transmits pressure equally to all brake shoes based on Pascal's Law which states that when any part of a confined fluid is subjected to pressure, the pressure is transmitted equally and undiminished to every portion of the inner surface of that container as shown in Fig. 19.5.

A pedal effort of 20 kg applied by the driver on the brake pedal is multiplied by a leverage of 15 cm giving {$20 \times 15 = 300$ kgf) on to the master cylinder piston. Let the master cylinder piston be of area 5 cm^2, then a pressure of $300/5 = 60$ kg/cm^2 is transmitted equally to all the four wheel cylinders in accordance with Pascal's Law. If the wheel cylinder piston is of 6 cm^2 area, the force on it would be ($6 \times 60 =$

360 kgf). Thus, the pedal effort is appreciably multiplied. The brake shoes move outwards to contact the brake drum in order to slow it down.

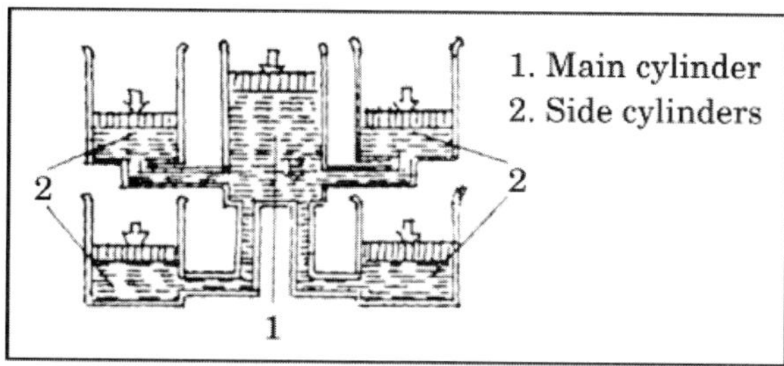

Fig. 19.5: Principle of Hydraulic Brake System.

19.7.3 Construction and operation of a hydraulic brake

The hydraulically operated brake comprises *(i)* master cylinder, *(ii)* wheel cylinder, *(iii)* brake drum, *(iv)* brake shoes and *(v)* brake fluid brake lines. Figure 19.6 shows a master cylinder and four wheel cylinder. Every wheel cylinder contains two pistons which move outward. The hydraulic fluid flows from the master cylinder to the four wheel cylinders through suitable lines. Springs are used to hold the brake shoes on all the four wheels. When the brake pedal is pressed, the piston in the master cylinder forces the liquid out of the cylinder. This liquid presses the two pistons in the wheel cylinders outward. These two pistons push the brake shoes outward. The brake shoes in turn press against the brake drums. This stops the motion of the brake drums. The wheels are thus stopped.

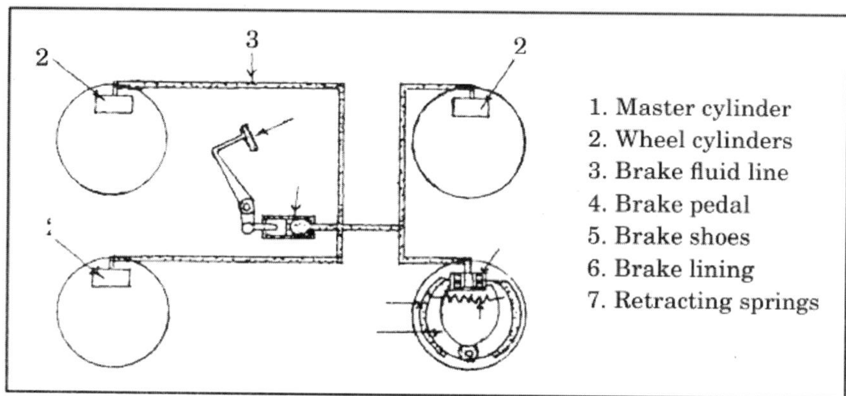

Fig. 19.6: Hydraulic Brakes System.

When the brake pedal is released, the master cylinder is pushed backward. This is done by a spring fitted in the master cylinder. The springs of the brake shoes bring the shoes closer. At this time, the two pistons in the wheel cylinder also come closer. The liquid in the wheel cylinder is pushed outward through brake fluid pipes. It returns through the pipes to the master cylinder. This is how the hydraulic system of the four wheels operates.

19.7.4 Master cylinder

The master cylinder is the heart of the hydraulic brake system. It consists of two main chambers; the fluid reservoir which contains the fluid to supply to the brake system and the compression chamber in which the piston operates. The reservoir supplies fluid to the brake system through two ports. The larger port is called feed hole or intake port and is connected to the hollow portion of the piston between the primary and secondary cups which act as piston seals. The smaller port is called the relief or bypass or compensating port which connects the reservoir directly with the cylinder and the brake fluid lines when the piston is in release position. The reservoir is vented to the atmosphere so that atmospheric pressure causes the flow through the filler port or feed port. The vent is placed in the top. The rubber boot covers the cylinder to keep it free from foreign material and dust. Figure 19.7 shows the cross-section of a master cylinder.

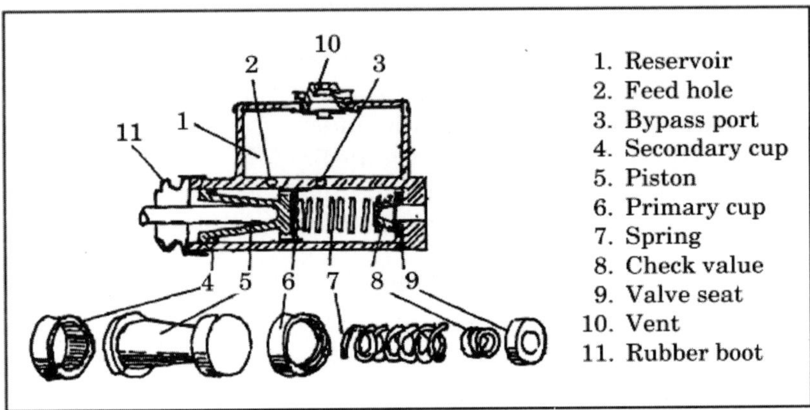

Fig. 19.7: Master Cylinder.

When the pedal is depressed, the master cylinder piston moves forward to force the liquid under pressure into the system. The piston in the process, covers the by-pass port. The fluid in the compression chamber is pressurized, opening the fluid check valve. Fluid under pressure flows to the wheel cylinders through the brake lines. The pistons in the wheel cylinder move apart allowing the brake shoes to come in contact with the brake drums, causing the wheel to stop.

When the brake pedal is released, the spring quickly forces the master cylinder piston back. As the fluid in the lines returns rather slowly, a vacuum is created in the compression chamber. This causes the primary cup to collapse to allow the fluid to flow from the reservoir through the feed hole past the piston to fill the vacuum. The system now comes to rest. When the pedal is in 'off position, the liquid may flow from the reservoir through the bypass port in the master cylinder, supply lines and wheel cylinders to make up for any fluid that may be lost. In this way, a complete column of fluid is always maintained between the master cylinder piston and wheel cylinder pistons.

19.7.5 Wheel cylinder

Wheel cylinder is the second important component of the hydraulic brake system. Fig. 19.8 shows a typical double-piston wheel cylinder. It consists of two pistons which can move in opposite directions by the fluid pressure. The dust cover protects the cylinder from foreign substance. Rubber cups fit tightly in the cylinder against each piston and seal the mechanism against leakage of the brake fluid. A spring serves to hold the cups against the piston when the pressure is decreased.

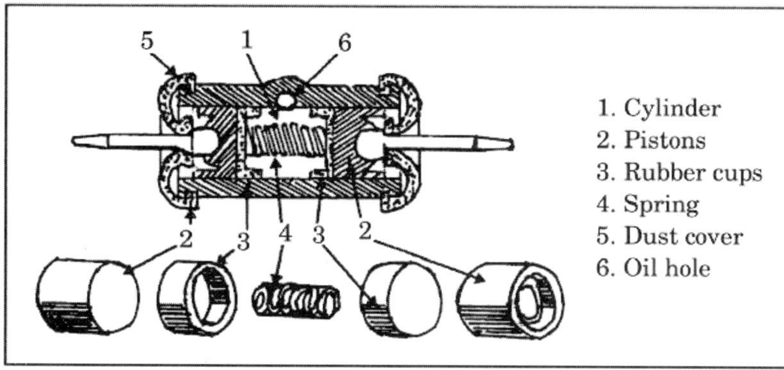

Fig. 19.8: Wheel Cylinder

When the brakes are applied, the brake fluid enters the cylinder from a brake line connection inlet between the two pistons. It causes to force out the two pistons in opposite direction. This motion is transmitted to the brake shoes to make contact with the brake drums. Once the pressure is released, the spring brings both the pistons back to their original position.

19.8 DISC BRAKE

The international B-275 tractor is fitted with double-disc type mechanical brakes mounted on the differential shafts. The motor vehicles are now being fitted with disc brakes instead of the conventional type drum brake. Figure 19.9 shows the construction and working of a disc brake. A disc brake consists of a rotating disc and two friction pads which are actuated by four hydraulic wheel pistons contained in two halves of an assembly called a caliper. The caliper assembly is secured to the steering knuckle in a front wheel brake and to the axle housing in a rear wheel brake. The road wheel is fastened to the outer surface of the disc. The friction pads ride freely on each side of the disc.

When the driver applies pressure on the brake pedal, hydraulic pressure pushes the pistons out of their housing. The pistons, in turn, press the brake pads against the moving disc faces, causing friction and hence slowing it down. Hydraulic pressure is equally applied by the hydraulic fluid to the floating pistons on either side. When the driver takes his foot off the brake pedal, hydraulic pressure on the friction pads is released, the pistons move inwards and break their contact with the disc.

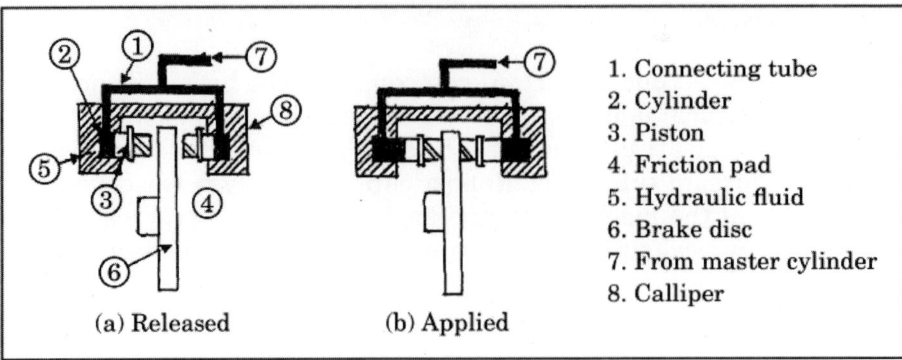

Fig. 19.9: Working Disc Brakes.

19.8.1 Advantages of disc brakes over drum brakes

(i) Disc brakes provide better stability since these have uniform pressure distribution over the pads than that of the brake linings in the case of drum brakes.

(ii) Increased temperature does not affect the disc pads much compared to the brake linings of the drum brakes.

(iii) Maintenance and repairs of disc brakes is easy.

(iv) The design of the brake adjusters becomes simple because when hot, the discs expand towards the pad causing no loss in pedal travel.

19.8.2 Disadvantages

(i) Disc brake assemblies are costlier than drum brakes.

(ii) The pads wear off fast compared to brake shoe linings of drum brakes. Disc brakes have higher brake pressure.

(iii) The high temperature operation of disc brakes causes evaporation of the brake fluid and deterioration of seals.

19.9 BRAKE FLUID

It is a special type of fluid named SAE-1703J and must meet the following requirements.

(i) It must have a high boiling point and low freezing point.

(ii) It must be chemically stable

(iii) It must remain fluid at low temperatures but must retain good film strength at high temperatures.

(iv) It must have good lubricating properties

(v) It must be non-corrosive and must not attack rubber or metallic parts.

These fluids are based on glycol or polyglycols with additives like castor oil in alcohol with a suitable neutralizer to neutralize the effect of acids in alcohol.

Chapter 20

Hydraulic System

20.1 INTRODUCTION

Tractor is equipped with hydraulic system in order to lift and lower the implements attached to it. This system is controlled by the tractor driver from his driving seat in the tractor cab. It includes a hitch linkage and a hydraulic lift mechanism. The hitch linkage serves to attach agricultural implements to the tractor. It consists of several links, rods, and arms, usually installed at the rear of the tractor. Hydraulic lift linkage system is necessary to utilize the useful power of tractor for the operation of various farm machines or implements attached to it. Before discussing the individual components used in hydraulic system, it is necessary to know the fundamentals of hydraulics on which the hydraulic system works.

20.2 BASIC CONCEPTS OF HYDRAULICS

20.2.1 Hydraulics and Hydraulic System

Hydraulics is the use of a liquid under pressure to transfer force or motion or to increase an applied force. Hydraulic system is a circuit in which force and power are transmitted through a fluid, generally an oil. Hydraulic systems may be divided into two groups, the hydrostatic and hydro-kinetic.

The hydrostatic system is that in which the primary function of hydraulic fluid is the transmission of forces and power by pressure. A hydrostatic system consists of two basic elements, a pumping unit to convert mechanical work into hydraulic energy and hydraulic motor of reciprocating or rotary type to convert fluid energy into mechanical work. A lead which forms a circuit connects the two main components. The pumping unit transmits fluid pressure where as hydraulic motor receives force and power by means of fluid pressure to do mechanical work.

The purpose of a hydro kinetic system is to transmit power and the required effect is obtained primarily by virtue of changes in velocity of flow, of the working media and changes of pressure are avoided as far as possible. A hydro-kinetic transmitter essentially consists of a centrifugal pump or impeller mounted on the driving shaft and an oil turbine or runner mounted on the driven shaft. Power is transmitted from the driving to the driven shaft through circulation of oil between the impeller and the runner. Examples of a hydro kinetic system are hydraulic coupling and torque converter.

Reciprocating, centrifugal and other types of pumps are usually employed for water lifting devices where large volumes have to be handled. In oil hydraulic field, however, rotary pumps of positive displacement type are used, for they develop very high pressure.

20.2.2 Incompressibility of Liquids

Increasing the pressure on a gas compresses the gas into a smaller volume. However, increasing the pressure on a liquid does not reduce its volume. A liquid is incompressible. This is because, the molecules of the liquid are relatively close together as opposed to a gas in which the molecules are relatively far apart. Putting pressure on a gas can squeeze out some of the space between the molecules. But in a liquid, there is no extra space that can be squeezed out by the application of pressure. The molecules are already as close together as possible. Therefore, putting pressure on the molecules of a liquid cannot force them, closer together.

20.2.3 Transmission of Motion by Liquid

Liquid can transmit motion as it is not compressible. Figure 20.1 shows two pistons in a cylinder, with a liquid separating them. When the applying piston is pushed into the cylinder 200 mm, the output piston will be pushed along the cylinder for the same distance or 200 mm.

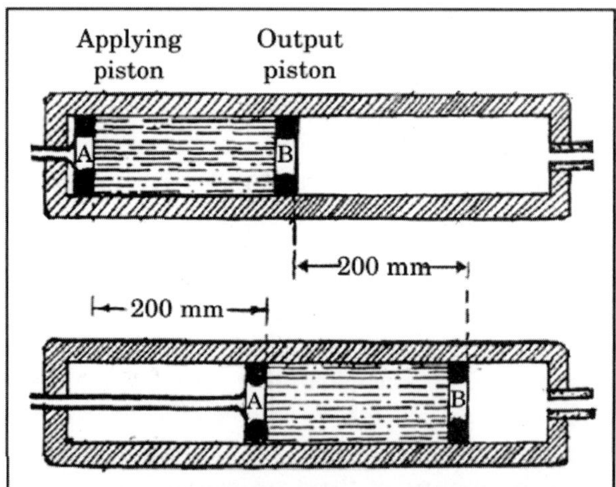

Fig. 20.1: Force and Motion can be Transmitted by a Liquid. When the Applying Piston is moved 200 mm. In the Cylinder, The Output Piston is also moved 200 mm.

Motion can also be transmitted from one cylinder to another by a tube or pipe. In Fig. 20.2, the applying piston is in cylinder A and the output piston is in cylinder B. As the applying piston is moved into its cylinder, liquid is forced from cylinder A into cylinder B. Thus causes the output piston to be moved in its cylinder. This is how the hydraulic system works.

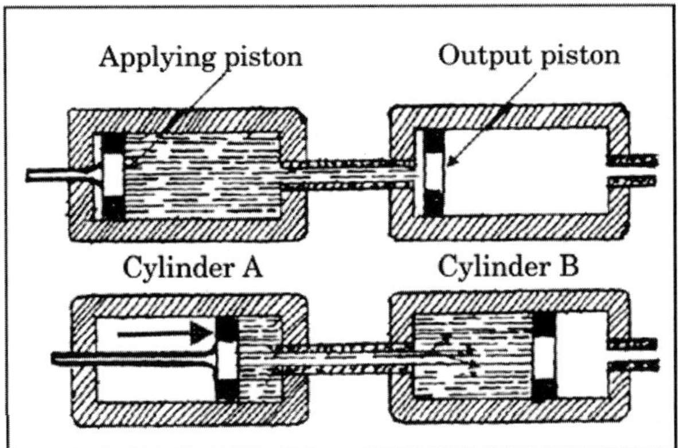

Fig. 20.2: Force and Motion can be Transmitted through a Tube from One Cylinder to another by *a* Liquid, that is by Hydraulic Pressure.

20.2.4 Working Principle of Hydraulic System

The hydraulic system used in tractor basically works on the fundamental law of hydraulics defined by Blaise pascal in 1653 which is as follows:

Pressure applied on enclosed fluid transmitted equally in all directions with every part of fluid and the surface of the container (Fig. 20.3).

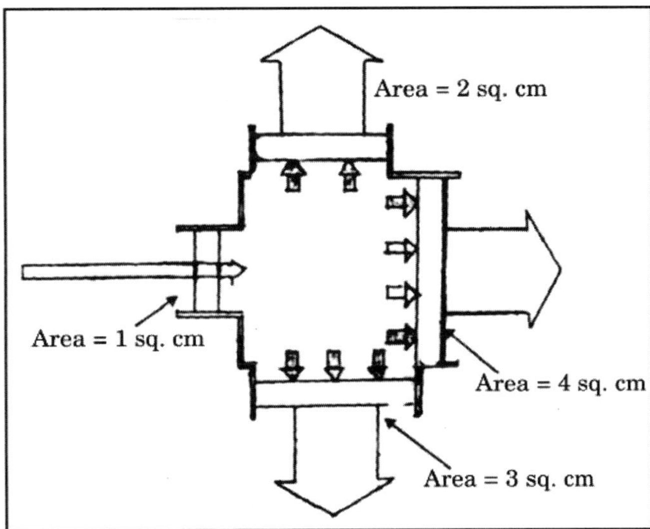

Fig. 20.3: Figure explaining Pascal's principle

Pascal's principles, together with the fact that the pressure force applied by a fluid at a surface is proportional to the surface area, has been the source of important technological innovations. It has resulted in many inventions that impacted many aspects of ordinary life such as hydraulic press, hydraulic brake, hydraulic jacks, hydraulic lift etc. If we take two cylinders, one having an area of 1 sq.cm. and the other having the area of 5 sq. cm, connected with a suitable pipe as shown in Fig. 20.4. Fill these cylinders with water or any other liquid and attach the matching pistons with both the cylinder. Then press the smaller cylinder with 10 kg weight by 10 cm. The bigger cylinder being interconnected will raise its piston by 2 cm and would carry a load of 50 kg. From this example, it is clear that by exerting 10 kg force, we are able to lift a weight of 50 kg easily or in other words, applying this principle, we can lift heavy weight with less force.

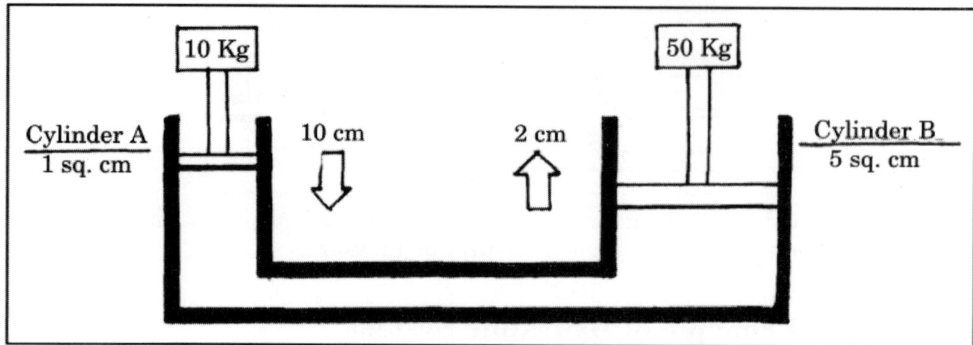

Fig. 20.4: Figure explaining Pascal's principle

The mechanical advantage obtained by the practical application of Pascal's law of transmission of liquid pressure can be explained with the help of Fig. 20.5. Noting that $P_1 = P_2$ (both pistons are at the same level and are enclosed by the liquid), the ratio of output force to input force is determined to be

$$P_1 = P_2 \rightarrow \frac{F_1}{A_1} = \frac{F_2}{A_2} \rightarrow \frac{F_2}{F_1} = \frac{A_2}{A_1} \rightarrow F_1 = F_2 \frac{A_1}{A_2} \tag{20.1}$$

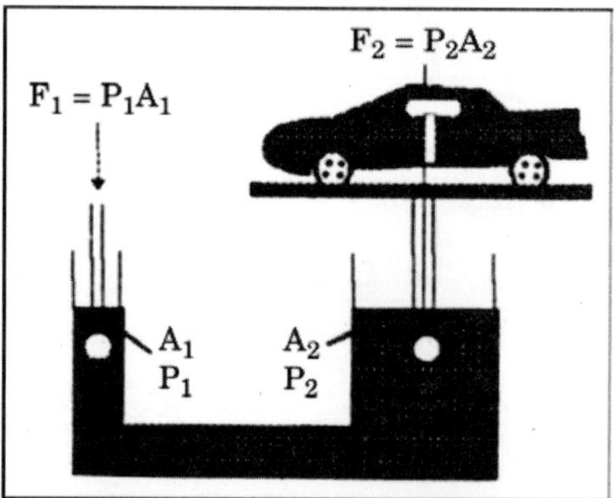

Fig. 20.5: Lifting of Large Weight by a Small Force by the application of Pascal's Principle.

The area ratio is called the mechanical advantage of the hydraulic A2 system. Using a hydraulic car jack with a piston area ratio of A =5, for example, a person can lift a 500 kg car by applying a force of just 100 kgf (= 980 N).

20.2.5 Advantages of Hydraulic System

The hydraulic system is preferred to the mechanical ways of power transfer due to the following reasons.

(i) It is easy and simple in design

(ii) The hydraulic working units can be fitted any where and in any position.

(iii) Speed can be easily controlled

(iv) There is practically no wear and tear as these parts always run in hydraulic oil and get lubricated.

(v) Hydraulic system is safer than a mechanical system as the moving parts are minimizcd.

(vi) It is easier to maintain

(vii) Heavier jobs can be done quickly with less force.

Though the hydraulic system is preferred but cannot be called a perfect system without any drawbacks. The efficient operation requires a regular cleanliness to avoid rusting, corrosion, dirt and other foreign materials.

20.3 OPERATION OF HYDRAULIC SYSTEM

The basic hydraulic system consists of a hydraulic pump which takes the hydraulic liquid from the reservoir and pumps it through the cylinder where hydraulic oil forces the piston out and work is performed. The other components are relief valve, control valve and filter. The hydraulic system relies for its operation on the pressure of oil forced by the oil pump (hydraulic pump) into the cylinder. During operation, the pump delivers oil, under high pressure to the hydraulic control valve. Depending upon the position of the hydraulic control lever, the oil is directed either to enter into the cylinder or to flow from cylinder to the reservoir or else is returned to the oil reservoir. Relief value protects the system from high pressure. It is set for slightly higher than the working pressure. In case the pressure increases beyond the working pressure, the relief valve opens allowing the oil to pass into the reservoir. The cylinder converts the hydraulic power to mechanical work for doing various jobs. The hydraulic pump is operated by suitable gears which get their motion from the tractor engine. The bypass line of the control valve is connected to the oil tank through an oil filter. The schematic diagram of a hydraulic system for doing various operations is shown in Fig. 20.6.

20.4 HYDRAULIC SYSTEM COMPONENTS

Followings are the components of a hydraulic system, *(i)* Reservoir or hydraulic tank or oil tank *(ii)* Hydraulic pump *(iii)* Pressure relief valve *(iv)* Hydraulic control valve *(v)* Hydraulic cylinder (or ram cylinder) *(vi)* Hose pipe and fittings *(vii)* Hydraulic control lever and *(viii)* Lifting arms.

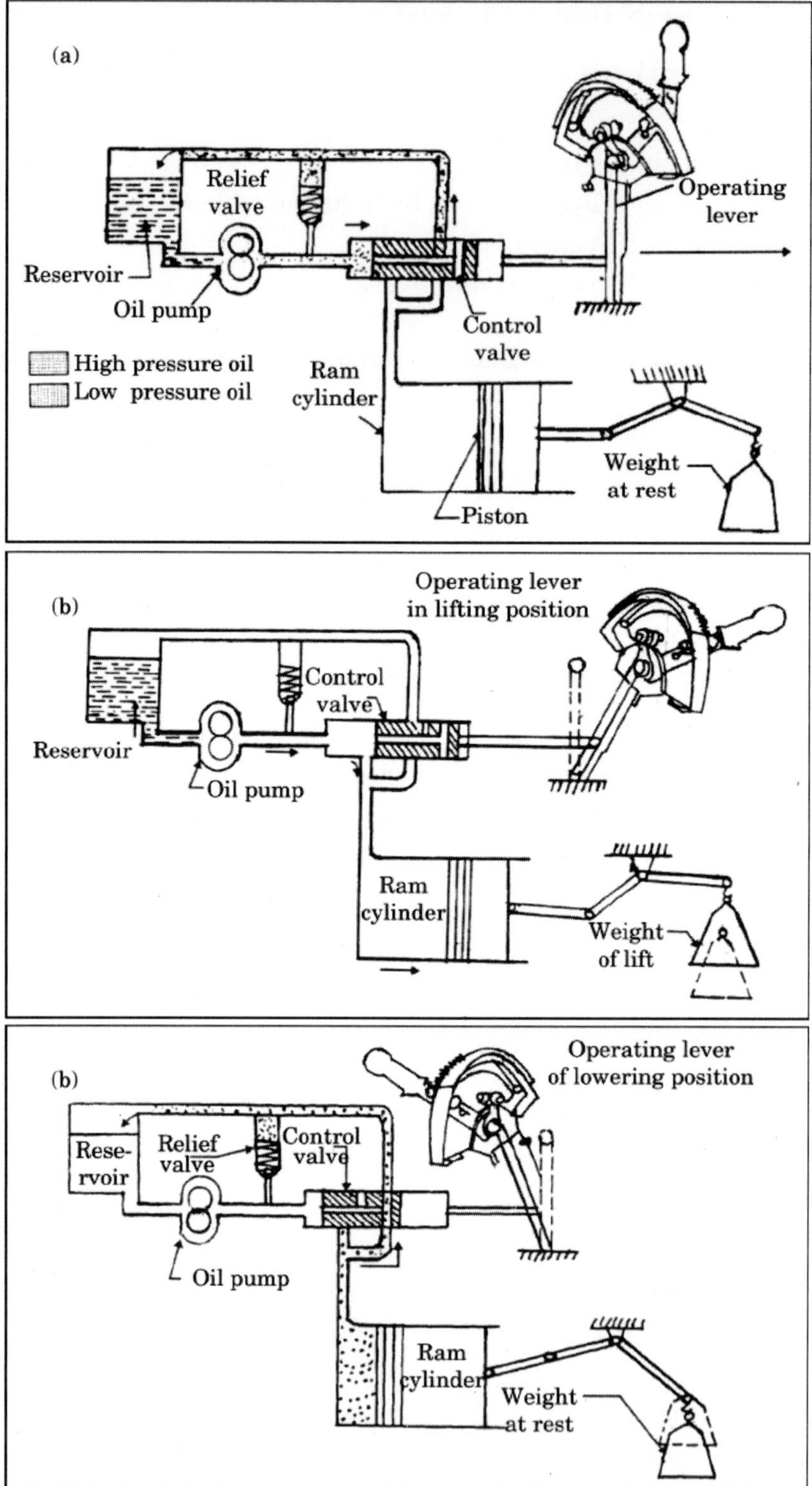

Fig. 20.6: Schematic diagram of hydraulic system

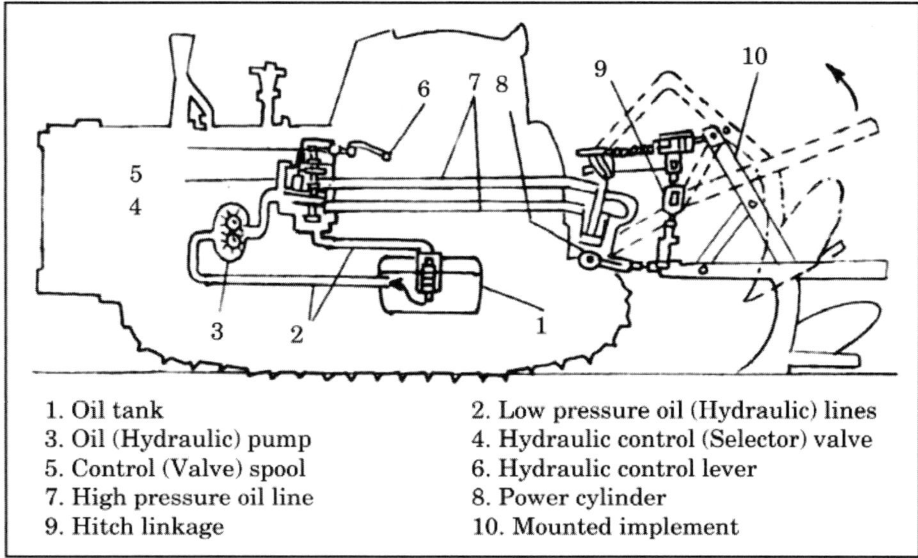

Fig. 20.7: Schematic Diagram of a Tractor Hydraulic System

The schematic diagram of a tractor with hydraulic system is shown in Fig. 20.7.

Hydraulic tank: Hydraulic tank is used for storing hydraulic oil for the system. It is a metallic tank welded with the tractor. The tank cover is equipped with the filler cap, breather (or air vent), oil level dipstick and oil filter. Two pipes are connected with the tank, one is the inlet pipe from where oil returns from the system and the other is the outlet pipe through which oil flows to the pump. Through the breather, clean air enters into the tank or air from the tank gets out when the tank is being filled with oil. At the bottom of the tank, a drain plug is fitted for the removal of oil at the time of its cleaning.

Hydraulic pump: Pump is the heart and the important component of the hydraulic system. The pump is a mechanical device which increases the pressure energy of a liquid. The lifting of water is done with the help of a rotodynamic (centrifugal pump, jet pump etc.) pump, but these pumps cannot be successfully used in handling oils because of their limitations of low pressure and high discharge volume. Therefore, in the field of oil hydraulics, positive displacement pumps are used. Positive displacement pumps are of two types, *i.e. (i)* reciprocating pump, *(ii)* rotary pump. Though reciprocating pumps are most efficient in hydraulic system, they occupy a lot of space and hence are not found suitable for the hydraulics of mobile machine. The rotary pump resembles in appearance to a centrifugal pump, but differs in action. The rotary pump may be a constant flow pump (gear pump, vane pump, screw pump etc.) or a variable flow pump (axial piston and radial - piston pumps).

The constant - flow pump delivers the same volume of oil for a given speed. The rate of flow changes with a change in speed. In a hydraulic system fitted with the constant flow pump, when the piston of the cylinder reaches its dead end or when the spool valve is brought to the neutral position, the oil delivered by the pump is diverted to the reservoir either through relief valve or spool valve. In case of the variable-flow pump, the volume of oil can be changed even at the same speed. These pumps supply oil only when it is needed. When the piston of the cylinder reaches dead end or the spool valve is put to neutral, the pump stops delivering further oil.

At present, most of the Indian tractors *i.e.*, International B-275, Escort 335 etc. are fitted with a gear pump, but Ford 3600 is an example where an axial piston pump is used.

Pressure relief valve: The oil pressure relief valve is one in which a ball and spring are used to release the oil pressure (Fig. 20.8). Suppose the pressure of the outlet oil is increasing, when it goes beyond a certain limit, the ball is lifted. The valve is thus compressed .The high pressure oil now passes through the by-pass passage. Finally it returns to the oil tank. The pressure of the oil is thus released and an uniform pressure of oil is maintained in the hydraulic circuit. To increase the pressure, the screw is tightened to increase the tension of the spring and to decrease the pressure, the screw is loosened. Hence an adjustment screw is fitted to change the tension of the spring.

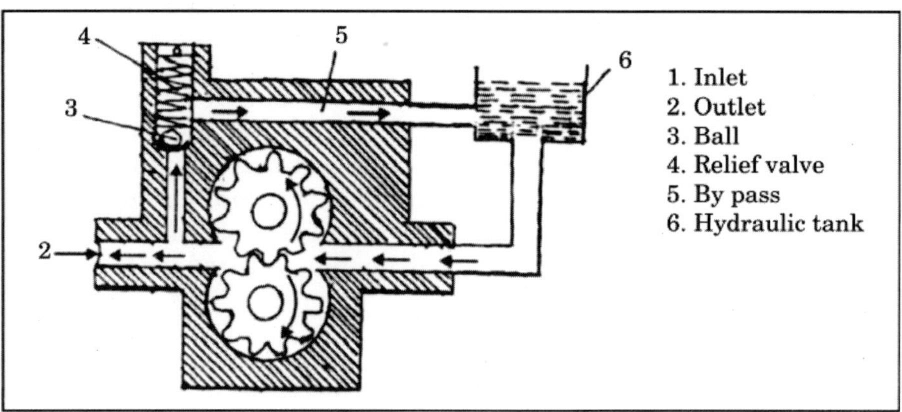

Fig. 20.8: Oil Pressure Relief Valve.

Hydraulic control valve: Hydraulic control valve is a type of direction-control valve, which directs and controls the flow of oil from the oil pump to the cylinder. Among a variety of direction-control valves (check valve, rotary valve, spool valve etc.) only spool valves are found on most of the Indian tractors.

The literal meaning of spool is a cylinder on which thread is wound. Hence, spool valve is a metal rod with cut away sections. The spool valve has at least three ports and two lands or outlets. A schematic diagram of a spool valve is shown in Fig. 20.9. The spool is a friction fit in the valve body, so that a momentary signal A from the bottom will cause the spool to move upwards and remain in the top position. An output, say, 83, will result without the application of further signals. The application of a signal B from top will similarly give another output, say S^ without further signal.

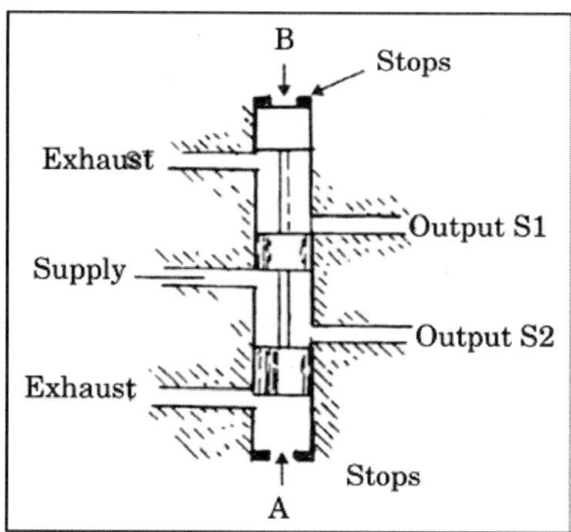

Fig. 20.9: Spool Valve

The control spools are fitted to have a very small clearance in their bores in the hydraulic control valve body. Each control spool is operated by hydraulic control lever of its own and controls the operation of the power cylinder .Each hydraulic control lever can place its associated control spool in different position to register the annular grooves in the spool with the appropriateness in the control valve body. The central position of the control spool is neutral. Pulling back on the hydraulic control lever places the spool into the "LIFT" position. Pushing forward on the lever moves the spool to the "LOWER" position.

With the control spool in the "LIFT" position, one of the grooves in the spool connects the discharge chamber of the hydraulic control valve with the power cylinder. The movement of piston in the power cylinder is transmitted to the linkage rods for the lifting of the implements. When the control spool in the "LOWER" position, it connects the cylinder to the oil tank, resulting in the flow of oil into the oil tank through the exhaust passage of the control valve. The implement is lowered by its own weight. When the control spool is in the neutral position, both the outlet ports are covered. The oil forced by the oil pump into main inlet passage of the control valve returns to the oil tank through the exhaust passage of the control valve. In this position, the attached implement can neither be lifted nor lowered and the load remains lifted.

The spool valve is basically of two types *i.e.* (i) open-centre system and *(ii)* closed centre system. The main difference between open centre and closed centre valve is that in open centre spool valve, when it is kept in neutral position, it closes the outlet port, but inlet port is kept open and oil passes in the centre of open spool valve and to the oil tank (Fig. 20.10). The closed centre spool valve when kept in neutral position closes both the outlet ports and pump stops pumping oil. Open-centre type valves are always preferred in hydraulic circuit with constant-flow pump so that even in the neutral position of the valve, the oil has a way back to the reservoir. Closed-centre valve is used with variable flow pump so that when the valve is in neutral, the discharge of the pump diminishes to zero.

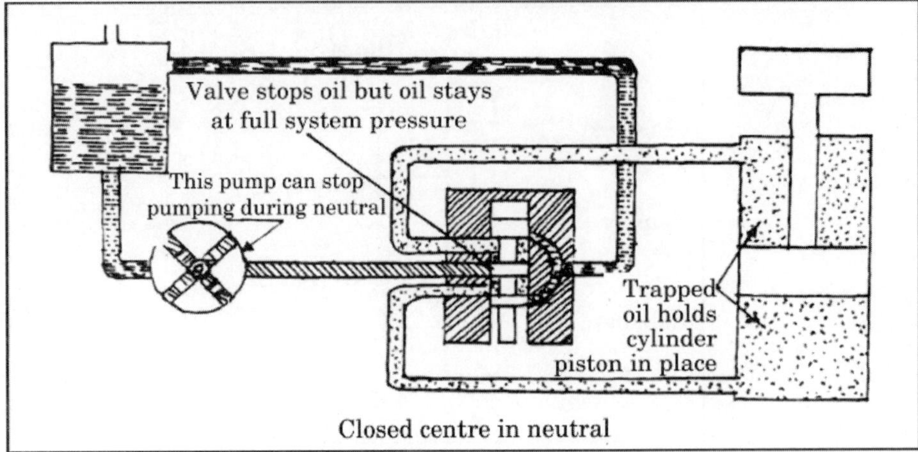

Fig. 20.10: Types of Hydraulic System

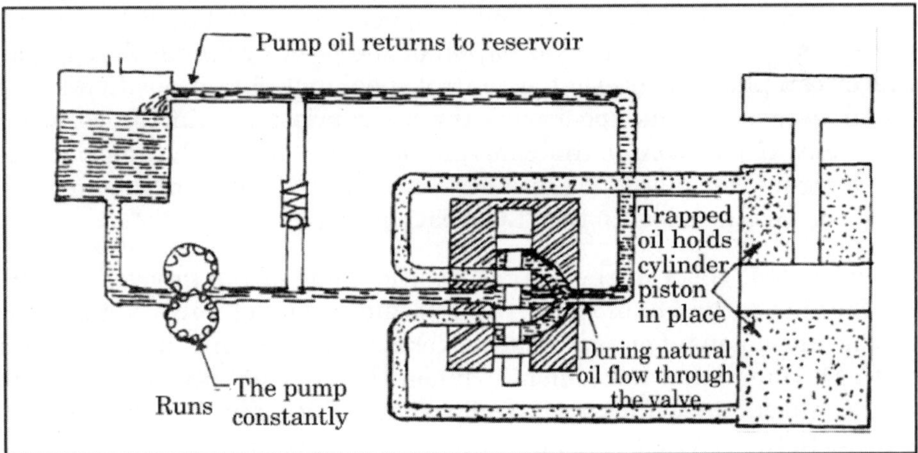

Fig. 20.10: (i) Open Centre in Neutral.

Hydraulic cylinder (or power cylinder or ram cylinder): The hydraulic, energy is converted to mechanical energy with the help of hydraulic cylinder. It consists of a cylinder, piston (or ram) and a connecting rod .The connecting rod transmits power from the piston to the lifting arms. Hence, hydraulic cylinders are designed to raise or lower an agricultural implement directly. Tractors are generally equipped with a single cylinder that is pivoted on the rear axle of the tractor and is intended to operate the hitch linkage.

Hydraulic cylinder may be either single acting or double acting. The single acting cylinder is powered only in one direction where as double acting cylinder is powered in both direction. The schematic diagram of single acting and double acting cylinder is shown in Fig. 20.11. In single acting cylinder, the piston fitted in the cylinder is lifted up due to pressure of fluid. When pressure is released, the piston comes back to its original position due to the weight of the implement or its own weight.

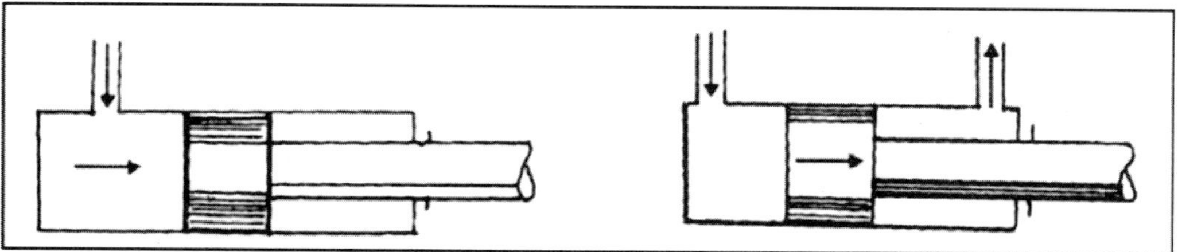

Fig. 20.11: Schematic Diagram of Single Acting Cylinder

Fig. 20.11: (i) Double Acting Cylinder

The cylinders are double-acting (two-way) type, *i.e.* oil can be delivered under pressure into the cylinder space both ahead and behind the piston. When we have to lift the piston in the cylinder, the hydraulic oil under pressure is sent to the lower part of the cylinder which makes the piston lifted up. At the same time, provision is made in the valve to drain back the fluid from the upper part of the cylinder to tank. Similarly when piston has to be lowered down, the upper chamber is connected to the pressure line while lower chamber to the oil tank. For keeping the cylinder in neutral position, valve closes both the upper and lower chamber of the cylinder causing the oil remains filled up in both the chambers and the piston remains in neutral position. To seal the piston in the cylinder, rubber seal is fitted in the piston .This seal does not allow the fluid to escape to the other chamber.

Hose pipe and fittings: Hose pipes are the fluid passages in the hydraulic circuit for the easy flow of oil. Pipes used in hydraulic system are of two types. High pressure steel pipe is fitted at such places where there is no moving part and the other is a high pressure flexible pipe, used, where flexible movement is needed. In some portions of the circuit, rubber pipes are also used. The rubber pipes are wire braided so that it can withstand high pressure prevailing in the system. Low pressure oil line connects the oil tank to the oil pump. The high pressure oil line connects the oil pump to the control valve and valve to the power cylinder.

Lifting arms: Lifting arms generally consist of control shaft, control lever quadrant, control lever and three point hitch linkage. The control shaft is ultimately connected to the piston rod for transmitting motion to the hitch linkage for lifting and lowering of the attached implement in the tractor.

Hydraulic oil: Hydraulic oil is the life of complete hydraulic system as blood in the human body. If something goes wrong with hydraulic oil, then hydraulic system does not function well. The hydraulic oil used in the hydraulic system is a petroleum by-product and is specially made and blended with certain chemicals so that its viscosity is kept under control, it does produce foam and lubricates the parts of hydraulic system. For efficient working of hydraulic system, oil should be maintained properly. The hydraulic oil should never be mixed with diesel oil, petrol and water. While filling in the tank it should be free from dirt, dust and other foreign materials.

20.5 HITCH LINKAGE IN A TRACTOR

The hitch linkage serves for attaching mounted and semi-mounted implements to the tractor and lifts and lowers the attached implements whenever required. The hitch linkage is arranged at the rear of the tractor and can be used for three-point hitch system. Three point linkage consists of one top link (Compression link) and two lower links (Tension link).

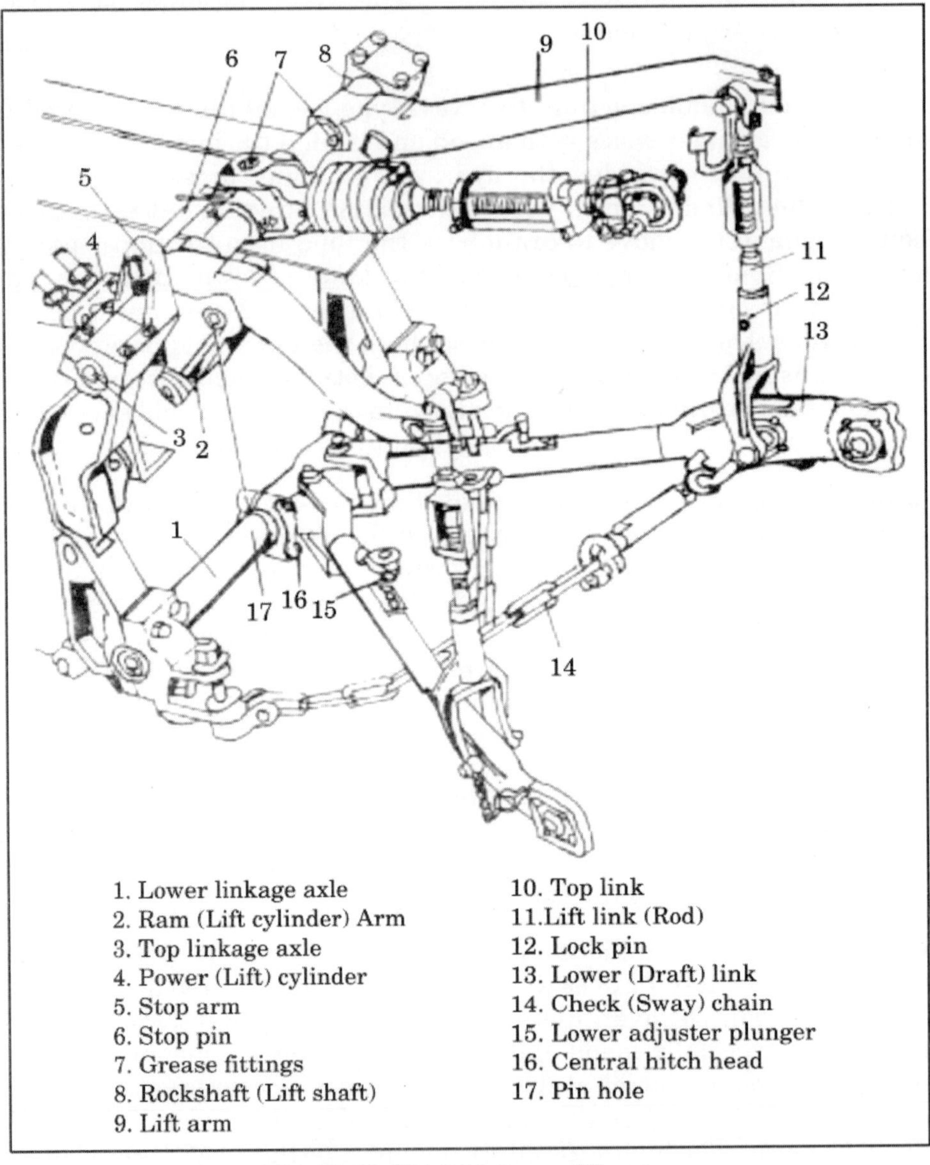

Fig. 20.12: Hitch Linkage of Tractor

Figure 20.12 shows a hitch linkage of a tractor. The hitch linkage consists of lower linkage axle *1* and top linkage axle, *3,* which are held to the tractor frame, top link *10,* two lift arms *9,* and two lower links *13,* that are hinged to the lift arms. The lift arms are splined to the ends of hollow rockshaft *8* that are supported by plain cast-iron bushings on the top linkage axle. Cylinder is attached to the left side of the rockshafts. Two lower links are at a same distance from the centre of tractor and also are at the same height from the ground level. As the rockshafts is connected to the piston rod, any movement of the piston is transferred to the bottom links. The top link is used for connecting the third hitch point of the implement and is adjustable for maintaining the implement level.

Mounted implements are attached to the ends of the top and lower links by ball joints. Placing the hydraulic control lever into the "LIFT" position, cause the power cylinder to extend and, with the piston rod acting on the ram arm, the rockshaft is turned together with the lift arms. The lift arms pull on the lift links which in turn pull on the lower links to lift the implement. Then placing the hydraulic control lever into the "LOWER" position, the power cylinder retracts resulting in the lowering of the implement due to its own weight.

Check chains are used to limit the sideways movement of the mounted implements. It holds the implement without swing when it is lifted or lowered. Load sensing for the draft control can also be done through top link which is spring loaded. In some tractors, the lower links are spring loaded for draft sensing. Depending upon the soil condition and type of operation, the mounted implement can be controlled either by position control or draft control.

20.6 CLASSIFICATION OF HYDRAULIC CONTROLS

The hydraulic control systems used on farm tractor can be classified into three system. These are *(i)* Nudging system (manually operated), (ii) Automatic position-control system (automatically operated), *(iii)* Automatic draft-control system (automatically operated).

Position control and draft control are the two important controlling factors in the hydraulic system. In most of the farming operations, like ploughing and sowing etc., it is required that the implement should penetrate into the soil to a certain depth. In position control system, the working depth of the implement is controlled by the adjustment of draft of tractor. In this system, the control valve can be operated directly by the driver to raise, lower or hold an implement, mounted on the linkage at any chosen height.

In draft control system, the implement is set for a particular draft (drawbar pull) rather than depth. In varied soil conditions, the implement automatically taking more or less depth to maintain the predetermined depth. Provision for draft control is made to ensure lifting of implement when it hits a hard ground. Failing which, there is every likelihood of its breaking or putting undue strain on the tractor.

20.6.1 Nudging-type Hydraulic Control

This system is used to raise, lower or position an implement, either mounted or trailed by moving a hand lever either forward or backward from its neutral position. If the control lever is moved (nudged), the hydraulic cylinder will move a complete stroke. If the lever is returned manually to its neutral position before the end of the complete stroke, the cylinder will stop and remain in that position, provided leaks do not exist in the system.

Figure 20.13 shows the raising of the implement by using manually operated lever. When implement hits the hard ground, operator has to raise it slightly. For raising the implement, he moves the control lever 1 to the back. As the lever 1 being pivoted to the cam follower 2, it is also pushed back and the cam follower also presses the valve rod. 3. When pushed back, the valve rod 3 operates the control lever 4. Due to this, the hydraulic oil enters into the hydraulic cylinder 5 and pushes the piston forward. When the piston rod is pushed, it pushes the shaft arm 6 and rotates the rocker shaft 7. Lift arm 8 which is fixed to the rocker shaft is lifted up, causing the implement to be lifted.

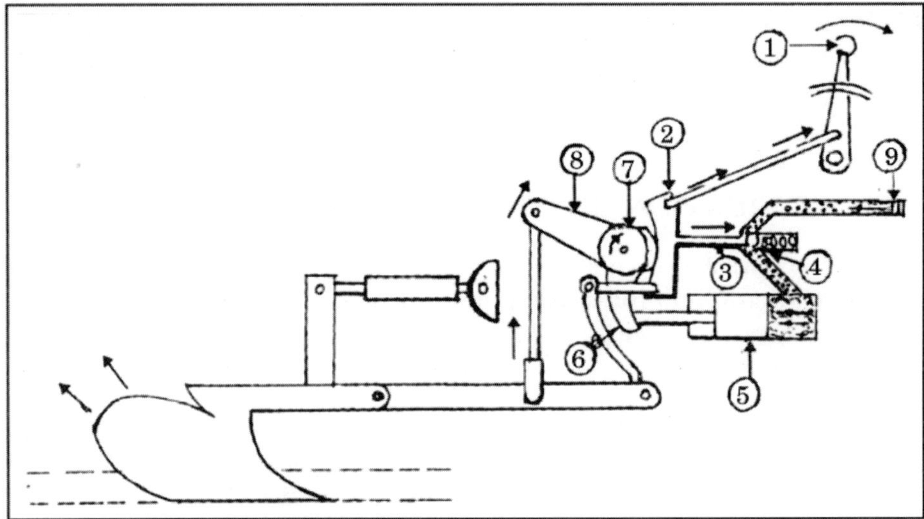

Fig. 20.13: Raising the Plough Using Lever (Not Automatic).

20.6.2 Automatic Position Control System on Tractors

Virtually all tractors with hitches for mounted implements have built-in lift systems with automatic position control. In this type of control system, any given position of the control lever in its quadrant represents a specific position or depth of the implement. Thus the operator can pre-select and identify an implement position by the location of the hand lever, and the cylinder will automatically move the implement to the corresponding position or depth.

The basic linkage for an automatic positions system is indicated in Fig. 20.14. The solid outlines repreaent a stabilized condition, the control valve being neutral.

If the control lever is then moved from A to A', joint C is moved to the left while joint E momentarily remains in the same position, thus moving D and the control-valve spool to the left. This actuates the power cylinder, which raises the implement until D is returned to the neutral position.

With the control valve again in neutral, the new position of the lift arms and connecting linkage is as shown by the broken-line outlines. Similarly, moving the control lever in the opposite direction results in lowering of the implement. In effect, D acts as a pivot point in determining the relative equilibrium positions of the control lever and the implement.

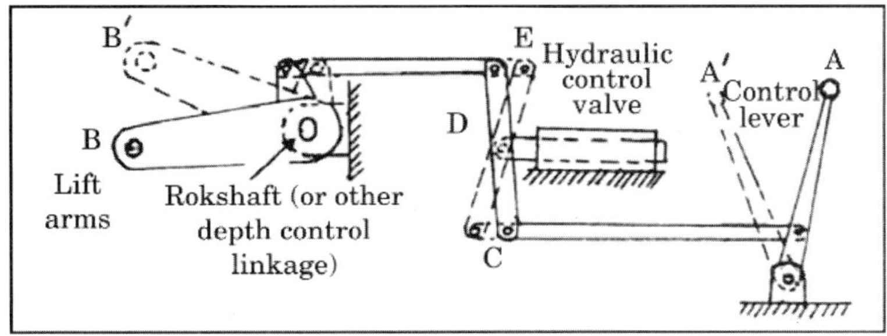

Fig. 20.14: Automatic Position Control

20.6.3 Automatic Draft-Control System

With automatic draft control, the position of the hand lever represents a given implement draft rather than a particular depth. Variations in draft cause the implement to be raised or lowered automatically as required to restore the draft to the pre-selected value.

As indicated in Fig. 20.15, the linkage for automatic draft control is similar to that for automatic position control except that the top end of equalizer link CE is moved in response to deflection of the load spring rather than by movement of the implement lift linkage. With the control lever in a given position, any change in draft changes the amount of deflection of the load spring thus moving the control valve from its neutral position, D. The hydraulic system then acts to change the implement depth as required to restore the draft and load-spring deflection to the pre-selected values and thus return the control valve to neutral.

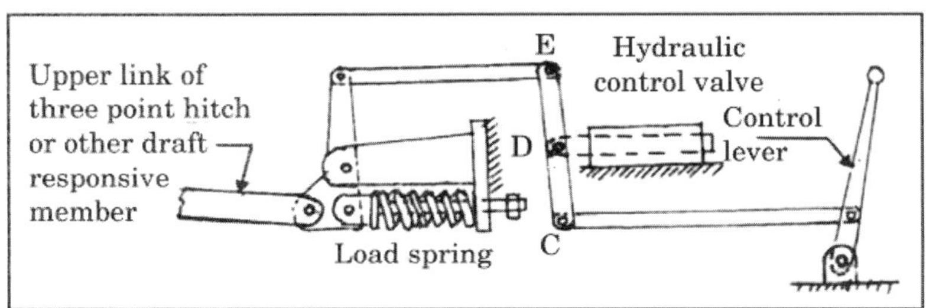

Fig. 20.15: Automatic Draft Control.

The load-spring, the sensing device, which tells the hydraulic system to lower or raise the hitch system, is located on either the lower links or the top link, depending on the size of the tractor. The position of the hand-control lever, in effect, establishes the draft to be maintained. The draft-control system on a tractor pulling an implement will raise and lower it to maintain a constant force on the sensing device.

Figure 20.16 shows how the implement automatically rises to maintain uniform draft while in operation. When the implement hits a hard ground 1, the load control arm 3 is pulled forward by draft link 2. As the load control arm 12 is linked with draft link, it is pushed backwards. Since control arm 12 is pivoted, its upper portion is pushed backward which in turn pushes the cam follower 4. The cam follower pushes rod 5 which in turn opens valve 6. As soon as the valve is opened, the hydraulic fluid with pressure enters into the cylinder 7 and its rod moves forward pushing arm 8 along with it. Since this arm is a part of rocker shaft 9, rocker shaft moves clockwise and the lift arm 10 is lifted upward. As lift arm is attached to the implement, the implement is raised slightly maintaining uniform draft. The implement will lower itself again when the hard surface is passed away. This happens as the tension of load control shaft 12 is partially released which allows cam follower to come back to its own position. This causes the hydraulic fluid to escape out from the cylinder resulting in the lowering of the implement.

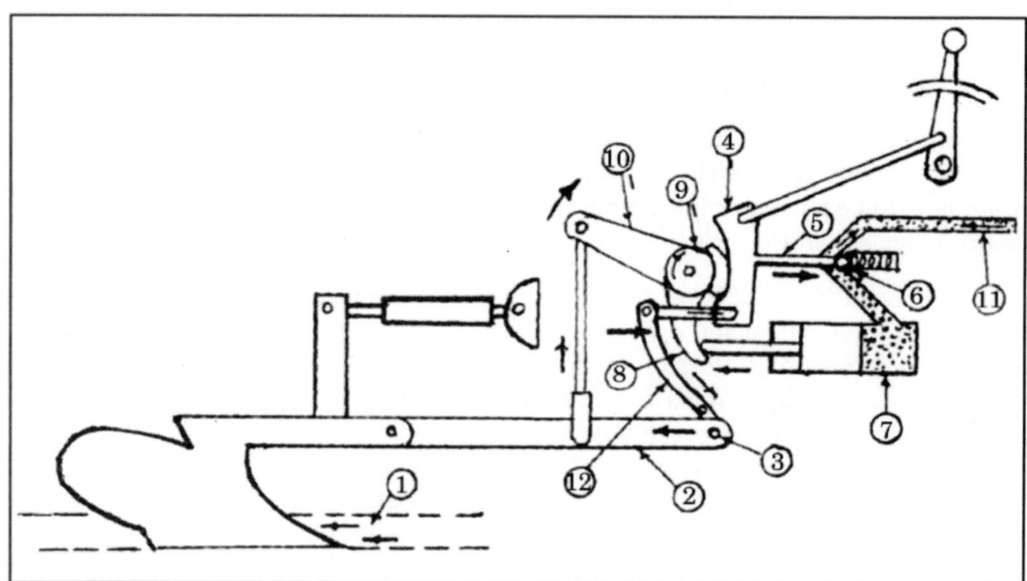

Fig. 20.16: Automatic Raising the Plough when Load Increases (Automatic)

20.6.4 *Mixed position and draft control*

This type of control system is a combination of position control and draft control. Position and draft control system is advantageous in that it eliminates the need for agricultural implements to have depth wheel fitted to it, so that the weight of the attached implement is transferred to the rear wheels of the tractor, resulting in the increase of traction. In some tractors *i.e.* HMT, Ford etc, there is the provision of suitably blending the response through an interlink mechanism so that a desired depth of ploughing is maintained within close limits and the draft control too is allowed to function for better traction.

Chapter 21

Electrical System

21.1 INTRODUCTION

Electricity is one of the forms of energy that are widely used in all tractors and automobiles. Every tractor or automobile of today contains an electric power plant under its hood, which produces and stores electric energy that is delivered either at low voltage or in the form of high voltage surges. Electrical equipment fitted on tractor or automobile is required to operate without failure for very long periods with little attention. Further, it is made to operate under widely varying climatic conditions. Hence, electricity does several jobs in tractors and automobiles. It starts the engine (where starter motor is used), ignites the compressed air-fuel mixture (in carburetor engines) and operates sound and light signaling devices *i.e.*, the lights, electric gauges and all sorts of auxiliary equipment.

Devices that supply electrical energy are referred to as power supply sources whereas those consuming electrical energy are called loads. The power supply sources used on tractors and automobiles are the battery and the generator (dynamo and alternator).The loads include the starter (cranking) motor, the ignition coil, signaling devices, the lights, electric gauges and other auxiliary electrical equipments. The power supply sources mentioned above convert chemical energy (the battery) and mechanical energy (the generator) into electrical energy and the loads serve to convert electrical energy into other forms of energy, such as mechanical, light (luminous), sound (acoustic) and heat (thermal) energy.

The heart of the electrical system is the battery, it supplies current to the starting system, lighting system and the ignition system. The generator keeps the battery charged. A regulator is used to control the output voltage and current. It also prevents the discharge of battery through the generator.

Wires of different sizes and colour codes connect the battery to the various components. These circuits are earthed to the vehicle's steel body as the body acts as a earth return. The various accessories and lamps are connected in parallel across the battery terminals. Each circuit has its own switch and earth connection. The negative or the positive terminal of the battery is earthed.

The electrical system in tractor or automobile is classified as : (i) Generation, storage and distribution unit *(ii)* Starting unit *(iii)* Ignition unit *(iv)* Lighting unit *(v)* Accessories like horns, signaling devices etc.

Generation unit: The generator is the primary source of electrical energy in generation system. It is driven by the V-belt which gets drive from the crankshaft. It converts mechanical energy obtained from the engine into electrical energy. Its main function is to recharge the battery. It also supplies electric current to the other accessories when the engine is running. Generators can be dc (dynamos) or ac (alternators). Alternators can give current output upto 35 amps whereas a dynamo output is limited to 12-14 amps. A control box or regulator controls the output voltage and current of the dynamo. Control system is required as at high engine speeds, the current output maybe increased to the values high enough to damage the battery and other accessories.

Storage unit: The battery stores energy in chemical form. The reaction in the battery starts as soon as any circuit is completed by the action of a switch. It supplies current up to 400 amps, required to crank the engine and a limited current to the accessories. It is continuously charged by the generator when the engine is running.

Distribution unit: The generator and the battery are connected to the various electrical units like lamps, horn, starter etc. by means of wires of different sizes, as shown in Fig. 21.1.

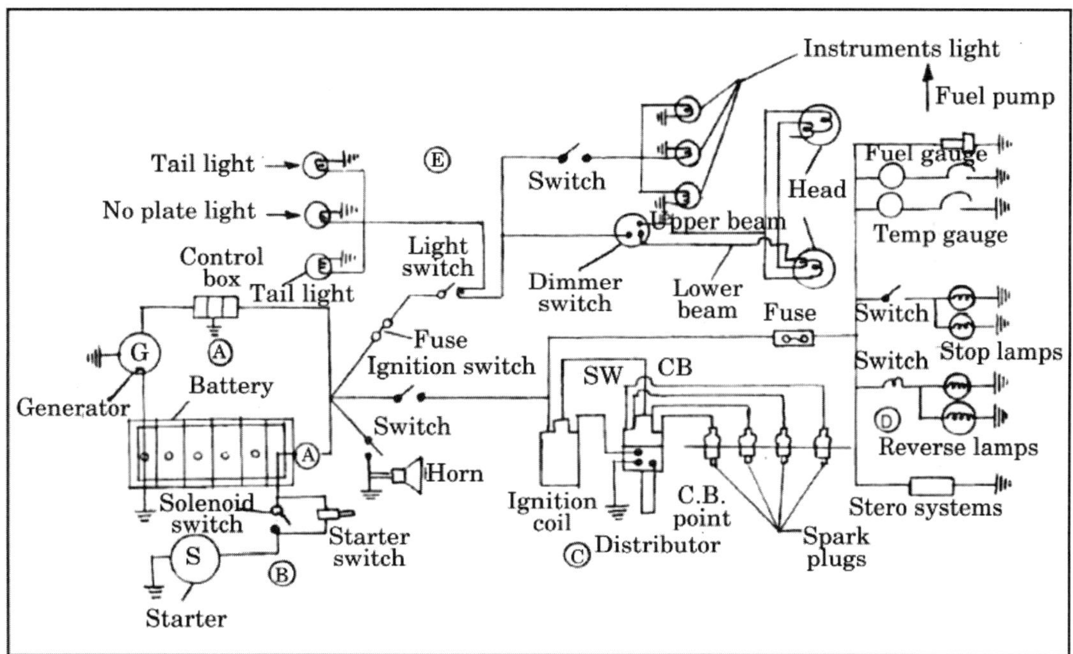

Fig. 21.1: Simple wiring diagram of electrical circuits.

Starting unit: The starter motor gets a heavy dose of current from the battery. It is a dc motor designed for intermittent service under great overload. When the circuit is completed by the solenoid switch, the pinion of the motor meshes with the crankshaft ring gear.The reduction ratio between these two is about 8:1 or 16:l. The crankshaft rotates to start ignition in the combustion chamber. As the engine speed increases, the pinion is thrown back to its initial position due to inertia effect. As shown in Fig. 21.1, the motor is earthed to the frame.

Ignition unit: The ignition unit plays an important role in the operation of petrol engines. The electric ignition unit is not deployed by diesel engines since the air temperature is sufficient at the end of compression stroke to cause burning of fuel which is injected at that time.

The ignition unit in petrol engine has to produce a high voltage spark (5000-20,000 volts) for a small duration at correct intervals of time, between the electrodes of the spark plug. The ways by which the sparks are generally produced are *(i}* coil ignition, *(ii)* magneto ignition. The coil ignition unit consists of a battery, high tension coil, distributor, spark plug and leads. The magneto ignition does not require battery current and therefore, it does not need any special low-voltage circuit. The magneto generates its own low-voltage current. It produces the high-voltage surges required for igniting the charge from this low-voltage current. It consists of an armature with windings, a permanent magnet and a contact breaker. The magnet rotates along with the engine and crankshaft. When the magneto shaft is rotated, it produces very rapid changes of magnetic flux linked with the armature windings resulting in a series of high-voltage surges. In the magneto ignition, the basic source of energy is the permanent magnet.

Lighting unit: The lighting circuit is shown in Fig. 21.1, It has a number of individual circuits for different lamps. Each has its own switch, wire connection and earth connection. The different lamps in tractor are head lamps, tail lamps, parking lamps, reversing light lamps, side indicator lamps, oil pressure and ignition warning lamps and instrument panel lamps.

Accessories like horns, dashboard gauges are connected parallel across the battery terminals. The dashboard gauges like oil pressure gauge, fuel lever gauge and temperature gauges do not have separate switches. These can be operated as soon as the ignition switch is on.

Earthed and Insulated return units: For electricity to flow, there must be a complete path, or circuit. The electrons or current must flow from one terminal of the battery or alternator, through the circuit and back to the other terminal. In the tractor or automobile, the engine and vehicle frame are used as the return circuit. Therefore, no separate wires are required for returns from electrical devices to the battery or alternator. The return circuit is called earth or ground and is indicated in wiring diagrams by the symbol .... So ground *i.e.*, engine and vehicle frame is the other half of the circuit. It is the return circuit from the electrical device to the source of electricity.

The use of fully insulated cables was made for electrical units in the wiring portions of earlier automobile, just like the domestic and industrial wiring units.

But presently, this method has been substituted by the single wire or earth return method. Figure 21.2 shows a simple circuit of a tractor engine cranking motor having insulated connecting cables. If the return current conductor AB is removed and the terminals of the battery and motor are connected to the metal member of the vehicle, the flow of the current will not be affected. By this method of connection, one set of insulated cables can be dispensed with. In this way, the various electrical units are connected.

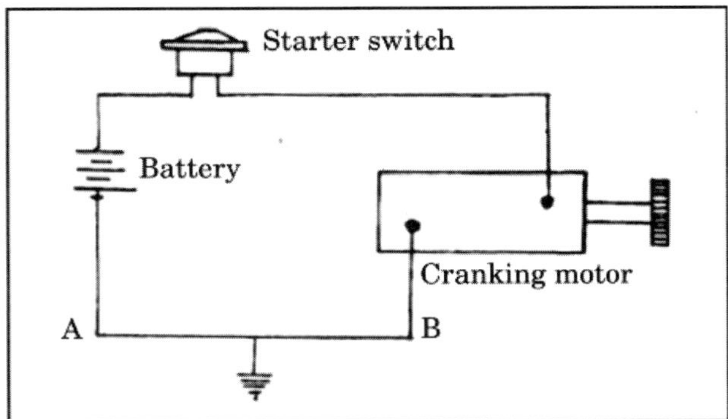

Fig. 21.2: A simple circuit of the starting system.

The advantages of such an earth return unit are as follows:

(i) Comparatively cheaper than insulated return unit as the length of the wire is greatly reduced,

(ii) The metallic earth return circuit provides negligible resistance whereas the insulated cable has a definite resistance

(iii) It is easier to trace out the faults in the earth return unit than in the insulated return unit.

The disadvantages of earth return unit are as follows :

(i) In case of short circuits, the whole electrical system fails whereas in insulated return unit, only the concerned unit fails.

(ii) The nut and bolt connections in the case of the earth return unit get loosened and corroded by vibrations and rust. This eventually leads to failure. In the case of insulated return unit, vibrations in chassis or engine do not affect the wiring.

Positive and negative earthing: It was a general practice to earth the negative terminal of the battery and allow its positive terminal to supply current. Now-a-days however, we earth the positive terminal of the battery. The advantages of earthing the positive terminal are given below.

(i) Lesser corrosion of the anode (positive plates) of the battery, gases corrode the positive plate if the negative terminal is earthed.

(ii) The central electrode of the spark plug has a longer life, being at lower potential than the metal electrode of the plug.

(iii) The central electrode being negative is hotter than the shell electrode giving a better spark at lower voltages. The voltages between the electrodes are also uniform.

(iv) The primary flux in the primary circuit voltage (High Tension, coil) is added to the secondary flux to produce high spark voltages.

If we use the alternator instead of a dynamo, we have to earth the negative battery terminal. This is so because the ac current rectifier has diodes and transistors.

21.2 THE BATTERY

The battery is an electro-chemical device for converting chemical energy into electrical energy and *vice versa*. The battery is the main part of the electrical system in an automobile or tractor. The main purposes of the battery are : to store electrical energy and to provide current for operating the cranking motor, inducing spark for ignition unit in petrol engine, provide heat in diesel engine and to energize other electrical units when the engine is idle or the speed of the generator is not sufficient to cope with the full load demands. The electricity supplied by the battery is limited. As the chemicals in the battery are 'used up', the battery runs down or is discharged and can be recharged. The 'used up' chemicals return to their original condition on recharging.

Two types of batteries are generally used in automobiles : the lead-acid type and alkaline type. The lead-acid type battery is usually used in tractor. The tractor is mostly provided with a 12 volt lead-acid battery. It consists of six cells and each cell produces approximately 2 volts.

21.2.1 Lead-Acid Battery

The chemicals in the battery are sponge lead (a solid, grey in colour), lead oxide (a paste, brown in colour) and sulphuric acid (a liquid). These three substances are made to react chemically to produce a flow of current. The lead oxide and sponge lead are held in plate grids to form positive and negative plates.

The main components of a lead acid battery are as follows : *(i)* container *(ii)* plates *(iii)* separators *(iv)* cell covers (v) electrolyte *(vi)* vent plugs. Figure 21.3 shows the various parts of a typical lead acid battery.

Container: The container is a single piece construction and is made of hard rubber or a bituminous material. It is divided into compartments by partitions for different cells. Bridges are formed at the bottom of each compartment on which the battery plates rest. The spaces between the bridge ribs are provided to collect sediment.

Plates: In the battery, several similar plates are properly spaced and welded or lead-burned to a strap to form plate group. The plates consist of perforated grids into which lead or lead peroxide is pressed. The grids are made of an alloy of lead and antimony, which makes them resistant to electrochemical corrosion and gives them strength and rigidity. There are two types of plate groups in each cell *i.e.*, positive plate group and negative plate group. The plate group connected to the positive terminal of the cell consists of grids filled with a paste of lead peroxide. The plate group connected to the negative terminal of the cell consists of grids filled with metallic lead. It is spongy and dull gray in colour. Each group of plates is held together by a post strap, to which each individual plate is welded. These straps are extended up through the cell cover to provide the cell terminals to connect one cell to the other. The plate groups are arranged on the cell so that the positive and negative plates alternate.

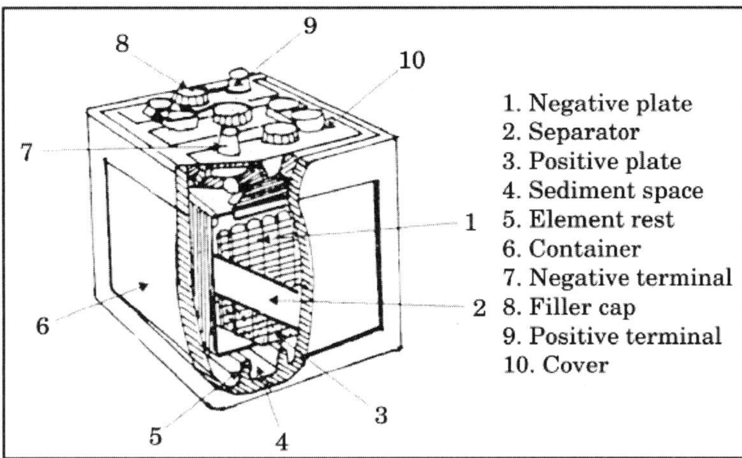

Fig. 21.3: Parts of lead acid battery.

Separators: In between negative and positive plates, separators made of wood or plastic are placed so that due to jerks while the vehicle is moving, these plates do not get short-circuited and both negative and positive plates, remain in their position. The separators must be porous enough to permit liquid to circulate between the plates. These are usually made of specially treated wood, hard rubber or resin impregnated fiber.

Cell cover: Each cell is sealed by a cover of hard rubber through which the positive and negative terminals project. Adjacent negative and positive terminals are connected by connector straps. Each cover has an opening through which liquid can be added. A filler cap is screwed on this opening. The filler cap has an air vent for escape of gas.

Electrolyte: The electrolyte used in the lead acid battery is the solution of sulphuric acid. It consists of 40 % sulphuric acid and 60 % distilled water. The level of the electrolyte in the container is about 10 mm over the tops of the plates.

Vent plug: Vent plug is provided in the cell cover for the purpose of pouring electrolyte and water whenever necessary. The vent plug or filler cap is screwed into the threaded holes provided in the cover. During battery action, water is lost, some by evaporation and some by conversion into hydrogen and oxygen. A hole is provided in the plug to permit escape of these gases. The vent plugs are provided with baffle plates. The gases emitted are likely to carry droplets of electrolyte along with them. These baffle plates trap the droplets of the electrolyte, thus avoiding loss of electrolyte.

21.2.2 Chemical Reactions in the Battery

The chemical reactions take place between the three chemicals *i.e.*, sponge lead, lead peroxide and sulphuric acid in the battery. Current is obtained due to the chemical action among the above three chemicals. Figure 21.4 shows the chemical action in a simplified form during discharge of the battery. The active material in the positive plate is denoted by the chemical symbol $Pb0_2$ and that of the negative plate by the chemical symbol Pb. The sulphuric acid in the electrolyte is denoted by the chemical symbol H_2SO_4, When a battery is being discharged, the chemical action starts. The sulphuric acid molecules split up into H_2 and SO_4. One $S0_4$ unites with the Pb of the positive plate, while the other unites with the Pb of the negative plate, forming $PbSO_4$ in each plate. This action sets free two atoms of oxygen from the positive plate and they get united with the hydrogen left behind in the electrolyte. The union of hydrogen and oxygen forms H_2O which is simply water. This means that during discharge, lead sulphate is formed on both the plates. While sulphuric acid is replaced by water.

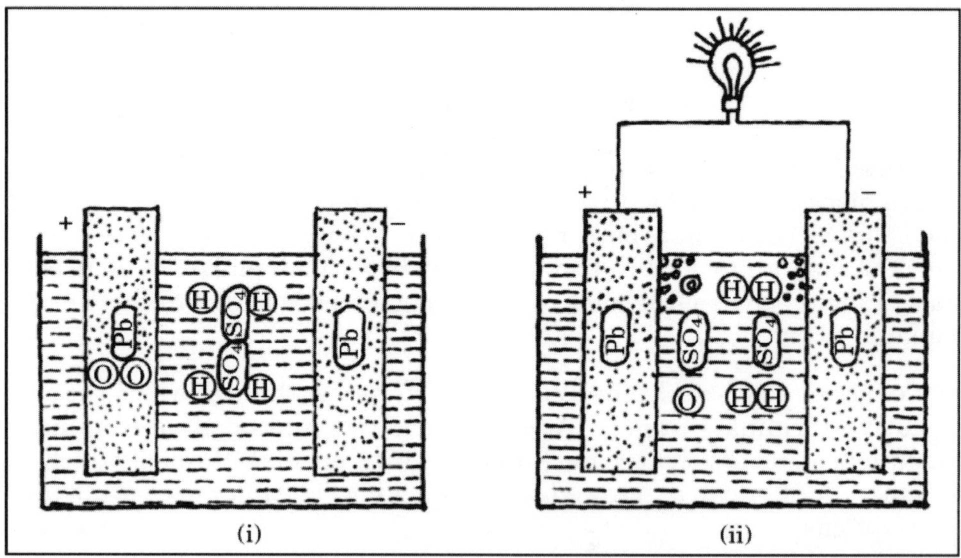

Fig. 21.4: Discharging the battery

While considering the chemical action, only a few molecules have been taken into account. In fact, the plates and the electrolyte contain billions of them and the battery is not considered discharged until most of them have been broken up and reunited.

Direction of current flow: Flow of current is nothing but the movement of electrons. It is produced by shifting $S0_4$ and O_2 to and from the plates. These parts of molecules are called ions. The electrons are carried by these ions from the positive plates and placed on the negative plate. It results in the massing of the electrons on the negative terminal and electron shortage on the positive terminal. On completion of the external circuit between the terminals, there is movement of electrons from the negative terminal to the positive terminal through the external circuit. This action continues so long there are sufficient number of H_2SO_4 and PbO_2 molecules yet to be split up and reunited into H_2O and $PbSO_4$ molecules. The battery is considered to be discharged when most of these molecules have split up and reunited.

Recharging a battery: Figure 21.5 shows the chemical action which takes place during battery recharge in a simplified form. In order to recharge a battery, it is essential that the electrons are forced into the negative plate and removed from the positive plate. To achieve this, an electric current must be forced through the battery in the opposite direction to that of the current flow during battery discharge. The water is broken up into hydrogen and oxygen during recharging. The SO_4 ion moves back to the electrolyte from the plates and it gets united with hydrogen forming sulphuric acid. The atoms of oxygen are driven back into the positive plates, thus forming lead peroxide. In this way, the sulphuric acid reappears in the electrolyte and lead sulphate disappears from both the plates, leaving lead on the negative plate and lead peroxide on the positive plate. The battery is said to be charged when most of the lead sulphate has disappeared from the plates.

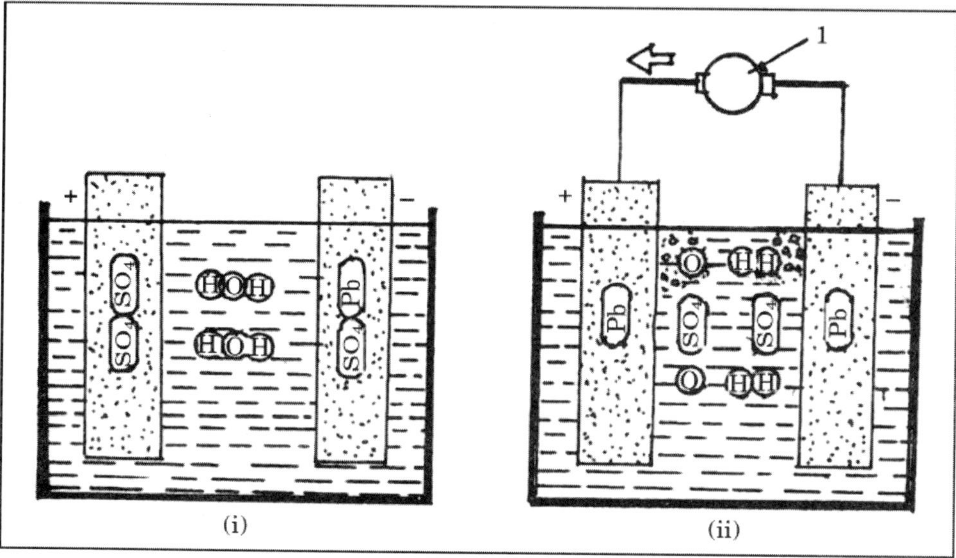

Fig. 21.5: Charging the battery.

The following chemical reactions take place while the battery is charged and discharged.

Charged				Discharged			
Lead peroxide	Sulphuric acid	Lead		Lead sulphate	Water	Lead sulphate	Energy
$PbSO_2$ +	$2H_2SO_4$ +	Pb	$\rightleftarrows$	$PbSO_4$ +	$2H_2O$ +	$PbSO_4$ +	Q
(+)	Electrolyte	(–)		(+)		(–)	

Thus battery is a means of converting electrical energy into chemical energy during charging and the chemical energy into electrical energy during discharging.

21.2.3 Battery Ratings

The battery rating is a measure of the energy stored in it. It is expressed in terms of the period during which the battery will give the rated current before it reaches the specified final voltage. The quantity of current that can be delivered by the battery depends upon the total area and volume of active material on the plates. It also depends upon the amount and strength of the electrolyte present in the battery. The main factors that influence the capacity of the battery are : the number of plates per cell, the thickness and size of the plates and the quantity of electrolyte.

The common ratings are : *(i)* 20-h rate *(ii)* 25-A rate. *(iii)* Cold rate.

20-h rate: It represents the capacity of the battery in terms of the amount of current, it can deliver in a period of 20 hours, holding the cell voltage about 1.75 V and starting with an electrolyte temperature of 27°C. If a battery delivers a steady current of 5A for 20 hours, starting with an electrolyte temperature of 27°C, it will be rated at 100 ampere-hours (Ah).

25-A rate: This rating gives an idea of how long a battery can carry the electrical operating load if the generator does not function. It is the measure of battery performance for the length of time in minutes that a fully charged battery at 27°C can deliver 25 amperes, holding the cell voltage not less than 1.75 V A typical rating would generally be 125 minutes.

Cold rate: It is also termed as short time rate. This rating is the number of minutes that a battery will deliver 300A of current at a starting temperature of –18°C before the cell voltage falls below 1.0 V It indicates the ability of battery during cold weather starting. A 100 Ah battery with a 20-hour rate may be in a position to deliver 300A for a period of 3.6 min, starting at –18°C, before the voltage drops below 1.0 V per cell.

21.2.4 Battery Efficiency

The ability of the battery to deliver current varies within wide limits. It depends on temperature and rate of discharge. At low temperature, chemical activities are greatly reduced. The sulphuric acid cannot work as actively on the plates. Therefore, the battery is less efficient and cannot supply as much current for as long a time.

High rates of discharge will not produce as many ampere-hours as low rates of discharge. At high discharge rates, the chemical activity takes place only on the surfaces of the plates. They do not have time to penetrate the plates and to use the materials below the plate surfaces. The chart below relates battery efficiency to battery temperatures.

Efficiency per cent	*Battery Temperature (°C)*
100	27
65	0
50	–17.8
10	–42.8

21.2.5 Effect of Temperature on Specific Gravity of Electrolyte

The state of battery charge is indicated by the specific gravity of electrolyte which can be measured by battery hydrometer. Hydrometers are calibrated to measure specific gravity correctly at an electrolyte temperature of 27°C. At low temperatures, the electrical resistance of the electrolyte increases and so does its specific gravity. The net effect is a fall in the battery capacity and hence the efficiency. Table 21.1 gives the state of charge of battery with respect to specific gravity of electrolyte.

Table 21.1

Specific gravity at 27°C	*State of charge of battery*
1.280	Fully charged
1.260	3/4 charged (75%)
1.220	½ charged (50%)
1.190	¼ charged (25%)
1.160	About to run down
1.110	Discharged

21.2.6 Procedure and Rate of Charging Battery

The following procedure is adopted for charging of a battery.

(i) Check the level of the electrolyte in the battery. If it is below the top edge of the plates, add more electrolyte so that the level is about 10 mm above the top edge of the plates. Also note the specific gravity and temperature of the electrolyte.

(ii) Connect the negative and positive terminals of the battery to the respective terminals of the battery charger. The battery charger is simply a source to supply direct current.

(iii) Adjust the value of charging current. This is usually kept half, in amperes, of the number of plates in the cell. For example, for a 17 plates battery, the charging current would be 8.5 amperes.

(iv) Continue the charging till the gassing begins, then decrease the charging current and continue till there is no further increase in the specific gravity of the electrolyte and cell voltage reading from three hours.

(v) If the temperature of the electrolyte during the charging process exceeds 45°C, discontinue the charging for some time so that it cools to normal temperature. Then again start the charging.

(vi) Check the specific gravity of the electrolyte hourly during the charging process.

(vii) Avoid overcharging which may damage the battery, particularly the ' positive plates.

A battery generally takes 12 to 20 hours for recharging, depending upon the charging rate and condition of the battery. The battery with sulphated. plates requires low charging current and hence takes longer time for charging. Several battery may be charged simultaneously if they are connected in series.

The battery may be charged either by slow rate or quick rate charging method. Slow rate charging method takes 12 to 20 hours for charging a battery, while quick rate charging method takes about half an hour only. Ordinary batteries are charged by slow rate charging method. It is safe method but takes long time for charging. Quick rate charging method takes much less time. High charging current is passed in the battery till it comes to about 80 % charge level, after which the current is reduced to a low value. While charging by this method, the battery requires a careful watch continuously so that it may not be damaged. However, if a battery is in good condition, it should not be charged by this method.

21.2.7 Battery Maintenance

The battery top should be kept clean. It should be kept dry to prevent surface current leakage. Filler caps should be free from dust, oil, battery acid or water. Similarly, cable connections should be kept clean.

The vent plugs should be removed and acid level should be inspected once every month. If the level is below the prescribed level, then a small quantity of distilled water should be poured in till the proper level is reached.

Given below are the service details that should be carried out: (i) check if the cable connections are loose. The remedy is to tighten the connections *(ii)* check if the cable connections are corroded. If they are so, then clean the terminals and clips by scraping with emery cloth. Smear vaseline or petroleum jelly on the outside *(Hi)* check whether the battery casing is cracked or distorted. If it is so, replace it with a new one.

21.2.8 Care of Batteries in Stock

1. If the battery has to be stored for one or two months, get the battery fully charged and store with its vent plug hole clean. Before reinstalling the stored battery, put it on charge.
2. If the battery has to be stored for longer period *i.e.* one year or more, then proceed as under:

(i) Fully charge the battery

(ii) After charging, drain the electrolyte completely.
(iii) Wash the element with fresh water.
(iv) All the water should be drained and the battery element should be dried.
(v) Store the battery in a room with their vent plug holes clean.
(vi) Before reinstalling, refill electrolyte and put the battery on charging.

21.3 THE CHARGING UNIT

21.3.1 The Purpose of Charging Unit

The charging unit has two jobs: *(i)* To put back into the battery, the current used to start the engine *(ii)* To supply the load for lights, ignition and other electrical equipment while the engine is running. The charging unit includes the dynamo or alternator as the case may be, regulator and battery with connecting wires. The dynamo or alternator works on the principle of electro magnetic induction.

There are two ways of inducing electromotive force by electromagnetic means, namely statically and dynamically. Transformers and ignition coils are the electrical apparatus which induce emf statically, while the generator is an electrical apparatus which induces emf dynamically. The generator is nothing but an electro-magnetic device that converts the mechanical energy received from the engine into electrical energy. The generator in a vehicle restores the current of the battery used in cranking the engine. It also supplies current to the various systems in the tractor such as lighting, ignition etc. at the operating speed.

The generator is generally mounted on the engine block in such a way that the engine fan-belt also provides a drive to the generator. The present-day generators employ the ventilation system. A fan is assembled in the drive end under the generator drive pulley, which sends a draft of air through the generator, carrying away the heat developed. The heat develops due to current flow through the armature and field conductors. This system of ventilation has considerably increased the amount of current that a generator of a given size can generate.

The use of alternating currents, in automobiles or tractors is becoming popular these days. These units are termed as alternators. Dynamo produces direct current and is called direct current generator. Alternator is the alternating current generator. Direct current produced by a dynamo flows in one direction only whereas alternating current produced by an alternator flows first in one direction and then in the other. Alternator uses diode which converts alternating current into direct current. The generators have regulators in the vehicle to keep the generator from producing excessive voltage and current.

21.3.2 Some Electrical Terms and their Definitions

There are many electrical terms which are frequently used while dealing with the electricity. Some of the important terms are given below.

Atoms and electricity: Atoms are the building blocks of the matter. They are extremely tiny. A single drop of water is made up of about 100 billion atoms. There are about 100 kinds of atom. The elements are different from each other with respect to the number of electrons that circle the centre or nucleus of the atom. When electrons break away from their atoms and are sent moving through a conductor *i.e.* a wire, there is a flow of electricity. This is called an electric current. It takes billions of electrons to make any noticeable amount of electric current. For example, it takes more than 6 billion electrons flowing each second to light a small one-ampere light bulb.

Measuring electricity: Electricity is measured in two ways. These are by the amount of current (number of electrons) flowing and by the push, or pressure that causes the current to flow. The push or pressure is, caused by the actions of the electrons. They repel each other. When electrons are concentrated in one place, their negative charges push against each other. If a path is provided for the electrons, they flow away from the area where they are concentrated.

The pressure to make them move is called voltage. If there are many electrons concentrated in one spot, we say that there is a high voltage. With high voltage, many electrons will flow provided there is a path or conductor through which they can flow. The more electrons that flow, the greater the electric current. Electric current is measured in amperes.

There are two measurements of electricity : *(i)* Voltage or the electrical pressure that causes the electricity - electrons to flow *(ii)* Amperes or the number of electrons, the amount of current that flows. To measure volts and amperes, two instruments are used a voltmeter and an ammeter.

Ampere: The rate of flow of electrons is called electric current. The unit of electric current is ampere (A) and represents a flow of one coulomb (6.24 X 10^{IS} electrons) through a given cross section of a conductor in one second.

Ampere hour: It is that quantity of electricity which is conveyed by one ampere current in one hour through a conductor.

Voltage or potential difference: Voltage is the pressure that causes the electric current to flow. Suppose one terminal of an electric source has a great many electrons and the other terminal has a great shortage of electrons. The pressure on the electrons to move from the "many electrons" terminal to the "few electrons" terminal is high. This is a high voltage. No current can flow when there is no potential difference. The unit of potential difference is volt.

Current: Flow of electrons in any conductor is called current. Its unit is ampere.

Electric Circuit: Electric circuit consists of battery or generator as source of electric supply, resistance, switch and the connecting wires.

Open circuit: In open circuit, the switch is open, so current does not pass through this current.

Closed circuit: In this circuit, the switch is in 'on' position, so current passes through it.

Short circuit: In this circuit, the resistance does not come into the circuit and the current takes the short path. Such a circuit takes more current than the normal current.

Sources of electricity: Battery, generator and thermocouple are generally the three important sources of electricity.

Effects of electric current: Electric current has five effects, (i) Physical effect, *(ii)* X-ray effect, *(iii)* Heating effect, *(iv)* Chemical effect, *(v)* Magnetic effect.

Physical effect: Our body is a good conductor. Current can easily pass through it. When the body touches the bare conductor carrying current, current flows from body to earth and body gets shock.

X-ray effect: If very high frequency voltage is passed through vacuum tube, the special type of rays come out, which cannot be seen. These rays are called X-rays. With the help of these rays, the photo of inner part of the body can be taken.

Heating effect: Whenever current passes through any conductor, the heat is produced in that conductor. Such effect is used in heater, press (iron), electric lamps etc.

Chemical effect: When current is passed through the electrolyte (acid mixed water), chemical action takes place. Such effect is used in battery for its charging and discharging.

Magnetic effect: It has been seen by experiments that whenever current passes through a conductor, a magnetic field is produced around the conductor. If this conductor (may be wire) is wound on an iron piece, it becomes a hard magnet.

Resistance: An insulator has a high resistance to the movement of electrons through it. A conductor has a very low resistance. Resistance is very important in all electric circuits. We want high resistance in some circuits so that too much current (too many electrons) will not flow. In other circuits, we want as little resistance as possible so that a high current can flow.

Resistance of a conductor is that property which opposes the flow of electrons (current) through the conductor. The unit of resistance is ohm (Q).

Conductor: The substance which offers least resistance to flow of electricity is called conductor.

Insulator: The substance which offers very high resistance to the flow of electricity are called insulator. It does not allow electricity to flow through it.

Conductance: The reciprocal of resistance is called conductance.

Alternating current: Current that flows first in one direction and then in the opposite direction at regular interval is called alternating current.

Direct current: The flow of electrons (or current) in one-direction only is called a direct current. Power supply source does not change its polarity.

Single phase system: An alternating current electrical system involving only a single coil in field and producing a single voltage and its associated current, is called single phase system.

Two phase system: The alternator producing two-phase supply has two windings (or phase) kept in such a way that the phase difference between them is 90°. It involves two voltages and their associated current.

Three phase system: The alternator producing three-phase supply has three windings (or phase) placed in such a way that the phase difference between them is 120°. It involves three voltages and their associated current. This system is used when large amount of electrical power is required by electric motors and other AC machines.

Kilowatt-hour : Kilowatt-hour is the amount of energy when one kilowatt energy is consumed for one hour. In commercial term, one kilowatt hour is called one unit. Electrical bills are usually paid according to this unit.

Fuse: A fuse is a small piece of metal connected in between two terminals, mounted on an insulated base. It forms a part of the electrical circuit. The function of the fuse wire is to carry safely the normal working current without heating the element. When the normal working current exceeds the predetermined limit, the fuse wire gets heated up to the melting point and the electric circuit is disconnected. The material used for fuse wire may be copper, aluminum, silver etc.

Circuit breaker: Circuit breaker and fuse are installed in the circuit to protect the electrical devices in the circuits. The circuit breaker has a small winding that carries the current in the circuit. When the current is too high, the win ding magnetism opens points to open the circuit. The advantage of the circuit breaker is that it keeps resetting itself. Therefore, it gives a warning of trouble but does not completely open the circuit. For example, if excessive current starts to flow in the headlight circuit, the circuit breaker will operate. The lights will flash on and off, warning the driver regarding the trouble.

Magnet: Magnet is the substance having the properties of attracting iron and its alloys. Magnet may be divided into two general classes, *(i)* Natural magnet, *(ii)* Artificial magnet. The magnet found in nature is known as lodestone or leading stone as it was used by navigators in sea etc. The natural magnet has chemical composition of $Fe_3 0_4$. The artificial magnets are prepared by artificial methods and are classified as (a) permanent magnet (b) temporary magnet or electromagnet. The magnet which retains the magnetic properties for a long period (indefinitely) is known as permanent magnet. The permanent magnets are available in different shape as *(i)* Bar magnet *(ii)* U-shaped *(iii)* Horse shoe magnet *(iv)* Compass needle etc. The magnet which loses it properties as soon as the magnetizing force is removed, is generally known as electromagnets.

Electromagnets act like magnets. An electromagnet can be made by wrapping wire around a tube. Then we connect the ends of the wire (through a switch) to the battery. Now when the switch is closed, there is a strong magnetic field around the coil of wire. With current flowing through the winding, the winding acts just like a bar magnet. One end of it will either attract or repel a pole of a bar magnet. We can change the poles of electromagnet by reversing the leads to the battery.

An electromagnet that is made by winding wire around tube is called a solenoid. It is used in several places in the electrical system of the automobile or tractor.

Strength of electromagnet can be varied but strength of permanent magnet cannot be varied. The cost of electromagnet is more and is suitable in the case of motors and generators of large size. The cost of permanent magnet is less and not suitable for larger size motors and generator. Only used in magnetos.

Magnetism is a property of magnets. Magnet can produce electricity and electricity can produce magnets.

Electromagnetism: The part of science which deals with the relation between electricity and magnetism.

A straight wire which carries current creates a magnetic field around itself. The lines of force form concentric circles around the wire. The magnetic field of a current-carrying conductor behaves in the same manner as the magnetic field of a permanent magnet, as long as current flows.-When the direction of current flow is changed, the direction of the lines offeree also changes. When the current flow in the conductor is increased, the magnetic lines of force also increase proportionately.

If a current carrying conductor is placed in the magnetic field of a permanent magnet (or electromagnet), the magnetic fields of the conductor and the magnet interact with the result the conductor will tend to move across the field of magnet, the direction of this movement depending on the direction of the current flow in the conductor. In this case, electrical energy is converted to mechanical energy. It is exactly this phenomenon that the operation of electric motors depends on. Figure 21.6 illustrates principle of operation of an electric, motor.

Electromagnetic Induction: Electromagnetic induction relates to the generation of electro motive force. If a conductor is moved between the poles of a magnet (or electromagnet), so that it cuts the magnetic lines of force or if a magnet is moved over the conductor so that the magnetic lines of force cut the conductor, a voltage (emf) is induced in the conductor. The electromotive force (emf) obtained in this way is said to be an induced emf and the process is called electromagnetic induction.

The polarity of induced emf depends on the direction in which the conductor moves to cut the magnetic lines of force. When the conductor is moved parallel to the magnetic lines of force, no voltage is induced. The polarity of the induced voltage also changes when the polarity of the magnetic field is changed, provided the conductor maintains its direction of movement. Now if the conductor is made to form the part of a complete circuit, the induced emf will cause current to flow through the circuit. The phenomenon of electromagnetic induction is used to convert mechanical energy to electrical energy by means of electrical machines called generators. Figure 21.7 illustrates the principle of operation of a simple direct-current (d.c.) generator.

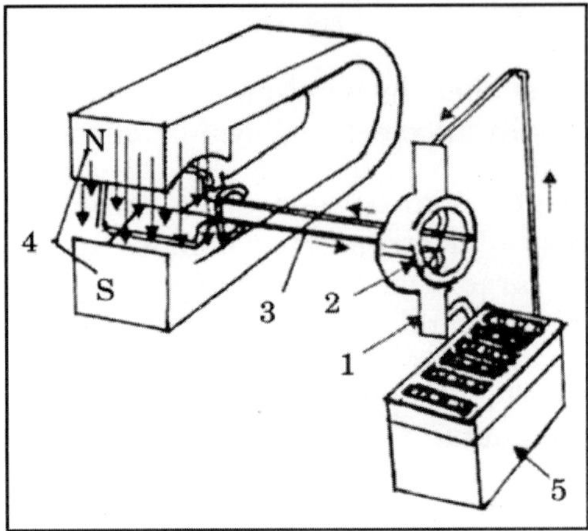

Fig. 21.6: Electric motor

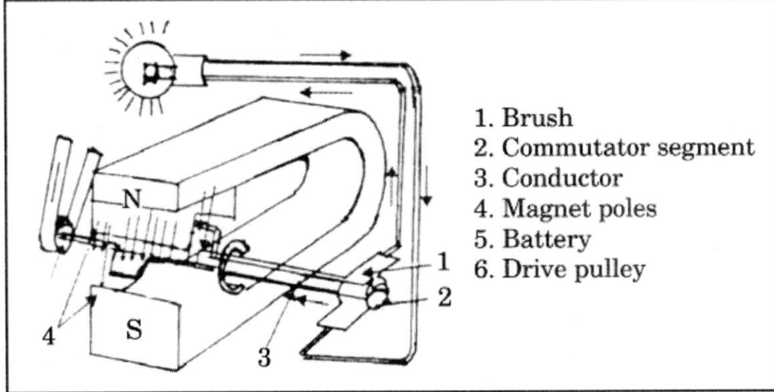

Fig. 21.7: Electric generator

Diode: The diode is an one-way valve for electrons or electric circuit. Automotive or tractor electrical systems use only dc current. The alternator produces ac current. Diodes convert this ac into dc, so the automotive electrical system can use it. A diode allows current to flow one way. But the diode does not allow current to flow in the opposite direction.

21.3.3 Generation of Electricity

Generation of electricity is based upon Faraday's laws of electro-magnetic induction. In 1831, Faraday discovered the basic principle of generation of electricity and explained the following laws.

First law: Whenever any conductor is made to rotate in a magnetic field and hence cut the magnetic lines of force or flux, e.m.f. will be induced in that conductor.

Second law: The magnitude of induced e.m.f. is directly proportional to the rate of change of flux linked with the conductor.

Figure 21.8 demonstrates how current will be generated in the conductor if it is moved in the magnetic field. Here the conductor is connected to the galvanometer and it is then pushed back and forth between the poles of the magnet. When it moves through the magnetic field, the meter needle will deflect, indicating the flow of current. The current flow will also be indicated if the magnet is moved with respect to the conductor. This' effect is termed as electro-magnetic induction. The current is induced as long as there is relative movement, so that the lines of foree are cut. In case of a dynamo (dc generator), the conductor is moved and in alternator (ac generator), the magnet is moved.

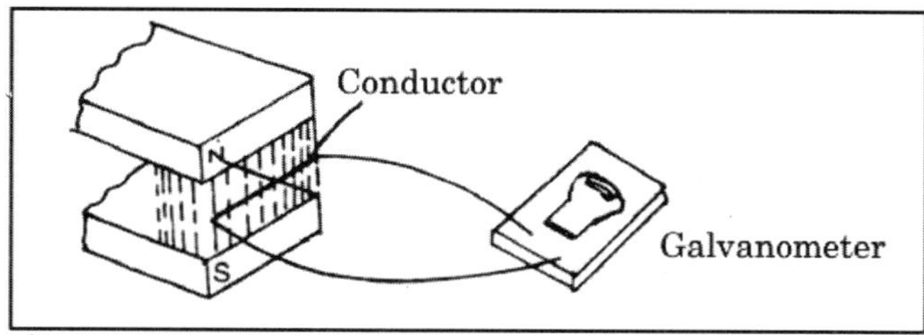

Fig. 21.8: Electromagnetic induction.

21.3.4 Principle of Generation of D.C. Voltage and A.C. Voltage

The direction of induced voltage depends on the way, the lines of force are being cut. Slip ring or collecting ring used in generator helps in producing A.C. voltage where as commutator generates D.C. voltage. The metal rings fitted on the armature shaft are called slip rings. The carbon brushes rest on these slip rings are to collect current in the machine. The terminals of armature winding terminates on these rings and are soldered with these rings. These rings rotate as fast as coils on the armature and give out the same voltage as is induced in the coils.

The commutator is a copper ring type but is the combination of small pieces. The one piece is called segment. The segments are insulated from each other and are combined together tightly and fitted in the shaft of the armature. This is called commutator. The terminals of the winding coils are soldered on the segments. Carbon brushes resting on this commutator are to collect the supply from the armature. The commutator rotates but the brushes are fixed. The work of commutator is to change the A.C. voltage produced in the armature conductor to unidirectional D.C. voltage.

Figure 21.9 shows the simple two pole and one coil generator. In Fig. 21.9A, the two slip rings are connected with the two terminals of the coil (armature). In Fig. 21.9B, the coil is at 90° to the field. When the coil rotates to right hand the coil sides become parallel with the field, so there will be no cutting of flux and hence no emf is induced.

In Fig. 21.9C, the coil sides are in parallel with the field after rotating 90° of the previous position. When coil rotates to right hand, it cuts maximum flux and hence maximum emf will be induced. With the help of Fleming's right hand rule (Fig. 21.10), the direction of current is shown in sides AB and CD.

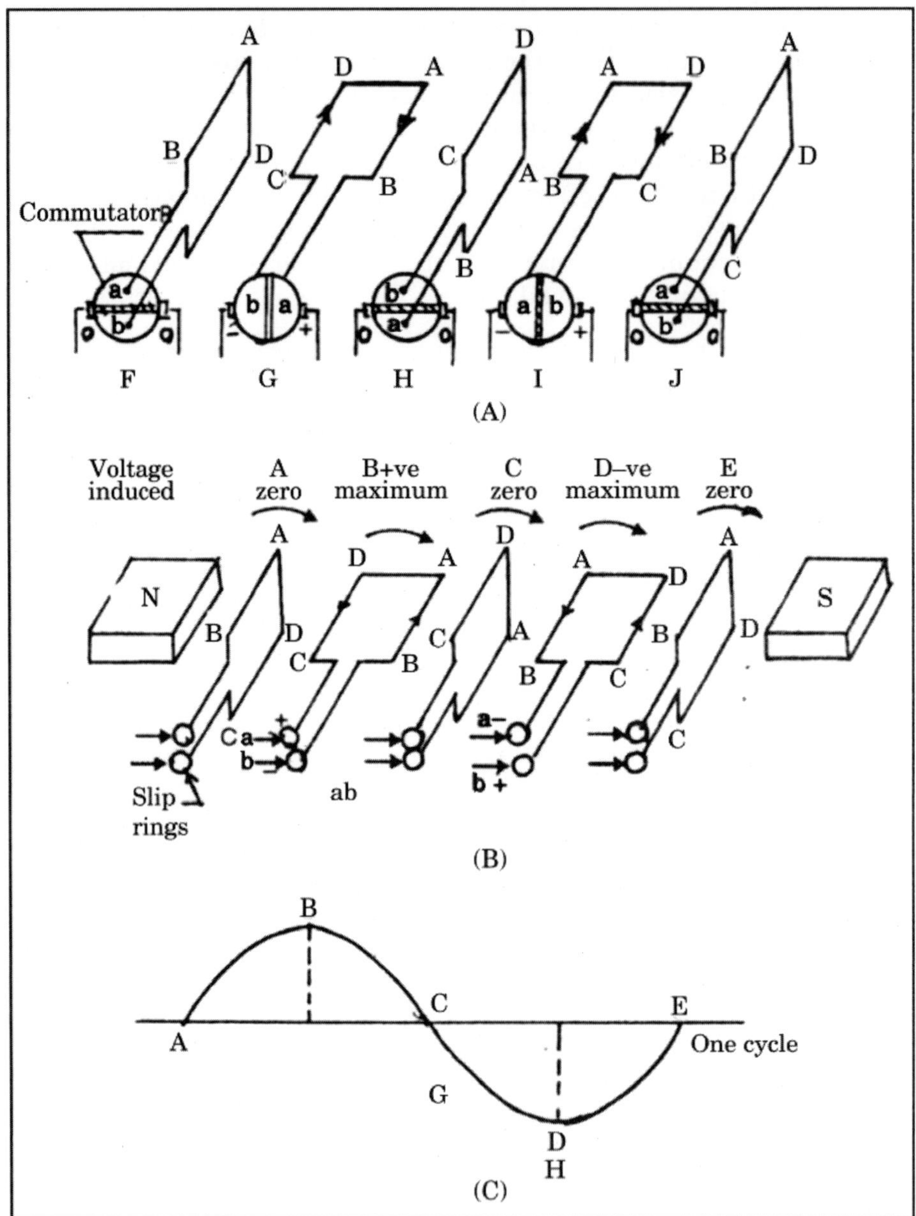

Fig. 21.9: Simple generator.

In Fig. 21.9C, the coil is in the same position as Fig. 21.9A, but AB side is downward and CD is upward, even then no voltage is induced as there occurs no cutting of flux.

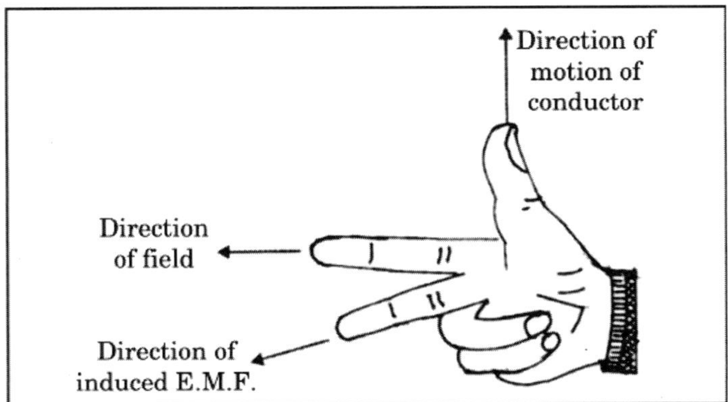

Fig. 21.10: Fleming's right hand rule.

Fig. 21.9D is same as Fig. 21.9B, but coil sides are reversed. But here sides cut maximum flux, the voltage induced is maximum but in reversed direction.

It is seen that in Fig. 21.9B the slip ring 'a' is positive and 'b' is negative but in Fig. 21.9D, the slip ring 'a' is negative and 'b' is positive. In this way a cycle of induced e.m.f. is complete. This is called alternating voltage and the associated current is alternating current.

D.C. voltage: In Fig. 21.9 F, G, H, I, J.,' the commutator is shown which rotates as the coil rotates as the brushes are fixed. Similar to 21.9 A, the voltage induced in the Fig. F is zero, in G maximum, in H zero, I maximum, and again in J, the voltage induced is zero. But it is seen that one brush is always positive and the other remains negative (Fig. 21.9 G and I). On the positive brush, where side AB is positive, segment 'a' is on positive brush and when side CD (Fig. I), is positive then its segment 'a' is at positive brush again. In this way unidirectional voltage is obtained with the help of commutator.

21.3.5 Dynamo

Dynamo is a dc generator. Dynamo is a machine which converts the mechanical energy from the automobile/tractor engine into electrical energy. It replaces the battery current used in starting the engine and also supplies current for operation of electrical devices such as ignition unit, light etc. Dynamo produces direct current, because the use of battery requires direct current.

The dynamo is usually mounted on the side of the engine block. It is driven by the engine fan belt. As the engine speed is subjected to large variations the generator speed also varies. But the necessity demands that the voltage should remain nearly constant. When dynamo is driven by the engine, it supplies electrical energy for all the circuits of the vehicle and keeps the battery fully charged. Generally all automobiles and tractors use generators or dynamo to supply direct current. Many manufacturers use the commutator type shunt-wound, self-energizing generator units. However, in recent years, alternator type generator is also used, which employs diodes to convert the generated alternating current, into direct current.

Construction: Figure 21.11 shows the sectional view of a commutator type dc generator or dynamo. The dynamo consists of a cylindrical soft-iron frame, a pair of field magnets fixed to it, an armature and field coils. The frame is closed at both ends by a commutator end (bracket) assembly and a driving end bracket. These end covers are held to the frame or yoke by long through bolts. The commutator end bracket has a brush gear assembly.

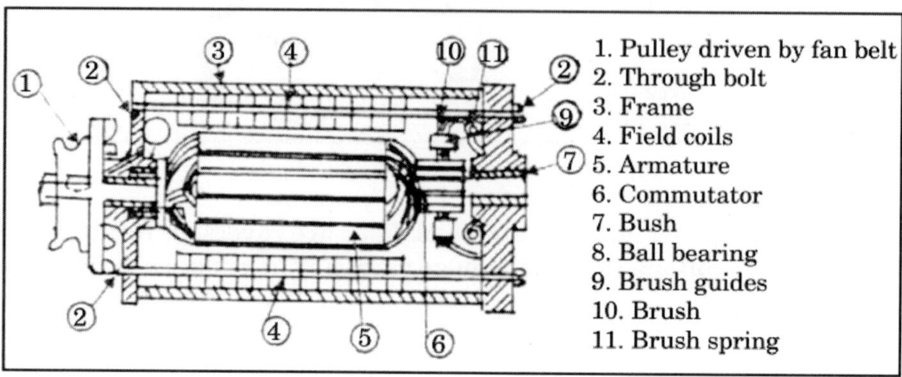

Fig. 21.11: The dc generator details.

Field magnets may be permanent magnets or electromagnets energized by the generator itself. Now-a-days electromagnets are used because higher outputs can be obtained and self-excited field coils can be controlled easily by regulating the field current.

The armature is supported at both ends on ball bearings. It has a core with longitudinal slots which house the coil windings. The ends of these windings are connected to the commutator.

Commutators are of two types *(i)* built-up and *(ii)* moulded. The moulded commutator is used in small generator but in-built types are generally employed now-a-days. The built-up type commutators have copper segments with radial sides. Mica sheets between copper segments provide insulation. The copper and mica segments are compressed and machined after being fixed to a shaft. The mica is ultimately under cut to avoid poor contact between the brush and the segments.

The brushes are of rectangular in shape. These brushes fit and slide in box type brush holders. A spring exerts pressure on the brush so that it may maintain good contact with the commutator in all conditions. Flexible leads are soldered with the brushes, the ends of which are attached to the terminals of the end plate.

The armature and commutator are designed to rotate together. They allow the generator to produce a flow of direct current, that is, the current continues to flow in the same direction. As the two ends of a conductor rotate and change position with respect to each other, the two segments of the commutator also changes positions so that the current continues to be fed to one brush in the same direction. A regulator controls the output current and voltage in order to save damage to the electrical units and to the generator itself.

21.3.6 Alternator

There has been a marked increase of the electrical load on the generators in the present day automobiles/tractors. The design of dc generators becomes cumbersome and presents commutation limitations. Now a days alternating current generators are used extensively as these can give current outputs as high as 40 amps.

The component parts are just the same as dc generators except that stator contains the generation coil and rotor of the field. Instead of commutator and the brushes, rectifier (or diode) is provided to convert ac into dc.

Advantages: (i) Alternators are smaller and lighter than dc generators for a given output *(ii)* Alternators require lesser maintenance because of the absence of brush gears and commutators and are more reliable, *(iii)* They can give higher output at low engine speeds. Batteries can be charged at engine idling speeds. Therefore, these are suitable for vehicles which have frequent stops *(iv)* These can run at higher speeds than dc generators producing higher current outputs (v) Cut-out relay is not required with alternators because the rectifier does not allow flow of reverse. The only disadvantage is the high initial cost.

Working principle: An armature magnet is shown in Fig. 21.12 *(i)*. A coil of wire is around this magnet. In this case, the north pole is at the top and the south pole at the bottom. When the armature magnet is rotated from top to bottom, that is, by half a revolution, it produces current in the wire in one direction.

In Fig. 21.12 *(ii)*, the south pole is at the top and the north pole at the bottom. When the magnet is rotated by another half revolution, the flow of current in the wire is reversed. Therefore, in one revolution of the armature magnet, the current flows in one direction and reverses in the opposite direction. This type of flow in both forward and reverse directions is called alternating current.

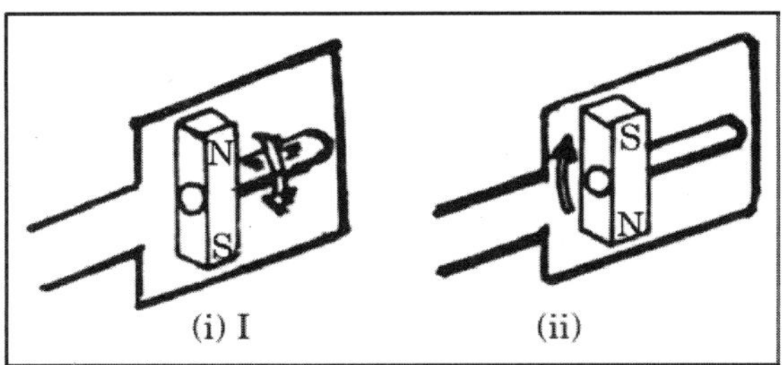

Fig. 21.12: Principle of production of alternating current.

Construction: Figure 21.13 shows various parts of an alternator. The alternator has a rotor and a stator. The rotor Is called the armature. It is mounted on bearings and is placed inside the stator. It is driven by fan belt. The armature contains one winding coil. This coil is connected to a separate slip ring. The current from the battery passes through the slip rings to this coil. This current.passes.through two small stationary carbon brushes. When the current flows through the armature coil, it turns into a magnet. One of its ends becomes the north pole while the other end becomes the south pole.

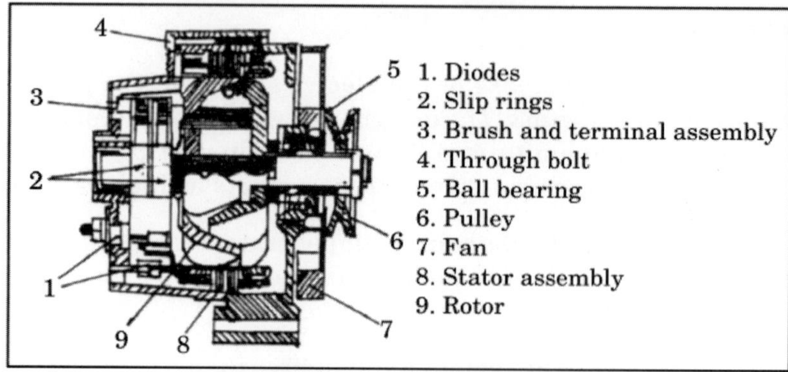

Fig. 21.13: Alternator

There are windings around the stator (Fig. 21.14 (i))- Whenever a magnet passes through each stator coil, current is generated. If there are more magnets passing through each coil in a given time, the current generated is more. The armature is a single magnet, but it acts like a set of magnets. For this purpose, the ends of the armature are formed into metal fingers [Fig. 21.14 *(ii)]*. Each finger becomes a magnet. These fingers inter lock, but do not touch each other.

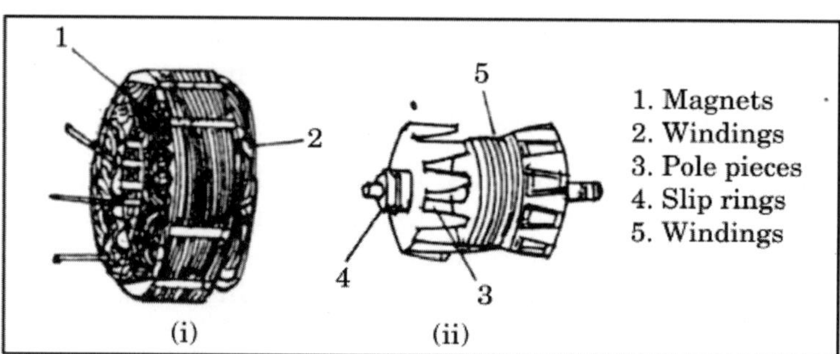

Fig. 21.14: (i) Startor (ii) Rotor.

The north and south poles of the magnets pass each stator winding. Due to this rotation, the current is generated in the winding, alternately positive and negative. This alternating current changes to direct current. Because only the direct current can change a battery. Diodes are built into the alternator for converting ac current to dc current.

The alternator limits its own current output. The rectifiers do not pass current in the reverse direction. Therefore, they do act like the cut-out. For this reason, the alternator only needs a voltage regulator.

21.3.7 Types of Field Coil of Generator

The performance of an electric generator depends upon the manner of exciting magnetic field.

Figure 21.15 illustrates the three ways of connecting field windings, namely series, shunt and compound wound. In series wound generator, the field coils are wound with few turns of thick wire of large cross-section and are connected in series with the armature. Under no-load or open circuit conditions, the voltage will be zero, since there is no flow of current in the field windings. The voltage of the generator will increase as the load current increases. In the shunt-wound generator, the field windings are connected in parallel with the armature conductors and are made with many turns of fine wire. The voltage of the generator is maximum on the open circuit and diminishes slightly as the load is increased. This is because of the following reasons : (a) voltage drop in the armature, (b) demagnetizing effect of the armature current, (c) the lower generator voltage causes reduction in field current.

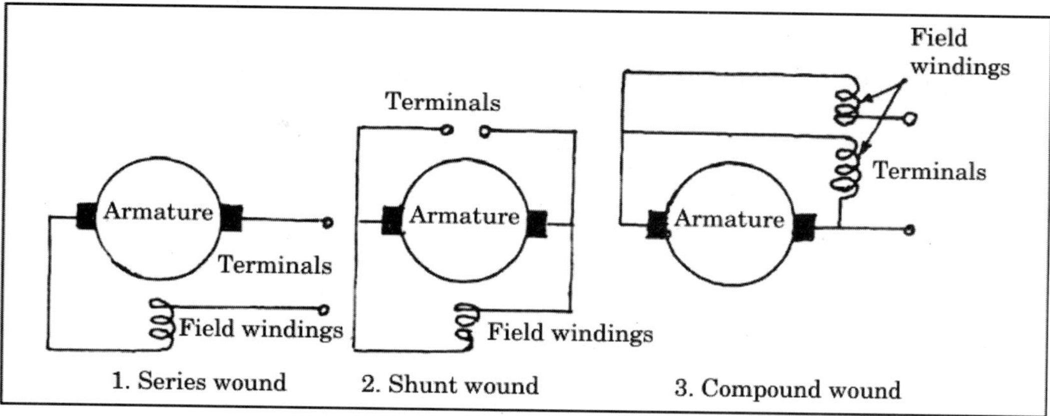

Fig. 21.15: Generator field windings connections.

The shunt-wound type of generator is suitable for steady load conditions where there is not much importance of voltage variations, for instance, battery charging.

The compound wound generator has both series and shunt field windings. The no load voltage is maintained by the shunt winding, while under load conditions, the series winding increases the field strength. If the series winding is suitably proportioned, the generator operation can be maintained over a wide load range without any appreciable variation in the voltage of the generator. This type of generator is most suitable for supplying power and lighting for industrial and domestic purposes.

21.3.8 Regulation of Dynamo

The output of dynamo is fully regulated with respect to the control of current and voltage being produced by it. The control unit *i.e.*, the regulator is a box comprising a voltage regulator, a current regulator and a cut-out element. The three are mounted on an insulated zinc plated steel base and are covered by a phenolic moulded cover. The terminals pertaining to dynamo field and battery are placed out of the base.

If the full charging current is generated at low engine revolutions, the charging current would be excessive at higher engine revolutions and may damage the battery and the general electrical system. The generator output must automatically adjust itself to the electrical needs of the vehicles. The charging rate of the generator must be adjusted to the condition of the battery *i.e.* a high rate for a low battery and a low rate for a fully charged battery. Therefore, it is necessary to introduce some means of controlling the voltage or current. By generator control (or regulation) we mean preventing the generator from producing excessive voltage and current without any control and generator would continue to increase its output as speed goes up until it would be producing much current that it would overheat and burn up. The voltage and current of a dc generator are controlled by using an external resistance.

21.3.8.1 *Cutout relay*

The cutout relay (or circuit breaker) closes the circuit between the generator and the battery when the generator is producing current. Also it opens the circuit so that the battery cannot discharge back through the generator when generator stops or slows down. DC generator is connected to the battery through a cutout relay.

When engine is running, the battery is charged by the dynamo. The speed of the dynamo is slow. At this time, there is a tendency for the battery to discharge the current to the dynamo. The dynamo then begins to run as a motor. This is not desirable. Therefore, there should be a device that would permit the dynamo to charge the battery, but at the same time, prevent the current from going back to the dynamo.

The device used is called a 'cutout'. It permits the dynamo current to flow to the battery. But it prevents any reverse flow from the battery to the dynamo. The cutout is nothing but a switch that operates electromagnetically. Figure 21.16 shows the schematic wiring diagram of a cutout relay. It has an armature of steel carrying the contact points. It has two windings, namely series and shunt winding. Shunt winding is connected parallel to the dynamo and series winding is in series with the dynamo and battery. These two coils are wound with soft iron core, There is also an L-shaped member with which the armature is hinged and is kept away from the windings by a spring when the generator is not running.

After starting of engine, the generator starts operating and a voltage is imposed on two windings of the cutout relay. It produces the magnetic field which attracts the armature. As soon as voltage produced by the generator is of sufficient value to be able to charge the battery, the magnetism is strong enough to pull the armature down after overcoming the spring tension. This action makes the two contact points to meet, thereby completing the circuit between generator and battery. The battery begins to get charged.

When the generator stops working, the current will begin to flow from the battery to the generator, changing the direction of current flow in the series winding. This results in reversal of the magnetic field in this winding, whereas the magnetic field in the shunt winding stays in the same direction. The two fields no longer help each other but instead oppose each other. It reduces the total strength of the magnetic field and the spring tension on the armature pulls the contact points away, thereby opening the circuit between the battery and the generator. Due to this, the battery current does not flow to the dynamo. In this way, the cutout relay provides protection to the generator and the battery.

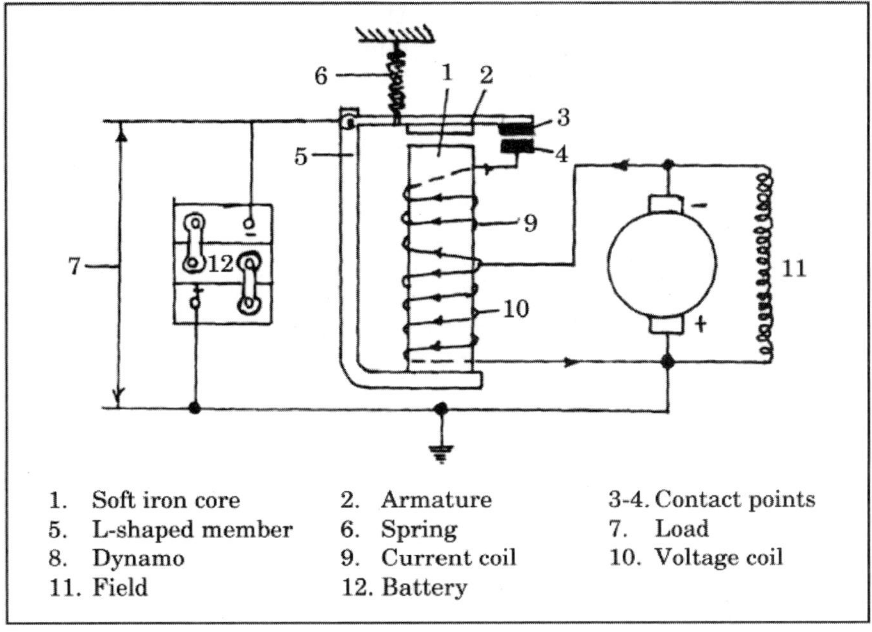

Fig. 21.16: Cutout

21.3.8.2 *Current regulator*

In the current regulator, there is a pair of contact points and a resistance. When the current output of the dynamo reaches pre-set value, it creates sufficient magnetic force in the core which attracts the armature to open the contact point. This action adds resistance to the generator field to cut down the generator output. When the generator current is reduced, the magnetic force is also reduced, the spring tension pulls the armature to close the contact points. The current begins to flow immediately and causes the generator current to increase again. In this manner, the contact points keep on opening and closing. This opening and closing takes place a number of times; about 250 times a second, causing the current regulator to limit the current output of the generator to the rated value. This prevents the flow of large amounts of current into the dynamo.

21.3.8.3 *Voltage regulator*

In voltage regulator, there is a shunt coil with a dynamo and a resistor (or a resistance). There is also a pair of contact points. When the battery is in a low stage of charge, the regulator does not operate. When the generator voltage reaches a predetermined value, the winding builds up sufficient magnetic strength that pulls down the armature against spring tension. Thus the contact points of the voltage regulator open, collapsing the magnetic field of the winding. The resistance comes into action. This increased resistance in the generator field circuit reduces the current through it and hence reduces the magnetic strength. This cuts down the generator output. As the magnetic field strength is reduced, the armature is again pulled back making the contact points closed. The value of the voltage of the dynamo now again increases. The entire sequence *i.e.*, opening and closing of contact points is repeated many times a second, causing the resistance to be inserted into and removed from the generator field circuit as many as 200 times a second. This holds the voltage to a constant value.

Figure 21.17 shows various parts of the regulator. It consists of the cut-out relay, current regulator, voltage regulator, dynamo and field coil. When dynamo is started, the voltage builds up. When sufficient voltage is reached, the cut-out relay comes into action. By this, armature (4) is pulled towards the soft iron core of the cut-out relay. This gives a connection between the dynamo and battery.

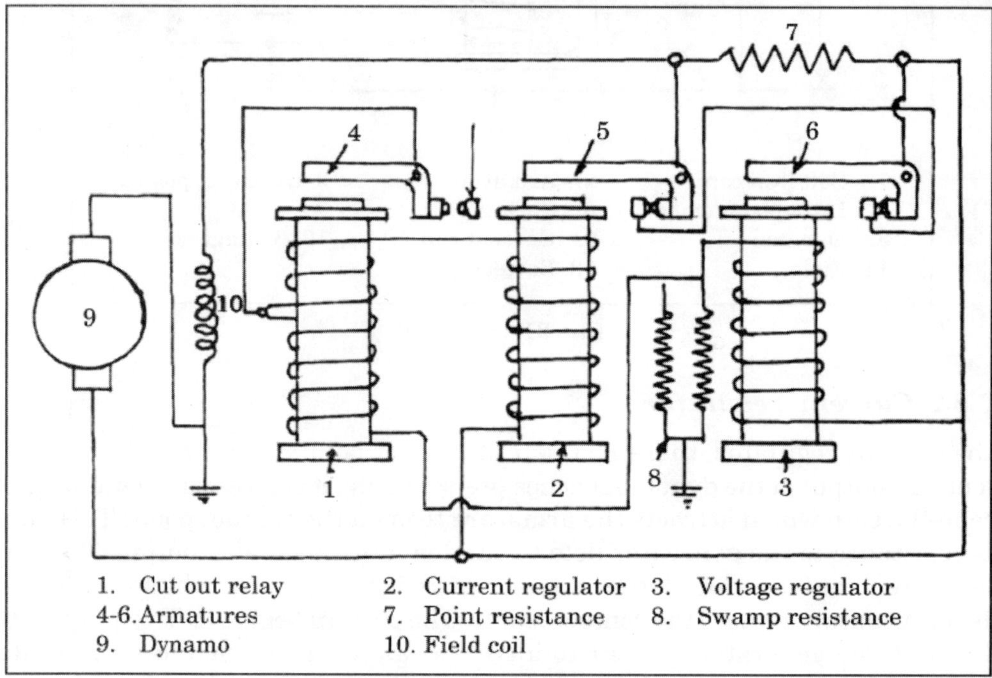

Fig. 21.17: Regulator

When dynamo starts running at a higher speed, the voltage regulator immediately comes into action. Armature (6) is pulled away from the contact points of the voltage regulator. The contact points are now opened. The resistance (7) is then introduced in the field coil, thereby bringing down the voltage.

Sometimes the voltage of the dynamo fails to reach the operating value of the voltage regulator. This is due to the discharge of the battery. This happens when there is large electrical loading. At this time, the current regulator comes into operation. If the current output is increasing, the current regulator pulls armature (5) closer. When the contact points of the current regulator open, the resistance (7) is brought in the field circuit. Thus current limit is maintained. By combining the voltage and current regulators, the optimum specified values of the voltage and current outputs are maintained.

21.3.8.4 *Alternator regulation*

Regulators for alternators operate in the same manner as regulators for generators. The regulation is achieved in both systems by varying the amount of resistance in the field circuit of the alternator/generator. It is not essential to employ an external device for limiting the current in order to control the output of the alternator. Alternators are designed and constructed in such a way that they have a built-in current limiting action. As the current increases, it produces stronger magnetic fields in the stator windings which counteracts the rotating field. Thus, when current reaches to its limiting value, the stator and rotating fields balance each other that prevents higher output.

The cutout relay is not required since the diodes permit the flow of current only in one direction- that which is required to charge the battery. When the alternator slows down or stops, the battery cannot discharge back through the alternator even though the two still remain connected to each other because diodes do not permit this flow of current.

Alternator has only voltage regulation system. The voltage regulator works like that of generator *i.e.*, as the voltage reaches a predetermined value, the regulator contact points open to insert resistance into the field circuit of the alternator, preventing excessive voltage.

21.4 IGNITION UNIT

The mixture of fuel and air in an engine cylinder in spark ignition engine at the end of compression stroke cannot produce power, unless it is ignited in every cycle at the proper instant. The spark ignition (S.I.) engines require some device to ignite the compressed air-fuel-mixture inside the cylinder at proper time for useful work. Arrangement of different components for providing such ignition at proper time in the engine cylinder is called ignition unit. The basic methods for igniting the fuel charge (i) Ignition by an electric spark («) Ignition by compression.

In a carburetor engine, a mixture of fuel and air is compressed in the cylinder. Then a high voltage spark jumps across the spark plug electrodes gap to fire the charge. The spark must be strong enough to jump across the gap and at the right time. If the spark occurs too early, *i.e.*, before the piston has reached TDC, then the engine tends to run in the reverse direction. At full engine speed, however the spark must occur earlier, so that proper pressure from the burning fuel charge is built up, just as the piston starts moving down from TDC to BDC. The ignition unit is so designed that the spark is retarded for starting the engine and as the engine picks up speed, the spark is advanced. The ignition unit is a part of electrical system which carries the electrical current to spark plug which gives spark to ignite the air-fuel mixture at the correct time. The electrical system which provides the spark, makes use of either a six or twelve volt battery or a magneto.

Compression ignition engines (diesel engines) do not have such ignition systems. In a compression ignition engine, only air is compressed in the cylinder and at the end of the compression stroke, the fuel is injected which catch fire due to the high temperature and pressure of the compressed air. Compression ignition is the chief characteristic of diesel engines, due to which they are often termed as compression ignition engines. It uses the heat developed due to compressing the mixture to a very high ratio. A pressure of nearly 35 to 45 kg/cm^2 developed by a compression ratio of 14:1 to 22:1 producing a temperature of nearly 500°C, which is high enough to explode the fuel. In this method, no electrical accessory is needed for ignition purpose. There is no external agency to produce a spark in the cylinder. The ignition is due to the heat of the compressed air.

21.4.1 Types of Ignition Unit

The ignition in petrol engines is basically of two types, *(i)* Battery ignition unit, (ii) Magneto ignition unit.

Both the ignition units are based on the principle of mutual electromagnetic induction. The battery ignition system is mostly used in passenger cars and light trucks. The magneto ignition system is generally used in small engines such as motor cycles, scooters, moped or portable engines. In battery ignition system, the current in the primary winding is supplied by the battery where as in magneto ignition system, the magneto produces and supplies the current in the primary winding.

21.4.2 Mutual Electromagnetic Induction

If two coils or circuits are placed near to- each other and if one is connected to electric supply, the e.m.f. will be induced in the other coil. The amount of emf. induced depends on the number of turns on the second coil.

So the mutual induction is the electromagnetic induction produced by one coil (circuit) to the nearby coil due to the variable flux of the first coil cutting the conductor of the second coil. That means when two coils are kept near to each other and if current is given to one coil and it is changed, the flux produced due to that current which is linking both the coils, cuts both the coil, an emf will be produced in both the circuits. The production of e.m.f. in second coil -is due to the variation of current in first coil known as mutual induction. Both the circuits or coils have no electrical contacts with each other but magnetic linkages. This principle is applied in transformers and induction coil of the automobile engine. The transformer is mostly used to increase the voltage and thus decrease the current and *vice-versa*, keeping' the power same.

21.4.3 Transformer

The transformer is a static piece of electrical apparatus, which converts electrical power from one circuit to the other circuit at the same frequency. It can increase or decrease the voltage with corresponding decrease or increase of current keeping the power same. Power (watt) = VI (voltage in volt x current in ampere). This transformation of energy is done due to the Faraday's Laws of electromagnetic induction through two windings, primary and secondary.

In its simplest form, a transformer consists of two coils, one called the primary winding and the other, the secondary winding. The coils usually are wound on a laminated core of magnetic material and the transformer is then known as an iron-core transformer. If the two coils are placed near to each other and if one is connected to the electric supply, the e.m.f. will be induced in the other coil. The e.m.f. induced will be according to the turns in the second coil. If the secondary winding has more turns than the primary winding, the voltage is increased and the current is proportionately decreased. Such a transformer is known as a step-up transformer. If the secondary winding has less turns than the primary winding, the voltage is decreased and the current is increased in the same proportion. The transformer in this case is called a step-down transformer.

The transformer is mostly used, in A.C. supply not D.C." supply to increase the voltage and thus decrease the current and keep the power same for transmitting the high voltage and to decrease the voltage at load point. Transformation ratio (K) is defined as follows :

$$\frac{N_s}{N_p} = \frac{V_s}{V_p} = \frac{I_p}{I_s} = K;$$

Where N_s = Secondary winding turns; N_p = Primary winding turns; V_3 = Secondary voltage turns; V_p = Primary voltage turns; I_s = Secondary current turns; I_p = Primary current turns.

When alternating current has to be transmitted over long distance, its voltage is stepped up to thousands volts, so the current becomes low and therefore, thin wire can be used in transmission. This saves cost and thus electricity becomes cheaper. When the supply is to be given to the consumers, the voltage is stepped down to 440 volts by transformer. The first coil to which energy is supplied is called the primary winding and the second coil from which energy is given out is called the secondary winding.

21.4.4 Condenser

Condenser can be defined as two conductors separated by an insulating dielectric. The insulating medium is called a dielectric. It is a device to store electrical energy and to release it when required. The condenser is also called capacitor. The conducting surfaces may be in the form of either circular, rectangular or spherical.

A parallel plate capacitor is shown in Fig. 21.18. One plate is joined to the positive end of the supply and the other to the negative end or is earthed. It is experimentally found that in the presence of an earthed plate B, plate A is capable of withholding more charge than when B is not there. When such a capacitor is put across a battery, there is a momentary flow of electrons from A to B. As negatively charged electrons are withdrawn from A, it becomes positive and as these electrons collect on B, it becomes negative. Hence a potential difference (p.d.) is established between plates A and B. This transient flow of electrons give rise to a charging current. The strength of the charging current is maximum, when two plates are uncharged, but it then decreases and finally ceases when p.d. across the plates becomes slowly and slowly equal and opposite to the battery e.m.f.. In fact charging of a capacitor is in many respects similar to the changing of a water tank T from a reservoir R (Fig. 21.19). On opening the valve, the water rushes through the connecting pipe from R to T. At first, the rush would be great and then flow of water will go on decreasing as the level of water in T goes on increasing. Obviously, the water flow will be stopped when the water levels in R and T become equal. In the same way, when capacitor is fully charged *i.e.*, when p.d. across its plates becomes the same in magnitude as the charging e.m.f. then no current flows.

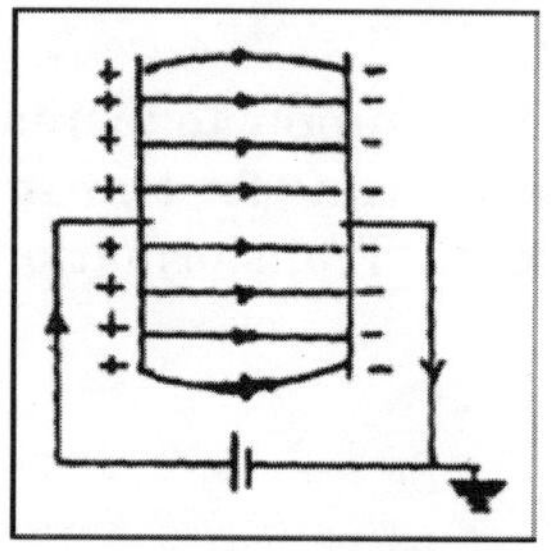

Fig. 21.18: Figure explaining working of a capacitor

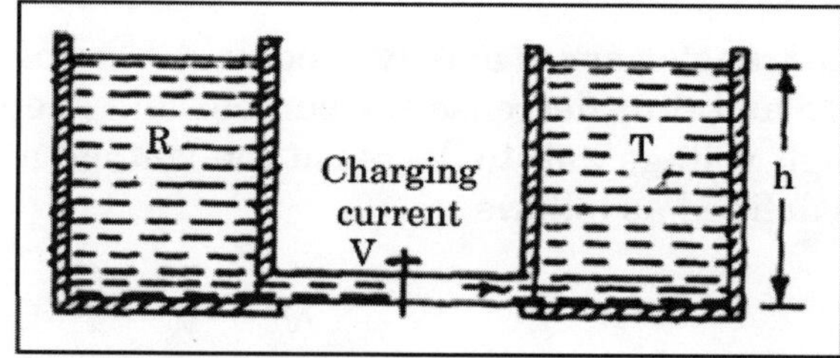

Fig. 21.19: Figure explaining working of a capacitor

The battery itself does not create electricity on plates A and B. Its function is merely to cause the transfer of electrons from A to B and hence create a p.d. between them. Its function is analogous to a pump. When capacitor is connected across a battery, no continuous current can flow through the capacitor. There is only a momentary shift of electrons from one plate to another.

Capacitance is the amount of charge required to create an unit potential difference between its plates

$$C = \frac{Q}{V} = \frac{Colomb}{Volt}(Farad)$$

21.4.5 Battery Ignition

Battery ignition unit includes two circuits (a) Low voltage (Primary circuit) (b) High voltage (Secondary circuit). The low voltage circuit consists of battery, ignition switch, primary winding, contact breaker, condenser. All are connected in series except condenser. The condenser is connected in parallel to the contact breaker unit.

The high voltage circuit consists of secondary winding, distributor rotor, high voltage wiring and spark plugs. They are also connected in series. The negative terminal of the battery, condenser and one terminal of each plug are all earthed to the metal body of the engine. Figure 21,20 shows the battery ignition unit for a four-cylinder engine.

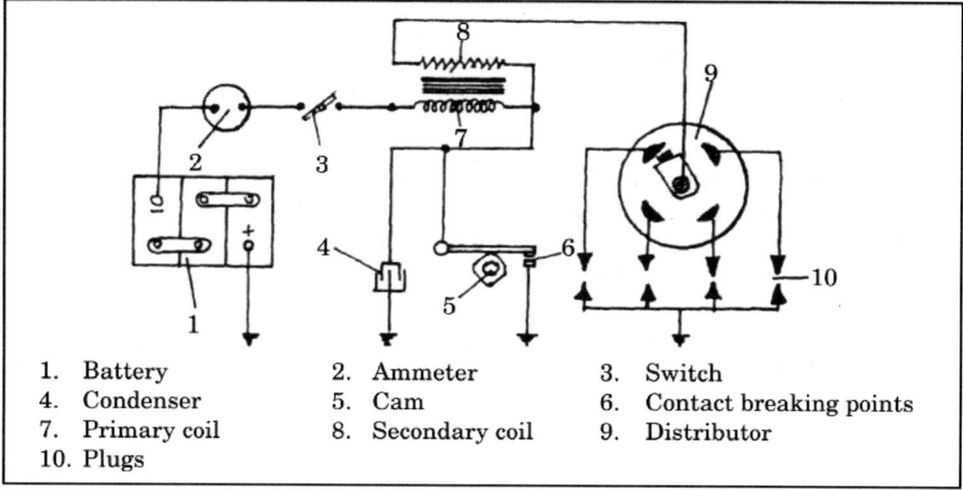

Fig. 21.20: Simple ignition circuit for a four cylinder engine.

The current is taken from battery to the ignition switch, from ignition switch to the ignition coil, from ignition coil to the primary circuit and to contact breaker and to the condenser and is earthed. From secondary circuit, a high tension lead is taken from the ignition coil to the distributor cap central terminal. From this, terminal current is taken to the rotor and distributed to various segments depending upon the number of cylinders and from these segments of distributor to various spark plugs.

When the ignition switch is on, the current flows from the battery through the primary winding. It produces magnetic field in the coil. When the piston is at the end of compression stroke, the contact breaker point opens. Thus the flow of current in the primary winding causes the magnetic field to collapse. As a result, an induced e.m.f. is generated in the primary winding. That induced e.m.f. tends to prevent the breakdown of the magnetic field. Hence, current flows in the same direction as the battery current. This current having no other exit, charges the condenser. The charge on the condenser builds up and discharges back into the primary winding breaking down the magnetic field. As the magnetic field collapses, its lines of force cut the wire turnings of the secondary winding. The secondary ignition circuit is not connected electrically to the primary ignition circuit. As the secondary winding has many more turns (about 20,000 turns) of fine wire than the primary winding, the voltage increases up to 30,000 volts due to step-up transformer principle. The primary winding consists of 200-300 turns of thick wire. About 15,000 volts are necessary to make the spark jump at 1 mm gap. The distributor then directs this high voltage to the proper spark plug when it jumps the gap, producing a spark which ignites the combustible mixture in the cylinder. The distributor automatically advances or retards the timing of the spark with regard to the engine requirements. This is accomplished by the centrifugal or vacuum advance mechanism within the distributor unit.

21.4.6 Components of Battery Ignition

This unit consists of a number of components such as (1) Storage battery (2) Ignition switch (3) Ignition coil (4) Contact breaker (5) Condenser (6) Distributor (7) Spark plug (8) Dynamo.

Storage battery: Storage battery is a device for converting chemical energy into electrical energy. It gets the charge from the dynamo driven by the engine. It provides the current for starting the engine. It also provides the current for primary winding of the ignition unit. The voltage of the battery is usually 6 or 12 V It supplies low voltage current.

Ignition switch: A switch provided in the primary circuit for starting and stopping the ignition unit (or the engine) is called ignition switch. It may be push-pull type or key type. It is placed within easy reach of the driver.

Ignition coil: The ignition coil is simply a transformer to step up the voltage in the ignition unit. It consists of a soft iron core, primary winding and secondary winding. The primary winding consists of 200-300 turns of thick wire and the secondary winding 15,000-20,000 turns of fine wire. The core is formed by lamination of soft iron. The primary winding is wound around the iron core. The secondary winding is wound outside the primary winding. The core assembly is placed in a steel casing which is fitted with a cap of moulded insulating materials. The terminals are provided in the cap. The windings are immersed in oil to improve insulation and to reduce moisture effect.

When the ignition switch is turned on, the current flows from the battery through the primary winding. When the contact breaker points open, the primary circuit breaks and the magnetic field collapses. Rapid collapsing of magnetic field induces high voltage in the secondary winding due to electromagnetic induction. The high voltage current flowing through the secondary winding goes to the distributor from where it is distributed to the proper spark plug through the high tension lead, where it jumps the spark plug gap producing the spark.

Contact Breaker: The contact breaker is a mechanism to make and break the primary circuit of the ignition unit for obtaining high voltage current in the secondary circuit. It consists of two points known as contact breaker points, one is fixed and the other is movable. The movable point is operated by the rotation of a cam by which the two points come in contact and separate, thus making and breaking the primary circuit. When the cam rotates further, the arm again comes into contact with the other arm. Now the contact points are in touch. This is due to the spring action in the system. Suppose, there was an ignition circuit unit for a four-cylinder engine, then the same cam would have four lobes. The contact breaking arm would be broken four times at suitable intervals. The operation of contact breaker has been shown in Fig. 21.21.

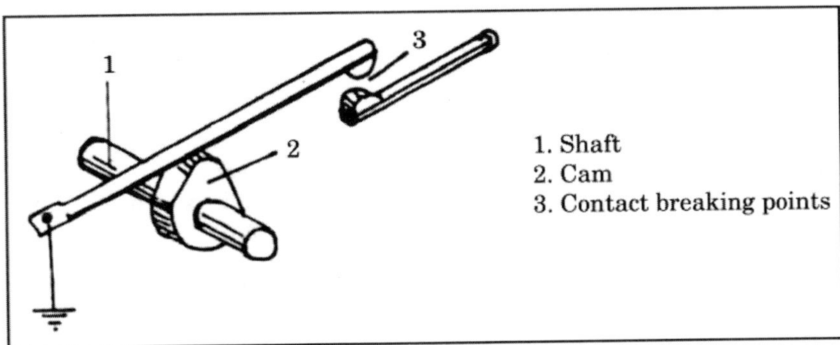

Fig. 21.21: Opening of contact breaking points

Condenser: Condenser is a device for storing of electric charge for a short time. As it is connected across the contact point - when these points open, condenser momentarily provides a place for current to flow in and get fully charged and send back the current when points close again. It also helps preventing arcing in the contact points thus increasing their life. If the condenser were not there in the circuit, the current would continue to flow across the points after they begin to open thus producing an arc. This arc would not only burn the points but also reduce the voltage. As such the condenser acts as storage of electricity and stops the flowing of current quickly.

A condenser consists of a pair of flat metal plates, separated by air. The most common type of condenser is of metal foil strips, separated by wax impregnated paper.

Distributor: Distributor is a component or device used to make and break the primary circuit and distribute high voltage current to various spark plugs of the engine cylinder in proper sequence and in proper time. The exploded view of a distributor is shown in Fig. 21.22.

The main functions of distributor are :

(a) It closes and opens the primary circuit with the battery.

(b) It distributes the resulting high voltage current from the secondary circuit to the spark plugs.

(c) It automatically advances or retards the timing of the spark with regard to engine requirements.

The distributor consists of : (a) housing (b) drive shaft (c) breaker cam *(d)* contact breaker points *(e)* governor (fly weights) (f) condenser *(g)* rotor (h) advance mechanism (i) cap. Distributor gets its drive from the cam shaft through gear and shaft (Fig. 21.23) and runs at half the speed of the crankshaft in 4-stroke engine. But in 2-stroke engine, it runs at the same speed of engine. When the shaft moves, it opens and closes the breaker points with the help of breaker cam. The breaker

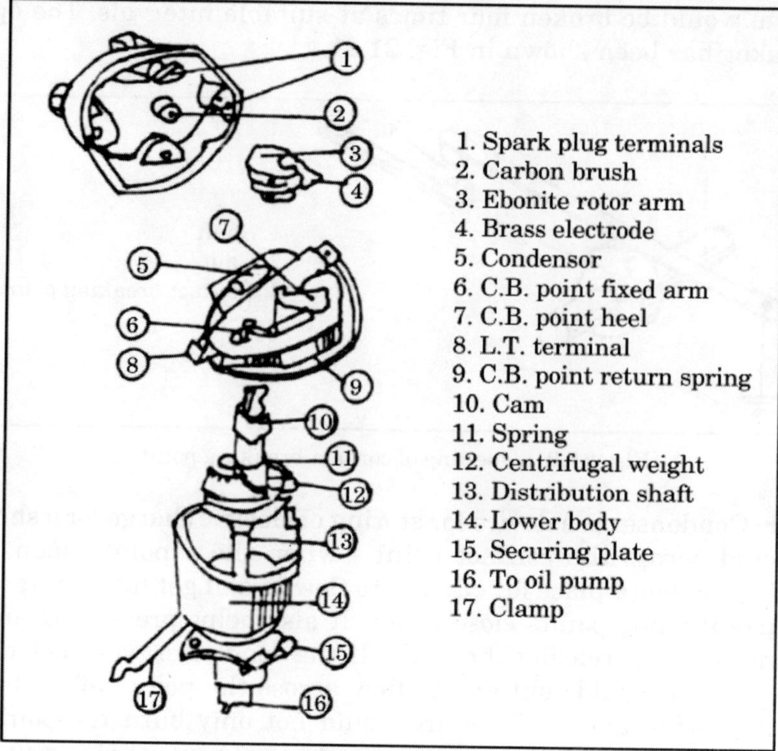

Fig. 21.22: Distributor

cam contains the same number of lobes as the number of cylinders in the engine. A rotor is mounted on the b'reaker cam which is carried by the drive shaft. The rotor connects the centre-terminal of the cap with each outside terminal in turn, so that high voltage current from the secondary coil is directed first to one spark plug and then to another, according to the firing order. High tension cables connect the terminals in the distributor cap to the spark plug. The centre of rotor is connected with the high tension lead from the secondary winding.

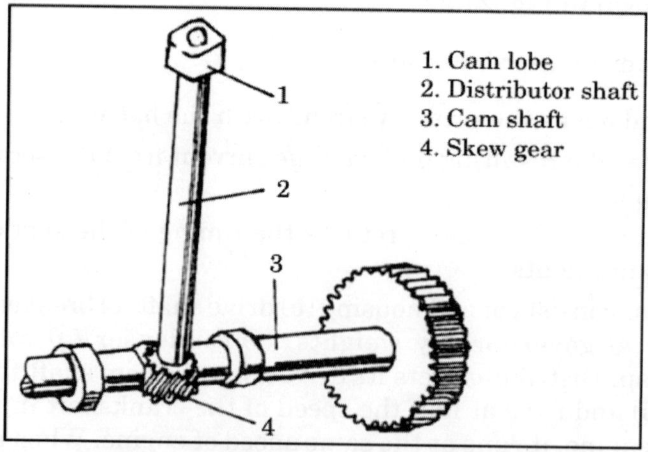

Fig. 21.23: Rotation of distributor shaft.

Spark plug: Spark plug is a device to produce electric spark to ignite the compressed air fuel mixture inside the cylinder. The spark plug is screwed in the top of the cylinder so that its electrodes project into the combustion chamber. The charge is ignited when high voltage current is made to jump from central electrode to the side electrode. When the current jumps from one electrode to another, there is spark. Due to this spark, compressed charge gets ignited. The spark should be of high voltage so that it could jump from one electrode to another specially in highly compressed air which is a non-conductor of electricity.

The spark plug mainly consists of three parts, (a) Centre electrode (6) Ground electrode (c) Insulation separating the two electrodes (Fig. 21.24).

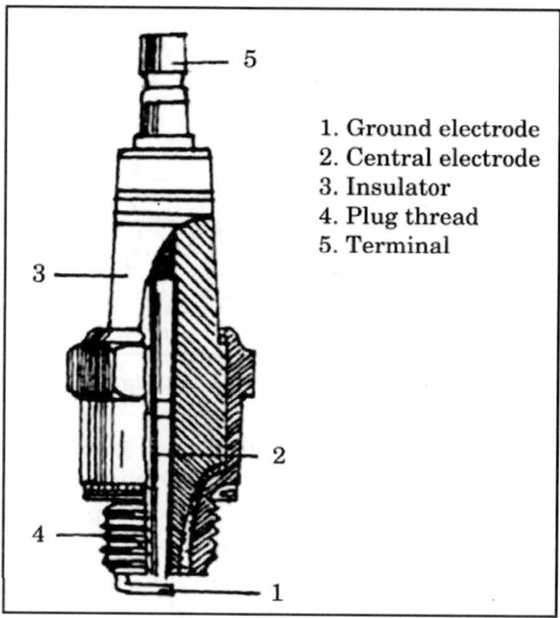

Fig. 21.24: Section of spark plug

The central electrode is connected to the high tension wire coming from the distributor. The ground electrode is earthed to the engine metal body. The central electrode surrounded by insulator carries positive current while the negative electrode is grounded and extended from the side of the plug.

The upper portion of the spark plug is surrounded by the porcelain insulator. The lower half portion of the insulator is fastened with a metal shell. The lower portion of the shell has a short electrode attached to one side and bent in towards the centre electrode, so that there is a gap between the two electrodes. The two electrodes are thus separated by the insulator. The sealing gaskets are provided between the insulator and the shell to prevent the escape of gases under various temperature and pressure conditions. The lower part of the shell has sorew threads and the upper part is made in hexagonal shape like a nut, so that the spark plug may be screwed in or unscrewed from the cylinder head.

The gap between the centre electrode and the ground electrode is called the spark plug gap. This gap is adjusted to recommended specifications by bending the ground electrode. It generally varies from 0.4 mm to 1.0 mm. Too large or too small gap reduces the efficiency of the entire ignition system, which in turn causes losses in engine power and operating efficiency. It is therefore of great importance to maintain proper spark plug gap.

Spark plug type: Spark plugs are of two types according to heat range.

1. Hot spark plug
2. Cold spark plug (Fig. 21.25)

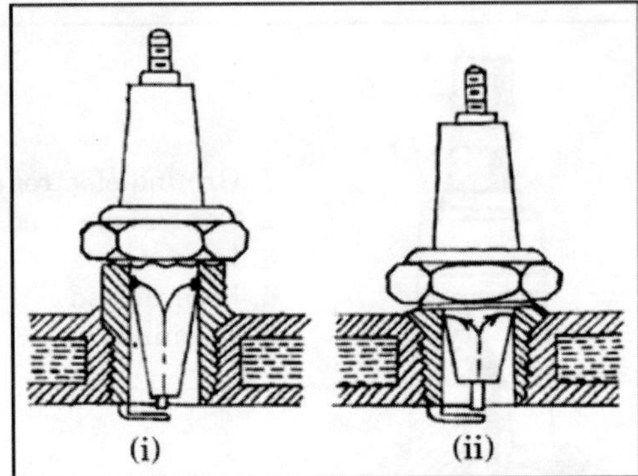

Fig. 21.25: Two types of plugs. (i) Hot plug, (ii) cold plug.

Heat range is a means of designating how hot a plug will run in operation. It refers to the ability of the spark plug to heat transfer from the firing tip of the insulator to the cooling system of the engine. The temperature that a spark plug will attain depends upon this distance through which the heat is transferred. If the path of heat travel is long, the plug will run hotter than if the path is short. A hot spark plug has longer path of heat travel and runs hotter than the cold spark plug which has the shorter path of heat travel and runs cooler.

Hot plug: It has comparatively longer insulator so the heat has to pass through a longer path to reach the cooling water and hence the heat is not dissipated quickly.

Cold plug: It has a short insulator, extending into the cylinder. It conducts the heat away from the point rapidly allowing it to be cooled by the cylinder jacket. The short path dissipates heat quickly, so that it is named as cold plug.

Cold spark plugs are used in heavy duty or continuous running high speed compression engines running at high temperature. Hot spark plugs are used in low speed, medium duty, low compression engines which are running at colder operating conditions.

Dynamo: The purpose of the dynamo is to keep the battery charged and to supply current for ignition, light and other electrical accessories. The dynamo supplies direct current to the battery and keeps it fully charged. It consists of a frame, soft iron pole shoes, field winding armature winding, commutator, brushes, brush holder, spring, terminals and drive pulley.

21.4.7 Magneto Ignition Unit

Magneto ignition unit (Fig. 21.26) consists of a magneto, instead of a battery, which produces and supplies current in the primary winding. The remaining arrangement in this unit is the same as that in the battery ignition unit. The magneto consists of a fixed armature having primary and secondary windings and a rotating magnetic assembly which is driven by the engine. When magnets rotate, current flows in the primary winding. The secondary winding gives high voltage current to the distributor, which distributes it to the respective spark plug. In a magneto, the magnetic field is produced by means of permanent magnet. The permanent magnet is of horse shoe type of magnet. The magnet is rotating while the armature pole pieces are stationary. This type of ignition system is mostly used in motor cycle. Some high speed engines have magneto built in the fly wheel.

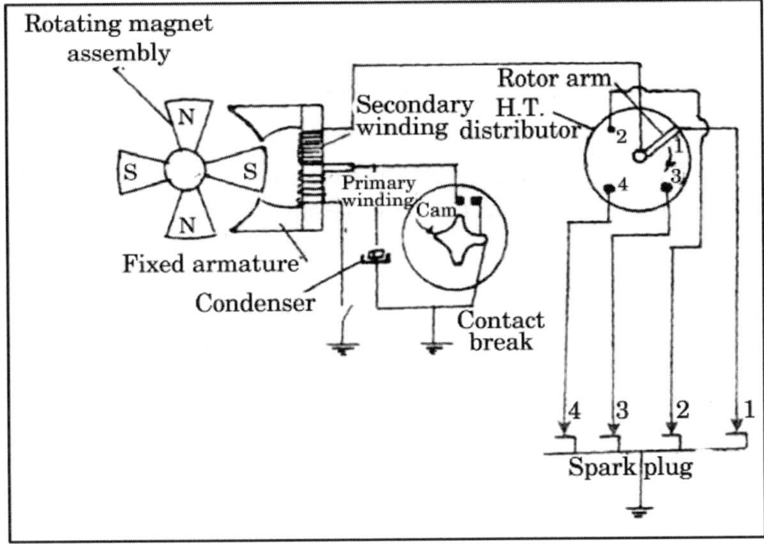

Fig. 21.26: Magneto ignition unit for a four cylinder engine

Table: Comparison of battery ignition and magneto ignition system

SI. no.	*Battery ignition system*	*SI. no.*	*Magneto ignition system*
1.	Current is obtained from the battery	1.	Current is generated by the magnets
2.	Sparking is good even at low speed	2.	Poor sparking at low speed
3.	Starting of engine is easier	3.	Difficult starting
4.	If the battery is discharged, the engine cannot be started	4.	No such difficulty, as battery is not needed
5.	Occupies more space	5.	Occupies less space
6.	Complicated wiring	6.	Simple wiring
7.	Less costly	7.	More costly
8.	Used in cars, buses, trucks etc.	8.	Used in motor cycles, scooters, racing car etc.

21.4.8 Spark Advance Mechanism / Ignition Advance

The high tension coil requires little time to build up current and spark plug also requires a short time to burn the charge. This total time required is called time lag. Let us assume that the time required for high tension coil to build current and also for the spark plug to start burning the charge *i.e*, time lag is 1/400 second. If the engine is running at 1000 rpm, the degree of flywheel for time........................ So ignition process should start 15° before piston reaches TDC in compression stroke, causing the charge ready for expansion. From the above explanation, it is clear why a spark advance is required.

As the engine speed increases, the spark must occur in the combustion chamber earlier in the cycle. If the engine is idling, the spark must occur just before the piston reaches TDC on the compression stroke. At low speed, the spark has enough time to ignite the compressed air-fuel mixture. The pressure rise due to combustion approaches maximum as the piston passes TDC and starts down on the power stroke.

When the ignition occurs early in compression stroke, before TDC, the ignition or spark is said to be advance. A retarded spark occurs when the compression stroke is more nearly complete at the time ignition occurs.

If the spark is advanced too much, the fuel will ignite completely before the piston reaches to TDC and the maximum pressure will be exerted on the piston when it is moving upwards during the compression stroke. It will cause the engine to stop suddenly. Under certain operating conditions, an advanced spark might also cause the fuel to explode or detonate.

If the spark is retarded too much, the fuel will ignite completely too long after the piston has begun to move downward during the power stroke. Because the combustion space increases by the downward movement of the piston, the pressure will also be reduced on the piston.If the spark is still further retarded, the fuel may not be burnt completely by the time exhaust valve opens. It will cause loss of power, overheating, carbon deposits and probably exhaust valve.

To change ignition advance to suit various speed and operating conditions, two systems are employed.

1. Centrifugal advance mechanism
2. Vacuum advance mechanism.

Centrifugal advance mechanism: It consists of two fly weights (advance weights), cam, spring and base plate (Fig. 21.27).The flyweights are carried by the distributor drive shaft through the base plate which is fixed to the drive shaft. The flyweights are pivoted on the base plate and attached to the cam through springs. The cam is connected with the distributor drive shaft through the springs, flywheel and base plate. When the engine speed increases, the weights move out due to centrifugal force rotating the plate and cam in the anti clockwise direction. This movement affects the desired advance. This timing of spark varies from no advance at low speed to full advance at high speed.

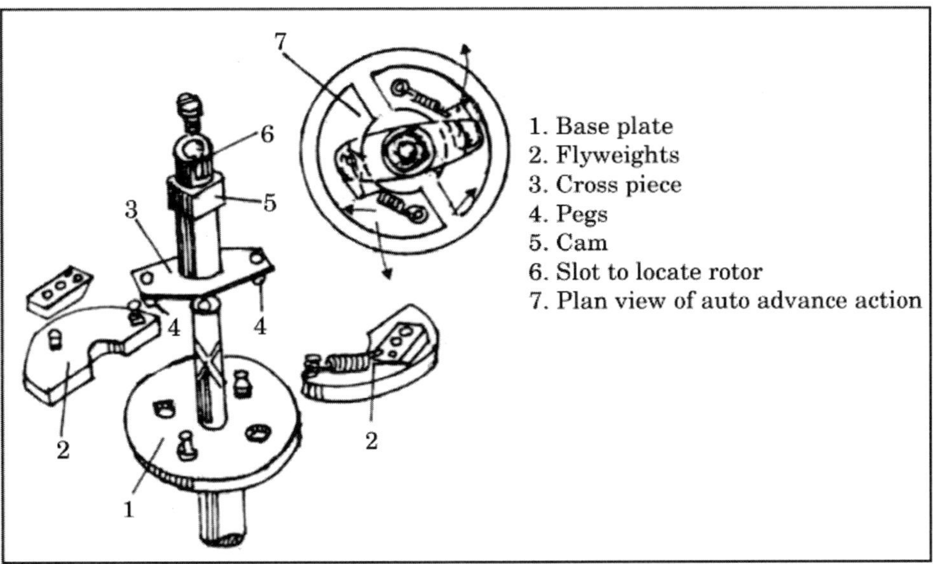

Fig. 21.27: (a) Automatic centrifugal timer.

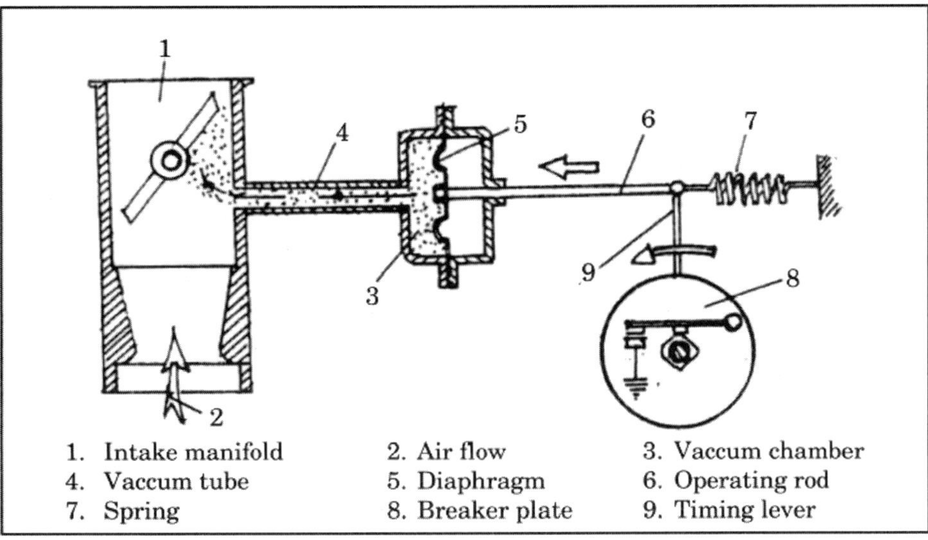

Fig. 21.27: (b) Construction of vaccum control unit.

Vacuum advance mechanism: Vacuum advance mechanism (Fig. 21.27(6)) consists of a diaphragm which automatically advances and retards the ignition timing according to the engine speed and operating condition. The diaphragm forms two chambers, one of which is connected to the induction manifold and the other is open to the atmosphere. A linkage connects the diaphragm to the distributor. When the diaphragm is in its normal position, the contact breaker is held fully retarded. When the engine speed increases, the induction manifold depression is high and the diaphragm is pressed up by atmospheric pressure. The movement of the diaphragm moves the contact breaker in the opposite direction to rotate and thereby advancing the ignition. A decrease in vacuum allow the diaphragm to return back to its original setting, retarding the ignition.

The vacuum advance mechanism relates the throttle opening. When the throttle is partly open, vacuum develops at the vacuum tube point, diaphragm is pressed up and ignition is advanced.

When engine is running idle, there is only a small throttle opening. At this time, the induction manifold depression is very high. This develops the advance of ignition timing.

But when the throttle is fully open, due to the maximum load, there is no vacuum effect. The lever is not operated. Therefore, there is no advance ignition timing at full load. When there is no vacuum, the diaphragm returns to its original position. This causes the timing to retard in relation to the load.

It is seen that centrifugal advance mechanism takes much care of speed only and not the load conditions, whereas the vacuum advance mechanism takes much care of load conditions. Therefore, a combination of the two mechanisms is applied on the distributor. In this design, part of the ignition advance is due to the centrifugal force and part to the vacuum produced in the intake manifold. The combination of the two mechanisms gives the practically perfect spark timing for all driving conditions.

21.5 STARTING UNIT

Internal combustion engines are not self starting and need to be rotated at a certain minimum speed in order for the engine to start running by the fuel supply. The automotive engines (both spark ignition and diesel) are cranked by a small but powerful electric motor. This motor is called the cranking motor, starter or starting motor.

For a carburetor engine to start reliably, its crankshaft must be turned at a speed of 60 to 80 rpm. To manually crank the engine with such a speed requires much effort. Therefore electric starting systems are now universally used to provide the same amount of effort or torque.

The starting motor is a direct current motor which cranks the engine for starting. It is used to replace the laborious hand cranking method. Cranking the engine means to rotate the crankshaft by applying torque on it so that the piston may get reciprocating motion. The starting motor is mounted on the engine flywheel housing. It is series wound and designed to operate on large currents at low voltage. It must be capable of exerting a very high torque when starting at low speeds. The armatures and fields are built with thick wire to keep the resistance low and to enable them to carry large currents without overheating. For starting, engine requires more torque for a short time. To prevent overheating of motor, it is not recommended to be used continuously. It should only be used for 30 seconds and if an engine fails to start during this period, a rest of few minutes is permitted before making another attempt to start the same.

Although the mechanical construction of motor is very similar to generator, their functions are entirely different whereas function of the generator is to generate voltage when conductors are moved through a magnetic field, the function of the motor is to develop a torque or twisting effort when the current is passed through the conductors held in a field. The generator is shunt-wound but the starting motor is series wound.

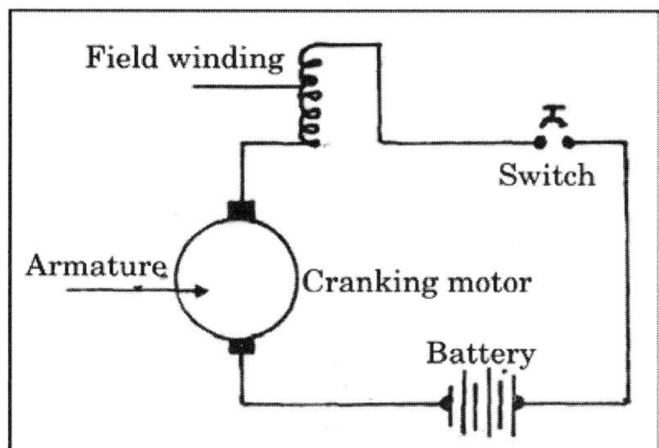

Fig. 21.28: Simple connection diagram of a cranking motor.

The battery sends current to the starting motor when the driver turns the key switch or starting switch to start. This causes a pinion gear in the starting motor to mesh with teeth on the ring gear around the engine flywheel. The starting motor then rotates the engine crankshaft for starting. Figure 21.28 shows the connection diagram of the cranking motor. The typical starting system includes the battery, starting switch, solenoid switch, starting motor, drive pinion gear mechanism with pinion and the wiring which connects these components. Starting unit circuit in operation is shown in Fig. 21.29 when the starter switch is activated. Each of the starting unit components is described in following sections.

21.5.1 Starting Motor

Starting motor is an electric motor which is a machine to convert electrical energy into mechanical energy.

Motor principle: The action of a motor is based on the principle that if a current carrying conductor is placed in a magnetic field, a mechanical force is experienced by a conductor, the direction of the force is given by Fleming's left hand rule [Fig. 21.30 (a and b)].

Constructionally, there is no basic difference between a d.c. generator and a d.c. motor (Fig. 21.31). In fact, the same d.c. machine can be used interchangeably as a generator or as a motor. D.C. motors are also, like generators., shunt-wound or series wound or compound wound.

In figure 21.32, a part of multi polar d.c. motor is shown when its field magnets are excited and its armature conductors are supplied with current from supply mains, they experience a force tending to rotate the armature.Armature conductors

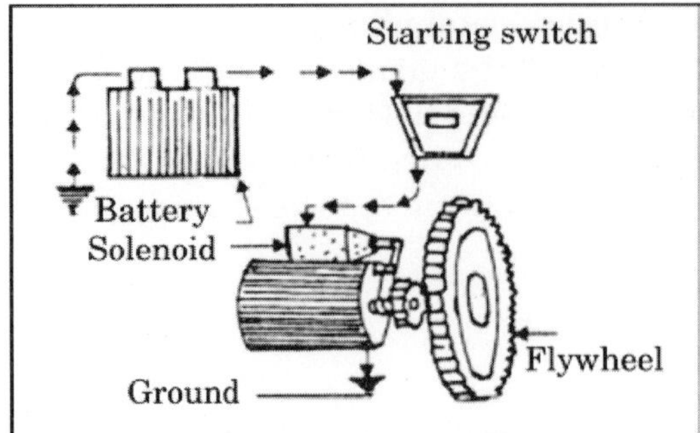

Fig. 21.29: (a) Starting circuit in operation when the starter switch is activated

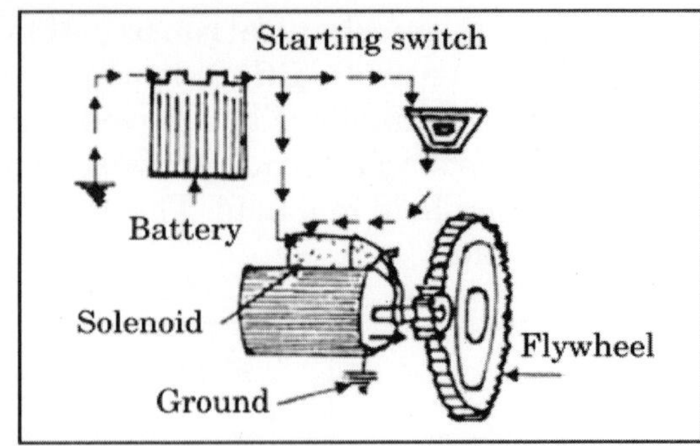

Fig. 21.29: (b) Starting circuit in operation when starter motor engages with flywheel

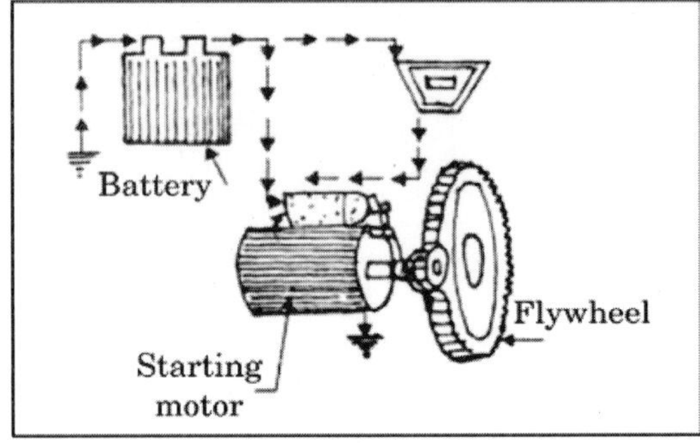

Fig. 21.29: (c) Starting motor cranks engine

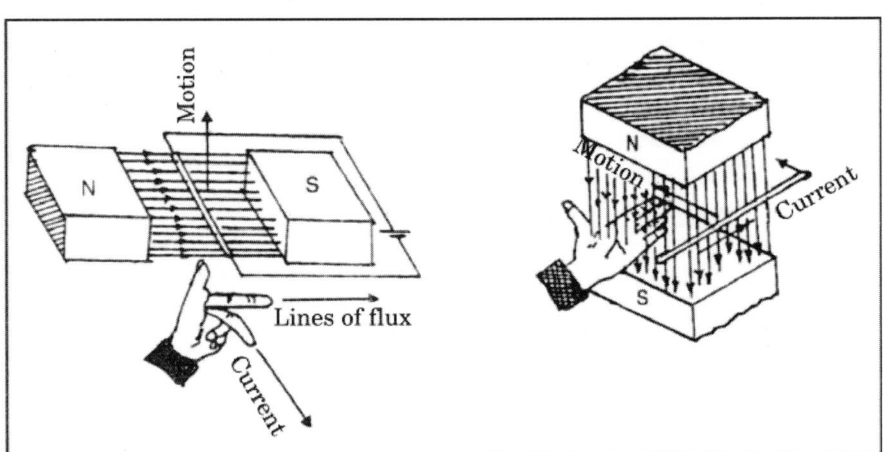

Fig. 21.30: (a) Working of DC motor

Fig. 21.30: (b) Working of DC motor

under N-pole are assumed to carry current downwards (crosses) and those under S-poles to carry current upwards (dots). By applying Fleming's left hand rule, the direction of the force on each conductor can be found. It is shown *by* small arrows placed *above* each conductor. It is seen that each conductor experiences a force F which tends to rotate the armature in clockwise direction. These forces collectively produce a driving torque which sets the armature rotation. It *may* be said that *any* current flowing in *a* loop or coil of wire produces a magnetic field. This produces a second magnetic field with north and south poles, perpendicular to the armature loop but the north pole of main magnetic field attracts the south pole of the armature and sinde the loop is free, it starts revolving.

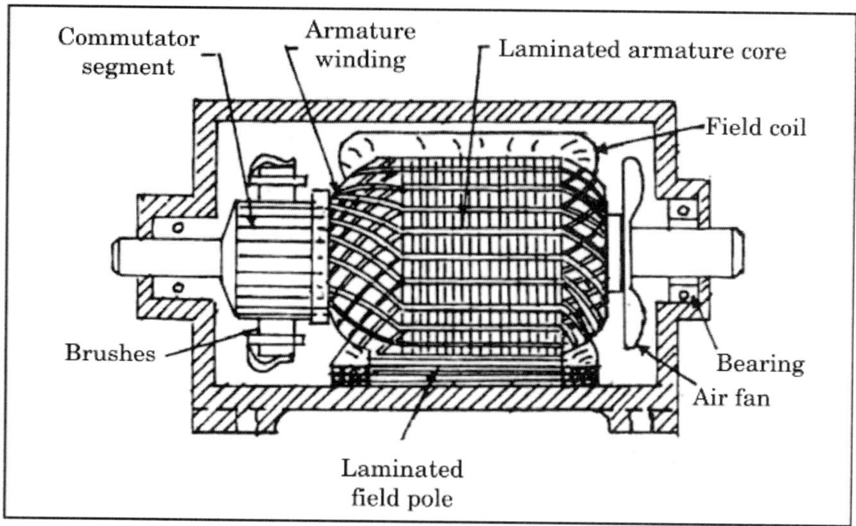

Fig. 21.31: DC motor

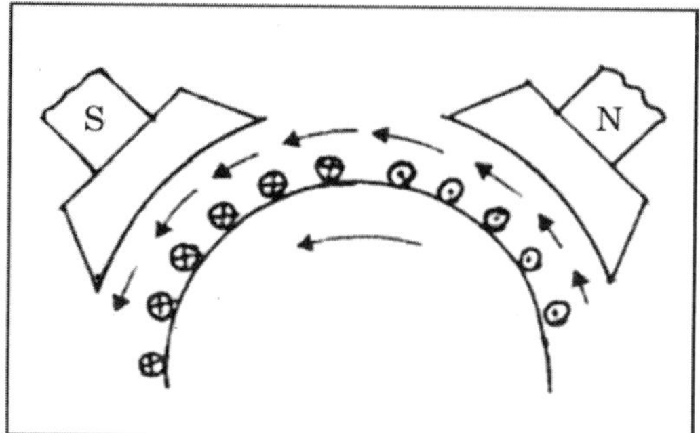

Fig. 21.32: A multi-polar DC motor

In the motor or starting motor, there should be some means of connecting the loop with the battery, so that, it receives the supply of current and thus rotates. The loop-ends are connected to two halves of a split ring called the commutator. Two brushes rest on the split ring. Current flows from the battery through one brush-through the loop and the other brush-and back to the battery as shown in Fig. 21.33. It may be noted that the commutator rotates with the loop. Each segment of the commutator gets disconnected from one brush and connected to the second brush when the loop rotates through half a turn, thus maintaining the supply of current in the same direction. This results in the same direction of rotation of the loop. Hence the function of the commutator in the motor is the same as in a generator.

By reversing current in each conductor as it passes from one pole to another, it helps to develop a continuous and unidirectional torque.

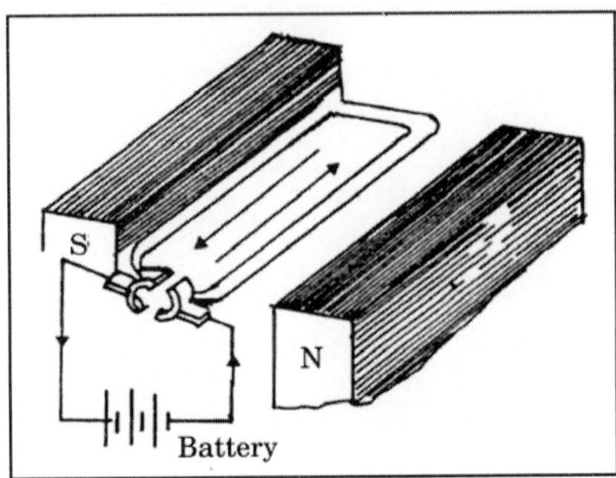

Fig. 21.33: Connections of the loop with the commutator and the battery

Construction: The construction of the starting motor is similar to that of the generator but the windings and brush terminals are heavier to deal with heavy currents. The brushes are made of low resistance material such as copper instead of carbon as in the case of generator. The main parts of the starting motor are frame, armature, commutator, field windings, brushes, poles and terminals. A drive mechanism is provided at the end of the armature shaft, by means of which the motor starts the engine.

The starting motor uses either two field windings or four field windings. Figure 21.34 shows the exploded view of a four-pole, four-brush cranking motor. The motor is energized by means of a key push switch on the main panel. Small current flows through the panel board switch and the heavier current is confined to the solenoid starter circuit.

The pinion is carried on a barrel type assembly which is mounted on a screwed sleeve. This sleeve is carried on splines on the armature shaft and is arranged so that it can move along the shaft against a compression spring to reduce the shock loading at the moment engagement takes place.

When the motor switch is operated, the armature shaft and screwed sleeve rotate. Owing to the inertia of the barrel assembly, the latter is caused to move along the sleeve until the pinion comes into engagement with the flywheel ring, The cranking motor will then turn the engine. As soon as the engine fires and commences to run under its own power, the flywheel is driven faster by the engine than the motor. This causes the barrel assembly to be screwed back along the sleeve, thereby drawing the pinion out of mesh with the flywheel teeth. In this manner, the drive safeguards the cranking motor against damage.

A piston retaining spring is incorporated in the drive. This spring prevents the pinion from vibrating into the mesh when the engine is running. If any difficulty is experienced with the cranking motor not meshing correctly with the flywheel, it may be that the drive requires cleaning.

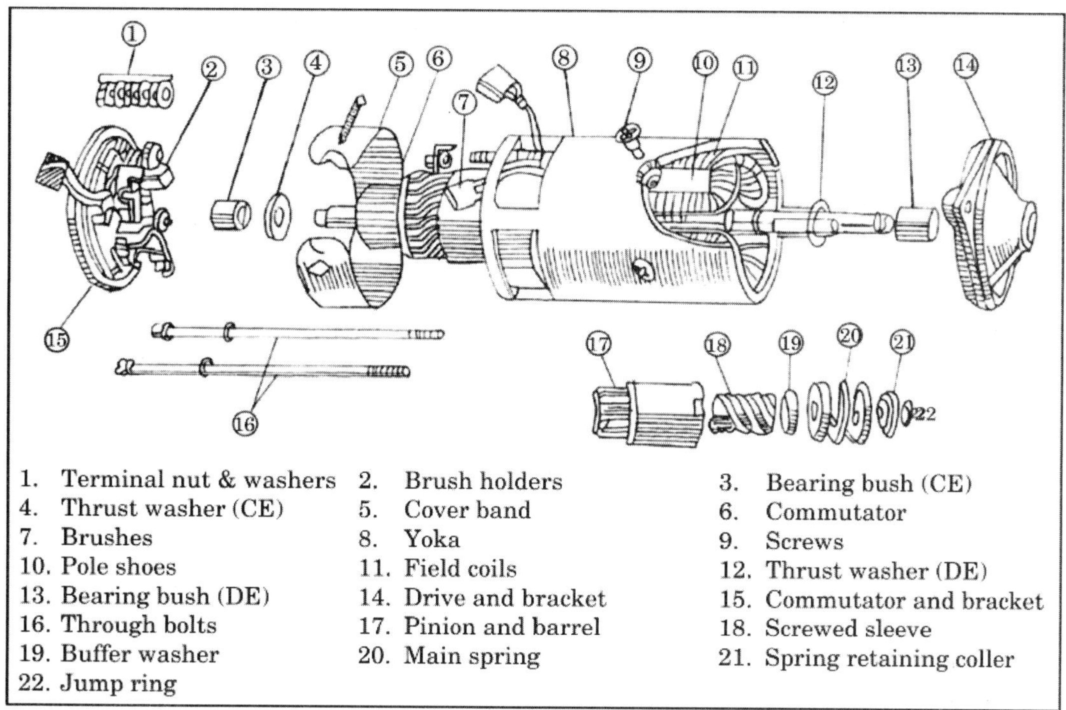

Fig. 21.34: Starter

21.5.2 Solenoid Switch

Different types of switches are used to connect the starting motor with the battery. A heavy duty foot operated switch was used in some early models. A magnetic switch, also known as solenoid switch or starter relay, is used in many present day automobiles/tractors.

The initial starting current of the motor is very high. An ordinary switch cannot bear high current. Therefore a solenoid switch is provided. This switch is able to take a very large current to flow through it. It works on electromagnetic principle. Battery is connected to one terminal of solenoid switch and the other is earthed to the self stator. To operate the solenoid switch, a starting switch is installed.

The solenoid switch consists of plunger, contact disc or contact bridge, winding, terminal and necessary connecting cables. The switch is connected between the battery and the starting motor as shown in Fig. 21.35. When the starter switch is on, a large amount current flows from battery through the solenoid coil. The current passing through the solenoid coil develops a magnetic effect in the hollow core. This pulls the sliding plunger. The plunger moves forward and the contact disc makes the contact between the terminals. These terminals are connected to the battery and the stator motor. A heavy current of say 300 to 400 ampere flows through the circuit, When the starter switch is off, the current does not flow in the solenoid coil. The solenoid core then loses its magnetic field. The sliding plunger is brought back to its original position by a spring. Thus the connection between the battery and the starter is disengaged.

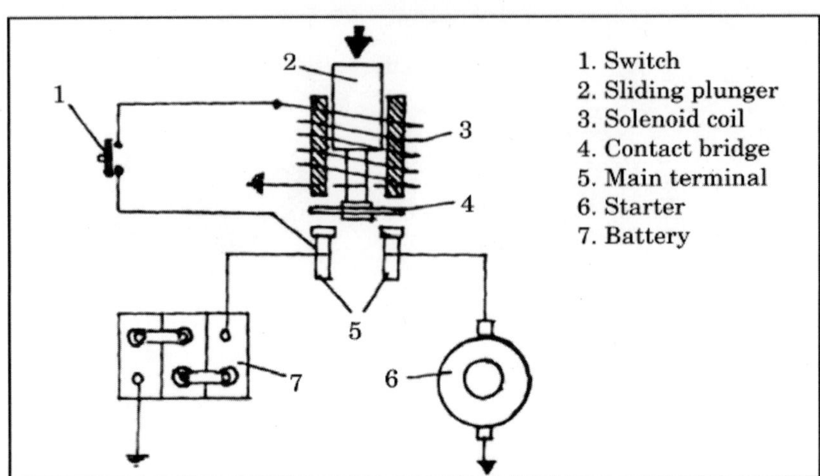

Fig. 21.35: Principle of a solenoid switch

21.5.3 Drive Mechanism

The starting motor is linked to the engine flywheel through a set of gears. A pinion gear is attached to the starter armature which drives a ring gear attached to the fly wheel. The arrangement is so made that the two gears engage to crank the engine until it starts and then disengage automatically when the engine is running. The gear ratio is about 15 : 1. The armature rotates 15 times to cause the fly wheel to rotate once. The armature may revolve at about 2000 to 3000 rpm when the cranking rotor is operated and hence the fly wheel will rotate as high as 200 rpm.

When the engine starts, its speed may increase to about 3000 rpm. If the pinion is still in mesh with the fly wheel, it will revolve the armature at about 45000 rpm, which is a very high speed. At this speed, the centrifugal force would cause the conductors and commutator segments to be thrown out of the armature causing damage to the motor. Hence the pinion must be disengaged from the flywheel, after the engine has started. The automatic engagement and disengagement of the motor with the engine fly wheel is obtained with the help of drive arrangement.

There are two basic methods through which starter pinion is engaged with the ring gear of fly wheel. (A) Inertia drive (B) Electromagnetic drive.

Inertia drive : Bendix drive works on the principle of inertia force. Figure 21.36 shows bendix drive for starting motor. Bendix starter drive arrangement is provided at the end of the armature shaft. The pinion gear having internal threads, is mounted on the threaded sleeve, just like a nut on bolt. The sleeve is not connected directly to the shaft of the starting motor, but uses it only as a bearing. A spring is attached to the drive head and also to the sleeve.

When the starting motor is at rest, the pinion gear is not engaged with the fly wheel. When the starting motor is switched on, the armature begins to rotate. This causes the sleeve to rotate because the sleeve is fastened to the armature shaft through a spring. The pinion, because of its inertia of rest and its unbalanced

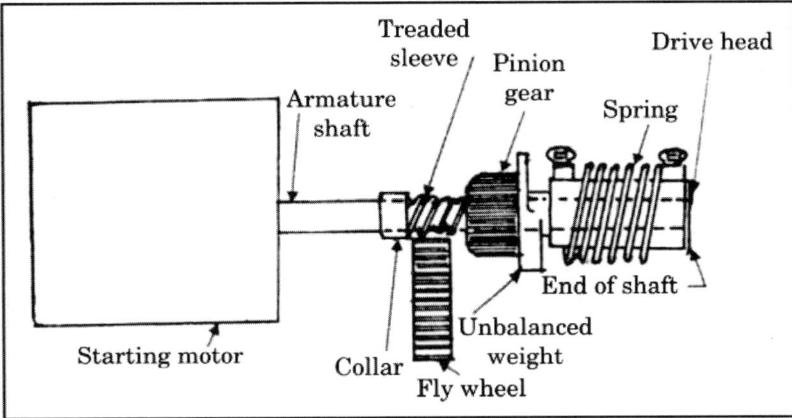

Fig. 21.36: Bendix Drive

weight, turns very little but it moves forward on the revolving sleeve until it engages with the teeth of the fly wheel. But there is a certain restriction on its movement. This is provided by the thrust face or collar. When the pinion gear strikes with the collar, it begins to turn with the sleeve, causing the fly wheel to run with it. When the fly wheel turns, the crank shaft also turns and the engine starts. Once the engine starts, the speed of the fly wheel increases. The speed of the starter pinion is less than the flywheel. Therefore, starter pinion returns along the screwed sleeve. Thus the pinion is disengaged. It is important that when the engine starts, the starter switch is off. Hence this drive works on the principle of inertia forces.

Electromagnetic drive: The pinion gear is shifted in or out of mesh by the magnetic field of the solenoid switch. Figure 21.37 shows the shift drive starter system. The first step is to operate the solenoid by switching on the starter switch. The plunger is then moved to the right. The shift lever is pivoted to the plunger. The movement of plunger operates the shift lever. The lever pushes the pinion to

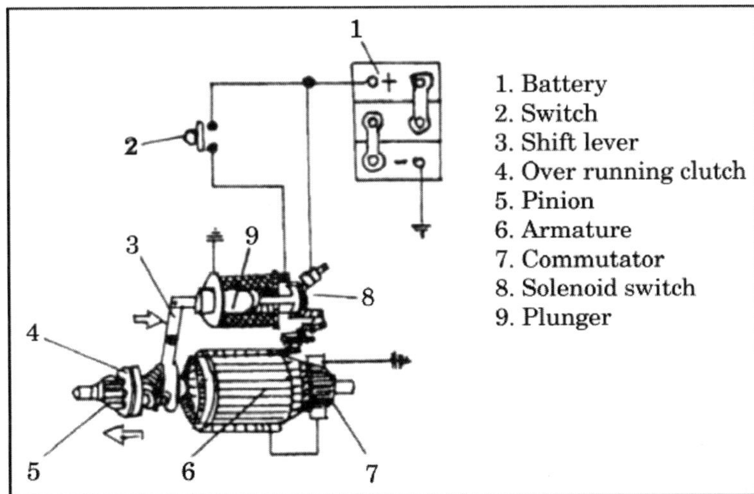

Fig. 21.37: Shift drive starter solenoid operated starting mechanism

the left. By this movement, the pinion comes in contact with the ring gear of the fly wheel. This is the first step.

The second step is the operation of the solenoid so as to give connection to the solenoid switch. By this contact, the starter motor gets the current from the battery. Therefore, the starter motor begins to rotate. At this stage, the pinion gets the torque from the shaft of the starter. Then it rotates the fly wheel of the engine. Once the engine is started, the switch is put off.

After switching off, the shift lever now pulls back as plunger goes back due to the demagnetization of solenoid coil. The pinion and flywheel then get disengaged.

21.6 LIGHTING AND AUXILIARY ELECTRICAL EQUIPMENTS

Lights are used in modern tractors for various purposes. The main lights are : (i) Head lights *(ii)* Parking lights *(iii)* Direction signal lights *(iv)* Stop lights (v) Backup lights *(vi)* Tail lights *(vii)* Interior lights *(viii)* Brake working light.

Head lights are required to illuminate the road sufficiently to permit safe night driving of the tractor. These are usually provided with two beams, one gives maximum illumination for night driving and the other gives deflection to the ground and to the side of the road to minimize glare when passing other vehicles on the road. Sometimes a third beam is used which is of low intensity while driving sometimes in city.

For parking the tractor during dark, the parking lights are kept on to provide a signal for other moving objects and thus avoid the accident.

Directional signal lights are used to indicate the direction in which the vehicle is to turn. These lights give signal to the vehicles coming from the front or rear.

Blinker lights provide a means of signaling when the vehicle is stalled on the highway or has pulled off to the side. The blinking is much more noticeable than a steady light and provides a warning to the approaching vehicle.

Stop lights are also provided at the rear of the vehicle and become on when brakes are applied.

Backup lights are operated when the driver shifts to the reverse direction. This closes a switch linked to the selector lever which connects the backup light to the battery.

Tail lights are required under driving conditions and when the vehicle/ tractor is running at night. It contains the miniature bulbs with filaments which have reflectors and fluted covers to disperse the light.

Interior lights include instrument panel lights, various warning indicator lights, compartment lights etc.

Brake warning lights are provided to give indication to the drivers at the rear to slow down. It operates as soon as the brake pedal is depressed.

21.6.1 Wiring Circuit

The wiring circuit of a tractor is shown in Fig. 21.38. It is a simple circuit system using one wire and ground. Electrical equipment of modern tractors and automobiles is a complex system of numerous electrical components interconnected by a multitude of wires making up a system of their own.

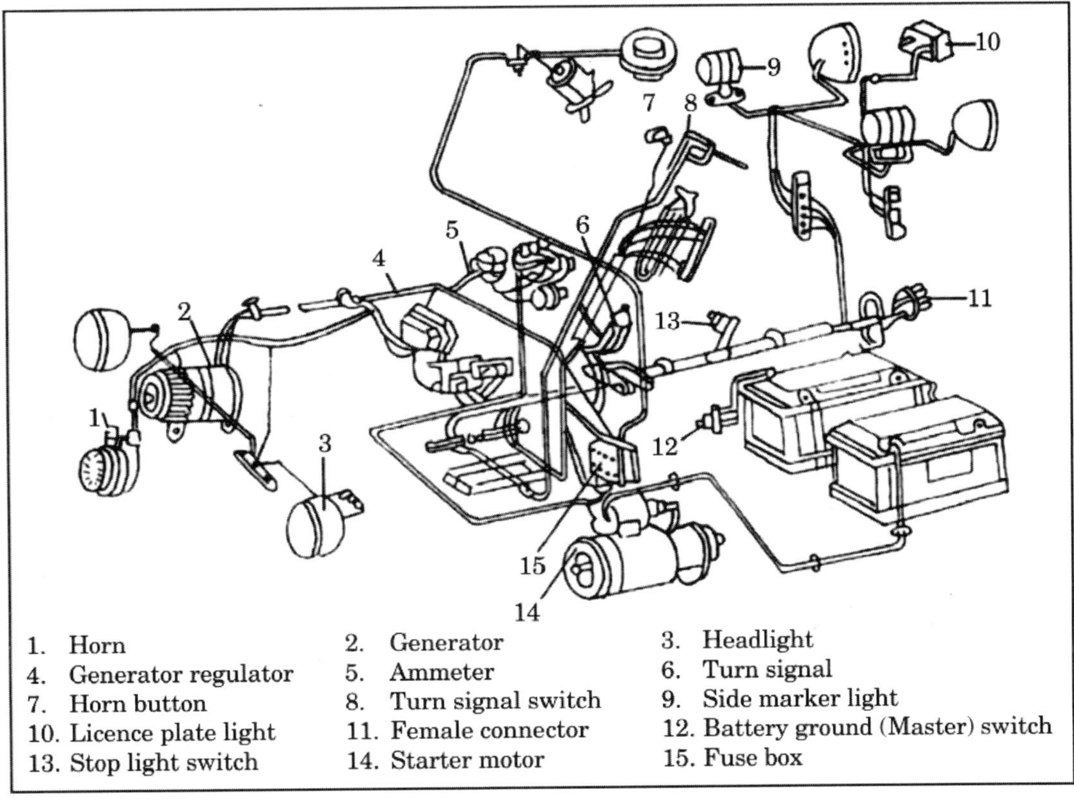

Fig. 21.38: Tractor wiring layout

The electrical units are connected by wires of different sizes. The size of each wire depends on the amount of current the wire must carry. The heavier the current the larger the wire must be. The wires are gathered together to form wiring harnesses. Each wire is identified by the colour of its insulation. For example, wires of light green, dark green, blue, red, black with a white tracer and so on. The operator's manual has illustrations that show the various wires and their colors.

All tractors use one-wire system in which vehicle frame and other metal parts are employed as the second (ground or return) wire. One-wire systems are simpler and less expensive than their two-wire counterparts, but require very careful handling and systematic inspection of the condition of wiring insulation. The most common problems with the wiring system are open circuit, short circuit and ground faults. Open circuits may be caused by loose, badly corroded or soiled connections or by broken wires. Short circuits and ground faults are primarily due to damaged wiring insulation.

21.6.2 Electric Horn

Horn is a sound creating device. Electrical horns are used in all the automobiles as well as tractors. The note emitted by the horn is neither musical nor hoarse. The horn works on electro-magnetic make and break system. The diagram of horn circuit has been shown in Fig. 21.39. It has a core E having an enameled wire wound on it. The diaphragm with an armature is secured on the horn body. Plunger which operates contact points is fixed with armature and diaphragm. When current in passed in the core E through horn button, the core gets magnetized and pulls the armature. As the armature is fixed with the diaphragm, it also gets pulled in along with plunger rod. This plunger rod, while being pulled in, presses the contact points open. When the contact point is opened, electric current going to- E core is stopped and the core gets demagnetized resulting in the armature and diaphragm coming back to its original position. As soon as the diaphragm armature and plunger move to their position, contact points are again closed and current starts flowing into core E which again attracts the armature. In this way, the diaphragm and armature goes on vibrating. This vibrating action of diaphragm produces sound. The condenser, placed in parallel with the contact points protects them against burning as a result of arcing. The cycle of opening and closing of circuit is repeated rapidly causing rapid movement of diaphragm to produce sound.

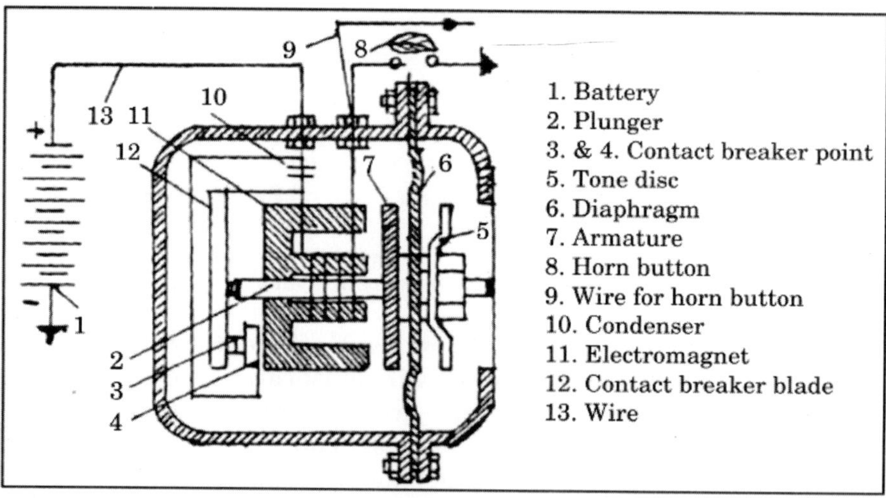

Fig. 21.39: Electric horn vibration type

21.6.3 Dashboard Instruments

The various dashboard instruments are oil pressure gauge, temperature gauge, fuel gauge, an ammeter, a speedometer etc. These are all indicating and warning devices provided in automobiles/tractors in order to keep the driver informed about the correct operation of the engine and other systems. They are usually operated electrically. These devices has been described in this section.

Oil-pressure gauge: This device reads the pressure of a vehicle's engine lubrication system and serves as a warning device to the driver against any likely damage to engine parts due to insufficient lubricating oil. Generally balancing coil type of oil pressure gauge is used in modern automobiles/tractors. Figure 21.40 shows the wiring diagram of oil pressure gauge.

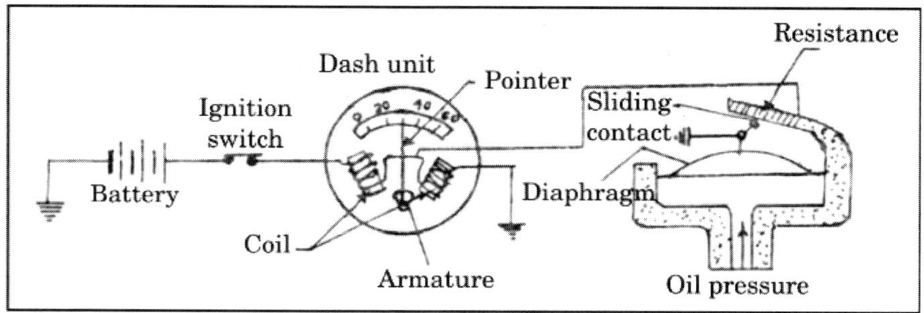

Fig. 21.40: Wiring diagram of balancing coil type of oil pressure gauge.

The oil pressure gauge consists of two units, namely the dash unit and the engine unit. A variable resistance is incorporated in the engine unit. An increase in the oil pressure causes the diaphragm to get pushed outward. This results in increase in the resistance at the engine unit, thus making the right-hand coil of the dash unit relatively magnetically stronger than the left-hand coil. Consequently, the armature and the pointer swing towards the right to indicate a higher oil pressure.

Water temperature gauge: This gauge keeps a constant check on the vehicles' engine cooling system. Figure 21.41 shows the schematic wiring diagram of the balancing coil type of temperature gauge. This gauge has a head unit containing a semi-conducting element in the form of a pellet. It is made of special material whose electrical resistance increases when temperature is lowered and reduces when temperature is increased. This phenomenon of resistance reduction with increase in temperature is made use of for indicating the temperature of the cooling water.

The operating current is supplied from the battery though the ignition switch. In this case, throughout the operation of the gauge, the current flowing through the left coil is constant, whereas the current flowing through the right coil changes, depending upon the resistance of the pellet. When the water is cold, the battery current flows to. the earth through the left coil. This causes the pointer and armature to swing to the cold side of the temperature scale.

When the water begins to heat up, thus heating the engine pellet, its resistance diminishes, thereby increasing the current through the right coil. This results in a stronger magnetic field. The pointer along with the armature moves to the hot end of the scale.

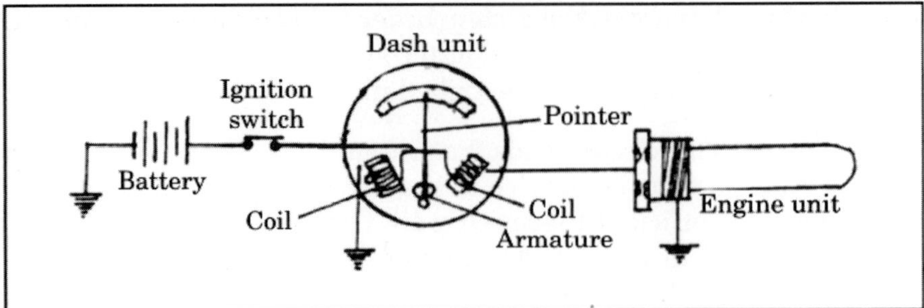

Fig. 21.41: Wiring diagram of a balancing coil type temperature gauge.

Fuel gauge: The balancing coil type fuel gauge is generally used in the automobiles/ tractors. Figure 21.42 shows the wiring diagram of a balanced coil type of fuel gauge. It has two units, the dash unit and tank unit. These are connected in series by a suitable wire to the battery through the ignition switch. When the ignition switch is turned on, the current from the battery flows through both the units.

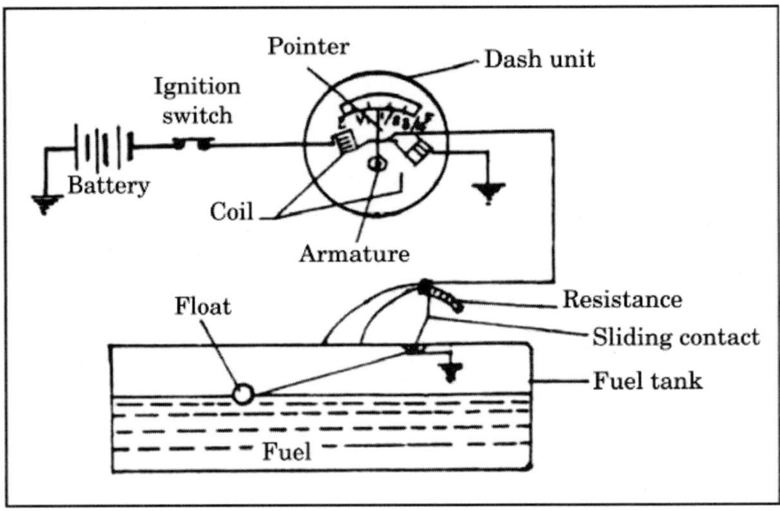

Fig. 21.42: Wiring diagram of balancing coil type fuel guage.

The tank unit consists of a float mounted at one end of the hinged arm and a sliding contact at the other end. The sliding contact moves along the resistance. The float lever moves up or down when the changes in fuel level in the tank take place. When the fuel level in the tank begins to empty, the sliding contact moves to the left. Thus more current flows through the left-hand coil of the dash unit and a little of it flows through the right-hand coil. This results in the left-hand coil being magnetically stronger than the right-hand one. The armature along with the pointer is moved towards the left side thus indicating a low fuel level in the tank.

On the other hand, when fuel level in the tank is high the float moves up thus making the sliding contact to insert most of the resistance into the circuit. Now most of the current that flowed through the left-hand coil also flows through the right hand coil. The right-hand coil is relatively stronger and this causes the armature and pointer to swing to the right, thereby indicating a high fuel level in the tank.

Ammeter: It shows the amount of current that flows to or from the battery. Figure 21.43 shows the construction of an ammeter.

In the ammeter, the conductor is connected at one end to the battery. The pointer is mounted on a pivot. There is a small piece of iron, oval shaped, mounted on the same pivot. This oval-shaped piece of iron is called the armature, A permanent magnet, almost circular in shape, is placed so that its two ends are close to the armature. The permanent magnet attracts the armature and tends to hold it in a horizontal position. In this position, the pointer or needle points to zero. Nothing is happening. Now suppose, the alternator starts sending.current to the battery. This current passes through the conductor. The current produces magnetism. This magnetism attracts the armature and causes it to swing clockwise. This moves the pointer to the 'charge' side. The more current that flows, the stronger the magnetism and the farther the pointer moves. The meter face is marked off to show the number of amperes flowing. If current flows from battery, the direction of flow of current is reversed. The armature is attracted in the opposite direction and it swings counterclockwise. This moves the pointer to the "discharge" side". The more current being taken out of the battery, the farther the pointer moves.

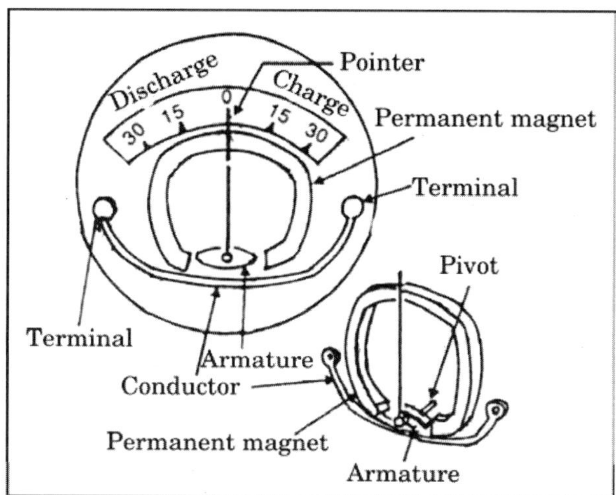

Fig. 21.43: Simplified drawing of ammeter showing its internal construction

Speedometer and Odometer: The mechanical speedometer and odometer are not electrical device. These are fitted on automobiles and are generally mechanically operated with the help of a flexible cable fitted between the instrument and the gear box. The electrical speedometer and odometer are also available, but mechanical operated devices are mostly used in tractors, The speedometer tells the driver how fast the vehicle is moving. The odometer tells the driver the distance the vehicle has traveled. Figure 21.44 shows the cut away view of the assembly.

There is a small magnet mounted on a shaft inside the speedometer. The magnet is driven by a flexible cable from the transmission. The faster the vehicle moves, the faster the magnet spins. This action produces a rotating magnetic field that drags on the metal ring surrounding the magnet. The faster the spinning, the more drag on the ring. The spinning causes the ring to swing around against the tension of a spring. This, in turn moves a pointer attached to the ring, which indicates the vehicle speed.

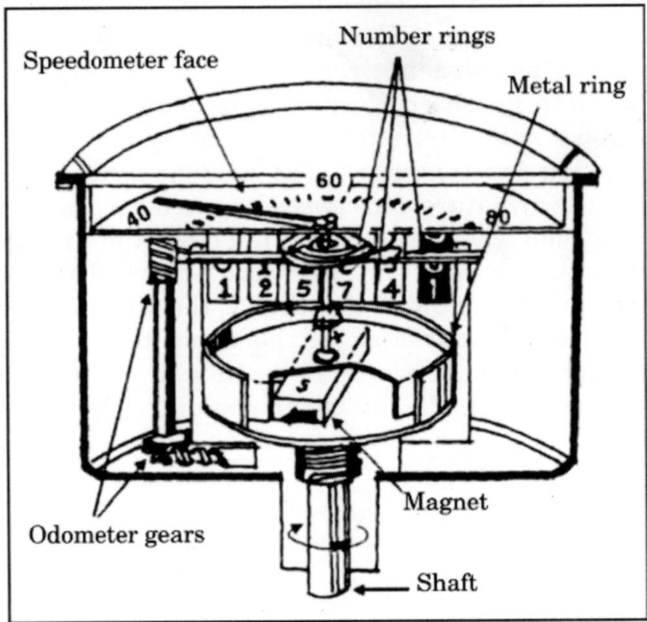

Fig. 21.44: A speedometer- odometer assembly

The odometer is operated by a pair of gears from the same rotating flexible cable that drives the speedometer. The motion is carried through the gears to the mileage rings on the odometer indicator. These rings turn to show how many miles the vehicle has been driven.

The cable is usually driven from a pair of gears in the rear extension housing of the transmission. One of these gears is on the main shaft of the transmission. The other is on the end of the flexible cable.

Chapter 22

Traction and Mechanics of Tractor Chassis

22.1 TRACTION

Traction is a term applied to the driving force developed at the point of contact between the drive wheel tyres and the ground surface for propelling the tractor with or without an attached load. The traction on the other hand is the ability of drive wheel tyres to transmit the driving force without slipping.

The literal meaning of traction is the action of pulling or hauling. Tractor is a vehicle for pulling or hauling load. Hauling means pulling with effort. Hence the term tractor is a synonym for traction engine which transforms the torque acting on the axle into linear drawbar pull. Transmitting engine power to the drawbar in agricultural and other off-highway vehicles is achieved through traction devices namely wheels or tracks. Track consists of two heavy endless metal chains moving on two iron wheels and is used in crawler tractor for providing large area of contact with the ground. Because crawler tractor is most suited for heavy work, especially for earth moving and land reclamation work.

Basic configurations of wheel are modified to improve the pulling force of the tractor. The predominate traction device is the pneumatic tyre. The traction of the tyre is achieved by its good grip with the surface over which tractor moves. The modern wheel type tractor is provided with steel rims and pneumatic tyres as the ground drive component. Pneumatic tyres, mounted on the rear axle of the tractor are so designed that they grip the ground by penetration as well as by adhesion. It is very important that the tyres are mounted in the proper way or under correct inflation pressure so that better traction is achieved when the tractor is moving in the forward direction to exert pull at the drawbar. The tractor tyre is also provided with tread bars which penetrate into the soil and result in better traction.

The power developed in the engine finally goes to wheels or tracks which move the tractor with or without an attached load. During movement of wheels, some slip occurs, which causes reduction in speed. Reduction of slip can be achieved by increasing traction. Traction depends upon *(i)* tyre size *(ii)* tyre tread *(iii)* tractor weight, *(iv)* ground surface etc and can be increased *(i)* by using rubber tyres with grooves and *(ii)* by putting lugs, cleats or grousers on the wheel rim.

22.2 FRICTION AND LAWS OF FRICTION

Whenever, there is a relative motion between two bodies which are in contact with one another, an opposing force arises to prevent relative motion of the two bodies. This opposing force is called force of friction. It depends on the nature of the two surfaces and materials of the bodies. Friction is helpful in our day-to-day's life. We are able to walk because of friction between the foot and the ground. In the absence of friction, as on perfectly smooth surface, we cannot walk. We slip and fall down. The friction experienced by a body at rest is called static friction and by a body in motion is called dynamic friction. The dynamic friction is less than the static friction.

22.2.1 Limiting Friction or Maximum Frictional Force

Limiting friction is also called as the limiting force of friction. Let us pull a body gradually by applying increased horizontal force. The magnitude of force of friction also increases and maintains the body at rest until at a certain stage, the force of static friction reaches its limiting value (maximum value). When the applied force to the body just exceeds the maximum force of static friction (developed in opposite direction), the body starts moving. This maximum value of frictional force, which comes in to play, when a body just begins to slide over the surface of the other body, is known as limiting force of friction or maximum force of friction.

Laws of friction: 1. The maximum force of static friction is almost independent of the area of contact. 2. The maximum force of static friction is proportional to the normal force.

If F is the maximum force of friction or shear force and W is the normal reaction, then F/W = u_s, where μ_s, is called the coefficient of static friction.

22.2.2 Limiting Angle of Friction or Internal Friction ϕ

$$\text{Tan}\ \phi = F/W = \mu_s.$$

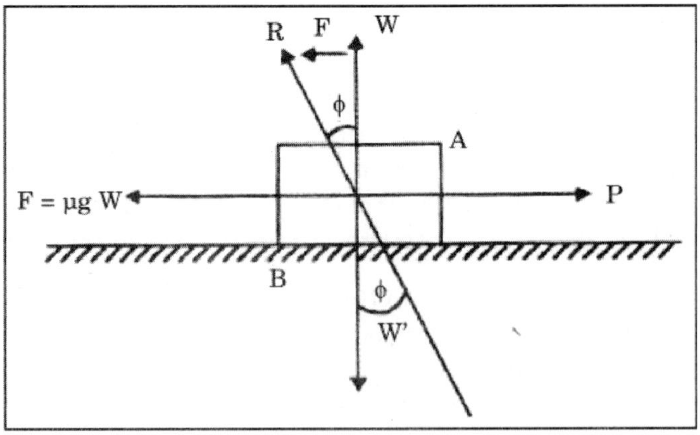

R F W

ϕ

A

F = µg W

P

B

ϕ

W'

The forces acting on the body are; 1. Weight of the body (w'); 2. Applied horizontal force (p); 3. Reaction force or resultant force (R) between the bodies A and B.

The reaction R must be equal and opposite to the resultant of W and P and will be inclined at an angle ϕ to the normal reaction W. This angle ϕ is known as limiting angle of friction or internal friction. It is defined as that angle which the resultant reaction R makes with the normal reaction W

22.3 MOHR-COULOMB FAILURE THEORY

Mohr-Coulomb failure theory considers the relationship between shear force (or frictional force) exerted on the soil and the normal load on it, corresponding to its failure. Of the many theories of failure, that have been proposed, Mohr-Coulomb theory is useful in case of soils. The essential points of this theory are that (i) Soil fails by shear. The critical shear stress causing failure depends upon the properties of the soil as well as on normal stress *(ii)* The ultimate strength of soil is determined by the stresses in the potential failure plane (or plane of shear).

When soil is subjected to "load, shearing stresses are induced in it. When the shearing stresses reach a limiting value, shear deformation takes place, leading to the failure of the soil mass. The failure may be in the form of sinking of the object. If soil fails, tractive force cannot be developed by the traction device. The failure conditions for a soil maybe expressed in terms of limiting shear stress, called shear strength. The shear strength of soil is the resistance to deformation by continuous shear displacement of soil particles or on masses upon the action of a shear stress. All stability analysis of traction device involve a basic knowledge of the shearing properties and shearing resistance of the soil. The shearing resistance of soil is constituted basically of frictional resistance or cohesion and adhesion between the surface of the soil particles. The shear strength in cohesionless soil results from intergranular friction alone while in all other soils, it results both from internal friction and cohesion.

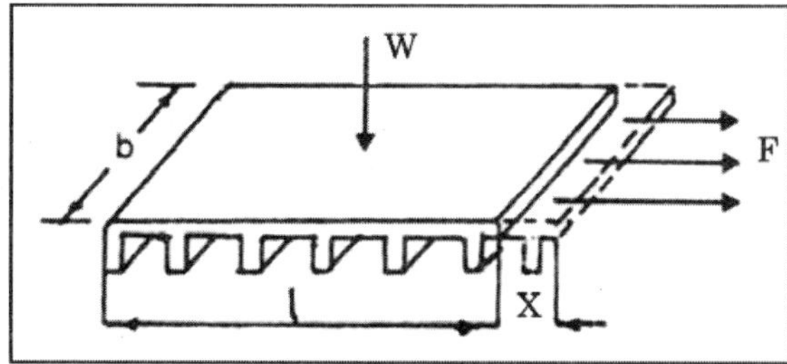

Fig. 22.1: A Soil plate under shear force and normal force

If a plate of width b and length 1 is equipped with lugs, such that an area A = bl shears off as shown in Fig. 22.1, then the force required is usually dependent upon both the normal force (load) and the area.

If we plot (Fig. 22.2) maximum value of ϕ against normal force (W) corresponding to failure for soil having cohesion *c* and internal friction ϕ, we get the values of c and ϕ. This plot or the curve is called the strength envelope. The equation for such a curve as given by Coulomb is

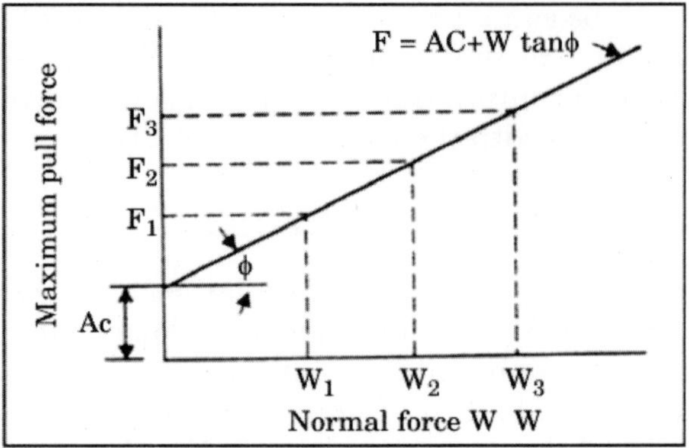

Fig. 22.2: Determination of Soil Parameters c and ϕ from the Piot of Maximum Shear Force and Normal Force.

$$F = Ac + W \tan \phi \tag{22.1}$$

Or,
$$F = A (c + W/A \tan \phi) = A (c + p \tan \phi) \tag{22.2}$$

Where p = W/A is the average normal soil pressure and A is the contact area of soil. The empirical constants c and ϕ represent respectively, the intercepts on the shear axis (or F) and slope of the straight line of Eq. (22.2).

For a track (Fig. 22.3a)

$$p = W/bl \tag{22.3}$$

Where b is the width of the track and 1 is the length of the track in contact with the soil. Here uniform pressure is also assumed.

For a rubber tyre (Fig. 22.3b), the shape of the contact area with soil is similar to an ellipse, for which,

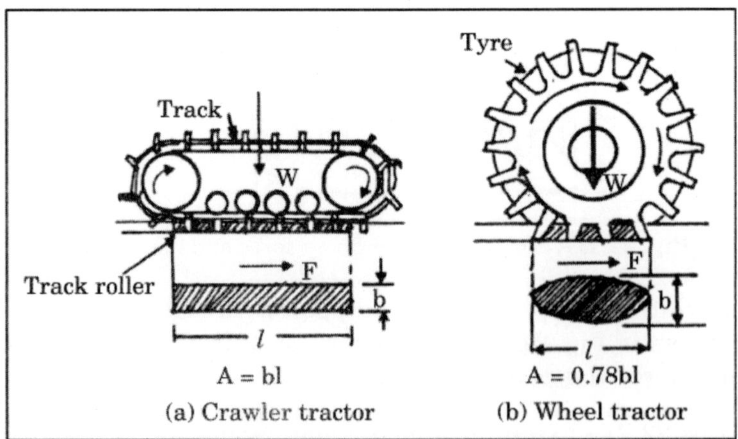

Fig. 22.3: Soil Thrust from a Traction Member by Both Shear Area and Weight

$$p = \frac{W}{0.78bl} \tag{22.4}$$

By knowing the soil values c and ϕ, the maximum soil thrust or maximum tractive force can be calculated by Eq. (22.2).

Mohr-Coulomb failure curve can be considered to be straight, if the angle of internal friction *(p* is assumed to be constant. Depending upon the properties of a material, the failure envelope maybe straight or curved and it may pass through the origin or it may intersect the shear force axis.

Example 1: Table below gives observations for normal load and maximum shear force for the specimens sandy clay tested in the shear box, 36 cm^2 in area under undrained conditions. Plot the failure envelope for the soil and determine the values of apparent angles of shearing resistance and the apparent cohesion.

Normal load (N)	*Maximum shear force (N)*
100	110
200	152
300	193
400	235

If we plot between shear force (F) and normal load (W), we get $\phi = 22°$ and total cohesive force = 70 N; unit apparent cohesion c_u = 70/36 = 1.95 N/cm^2.

22.4 SOIL TRACTION TYPES

Soils involved with traction devices are generally classified as follows :

(i) Purely cohesive soils are those which possess only a cohesive (non-frictional) soil strength component; example'- water-saturated clay.

(ii) Purely frictional soils are those which possess on a frictional (zero cohesion) soil strength component; Example - Dry sand.

(iii) Cohesive - frictional $(c - \phi)$ soils are those which possess both cohesive and frictional strength components; Examples - partially saturated clay, loams, sands etc.

The agricultural, earthmoving and forestry equipments generally work in purely cohesive and cohesive-frictional soil type.

22.5 SOME TERMINOLOGIES RELATING TO TRACTION

Traction device: It is a device for propelling the vehicle using the traction forces from the supporting surface.

Coefficient of traction: It is the ratio of total force output of the traction device in the direction of travel to the normal dynamic load on the traction device.

Dynamic load: Total force exerted by the traction device normal to the supporting surface under operating conditions.

Static load: Total vertical force exerted by the traction device when stationary, on a level surface and with zero drawbar pull.

Net traction coefficient (μ)**:** It is the ratio of net pull produced to the dynamic normal load on the traction device.

Tractive efficiency (TE): It is the ratio of output power to input power for a traction device. It is the measure of efficiency with which the traction device transforms the torque acting on the axle into linear drawbar pull. In other words, it can be expressed as TE = Drawbar power/PTO power; where drawbar power = drawbar pull x speed. Drawbar pull (P) = Tractive force (F) - resulting resistance force (Σ R).

Hence tractive efficiency $TE = \frac{(F - \Sigma R)}{\text{PTO Power}}$; the resistance force Σ R are due to.

(i) Rolling resistance *(ii)* Transmission resistance *(iii)* Air resistance *(iv)* Inertia force *(v)* Gradient resitance etc. Approximately 30% of engine power is lost in transmission for overcoming motion resistance and vehicle slip.

However, the difference between tractive efficiency and traction coefficient should be recognized, the traction coefficient refers to the ratio of forces where as tractice efficiency refers to the conversion of energy. Therefore,

$$\text{Coefficient of traction} = \frac{\text{Drawbar pull}}{\text{Dynamic normal load on wheels}}$$

Power input: Power input to traction device is calculated from input torque and angular velocity of the driving axle.

Power output: Power output of a traction device is calculated from net traction force and velocity in the direction of motion.

Rolling resistance or motion resistance: It is force required in the direction of travel of a vehicle in a given surface to overcome the resistance of motion.

Coefficient of rolling resistance or motion resistance ratio *(r)*. It is the ratio of the rolling resistance force to the normal load on the traction device.

Rolling radius (r): It is the distance traveled per revolution of the traction device divided by 2π when operating at the specified zero condition. Zero condition refers to the vehicle operating in a self-propelled condition on a hard surface, such as a smooth road with zero drawbar pull. Rolling radius is more in case of hard soil and less in case of soft soil.

Slip: It is the relative movement in the direction of travel at the mutual contact surface of the traction device and the surface which supports it. Mathematically, it can be expressed as:

$S = I - \frac{V_a}{V_t}$, where S = wheel slip or travel reduction V_a = actual travel speed and V_t is the theoretical wheel speed = r ω and ω is the angular velocity of wheel. The slip can also be expressed as $S = \frac{N_1 - N_0}{N_1}$ where N_1 = number of revolution of driving wheels for a given distance under load and N_0 is the number of revolution of the driving wheel for the same distance at no load. Also it can be expressed as the following expression :

$$\text{Slip} = \frac{\begin{pmatrix}\text{Distance travelled per} \\ \text{revolution of traction} \\ \text{device with no pull}\end{pmatrix} - \begin{pmatrix}\text{Distance travelled per} \\ \text{revolution of traction} \\ \text{device with pull}\end{pmatrix}}{(\text{Distance travelled per revolution of traction device with no pull})}$$

Thrust: Thrust is nothing but the tractive force developed in the direction of motion and in the horizontal direction. It is the net traction plus rolling resistance of tractions device while pulling.

Weight transfer: The change in normal forces on the supporting devices (traction and transporting) on a vehicle under operating conditions, as compared to a static vehicle on a level surface.

Cone index: Cone index is a measure of soil strength. It is the average force per unit base area required to force a cone shaped probe into a soil at a steady state.

22.6 SOIL MECHANICS APPLIED TO TRACTION

M.G. Bekker* states the following seven parameters of soil with the help of which, the performance of a traction device can be reasonably predicted for F (thrust or tractive force), R (rolling resistance) and P (Pull), c = cohesion of soil (kg/cm^2); ϕ = Internal factional angle of soil; Kc = cohesive modulus of soil deformation (kg/cm^{n+1}); Z = wheel sinkage (cm), n - coefficient of Z, b = smallest dimension of traction area (cm); 1 = length of the traction area. F = Ac + W tan ϕ where A = Ground contact area and W is the normal weight (load) on traction wheel.

An empirical expression for pressure P_z (direction of sinkage Z) required to sink a plate into the soil is

$$P_s = \left(\frac{K_c}{b} + K_\phi\right) Z^n \tag{22.5}$$

*M.G. Bckker, ofT-the-Road Location. The University of Michigan Press 1961.

Rolling resistance is also given by

$$R = \frac{2}{(n+1)(K_c + bK_\phi)^{1/n}} \left(\frac{W}{2l}\right)^{\frac{n+1}{n}} \tag{22.6}$$

The effective pull of a traction member is seriously affected by the rolling resistance, which is expressed for a simplified condition in Eq. (22.6). The rolling resistance will decrease as b and *l* increase, but the effect *of I* is more pronounced than b. Also rolling resistance decreases if K_c and K_ϕ increase. But K_c and K_ϕ increase as the moisture content of the soil decreases.

22.6.1 Calculation of K_c, K_0 and n

Equation (22.5) is a modification of the classical expression for the sinkage of a foundation footing with allowance for the fact that pressure does not increase linearly as Z becomes large. The sinkage of a traction member, is of course, large in comparison with the sinkage of a foundation footing of a building.

In order to determine the soil parameters K_c, K_f and n. an apparatus similar to that shown in Fig. 22.4 is used. For any given test, b and *I* will be fixed. The force W required to sink the plate to any given depth Z can be recorded continuously as in Fig. 22.4.

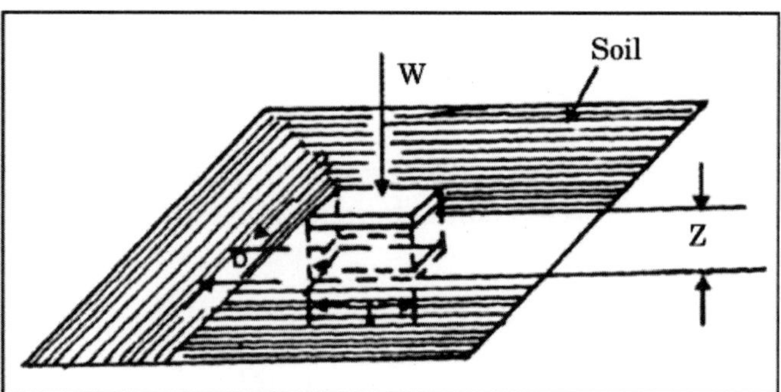

Fig. 22.4: Method of determining K_c, K_ϕ and n

For any given test, there will be three unknown values in Eq. 22.5. By writing the equation for three different conditions of *b, p* and Z, there will result three equations from which the three unknown values of K_c, K_ϕ and n can be determined.

22.7 TRACTION PERFORMANCE EQUATIONS

The analytical and empirical relationships for the tractive performance of the wheeled vehicles have been discussed in this section. The tractive ability is affected by the vertical soil reaction against the traction wheels. Weight transfer, from drawbar pull, decreases the soil reaction against the front wheel and increases the reaction against rear wheel, thus adding to the maximum drawbar pull for a two-wheel-drive tractor. Any means of increasing the rear wheel reaction will increase the traction of the rear wheel if the soil has sufficient strength and if sinkage does not limit the traction.

The different conditions under which a wheel comes into action are (i) towed condition, *(ii)* self propelled condition, *(iii)* driving condition and *(iv)* braked condition. The transition point between the braked and driven force state is the towed wheel condition. A towed wheel is unpowered and axle torque is zero. The transition point between the driven and driving force states is the self-propelled wheel condition. For a self-propelled wheel, pull is zero, with the applied torque simply overcomes the motion resistance of the soil.

Self propelled wheel: It is a powered wheel, power given to the wheel is utilized to overcome rolling resistance. Self propelled wheel develops traction exactly equal to rolling resistance but not useful work. Example is the drive wheels of automobile vehicle. For self propelled wheel pull, $P = 0$ and wheel axle has torque *i.e.* $T \neq 0$

Traction wheel or driving wheel: Traction wheel develops enough traction that overcomes rolling resistance and does useful work. Example is that tractor rear wheels are the driving wheel. In this case Pull $P > 0$ and wheel axle has torque.

Towed wheel: In case of towed wheel, the wheel is not powered. In order to pull it, we require towing force *i.e.* pulling force to move it forward. The front wheel of tractor is towed wheel. In this case $T = 0$.

Brake wheel: In braked wheel, $T < 0$ and pull direction is opposite to use that in driving wheel.

For agricultural and off road vehicles, the towed wheel condition and driving wheel condition are particularly important. Free body diagrams of the wheel for these two conditions are shown in Figs. 22.5(a) and (6).

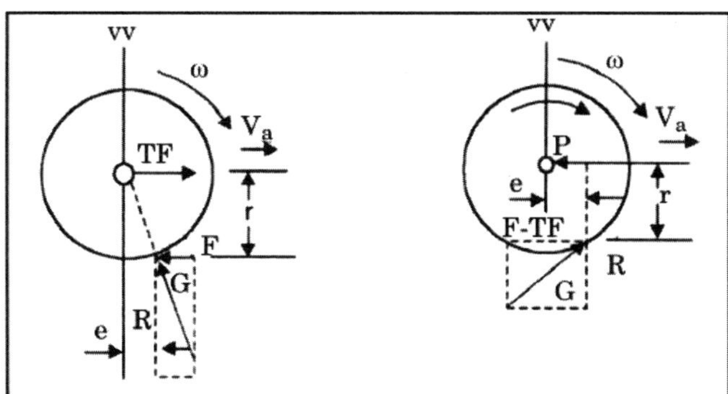

Fig. 22.5: (a) Pull-Torque Slip Relation for Wheels on Soil, (b) Free-Body Diagram of a Towed Wheel, (c) Free-Body Diagram of a Driving Wheel.

In Figure 22.5 (a) for the towed wheel, the soil reaction G is resolved into a horizontal component (which from equilibrium, considerations must be equal and opposite to the towed force TF exerted at the axle centre) and a vertical component R (which must be equal and opposite to the wheel load W). The horizontal component of the soil reaction is assumed to act at a distance r below the wheel centre. Since there is no axle torque acting on the towed wheel. (TF)r – Re = 0.

or
$$e = \left(\frac{TF}{W}\right)r = \left(\frac{TF}{R}\right)r \tag{22.7}$$

But motion resistance ratio $\rho = \frac{TF}{W}$, Hence $e = pr$

In Figure 22.5(b), for the driving wheel, the soil reaction G is again resolved into horizontal and vertical components. The horizontal component is again assumed to act at a distance *r* below the wheel centre and is now divided into two forces : a gross traction force F and a motion resistance force TF. Because motion resistance force acting in the driving wheel is considered to be the same as the towed force for the wheel.

Summing forces in the horizontal direction,

$$P = F - TF \tag{22.8}$$

But recalling gross tractive coefficient $\mu_g = \frac{F}{R} = \frac{F}{W}$ net traction coefficient $\mu = \frac{P}{R} = \frac{P}{W}$ and dividing Eq. (22. 8) by W, we get,

$$\frac{P}{W} = \mu = \frac{F}{W} - \frac{TF}{W} = \mu_g - \rho \tag{22.9}$$

Taking moments of all the forces about the centre of wheel and summing, we get

$$T - (F - TF)\, r - Re = 0 \tag{22.10}$$

With the help of Eq. (22.7),

$$T = (F)\, r \tag{22.11}$$

Thus wheel torque T is assumed to be equal to the gross tractive force F acting at a moment arm equal to the rolling radius *r*.

22.8 FORCE TRIANGLE ANALYSIS OF TRACTION WHEEL AND TOWED WHEEL

Two kinds of soil reaction forces are acting on a traction wheel (Fig. 22.6a). One is tangential force and the other is radial force. Tangential force is generated in opposite direction of torque applied and acts in the contact zone. Radial forces are passing through the centre of the driving wheel.

If $F_{(t)}$ = Resultant of all tangential force and R_r = Resultant of all radial forces, then from the force triangle $F_h = P + R_h$ and $W = F_v + R_r$; where F_h = Gross tractive effort; P = pull and R_h, is the rolling resistance, acting in opposite direction of movement of wheel.

In case of towed wheel (Fig. 22.66), there is no tangential force as no torque is provided, Only radial force is generated. From the force triangle, TF = R_h and W = R_v.

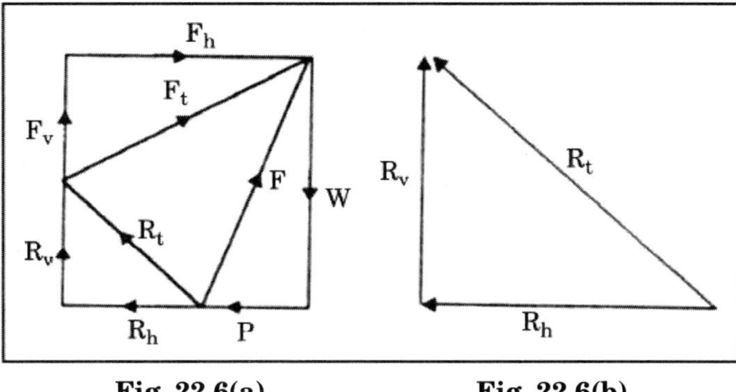

Fig. 22.6(a) **Fig. 22.6(b)**

22.9 PREDICTION OF TRACTION PERFORMANCE FROM DIMENSIONAL ANALYSIS

The prediction of performance of traction device is not so easy because of the difficulties in finding out the soil properties for cohesion and internal friction. Dimensional analysis is used to simplify the prediction equations for the multivariable system. The number of variables for dimensional analysis is nine and are as follows:

Cone index (CD) Tyre section width (b); Overall tyre diameter (d); Tyre rolling radius (r); Load (W); Towed force (TF); Pull (P); Gross tractive force (F) and slip (s).

The above nine variables are involved in traction equations and seven dimensionless ratios are needed to formulate the prediction equation. Seven dimensionaless ratios are:

(i) $\rho = \frac{TF}{W}$ (ii) $\mu = \left(\frac{P}{W}\right)$ (iii) $\mu_g = \left(\frac{P}{W}\right)$

(iv) $\mu_g = f\left(\frac{Clbd}{W}\right)$ (v) $\mu_g = f\left(\frac{b}{d}\right)$ (vi) $\mu_g = f\left(\frac{r}{d}\right)$

(vii) $\mu_g = f(s)$

Towed force: The towed force or motion resistance of a pneumatic tyre is dependent on load, size, inflation pressure and soil strength. For soils that are not very soft and tyres that are operated at nominal tyre inflation pressures; the towed force can be predicted from

$$\rho = \frac{TF}{W} = \frac{1.2}{C_n} + 0.04$$

where, C_n = wheel numeric $= \frac{Clbd}{W}$ (22.12)

The above equation has been developed for $\frac{b}{d} = 0.3$

Gross tractive force: The variations of the gross tractive force with soil strength and slip have been incorporated into a relation including the effect of wheel load and type size

$$\mu_g = \frac{F}{W} = \frac{T}{rW} = 0.75\left(1 - e^{0.3C_nS}\right) \quad (22.13)$$

where e = base of natural logarithms.

***Net traction coefficient*:**

$$\mu = \frac{P}{W} = \mu_g - \rho;$$

$$\mu = 0.75\left(1 - e^{-0.3C_nS}\right) - \left(\frac{1.2}{C_n} + 0.04\right) \quad (22.14)$$

Tractive efficiency: The pull, torque and slip characteristics of a driving wheel determine both the magnitude and efficiency of tractive performance.

Tractive efficiency (TE) = $\frac{\text{Output power}}{\text{Input power}}$;

So,

$$TE = \frac{PV_a}{T\omega} = \frac{PV}{T\left(\frac{V_t}{r}\right)} = \frac{PV_a}{PV_t}\left(\because T = (F)r\right)$$

$$TE = \frac{\left(\frac{P}{W}\right)V_a}{\left(\frac{F}{W}\right)V_t} = \frac{\left(\frac{P}{W}\right)}{\left(\frac{F}{W}\right)}(1-S) \tag{22.15}$$

From Eq. (22.5), it is clear that tractive efficiency increases at low slip and vice versa. Tractive efficiency is also affected by several factors such as (i) Tyre inflation pressure *(ii)* Soil condition *(iii)* wheel size *(iv)* Speed of travel (v) Slope of the land *(vi)* Height of the hitch *(vii)* Tread design of the wheel *(viii)* Shape and size of lugs etc.,

22.10 PRINCIPLE OF WEIGHT TRANSFER OF TRACTOR

Weight transfer that occurs from the drawbar pull decreases the soil reaction against front wheels and increases the reaction against the rear wheels, causing the increase in the tractive force of the drive wheels.

Prior to the evolution of three point hydraulic system on tractor, the implements were simply pulled by tractor and the depth was maintained with the help of depth wheel. The attached implements were lifted with the help of power driven dog clutch and was very cumbersome and time consuming. Later on the PTO mechanism was introduced to lift and lower the implement, but due to frequent troubles, it was soon replaced by a hydraulic system (1935). The hydraulic system though helped in lifting and lowering of the implement but on the other hand created the problem of front wheel lift, which was later counter balanced by front ballasting, causing ultimately the tractor more heavier and bulkier. This called for some device by which the tractor would be heavy when required and may become light when not required . This was achieved by the adoption of the process called weight transfer of tractor. The weight transfer was possible with the introduction of three point linkage.

When a tractor is pulling some load either at drawbar or connected to the three point linkage, the weight of the implement along with a part of the weight of the front axle is transferred to the rear axle. This shifting of weight to the rear axle is known as weight transfer, which adds to the tractive efficiency of the tractor.

The magnitude of the weight transfer depends on the configuration of the tractor, *i.e.*, its total weight, reactions of rear and front wheel, location of *e.g.* (center of gravity), wheel base, weight of the implement, soil reaction, soil weight and size of the implement.

Though higher weight of the implement or higher draft will result in more weight transfer and theoretically would add to the tractive efficiency, but on the other hand, the total reaction of the front axle should not be too low to make tractor overturn. Normally the weight transfer from the front axle to the rear axle is allowed to 30-40 % without affecting the steering ability.

The process of weight transfer of tractor works simply on law of mechanics. Let us consider a beam pivoted at one point with its short end exactly half the length of the long end (Fig. 22.7).

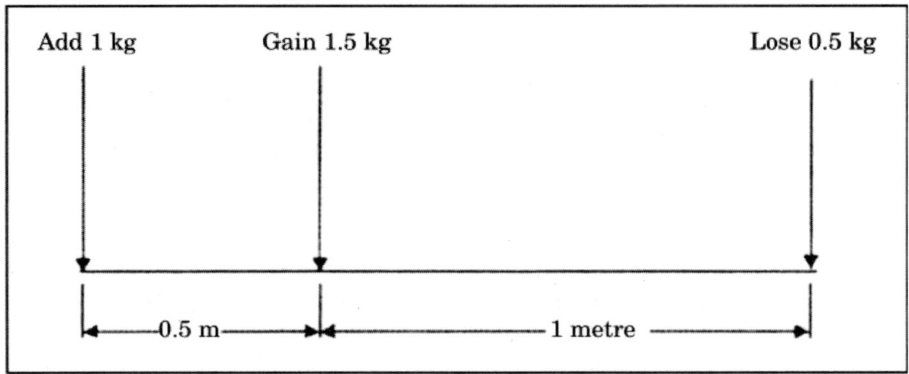

Fig. 22.7: Figure explaining weight transfer in tractor

Now if a load of 1 kg is applied at the far end of the short distance, the beam can be brought to the state of equilibrium by placing the weight of 1/2 kg of the far end of the longer distance. Also the pivot point has taken up a total load of 1.5 kg. Thus it is seen that by arranging a fulcrum in such a way, the rod is balanced by putting a weight of 1.5 kg so that the total weight of 1.5 kg acts at the fulcrum.

A tractor mounted with an implement works exactly in the same way. The wheel base serves the purpose of the long end of the lever where as the distance of *e.g.* of the implement from the rear wheel, as short end. The ground contact of rear wheel serves the purpose of pivot point. As shown in Fig. 22.7, the weight distribution changes as soon as downward load is applied on the short end. Exactly the same thing happens when an implement is carried at the rear of the tractor. The rear wheel gains all the weight of the implement plus part of the weight from the front axle, which is half the weight of the implement.

In actual practice, it is not only the implement weight, but there are other weights, such as soil resistance and soil weight.

We may take the case of a tractor, connected to a 2 bottom MB plough. The distribution of weight has been observed on weigh bridge as under and also be verified mathematically.

Tractor weight	*Rear axle (kg)*	*Front axle (kg)*
Tractor 1550 kg	900	650
Attach implement, 200 kg	+ 300	–100
Soil resistance, 100 kg	+ 150	–50
Soil weight, 50 kg	+ 75	–25
Resulting in total distribution	1425	475

From the above data, it is seen that load on rear axle changes from 900 kg to 1425 kg and on front axle, it changes from 650 kg to 475 kg as implement is put to work, which means significant gain by rear axle and hence better traction. Although appreciable weight is taken off from the front wheels, still there is enough left for adequate stability and effective steering (around 73 %).

By transference of weight, though the front end becomes lighter, but the lighter front end has the tendency to lift the front end or twist back around rear wheel. This is however, overcome by the compression force developed in the top link. The compression forces acting above the centre line work together to keep the front of tractor down. The calculation of weight transfer from front axle and rear axle of a tractor has been done in section 22.11.1.

22.11 MECHANICS OF TRACTOR CHASSIS

The tractor chassis presents varied and complex problems in mechanics because of the great number of combinations of designs that are possible with the varied operating conditions under which tractors are used. The following parameters, not of course complete, indicate the complexity of the mechanics of tractor chassis.

(i) Location of centre of gravity *(ii)* type and design of traction wheel *(iii)* design and location of hitch system *(iv)* resistance characteristics of load or implement being pulled, *(v)* traction surface *(vi)* power-take-off torque *(vii)* distribution of mass of the tractor *i.e.* mass moment of inertia about its three principal axes *(viii)* dynamic response of tractor; effect of terrain, damping etc *(ix)* grade; making angle to the direction of travel *(x)* acceleration; very gradual to very abrupt changes of speed *(xi)* gyroscopic effect of the engine upon the complete tractor etc. But to understand the main principle regarding the mechanics of tractor chassis, a simple approach is followed considering the main forces such as *(i)* gravitation *(ii)* soil reaction, *(iii)* traction force and *(iv)* drawbar pull etc.

22.11.1 Mechanics of Tractor Chassis Under Simplest Operating Condition

Figure 22.8 shows the forces acting on a wheel type tractor under simplest operating conditions. The analysis is based on the following simplifying assumptions.

(i) Rear wheel drive, *(ii)* Uniform forward motion on a plane and level surface, *(iii)* Line of action of drawbar pull P horizontal, midway between traction wheels and parallel to the direction of motion of tractor, *(iv)* Line of action of vertical soil reaction R^ against traction wheel, passing through the centre of rear axle O

(v) Line of action of vertical reaction, R_1 against front wheels, passing through centre of front axle O' *(vi)* Line of action of tractive force F is tangent to traction wheel at B *(vii)* Rolling resistance assumed zero.

Force analysis: On the basis of the assumptions listed above, gravitational force can be represented by the weight W_1 supported by the front wheels and the weight W_2, supported by rear wheels, when the drawbar pull is zero. Likewise, the soil reaction can, for the purpose of this approximate analysis be resolved into three components R_1 R_2 and F.

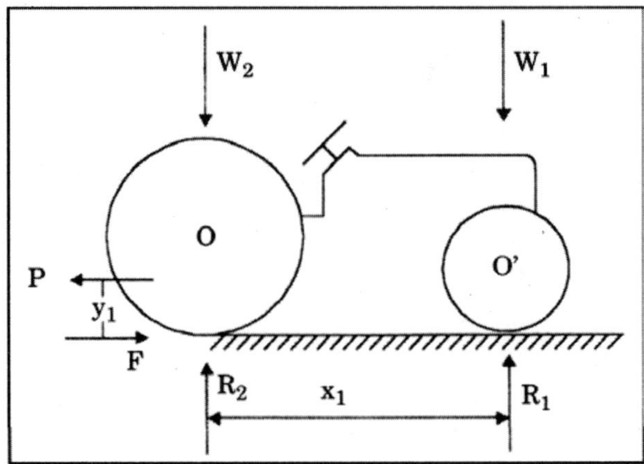

Fig. 22.8: Forces Acting on a Wheel Type Tractor Under Simplest Operating Conditions. Drawbar Pull is Parallel to Direction of Travel on a Level Surface.

If the tractor is considered as a free body, the algebraic sum of all forces acting parallel to the direction of motion must be equal to zero.

$$F - P = 0 \tag{22.16}$$

Similarly, the algebraic sum of all forces acting perpendicular to the direction of motion must be equal to zero.

$$R_1 + R_2 - W_1 - W_2 = 0 \tag{22.17}$$

The algebraic sum of the moments about any given axis must also be equal to zero. The problem can further be simplified by taking moments about B, the intersection of the soil reactions R_2 and F. The line of action of force W_2 passes through this axis. Hence taking moment about B,

$$W_1 x_1 - Py_1 - R_1 x_1 = 0. \tag{22.18}$$

From the above three equations, the values of the soil reaction can be easily calculated in terms of the tractor's weight and drawbar pull. Solving Eq. (22.18) for R_1,

$$R_1 = W_1 - \frac{Py_1}{x_1} \tag{22.19}$$

Likewise from Eq. (22.17) ;

$$R_2 = W_1 + W_2 - R_1 \tag{22.20}$$

Substituting the value of R_1 from Equation (22.19) for R_1 in Eq. (22.20)

$$R_2 = W_2 + \frac{Py_1}{x_1} \tag{22.21}$$

The stability of tractor is, to a great extent determined by R_1 and tractive capacity by R_2. The term $\left(\frac{Py_1}{x_1}\right)$ expresses the change in soil reactions R_1 and R_2 resulting from drawbar pull P. The soil reaction R_1 supporting the front wheels decreases as P increases and soil reaction R_2 increases. This relationship is true until P becomes large enough to cause $\frac{Py_1}{x_1}$ to become equal to W_1, which in turn causes R_1 to become zero. Any further increase in P will cause the front wheels to leave the ground.

Although there is no actual shift of the weight, but this change $\frac{Py_1}{x_1}$ in soil reaction R_1 and R_2 is generally known as weight transfer.

If zero is substituted for R_1 in Eq. (22.19) and the equation is solved for P, an expression is obtained for the value of the drawbar pull P, at which the soil reaction against the front wheel becomes zero.

$$P = \frac{W_1 x_1}{y_1} \tag{22.22}$$

From equations (22.19), (22.21) and (22.22), it is clear that by increasing wheel base length x_1, R_1 will increase and R_2 will decrease for any given drawbar pull and hitch height. It will also increase the drawbar pull required to lift the front wheels from the ground. That means that, if wheel base is lengthened, the tractor is to be designed to carry a greater portion of its weight on the rear wheels by considering other factors.

Similarly from the above equations, it is clear that by increasing the effective drawbar height y_1, the change in soil reactions between R_2 and R_1 will increase for a given drawbar pull P and decreases the drawbar pull at which the front wheels will be lifted from the ground. Thus, from the stability point of view, the line of action of the drawbar pull should be as low as possible. But there should always be a compromise between x_1 and y_1.

22.11.2 Mechanics of Tractor Chassis Under Nonparallel Pull Condition

The most common variation from the simple situation shown in Fig. 22.8 is operation of tractor with a hitch so arranged that the direction of the drawbar pull is not parallel to the direction of the motion of the tractor. This situation is shown in Fig. 22.9 where drawbar pull is at an angle a below the line parallel to the motion of the tractor. As in Fig. 22.8, point B is the interaction of the resultants of the supporting and traction components of the soil reactions against the rear wheels. In order to make the present situation more general and simpler, the total weight W, acting through the centre of gravity G, is used rather than the weights W_1 and W_2 supported at the front and rear wheels respectively. All other assumptions remain same as in Fig. 22.8.

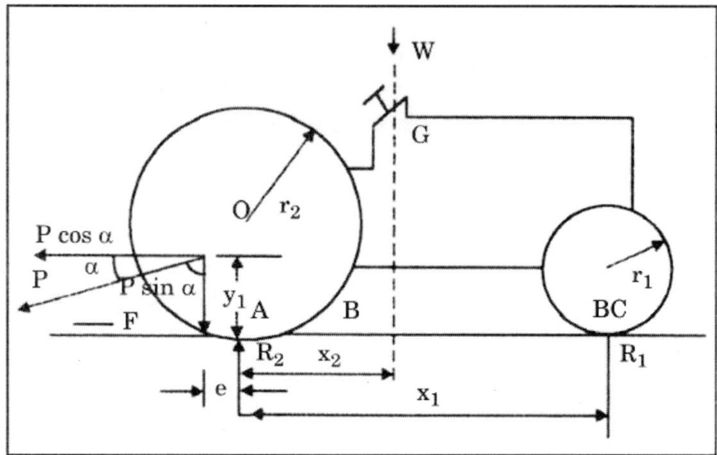

Fig. 22.9: Important Forces on a Tractor Chasis.

Referring to Fig. 22.9 and considering tractor as a free body, the algebraic sum of all forces acting parallel or perpendicular to the direction of motion as well as the algebraic sum of moments about any point must be separately zero. All forces acting parallel to the direction of motion is

$$F - P \cos \alpha = 0 \tag{22.23}$$

where F is the tractive force. Algebraic sum of all forces acting perpendicular to the direction of motion must be equal to

$$R_1 + R_2 - W - P \sin \alpha = 0 \tag{22.24}$$

Taking moments about the point B,

$$Wx_2 - R_1x_1 - y_1 P \cos \alpha - e P \sin \alpha = 0 \tag{22.25}$$

Solving equation (22.25) for R_1

$$R_1 = \frac{Wx_2}{x_1} - \left(\frac{y_1 P \cos \alpha + eP \sin \alpha}{x_1}\right);$$

Also $R_1 + R_2 = W \rightarrow R_2 = W - R_1$

or $$R_2 x_1 = \left[W - \frac{Wx_2}{x_1} + \left(\frac{y_1 P \cos\alpha + eP \sin\alpha}{x_1} \right) \right]$$

or $$R_2 x_1 = Wx_1 - Wx_2 + y_1 P \cos\alpha + eP \sin\alpha$$

$$= \frac{W(x_2 - x_1)}{x_1} + \frac{P(y_1 \cos\alpha + e\sin\alpha)}{x_1}$$

The stability of tractor is determined by R_1 and the tractive capacity by R_2. The term $\frac{P(y_1 \cos\alpha + e\sin\alpha)}{x_1}$ is known is weight transfer and indicates the change in soil reactions between R_1 and R_2. In order to avoid overturning of the tractor, this factor must be less than $\frac{Wx_2}{x_1}$. As soon as $\frac{P(y_1 \cos\alpha + e\sin\alpha)}{x_1}$ becomes more than $\frac{Wx_2}{x_1}$, then front wheels of the tractor leaves the ground. Hence the drawbar pull P at which the tractor becomes unstable is given by

$$\frac{P(y_1 \cos\alpha + e\sin\alpha)}{x_1} = \frac{Wx_1}{x_1}$$

Or, $$P = \frac{Wx_2}{y_1 \cos\alpha + e\sin\alpha} \tag{22.26}$$

From Eq. (22.26), it is clear that the change of pull depends on the weight of tractor, height of drawbar hitch, angle α etc.

22.12 CENTRE OF GRAVITY AND MOMENT OF INERTIA DETERMINATION OF TRACTOR

22.12.1 Centre of Gravity

Centre of gravity is the point on the body of tractor at which its total weight may be considered as acting. Since tractors are composed of many irregularly shaped parts, it is difficult to analytically find the centre of gravity of the assembled tractor. Therefore, experimental method of determining the centre of gravity is an easy way. It is necessary to determine only three planes, each containing the centre of gravity and each intersecting the other two .Exact location of the centre of gravity of tractor *can* be determined by the following method.

(i) Suspension method *(ii)* Balancing method- *(iii)* Weighing method

Suspension method: The tractor is suspended from any convenient point, strong enough to carry its weight. The centre of gravity will be in the vertical plane through the point of suspension. By repeating this operation with another point of suspension, another plane maybe located and its intersection with the previous vertical line locate the centre of gravity.

Balancing method: The balancing method can be applied for determining the centre of gravity of a track type tractor. The equipment needed is a plumb bob and a large timber equal in length to the overall width of the tractor and about 15 cm or more thick. If the tractor is driven so that the lug point positions of the two tracks match and it is then driven slowly on the timber, so that it can be balanced over the two lug points. The centre of gravity will be in the vertical plane through the lug points. By backing the tractor on the timber, another plane can be located for finding out the point of centre of gravity.

Weighing method: The weighing method of finding the centre of gravity is commonly used for a 4-wheeled tractor referring to Fig. 22.9, a vertical plane containing the centre of gravity can be determined by the following equation.

$$x_2 = \frac{R_1 x_1}{W} \tag{22.27}$$

If the front wheels of the tractor are raised a distance y_2 and R_1 is determined (Fig. 22.10), another plane containing the centre of gravity may be obtained from

$$x_2' = \frac{R_1' x_1'}{W} \tag{22.28}$$

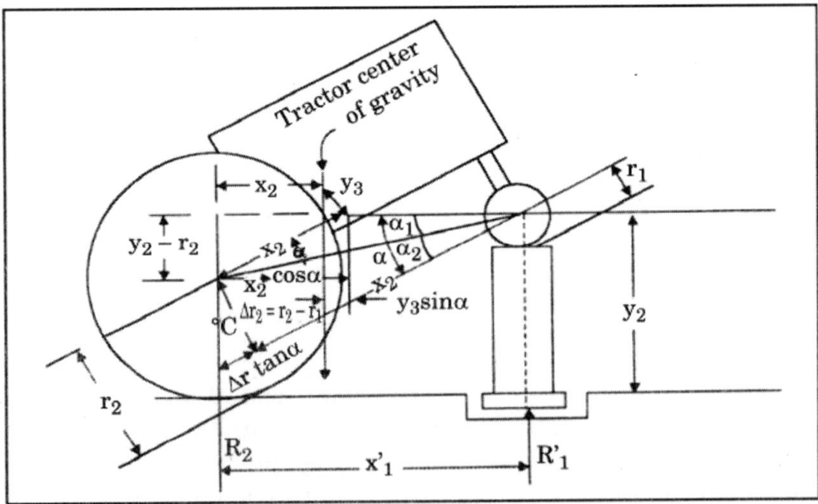

Fig. 22.10: Determination of the Vertical Location of the Center of Gravity using the Weighing Method.

R'_1 is the weight of the tractor on the front wheels which is measured. But from the geometry of Fig. 22.10, $x'_2 = x_2 \cos a - y_3 \sin a$

Thus $$\frac{R'_1 x'_1}{W} = x_2 \cos\alpha - y_3 \sin\alpha \tag{22.29}$$

But $$x'_1 = (x_1 + \Delta r \tan \alpha) \cos \alpha \tag{22.30}$$

Where $\Delta r = r_2 - r_t$; Substituting Eq. (22.30) in the Eq. (22.29), dividing by cos α and solving for y_3.

$$y_3 = \frac{(Wx_2 - R'_1 x_1)}{W \tan\alpha} - \frac{R'_1 \Delta r}{W}$$

The angle α is the only quantity in Eq. (22.31) that cannot be directly measured. However, $\alpha = \alpha_1 + \alpha_2$ where $\tan \alpha_1 = (y_2 - r_2)\, lx'_1$ and $\tan \alpha_2 = \frac{\Delta r}{x_1}$ An approximate value of x'_1 for determining the value of α_1 is also given by

$$x'_1 = \sqrt{x_1^2 - (y_2 - r_2)^2 + (\Delta r)^2} \tag{22.32}$$

In the above three methods of determining the centre of gravity of tractor, the fluid movement occurring in the fuel tank, coolant, crank case and transmission box is ignored.

22.12.2 Moment of Inertia Determination of Tractor

A body free to rotate about an axis has a tendency to oppose any change in its state of rest or uniform rotation. The body also possesses rotational inertia. This property of the body is expressed by saying that the body possesses moment of inertia about the axis of rotation. If the rotational motion of a tractor is being considered, it is necessary to know its mass moment of inertia about the axis of rotation in question. The tractor may rotate about any one of its three axes; longitudinal, transverse and vertical. Due to irregularly shaped parts of tractor, it is also difficult to analytically find out its moment of inertia about the three axes passing through the centre of gravity. Hence it is required to measure the moment of inertia experimentally.

If a tractor is suspended from a pivot as in Fig. 22.11, the tractor plus the sling will become a compound pendulum which oscillate with the time period, given as

$$T = 2\pi \sqrt{\frac{I_0}{WR_0}}$$

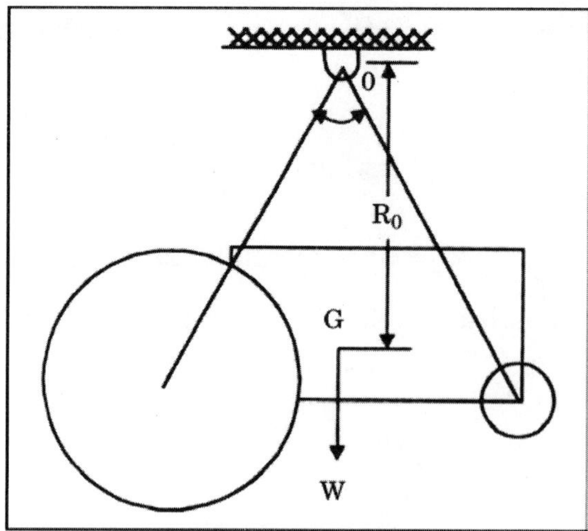

Fig. 22.11: Determination of the Pitch Moment of Inertia using the Pendulum Method.

Where T is the time period (second); time taken for one complete cycle of the body about the pivot. I_0 = moment of inertia of tractor plus sling about pivot O ; W = Weight of tractor plus sling; R_0 = Distance between point O and the centre of gravity of the tractor plus sling. $W = W_t + W_S$ where W_t is the weight of tractor and W_S is the weight of sling. For negligible weight of sling, Eq.(22.33) reduces to

$$T = 2\pi\sqrt{\frac{I_0}{W_t R_0}} \tag{22.34}$$

From Eq. (22.33); all the terms can be measured except I_0, which can be computed from this equation.

If the weight and moment of inertial of the sling can be assumed to be neglected with respect to the weight of the tractor, then moment of inertia of the tractor about its centre of gravity l_t can be computed using the parallel axis theorem.

$$I_t = I_0 - m_t R_0^2 \tag{22.35}$$

Where m_t = mass of tractor which is equal to its weight divided by g (acceleration due to gravity).

For accurate measurement of I_t, the distance R_0 should be made as small as possible.

Chapter 23

Tractor Tests and Performance

23.1 INTRODUCTION

With the increasing use of tractors in the farming operations, there is the requirement for manufacturers/buyers to test the tractor on its power output and performance. Testing helps in assessing the functional suitability, durability and performance characteristics under different operating conditions. The results of the test provide as a basis to decide the suitability of machine for different soil agro climatic conditions. In general, tests may be carried out with one of the following purposes:

(a) To justify the rating of the engine
(b) To get information which cannot be obtained by calculations
(c) To make the users confirmed regarding the data reported by the manufacturers
(d) To help, and guide the manufacturers and designers to improve their product or design.

The basis characteristics describing the performance of an engine are the engine torque, power, specific fuel consumption and efficiency. Some of the heat energy evolved in the process of burning of fuel in the cylinders of an engine is converted into mechanical energy. The gas pressure force acting upon the piston is transmitted through the connecting rod to the crank throw and thus applies torque to the crank shaft. Torque is the product of the force rotating the crankshaft and the crank radius. While an engine is developing torque, it performs work. The amount of work done in unit time is termed as power. The fuel economy of an engine in characterized by its specific fuel consumption which is determined by dividing the hourly fuel consumption of the engine by its useful power. The efficiencies of the engine are generally expressed by its mechanical efficiency and thermal efficiency. In addition to the engine performance, the other systems of the tractor are also tested to assess its overall working capacity.

Basically two types of tests are done in tractor, one is the confidential test and the other is the commercial test. The test conducted for providing confidential information on the performance of tractor before commercial production is called the confidential test. The test conducted for establishing performance characteristics of tractors that are ready for commercial production or already in production is called commercial test. The confidential test also provides special data that may be required by the manufacturer for rectification and modification in their products.

23.2 GENERAL GUIDELINES FOR PREPARATION OF TESTS

1. **Selection:** The tractor, if under production, should be selected at random from the production unit as directed by the testing station with its usual accessories and in a condition as generally offered for sale.
2. **Specification sheet:** The tractor manufacturer should supply the specification of the tractor. The manufacturer should also supply all literature, service manual and parts catalogue. These specifications should be verified by the testing station and should be reported.
3. **Running-in and preliminary adjustment:** The manufacturer should run-in the tractor before the test under his responsibility. The running-in should be carried out in collaboration with the testing authority. The running-in test report should state the place and duration of test. The manufacturers may make adjustments in fuel injection pump, governor, fuel injector and any other adjustments during the period the tractor is prepared for tests.
4. **Ballasted tractor:** The ballast mass, which are commercially available and approved by the manufacturer for use in agriculture may be fitted. The inflation pressure should be within the limits specified by the tyre manufacturer for field work.
5. **Repairs and adjustments during test:** All repairs or adjustments made during the test should be reported together with comments on any practical defects or short comings.
6. **Funds and lubricants:** Fuels and lubricants should be selected from the range of products commercially available.
7. **Dimensions checking:** For measurement of length, width and other dimensions, the tractor should be parked on a firm horizontal surface with steering wheels in a position to allow the tractor to travel in a straight line. The tractor should also be without any wear on tyres.
8. **Disconnecting the auxiliary equipments:** For all power tests, accessories such as hydraulic lift pump or any other power consuming device like air compressor may be disconnected only if it is possible for the operator to do so as a normal practice during work in accordance with the operator's manual without using any tool. If not, they should remain connected and operate at minimum load.

23.3 TYPES OF TEST

The various tests to be conducted on an agricultural tractor are of the two types; a) Laboratory tests, b) Field tests. The tests under laboratory test can be summarized as follows:

1. **Power test**
   (i) Engine test
   (ii) PTO performance test
   (iii) Belt-pulley performance test
   (iv) Drawbar performance test
   (v) Hydraulic power lift test.
2. **Safety test**
   (i) Brake test
   (ii) Centre of gravity
   (iii) Turning ability test.
3. **Ergonomical test**
   (i) Noise measurement
   (ii) Visibility from driver's seat
   (iii) Mechanical vibration test,
   (iv) Smoke level test.
4. **Miscellaneous test**
   (i) Air cleaner oil pull over test
   (ii) Component/Assembly inspection.

The tests under field test are as under:

5. Field evaluation with specific implements for a fixed number of hours
6. Haulage test

Power Test

(i) **Engine test:** In this test, sufficient data should be obtained pertaining to specific fuel consumption with power, torque and engine speed throughout the working period. The results should be presented graphically.

(ii) **PTO performance test:** This test gives the results of the data for the power available at the power take-off (PTO) of the agricultural tractors.

(a) **Test at maximum power:** This test is done with the help of a suitable dynamometer. The tractor is operated for a period of about two hours after the engine has warmed up and to reach the stabilized running conditions. The power, torque and fuel consumption readings are taken at almost equal intervals during the two hours test. The maximum power quoted in the report should be the average of at least six readings made at regular intervals during the 2 hour period. If the power varies by more than ± 2 per cent from the average, the test should be repeated. If the same condition continues, it should be recorded in the report.

(b) **Test at varying speed at full load:** The power, torque and fuel consumption of the tractor are measured with respect to varying speed at full load at approximately 10 per cent speed increments. With the recorded data, a suitable curves are plotted.

(c) **Test at varying load:** Keeping the governor position for maximum power, the power, speed and fuel consumption of tractor are recorded. Engine speed at no load, is also recorded. The readings are taken at no load, half load, three quarter load and full load corresponding to maximum power and at 85 % of the torque.

(iii) **Belt-pulley performance test:** The power available at the belt or pulley shaft of tractor is measured by connecting tractor pulley with the dynamometer by a flexible belt. The belt slip should not exceed about 2 %. The engine power and fuel consumption are recorded corresponding to engine speed.

(iv) **Drawbar performance test:** Drawbar pull is measured by means of suitable dynamometer, inserted at the hitch between the tractor and the pulling load. For draw bar testing, the load on the back of the tractor has to be varied and for this, a loading car is used. A loading car is device in which any amount of load can be applied. In this test, the parameters generally measured are (a) drawbar pull (b) wheel slip, (c) fuel consumption, (d) speed. Drawbar performance curves are obtained by applying a large number of known loads in each forward gear and measuring the distance traveled and the time taken for a fixed number of wheel revolution.

(v) **Hydraulic power lift test:** This test is done with an aim to find out the maximum static vertical force that can be exerted by the hydraulic lift at the lower hitch point and at a point 610 mm to the rear of the hitch points. The performance of hydraulic pump and relief valve is tested to see whether these items are functioning well according to the requirements or not. The specification of hydraulic fluid should conform the specification of the manufacturer.

Safety Test

(i) **Brake test:** The brake test is done to assess the performance of wheel brakes and parking brakes. To perform the wheel brakes, the tractor is allowed to move at its maximum designed speed under ballasted and unballasted conditions. Then the brakes are applied. The time and distance moved by the wheel are noted after using brake. The stopping distance for cold and hot brake should not exceed 7.6 m and 9.5 m, respectively. A brake is considered cold when its temperature is lower than 80°C. Similarly the performance of parking brakes is checked by parking the tractor on slopes.

(ii) **Centre of gravity:** To evaluate the effect of gradients on tractive effort and stability, it is necessary to know the position of centre of gravity of the tractor. It is determined through a method of weight transfer by lifting it. It is also determined with and without ballast with fuel tank full and the lubricating oil up to the desired mark. A weight of 75 kg is put on the seat of driver to replace the weight of driver. If the centre of gravity is below the rear axle of the,tractor, then it is better stabilized.

(iii) **Turning ability test:** Turning ability test determines the turning radius of tractor. It is the radius of circle on which outside front wheel moves when the front wheels are turned to their extreme outer position. This determines the extent of headland to be made available for tractor for field operation. This turning radius is checked as per the manufacturers' specification.

Ergonomical Test

(i) **Noise measurement:** Noise is measured with the help of a sound level meter. The results of the test provide information regarding the comfort of the operators against exposing themselves to noise level that put their hearings at risk. The average amplitude of mechanical vibration should not exceed 100 microns.

(ii) **Visibility from driver's seat:** This test is done to find whether sufficient visibility is there from the seat of driver or not.

(iii) **Mechanical vibration test:** The amplitude of mechanical vibration of the assemblies and components of tractor is measured with the help of vibration measuring device. The tractor is braked on a level concrete surface and operated at rated speed at no-load and at load corresponding to 85 per cent of maximum PTO power. The maximum horizontal displacement and vertical displacement in microns is measured by mounting the measuring device in related positions.

(iv) **Smoke level test:** The smoke level measured at 80 % of maximum power should not exceed the desired level.

Miscellaneous Test

(i) **Air cleaner oil pull over test:** The tractor is parked on a level ground. The air cleaner should be cleaned and filled with normal oil. The engine should then be operated at full governed speed for 15 minutes with alternate acceleration and decelerations made after every 30 seconds. The air cleaner assembly should be weighed before and after the test. The loss of mass of oil should be reported. If there is no pull-over with the tractor in level position, the test is repeated with tractor tilted 15° to either side and then 30° forward and backwards with direction of travel.

(ii) **Component/assembly inspection:** After the completion of all the laboratory, track and field tests, the tractor assemblies such as engine, transmission, brakes, front axle, starter motor dynamo etc are dismantled to check their specifications as given by the manufacturers.

Field Test

A range of implements, matched to the tractor should be supplied by the manufacturer. The tests should be conducted with these implements. The field operations with each matching implement should be carried out for a duration of at least 25 hours.

Haulage Test

The haulage test should be carried out with two wheel or four wheel trailers at the gross trailer loads recommended by the manufacturer. The haulage test is conducted on a level track having about 80 per cent of tarmacadam surface and 20 per cent of earthen surface and having gradients not exceeding 8 per cent at certain short length.

23.4 POWER MEASUREMENT METHODS

To understand power and its measurement, certain terms need to be defined and are as follows.

Power: It is the rate of doing work. Its unit is Newton meter per second (watt).

Indicated power: It is the power available at the piston inside the engine cylinder. It is the power developed in the cylinder without or auxiliary unit.

Brake power: It is the power output of the engine crankshaft.

Friction power: The power required to run the engine at any given speed without production of useful power. It represents the friction and pumping losses of an engine.

Indicated house power = brake power + friction power

Power-take off power: It is the power delivered by a tractor through its pto shaft.

Drawbar power: It is the power of a tractor, measured at the end of the drawbar. This power is used for pulling loads at the drawbar

Belt power: It is the power of the engine, measured at the end of a suitable belt, receiving drive from the pto shaft.

Pulley power: Power measured by coupling the pulley shaft directly to the dynamometer for pulley work.

Observed power: The power obtained at the dynamometer without any correction for atmospheric temperature, pressure or vapour pressure.

Corrected power: Corrected power is obtained by correcting the observed power to standard conditions of sea-level pressure and temperature.

Rated engine speed: The speed of engine in rev/min specified by manufacturer for continuous operation at full load.

Rated horsepower: The power stated by the manufacturer that the engine will generate. It may or may not be the maximum power generated by the engine. Generally manufacturer rates the power within a safe limit avoiding the overloading of the engine

23.4.1 Measurement of Indicated Horse Power

The power developed inside the engine cylinder is known as indicated horse power and designated as I.H.P The I.H.P of an engine at a particular running condition is obtained from the indicator. The recording of pressure and volume inside the engine cylinder in the form of a p – v diagram is done with the help of the indicator. By getting the indicator diagram, the indicated horsepower may be computed by measuring the area of the diagram, either with the planimeter or by the ordinate method and dividing by the stroke measurement to obtain the mean effective pressure. In case indicated horsepower cannot be measured directly, it is made possible by measuring the brake horsepower and also the engine losses.

23.4.2 Mechanical Indicator

The mechanical indicator consists of two parts *(i)* A pencil mechanism causes a pencil or stylus to rise in a straight line with the height that is proportional to the pressure in a small cylinder (indicator cylinder) at the base of the indicator *(ii)* the other main part is designed to provide a lateral motion which is proportional to the cylinder volume.

The mechanical indicator consists of a cylinder on to the end of which is fitted a coupling nut. This coupling nut is used for coupling the indicator on to a gas tap, which is fitted to the cylinder head of the engine under test. A gas passage exists from the combustion space of the engine cylinder to the gas tap, through the coupling nut into the indicator cylinder, A piston slides in the indicator cylinder. From the piston, a piston rod passes out at the top of the cylinder through a control spring and is then coupled to a straight line linkage at the top. The control spring is connected to the piston rod in order that it can control the piston movement when the piston is under pressure from engine cylinder. The straight line linkage is mounted in a swinging collar which can rotate on the top of the indicator cylinder. A brass stylus is attached to the end of a long link as shown in Fig. 23.1. This

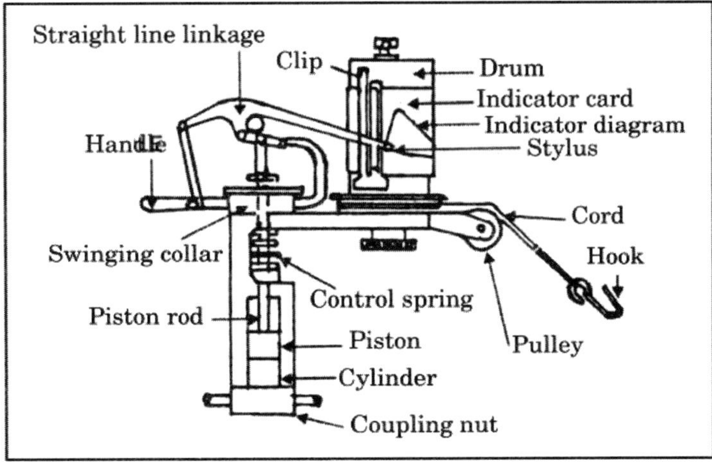

Fig. 23.1: Mechanical Indicator

stylus moves up and down in a vertical straight line as the piston moves up and down. The stylus movement is magnified through straight line linkage. A drum is mounted on a platform located at the top of the cylinder. This drum can rotate on a central spindle and its turn is controlled by a coiled spring, fitted over the central spindle and inside the drum. The paper indicator cord is wrapped round the drum. The one end of the paper is fastened to the clip fitted on to the side of the drum. The other end of the card is fastened to the bottom of the drum and is then wrapped round a channel at the bottom of the drum some two or three times. The card then passes away from the drum, over a pulley and a hook is fitted to its end. The hook is then connected to the reciprocating mechanism on the engine when indicator diagram is required. By connecting with the reciprocating mechanism, the drum oscillates and hence produces the horizontal component of the indicator diagram. The stylus makes a black line over the surface of the indicator card to get the p-v diagram of the engine cylinder.

The gas tap on the engine is opened and then indicator piston starts oscillating as the pressure varies in the engine cylinder. This oscillating appears as the vertical oscillations of the stylus. The straight line linkage is now rotated by means of the handle on the swinging collar until the stylus touches the indicator card on the osculating drum. The stylus is held in contact with the card for a small time and then is removed. The gap tap is shut and the hook is removed from the oscillating mechanism. The indicator card is removed and upon inspection, the indicator diagram is found to have been traced on its surface. The height of the diagram is controlled by the control spring of the indicator piston rod. This spring can be changed if required and a heavier spring may be used for the high engine loads. A typical p-v diagram taken by a mechanical indicator is shown in Fig. 23.2.

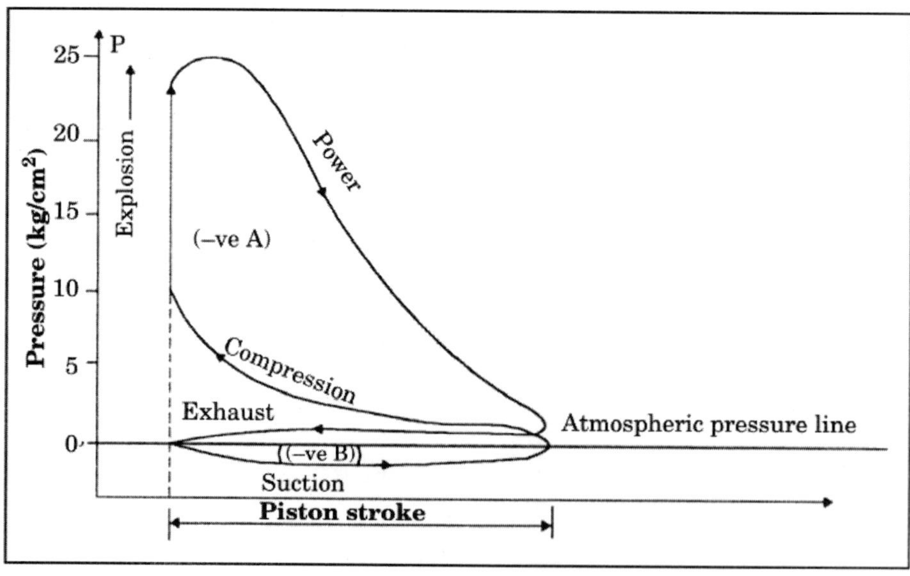

Fig. 23.2: P – v Diagram taken from Mechanical Indicator.

23.4.3 Measurement of Area for Indicator Diagrams

The net area of indicator diagram taken from mechanical indicator is proportional to the work done per cycle of the engine and is therefore, a measure of the indicated horsepower if speed is known. In the general type of four stroke cycle engine as shown in Fig. 23.2, the area A represents positive loop or positive work done by the engine and B represents exhaust-suction loop or negative loop or pumping loop or wasteful work done by the engine. The pumping work is required for admitting the fresh charge and for exhausting the burnt gases. This loss of work is known as pumping loss and power consumed for this is known as pumping horse power. The net work per cycle of the engine is therefore, represented by (A-B).

It is not necessary to measure the area of the indicator diagram, but only the mean height which is proportional to the indicated mean effective pressure. The usual method adopted is to divide the diagram into ten equal parts as shown in Fig. 23.3, by ordinates parallel to the pressure axis and to draw the dotted ordinates shown, such that they lie midway between the full lines. The suni of the lengths of the dotted ordinates divided by 10 gives the mean height of the diagram. The mean height represents the mean effective pressure of the engine cylinder.

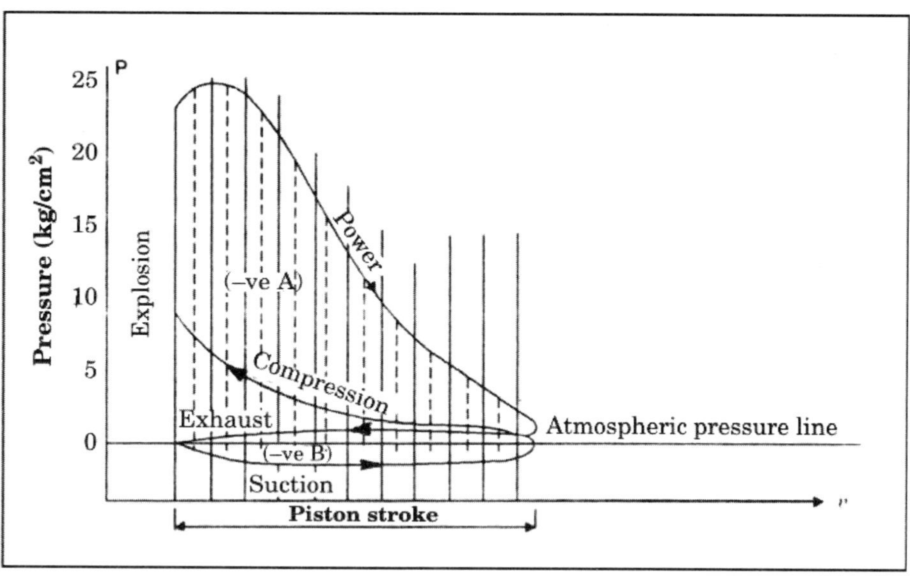

Fig. 23.3: Measurement of Indicator Diagram Area.

23.4.4 Calculation of Indicated Horse Power

After obtaining the mean effective pressure of engine from the above method, the indicated horse power can be calculated.

Let P = indicated mean effective pressure (kgf/cm^2); A = Area of the piston in cm^2; L = Length of stroke in metres; N = Number of power strokes per minute; Force on piston = P x A.; Work done per power stroke = Force x length of stroke = P x A x L.

Work done per minute = Work done per stroke x No. of power strokes per minute = P x A x L x N. This is usually written as (PLAN). Horse power (HP) = 4500 *m* - kg/min. Hence I. H.E = PLAN/4500. In case of four-stroke engine, N = revolution per minute/2, as there is only one power stroke in every two revolutions of the crank shaft. In case of two-stroke engine, N = revolution per minute as there are one power stroke per revolution of crankshaft.

23.4.5 Measurement of I.H.P. of Multi Cylinder Engine (Morse Test)

This method is used in multi-cylinder engines to measure I.H.P without the use of indicator. The BHP of the engine is measured by making each cylinder inoperative in turn. If the engine consists of let 4- cylinder, then the BHP of the engine should be measured four times by making each cylinder inoperative turn by turn. This is applicable to petrol as well as for diesel engine. The cylinder of a petrol engine is made inoperative by 'shorting' the spark plug where as in case of diesel engine, fuel supply is cut-off to the required cylinder.

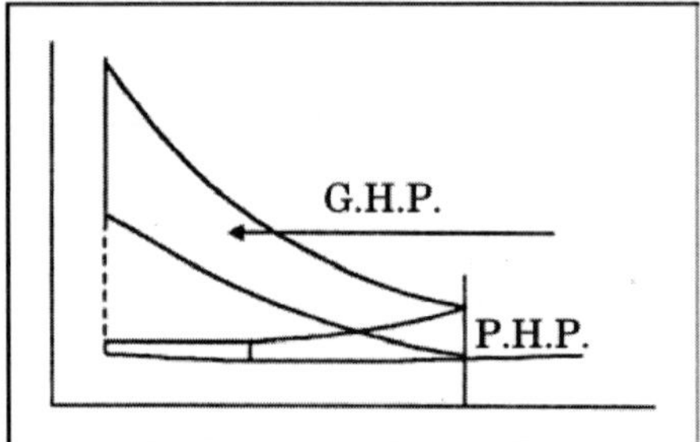

Fig. 23.4: Morse Test

In Fig. 23.4, the top area of the indicator diagram is a measure of the gross horsepower (G. H.P) developed by the engine, the bottom area the pumping horsepower (P.H.P). The net IHP per cylinder = (GHP-PHP). The net B.H.P per cylinder = (IHP - FHP) where F.H.P. stands for frictional horsepower with all the 'n' cylinders operating if there are *'n'* cylinders in an engine: The total BHP *i.e.*, $(BHP)_n - (GPH)_n - (PHP)_B - (FHP)_n$. If one cylinder is inoperative, then the power developed by that cylinder (IHP) is lost and the speed of the engine will fall as the load on the engine remains the same. The engine speed can be maintained its original value by reducing the load on the engine. This is necessary to maintain the FHP constant, because it is assumed that the FHP is independent of load and depends only on the speed of the engine. Hence by maintaining rpm of the engine constant, the FHP remains constant. By making one cylinder inoperative, the new BHP represents that of an engine having (n-1) cylinders minus the pumping and friction losses of the inoperative cylinder.

When one cylinder is cut-off, then $(BHP)_{n-1} = (GHP)_{n-1} - (PHP)_{n-1} - (FHP)_{n-1} - (PHP + FHP)$ of idle cylinder. Then the reduction in BHP by cutting-off one cylinder is

$[(BHP)_n - (BHP)_{n-1}] = [(GHP)_n - (PHP)_n - (FHP)_n] - [(GHP)_{n-1} - (PHP)_{n-1} - (FHP)_{n-1} - (FHP)_{n-1} - (PHP + FHP)]$

Or, $[(BHP)_n - (BHP)_{n-1}] = [(GHP)_n - (GHP)_{n-1}]$

As pumping losses are usually small, the reduction in BHP is usually regarded as the IHP of the inoperative cylinder.

The BHP of the engine by making each cylinder inoperative can be measured by any type of dynamometer. From the measured BHP, the I HP of inoperative cylinder can be obtained. Similarly IHP of all cylinder can be measured one by one and sum of IHP of all cylinders is the total IHP of the engine. This method of obtaining IHP of multi-cylinder engine is known as 'Morse Test'.

23.4.6 Measurement of Brake Horsepower

Measurement of the power output of an engine is the most important test carried out in the laboratories. The power output is the net power available at the crankshaft of the engine and is also known as brake horsepower, denoted as BHP Dynamometers are generally used to measure BHR Dynamometer is a device which measures the power of an engine. Dynamometers maybe classified as brake or drawbar dynamometer according to the manner in which the work is being applied. They are also classified as absorption or transmission dynamometer depending on the disposition of the energy. In absorption dynammoeters, the energy produced by the engine is absorbed by factional resistance of the brake and is converted into heat. In transmission dynamometer the energy is not wasted in friction but is used for doing work. The power is transmitted through the dynamometer from the source of power to the machines and hence power developed may be measured. In tractor, brake horse power and drawbar power need to be measured, for which, the working of some brake dynamometers and drawbar dynamometers have been discussed. Brake dynamometer measures BHP of the engine and drawbar dynamometer measures the drawbar power of the engine.

Brake dynamometer maybe of (a) prony brake type (b) hydraulic type (c) electrical type. Similarly drawbar dynamometer maybe of *(a)* spring type, (b) hydraulic type.

In brake dynamometer, the prony brake and hydraulic type of dynamometer are again coming under absorption type of dynamometer.

Prony brake dynamometer: The simplest form of an absorption type dynamometer is a prony brake and its arrangement is shown in Fig. 23.5. It consists of brake shoes, blocks made by wood, which are clamped on to the flywheel rim by means of the bolts. The blocks are set to grip the flywheel tightly by suitable screws. A helical spring is provided between the nut and the upper block so that the pressure between the blocks and the pulley remains constant. One of the blocks has a load lever attached to it and carries a weight W at its outer end. The magnitude of the weight is adjusted so that its moment balances the moment of the frictional resistance between the blocks and the fly wheel. A counterweight is placed at the other end of the lever which balances brakes(blocks) when unloaded. Two stops S, S are provided to limit the motion of the lever and to keep the load lever horizontal.

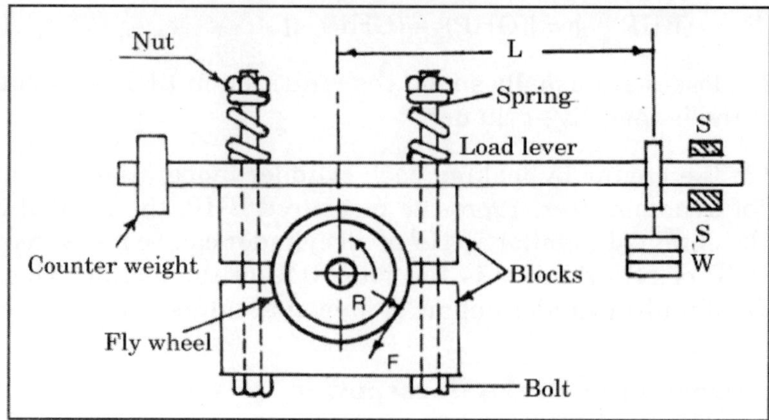

Fig. 23.5: Prony Brake.

Let W = Weight at the outer end of the lever (kg); L = Horizontal distance of the weight W from the centre of the flywheel (metre); F = Frictional resistance between the blocks and the flywheel (kg) ; R = Radius of the flywheel in meter; N = Speed of crankshaft in rpm. Taking moments about the centre of the flywheel, F × R = W × L = T; where T = torque of the crankshaft. Work done against frictional force in one revolution of the fly wheel = 2 r RF (kg-m). Also work done per minute = 2r WLN (kg-m).

$$\text{Brake horse power} = \frac{\text{Work done}}{4500} = \frac{2\pi WLN}{4500} = \frac{2\pi NT}{4500}\,\text{hp}$$

$$\text{Power (W)} = \frac{2\pi NT \times 9.8}{60}\ \text{Joules/sec (as 1 kgf} = 9.8\text{N)}$$

The energy supplied by engine to the brake is eventually dissipated as heat. Therefore, most of the brakes are provided with a means of supply of cooling water inside the run of flywheel. The prony brake is inexpensive, simple in operation and easy to construct. It is therefore, used extensively for testing of low speed engines. At high speeds, grabbing and chattering of the blocks occur and lead to difficulty in maintaining constant load

Hydraulic dynamometer: The hydraulic dynamometer (Fig. 23.6) also works on the principle of converting work into heat. The working medium, usually water, is circulated within the housing and it comes out at higher temperature by absorbing heat energy due to friction. Removal of the heat generated by this friction in the hydraulic brake is easily accomplished by providing a steady flow of water into and out of the dynamometer friction area. In this way, it is possible to avoid mechanical wear on a brake band or brake drum. In this dynamometer, the water fills the space between the rotor and rotor cover. The rotor is connected to the shaft of the engine. The rotor cover or outer case is not fixed but is free to rotate in trunnion bearings which are again mounted in the dynamometer bed. From the side of the rotor cover, a load bar (torque arm) extends on to the end of which is attached a weight hanger with weights. The stationary rotor cover is held and restrained from rotating by the torque arm.

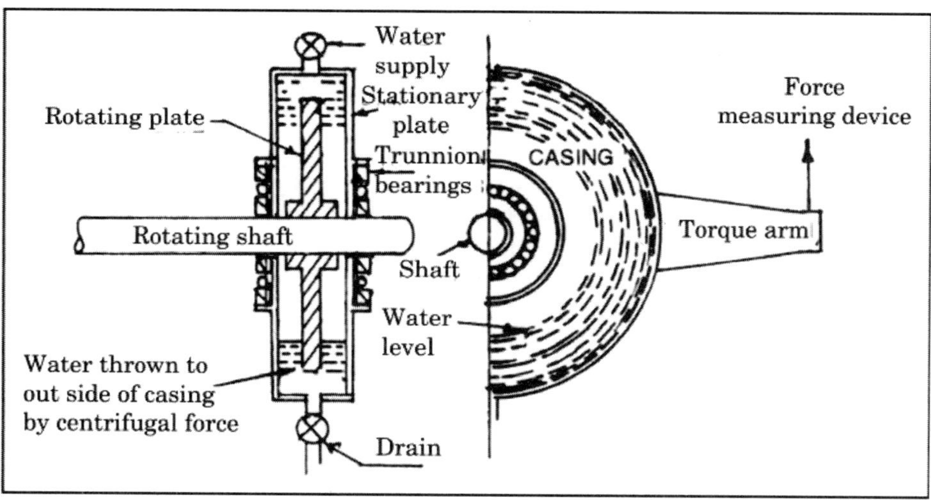

Fig. 23.6: Hydraulic Brake.

The principle of operation is that if the rotor is now rotated, a reaction will be set up with the water and the cover with the result that the cover will tend to rotate in the trunnion bearings with rotor. The cover rotation is prevented by hanging suitable weights on the weight hanger. Thus a torque reaction is set up and this torque, which is generated by the engine, can be measured.

If W = weight on weight hanger (kg); R = Distance from the centre of dynamometer to weight hanger *(m);* T = Torque generated by engine *(kg-m)*; Then T = WR.

$$\text{BHP} = \frac{2\pi \text{NT}}{4500} \text{hp} \text{ ; where N = speed of engine shaft in rmp.}$$

The amount of torque absorbed by the dynamometer can be varied by control of water. In some hydraulic dynamometer, there is fixed quantity of water in the cover, but in others, the cover is usually filled and means are provided for free flow of water through the cover.

Electric Dynamometer: The electric generator can also be used for measuring BHP of the engine, but the output of the generator must be measured by electrical instruments and correct in magnitude for generator efficiency. If the generated voltage = V volts and current output = I amperes. Then power output = VI watts *i.e.* VI/746 hp (*i.e.* bhp) of the engine. By neglecting electrical losses, this is generally very close to bhp of engine. This dynamometer consists of a generating unit (a dc generator) a resistance unit and a control board. The engine to be tested is mounted by the side of generator and its crankshaft is connected directly to the armature. The generator has an annular field frame. The field frame instead of being rigidly bolted to the base is supported in two large radial ball bearings on pedestals rising from a cast iron base. With the armature being rotated by the engine, it exerts a torque on the field frame and tends to carry the latter around with it. The pull so exerted is measured by means of a suitable balance. This type of dynamometer is costly especially in large sizes needed to test the automobiles. This type of dynamometer can also be used as a motor to start and drive the engine at various speeds.

Drawbar Dynamometer: Draw bar dynamometers are generally used to determine the drawbar pull of power units and to record the draft of the field implements. Followings are the type of drawbar dynamometer commonly employed for draft measurements for agricultural implements.

Spring dynamometer: It is a simple type of dynamometer which consists of a spring. The spring absorbs the pull and elongates under tension or shortens under compression, showing direct reading of load (Fig. 23.7). Such a dynamometer is suitable for rough measurement of forces; because of rapid variations in loads, such as are commonly observed with the agricultural implements.

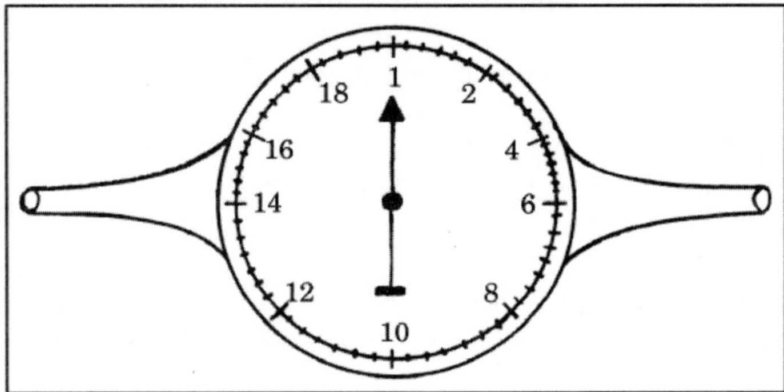

Fig. 23.7: Spring type dynamometer

Hydraulic drawbar dynamometer: This type of dynamometer is inserted in between the tractor and load. It consists of cylinder filled with hydraulic oil, a piston, connected to the pressure gauge. The drawbar pull causes an increase in oil pressure that is recorded in the pressure gauge. The gauge is calibrated in such a way that it shows the direct reading in kg. A hydraulic cylinder for measuring drawbar pull has an advantage over a spring dynamometer is that the momentary fluctuations in pull can be dampened with the use of hydraulic oil. It is a simple and accurate device for measuring drawbar pull.

23.5 FUEL CONSUMPTION MEASUREMENT

Fuel consumption of an engine may be expressed either in terms of volume or weight of fuel supplied in a specified time. The arrangement for measuring the fuel supplied to the engine is shown in Fig. 23.8 (a).

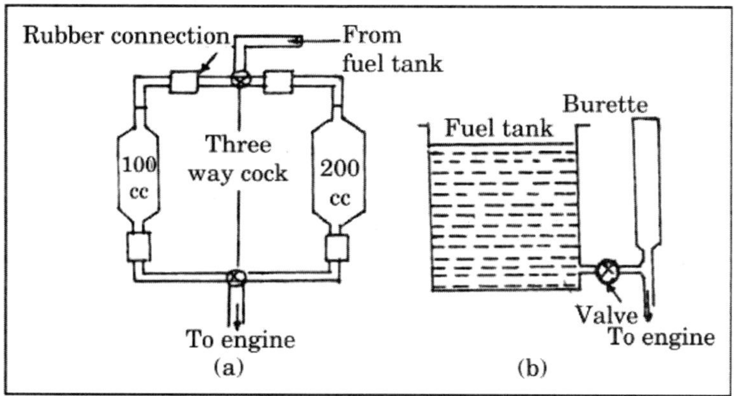

Fig. 23.8: Fuel Consumption

Two glass vessels of 100 cc and 200 cc capacity are connected in between the engine and main fuel tank through three way cocks. When one is supplying the fuel to the engine, the other is being filled. The time for the consumption of 100 and 200 cc fuel is measured with the help of stopwatch.

Another simple arrangement for measuring the fuel consumption rate is also shown in Fig. 23.8(b). A small glass tube is attached to the main fuel tank as shown in the figure when fuel rate is to be measured, the valve is closed so that fuel is consumed from the burette and is used for combustion in the engine. The time for a known value of fuel consumption can be measured and fuel consumption rate can be calculated with the help of the following expression.

$$\text{Fuel consumption (kg/hr)} = \frac{(X_{cc})(\text{specific gravity of fuel})}{(1000)(t)}$$

Where X is the quantity of fuel consumed in (cc); t is the time of hours.

$$\text{Specific gravity of a liquid} = \frac{\text{Density of liquid}}{\text{Density of pure water at 4°C}}$$

Density of pure water at 4°C = I gm/c.c, or 1000 kg/m^3. Putting the density of pure water as Igm/cc, the amount of fuel consumed will be gram. Dividing 1000, the amount will be in kg.

23.6 MEASUREMENT OF AIR SUPPLY

The measurement of air quantity is an important task in engine tests. The method commonly used in the laboratory for measuring the consumption of air is known as 'Orifice chamber method'. The arrangement of the system is shown in Fig. 23.9.

It consists of an air tight chamber fitted with a nozzle or plate orifice of known coefficient of discharge. The orifice is located away from the suction connecting to the engine. Due to the suction of engine, there is a pressure depression in the chamber which causes the flow through orifice for obtaining a steady flow. The volume of chamber should be sufficiently large compared with the swept volume of the cylinder; generally 500 to 600 times the swept volume. It is also assumed that the intermittent suction of the engine will not affect the air pressure in the air box as the volume of the box is sufficiently large and pressure in the box remains constant. The pressure difference causing the flow through the orifice is measured with the use of a manometer. Air passes through the orifice into the air box and from there into the air cleaner or intake manifold.

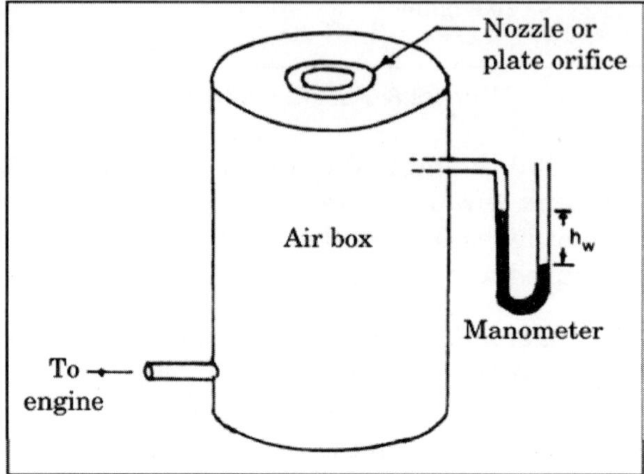

Fig. 23.9: Method of Measuring the Airflow into an Engine.

Let A = Area of orifice is m^2 ; h_w = head of water in cm. causing the flow; *Cfi* = coefficient of discharge for orifice; *d* = diameter of orifice in cm; p_a = density of air in kg/m³ under atmospheric conditions. Head in terms of air in meters is given by

$$H = \frac{h_w}{100} \times \frac{\rho_w}{\rho_c} \times \frac{h_w}{100} \times \frac{1000}{\rho_a} = \frac{10h_w}{\rho_a}$$

The velocity of air passing through the orifice is given by

$$v = \sqrt{2gH} = \sqrt{2g\frac{10/h_w}{\rho_a}} m/\sec$$

The volume of air passing through the orifice is given by

$$Va = A_0 v C_d = A_0 C_d \sqrt{2g\frac{10/h_w}{\rho_a}} = 14A_0 C_d \sqrt{\frac{h_w}{\rho_a}}\ m^3/\text{sec.}$$

$$840A_c C_d \sqrt{\frac{h_w}{\rho_a}}\ m^3/\text{min.}$$

Where A_O is the area of orifice. The mass of air passing through the orifice is given by

$$m_a = V_a\rho_a = 14A_0C_d\sqrt{\frac{h_w}{\rho_a}}\,\rho_a\,(\text{kg/sec})$$

$$m_a = 14 \times \frac{\pi d^2}{4\times100^2}C_d\sqrt{h_w\rho_a}$$

$$m_a = 0.0011\,C_d\,d^2\,\sqrt{h_w\rho_a}\ \text{kg/sec}$$

$$m_a = 0.066\,C_d\,d^2\,\sqrt{h_w\rho_a}\ \text{kg/min}$$

Where d is in cm ; h_w is in cm of water and ρ_a is in kg/m^3. The value of C_d varies from 0.55 to 0.65.

The measurement of air consumption by the orifice chamber method is used for the

(a) The determination of the actual A: F ratio of the engine at running condition.
(b) The weight of exhaust gases produced
(c) The volumetric efficiency of the engine at the running condition.

23.7 MEASUREMENT OF HEAT CARRIED AWAY BY EXHAUST GASES

Heat carried away by exhaust gases is given by the expression

$$Q_g = W_g\,C_e\,(T_e - T_0)$$

Where W_g = weight of exhaust gas; C_e = specific heat of exhaust gas; T_e = temperature of exhaust gases coming out from the engine; T_a = Ambient air temperature. Thus measurement of heat carried away by the exhaust gases involves determination of three quantities *i.e.*, (i) Determination of weight of exhaust gases *(ii)* Mean specific heat of exhaust gases *(iii)* Temperature of exhaust gases.

Determination of weight of exhaust gases: In case air and fuel consumption rates for the engine are known, the weight of exhaust gases can be directly calculated by the sum of two parameters *i.e.*, Weight of exhaust gases per minute = Air consumption per minute + Fuel consumption per minute.

In case volumetric analysis of the exhaust gases is known (by Orsat apparatus), the weight of exhaust gases can be calculated by the relation.

Weight of air supplied per kg of fuel,

$$W_a = \frac{N\times C}{33(C_1 + C_2)}$$

Where N is the percentage of nitrogen by volume in exhaust gases; C_1 is the percentage of CO_2 by volume in exhaust gases; C_2 is the percentage of CO by volume in exhaust gases; C is the percentage of carbon in fuel by weight. If $C_2 = 0$ then $W_a = \frac{N \times C}{33C_1}$ and the Air – fuel ratio. By knowing air-fuel ratio, the weight of exhaust gases per kg of fuel can be estimated.

The weight of exhaust gases per minute = (A + 1) x weight of fuel in kg per minute; where A is the air-fuel ratio.

Mean specific heat of exhaust gases: Specific heat of exhaust gases depends upon the constituent gases, vapours etc. In case composition of exhaust gas is known, mean specific heat can be estimated by allowing appropriate proportion of specific heat of each constituent.

Temperature of exhaust gases: Temperature of exhaust gases is measured by inserting pyrometer or mercury thermometer in the exhaust line as close to the engine as possible. The heat carried away by exhaust gases can also be measured with the help of exhaust gas calorimeter

Exhaust gas calorimeter: The exhaust gas calorimeter is a simple heat exchanger in which part of the heat of the exhaust gases is transferred to the criculating water. This calorimeter helps to determine the weight of exhaust gases coming out of the engine. The arrangement of the exhaust gas calorimeter is shown in Fig. 23.10.

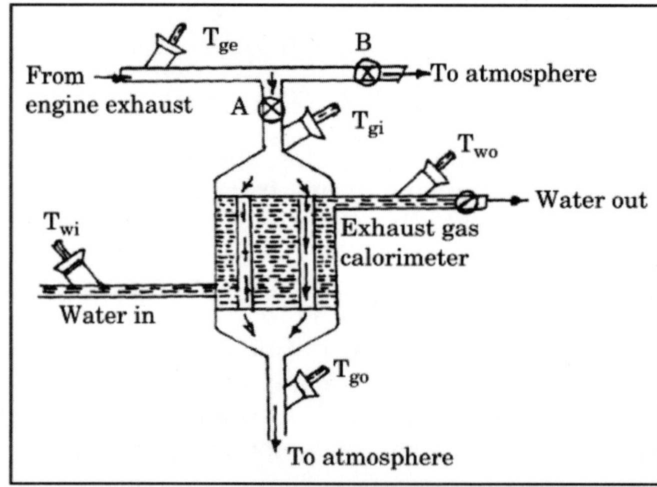

Fig. 23.10: Exhaust Gas Calorimeter

The exhaust gases from the engine exhaust are passed through the exhaust gas calorimeter by closing the valve B and opening the valve A. The hot gases are cooled by the water circulated in the calorimeter. As the calorimeter is well insulated and if it is assumed that there is no heat loss except by heat transfer from the exhaust gases to the circulating water, then, heat lost by exhaust gases = heat gained by circulating water

$$W_g C_e (T_{gi} - T_{go}) = W_w C_w (T_{wo} - T_{wi})$$

Where T_{gi} = Temperature of the exhaust gases entering the calorimeter

T_{go} = Temperature of exhaust gases leaving the calorimeter.

T_{wi} = Temperature of water entering the calorimeter

T_{wo} = Temperature of water leaving the calorimeter

W_w = Weight of water circulated through the calorimeter

W_g = Weight of exhaust gases (unknown)

C_e = Specific heat of exhaust gases

C_w = Specific heat of water ($C_w = 1$)

$$\therefore \qquad W_g = \frac{W_w}{C_e}\left(\frac{T_{wo} - T_{wi}}{T_{gi} - T_{go}}\right)$$

As all the parameters on the right hand side of above equation are known, then the weight of exhaust gas can be determined. Then the heat carried away by the exhaust gases is given by

$$Q_g = W_g C_e (T_e - T_a)$$

Usually a valve connections are provided as shown in figure so that the exhaust gases are exhausted to the atmosphere during normal operation by closing the valve A and opening the valve B. Only when the appratus is to be used, the valve A is opened and valve B is closed so that the gases pass through the calorimeter.

23.8 TORQUE CURVES

The torque curve or lugging ability is the one performance criterion of the engine. Torque is the turning force. When the piston is moving down on the power stroke, it applies torque to the engine crankshaft (through the connecting rod). The harder the push on the piston, the greater the torque applied. Therefore, the higher the combustion pressures, the greater the amount of torque. The dynamometer is normally used to check engine torque. Torque can be measured at the same time as horse power on the dynamometer. Engine torque is in N-m and engine speed is in rpm.

The torque that an engine can develop changes with engine speed. During intermediate speeds, volumetric efficiency is high. There is sufficient time for the cylinders to become fairly well "filled up". This means that with a fairly full charge of air-fuel mixture, higher combustion pressures will develop. With higher combustion pressures, the engine torque is higher. But at higher speed, volumetric efficiency drops off. There is not enough time for the cylinders to become filled up with air-fuel mixture. Since there is less air-fuel mixture to burn, the combustion pressures are not as high. There is less push on the pistons. Therefore, engine torque is lower. Figure 23.11 shows how the torque drops off as engine speed increases.

The basic requirements of a tractor engine are different from that of an automobile vehicle. It is estimated that a tractor engine requires to be operated at full load approximately 25 to 30 per cent of the operating time. Most of the time, the engine is required to be operated between 50 to 70 per cent of full load.

An increase in load, applied to a tractor by a hitched implement, demands from the engine a corresponding increased torque. Usually, the tractor is subjected to varying load conditions and in order to enable the engine to bear momentary overloads, it is designed to produce its maximum torque at a much lower engine speed than that at which maximum power is developed. The curves between bhp vs. speed, flip vs. speed and torque *vs.* speed of a tractor engine have been shown in Fig. 23.11. It may be seen that at the speed corresponding to maximum power at 1900 rpm, the torque is 88 % of the maximum torque. The maximum torque is produced at 1400 rpm. The maximum torque is 110 % of the torque at maximum power.

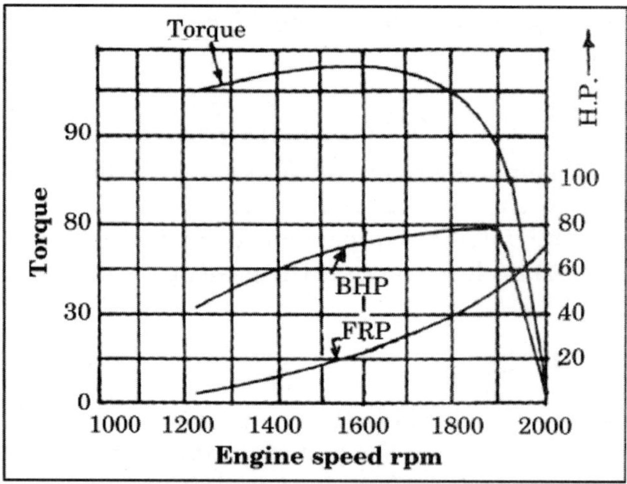

Fig. 23.11: Relation of Engine Speed with Torque, BHP and FHP.

The bhp curve of the engine is different from the torque curve. Figure 23.11 compares the bhp of the same engine for which the torque curve is shown. It starts low at low speed and increases until a high engine speed is reached. Then, at still higher engine speeds, bhp drops off.

The drop -off of bhp is due to reduced torque at higher speed and to increased fhp at the higher speed. It is also observed from the figure that the maximum engine torque and maximum engine output (power) is obtained at different engine speeds. This is because torque developed by the engine is maximum where volumetric efficiency or the amount of air sucked during the suction stroke is maximum where as engine output is maximum when amount of air sucked per minute is maximum.

A desirable torque curve is one that increases significantly as speed decreases and is therefore, stable. Such a torque curve results in a minimum of speed variation in the engine. It is also desirable for the torque curve to peak as far to the left (at lowest speed) as possible. A diesel tractor engine has normally less speed variation for a given change in torque than a comparable gasoline engine. This behavior of diesel engine is called slogging or lugging ability. The lugging ability, also called the reserve torque is the difference in the values of maximum torque and torque at maximum power of the engine. Better the lugging ability, the better is the tractor.

23.9 ENGINE PERFORMANCE

The performance of an engine is studied by finding out the variations of brake horsepower, torque, fuel consumption, frictional horse power and specific fuel consumption at different engine speeds. By drawing the curves with the above relations, we can compare the performance of the engine of different tractors and use these according to their service suitability.

The variations of above parameter with engine speed both for petrol engine and diesel engine have been shown in Fig, 23.12 and Fig, 23.13 respectively.

23.9.1 Torque *vs.* Engine Speed

The variation of torque with engine speed has been discussed in section 23.8. In this sub-section, the comparative study of torque curves for both the engines has been explained. From the figures, it is seen that *(i)* diesel engines give higher torque at low engine speeds as compared to petrol engines, *(ii)* diesel engines operate at nearly uniform torque over a wide range of engine operating speeds than petrol engines.

23.9.2 BHP *vs.* RPM

Comparing the bhp vs. rpm curves of petrol and diesel engine (Figs. 23.12 and 23.13), the following conclusions can be drawn, *(i)* At low engine speeds, petrol engines have higher bhp compared to diesel engine *(ii)* At high engine speeds, in diesel engine, there is improvement of bhp over petrol engines. The bhp curve in the case of petrol engines reaches a maximum value but it does not do so in case of diesel engine as the maximum speed is limited owing to heavier reciprocating weights.

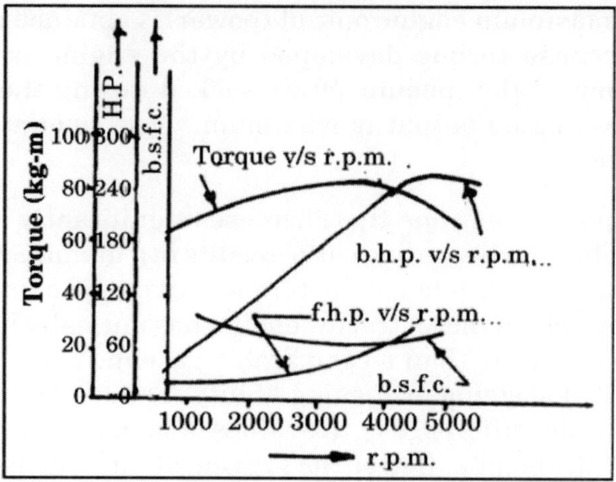

Fig. 23.12: Performance Curves (Diesel Engine)

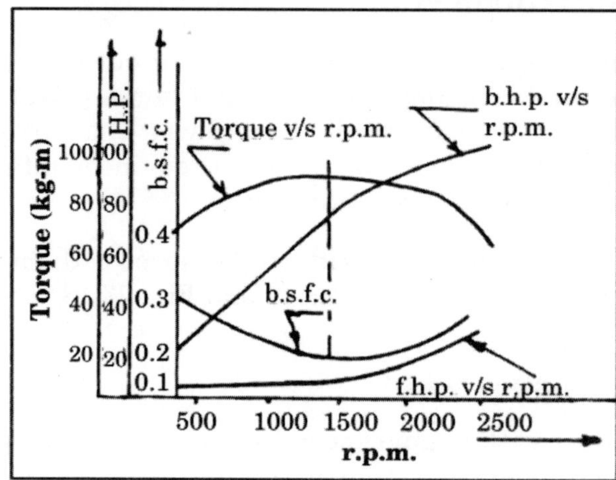

Fig. 23.13: Performance Curves (Diesel Engine)

23.9.3 FHP *vs.* RPM

Frictional horse power increases with increase in speed. Frictional losses are lower at low engine speeds. These losses increase considerably with the increase in rpm. The horsepower lost due to friction increases five times as the engine speed increases from 1000 to 3000 rpm.

The fractional losses in an engine may be due to friction between rings and cylinders, valves, timing gears, resistance of inlet and exhaust valves etc. In both petrol and diesel engines, the frictional horsepower curve rises rapidly with the increase of engine speed.

23.9.4 Specific Fuel Consumption *vs.* RPM

The specific fuel consumption decreases with the increase of engine rpm in both petrol and diesel engines of same swept volumes. The fuel consumption per hp-hour in case of diesel engines is less at all engine speeds. As shown in the figure, at engine speeds above 2,500 rpm, the specific fuel consumption increases in both petrol and diesel engines. This increase is more in the case of petrol engines than diesel engines of similar capacity. The best speed of the engine is the speed at which the specific fuel consumption per hp-hr is minimum. The minimum specific fuel consumption occurs at speed of 1400 rpm. It is the most economical speed for the engine.

23.10 ENGINE EFFICIENCY

Different types of engine efficiency have been discussed in chapter-2. In addition to mechanical efficiency, thermal efficiency and volumetric efficiency of an engine, there are some other engine efficiencies as explained below.

(i) **Air standard efficiency:** The thermal efficiency of an ideal air standard cycle is called the air standard efficiency. In an ideal air standard cycle, the working substance is assumed to be air. The petrol and diesel engine working on Otto and Diesel cycles use petrol and diesel fuel respectively with air.This air fuel mixture behaves like air before the combustion takes place. The properties of combustion product are also not different from those of air. Therefore, the efficiencies of petrol and diesel engines are calculated assuming them working on air standard cycle. The efficiency of the cycle is given by

$$\eta = \frac{\text{Work done}}{\text{Heat supplied}} = \frac{\text{Heat supplied} - \text{Heat rejected}}{\text{Heat supplied}}$$

(ii) **Relative efficiency:** It is the ratio of the indicated thermal efficiency to the corresponding ideal air standard efficiency. Thus relative efficiency

$$\eta = \frac{\text{Indicated thermal efficiency}}{\text{Air standard efficiency}}$$

(iii) **Scavenge efficiency:** For two stroke cycle engine, the concept of volumetric efficiency does not apply. Here another term called scavenge efficiency is used. It is the measure of the extent of which burnt gases are removed from the cylinder and the cylinder filled with fresh air charge.

where W_a = Weight air charge retained in the cylinder in kg; V_s = displacement volume in m^3 ; V_c = Clearance volume in m^3 ; w = specific weight of air charge in kg/m^3 at standard conditions of pressure and temperature.

$$\text{Scavemge } \eta = \frac{W_a}{(V_s + V_c)w};$$

(iv) **Overall efficiency:** The fuel enters the engine with a certain energy content, a certain ability to do work. At every step in the process, from burning of fuel in the cylinder to the rotation of drive wheels, energy is lost. Figure 23.14 illustrates these losses for one vehicle. As little as 15 per cent of the energy in the fuel remains to move the vehicle. This energy is used to overcome rolling resistance, air resistance, gradient resistance, power-train resistance and to accelerate the vehicle. All these resistances oppose the movement of vehicle.The resistance to the motion of the vehicle is known as tractive resistance. Tractive effort is on the other hand, the force available at the points of contact between rear wheel types and the road. The traction is the ability of rear wheels to transmit the tractive effort without slipping. Therefore, the useful tractive effort is always less than traction. If the tractive effort is less than the total tractive resistance, the vehicle becomes unable to move. In this situation, by keeping the speed of the engine same, the torque available on the drive wheel is to be increased by changing gear ratio by the use of gear box. The net energy available at the drive wheels to propel the vehicle after substracting energy to overcome all the resistances, is used to calculate the overall efficiency of the vehicle.

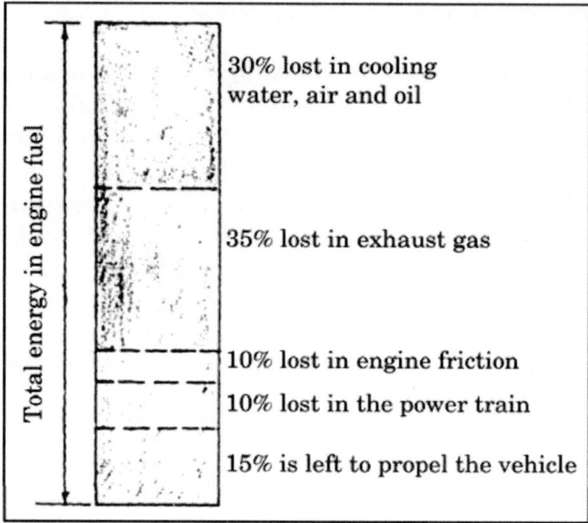

Fig. 23.14: Energy Lost from Cylinders to Wheels.

(a) **Rolling resistance:** This results from irregularities in the road over which the wheels move. It generally varies with the type of the road surface, load on each tyre, inflation pressure and type of type tread.

(b) **Air resistance:** Air resistance is the resistance of air to the passage of vehicle body through it. As vehicle speed increases, so does the air resistance. It depends on the speed of the vehicle, wind velocity, size and shape of the body of the vehicle.

(c) **Gradient resistance:** It is the force opposing forward motion of a vehicle up a gradient. It does not depend on the vehicle speed and is only the function of vehicle weight and gradient.

(d) **Power train resistance:** Some engine power is lost between the engine and the drive wheels. One reason is because of friction between moving parts in the power train which includes the clutch, transmission and drive axle.

(e) **Acceleration:** Power is required to increase car speed. The power applied to accelerate the vehicle overcomes its inertia.

Chapter 24

Power Tiller Engine System

24.1 INTRODUCTION

Power tiller is a tractor usually fitted with two wheels (pneumatic or steel) in which the direction of travel and its control for field operations are performed by operator walking behind it. It is also known as hand or walking type tractor. It may also be called as single axle walking type tractor, though a riding seat is provided in-certain designs. Power tillers are especially designed and developed for use on small (< 4 ha) and medium farms (4-10 ha) and under farming conditions where conventional four wheel tractors are either difficult or uneconomical for use.

The concept of power tiller came in the world in the year 1920. Japan is the first country to use power tiller in rice cultivation. In Japan, the first successful model of power tiller was designed in the year 1947. Then it became popular as a source of farm power on small and medium farms in China, Korea and other countries of South East Asia especially for rice cultivation. It was introduced in India in early sixties for rice cultivation only for preparing seedbed since very few matching implements were available during that period. The attachments available at the time were rotatiller, mould board plough, ridger and trailer. Getting enthusiastic with the initial performance of the power tiller in rice cultivation, six firms started indigenous production of power tiller. Manufacturing of several makes of power tillers like Iseki, Satoh, Krishi, Kubota, Yanmar and Mitsubishi were started in India after 1962.

24.2 CLASSIFICATION OF POWER TILLER

Power tillers may be classified in many ways depending upon the type of engine mounted on them, horsepower and type of tilling device etc.

Japanese Society of Agricultural Machinery (JSAM) has classified the power tiller as large (8-14 hp), medium (5-7 hp), small (3-5 hp) and very small (1.5-2 hp). However, the power tillers (garden tractors) in America are defined as the tractors weighing less than 545 kg. They are classified as riding or walking type tractors.

Bureau of Indian Standards (BIS) has given the following classification on the basis of utility of various types of power tillers.

24.2.1 General Purpose Type

The power tillers which can be used for number of operations are suitable for both rotary and traction work.

24.2.2 Pull Type (Traction Type Power Tiller)

The power tillers have the ability to pull various types of equipments. They are used for ploughing, harrowing, leveling, seeding and transport etc.

24.2.3 Tilling Type (Rotary Type Power Tiller)

The tilling type power tillers have an engine power driven tilling device such as rotary and crank or screw blades. The rotary tillers are fitted with 10-20 curved blades of different designs.

24.3 FEATURES OF POWER TILLERS

24.3.1 Advantages

(i) Power tillers are compact in construction and have good trafficability. The narrow wheel track enables them to go through the narrow path in the country side. These can be operated conveniently between the rows of plants in orchards.

(ii) The power tillers are light in weight and therefore, their sinkage is low on soft soils. The lightness and low centre of gravity facilitate easy operation on sloping fields and in forest areas and hills.

(iii) The power tillers have good maneuverability. Due to absence of two front wheels (as in case of four wheel tractors) and narrow wheel track, power tillers have short turning radius. Thus less land is left untilled during field operations.

(iv) The provision for adjustment of wheel track and narrow width of tyres of power tiller makes it suitable for interculture, spraying and dusting in crops, having row spacing 30 cm or more.

(v) The power tillers are simple in construction as there are no front wheels, hydraulic system etc.

(vi) The power tillers are capable of performing various operations with different matching implements for farming work.

24.3.2 Disadvantages

As compared with four wheel tractor, the walking tractors have following disadvantages.

(i) The power tillers have low horsepower, thus not suited for large farms and deep tillage.

(ii) Power tillers have low tractive efficiency. The drawbar power is usually 15 to 20 % of brake horse power and thus there is high energy requirement per unit area for tractive work.

(iii) Power tillers have high cost per drawbar power

(iv) Lack of seating arrangement in most power tillers available in country causes considerable fatigue to the operator.

24.4 CONSTRUCTION OF POWER TILLER

A power tiller consists of the following main parts:

(i) Engine,
(ii) Power transmission system,
(iii) Brakes,
(iv) Steering system,
(v) Rotary unit,
(vi) Control mechanism,
(vii) Chassis,
(viii) Wheels

Manufacturers supply counter weights, and ballast weights also as optional accessories for balancing the power tiller during operation and increasing the drawbar power respectively. The schematic view showing different components of a power tiller is given in Fig. 24.1.

24.4.1 Engine

A light weight diesel or petrol engine of small to medium size and medium speed range is used in power tillers to generate necessary power. Most of the engines used in power tillers are water cooled and have splash system of lubrication. The engine is provided in the front side of the power tiller. The power generated in the engine is transmitted from crankshaft to the clutch, gear and finally to the drive wheel with the help of belt or chain. The rotary unit is fitted in the rear part of the power tiller for tillage operation.

24.4.2 Power Transmission System

The function of power transmission system of a power tiller is to transmit the engine power to ground drive wheels and rotary tiller, to change the engine torque and speed into different driving force and speed combination required by ground drive components and rotary tiller and to reverse the direction of motion.

The engine power goes to the main clutch with the help of belt or chain. From main clutch, the power is divided in two routes, one goes to transmission gears, steering clutch and then to the wheel. The other part of power goes to the rotary tiller gear box and then to the rotary tiller shaft and to rotary tynes.

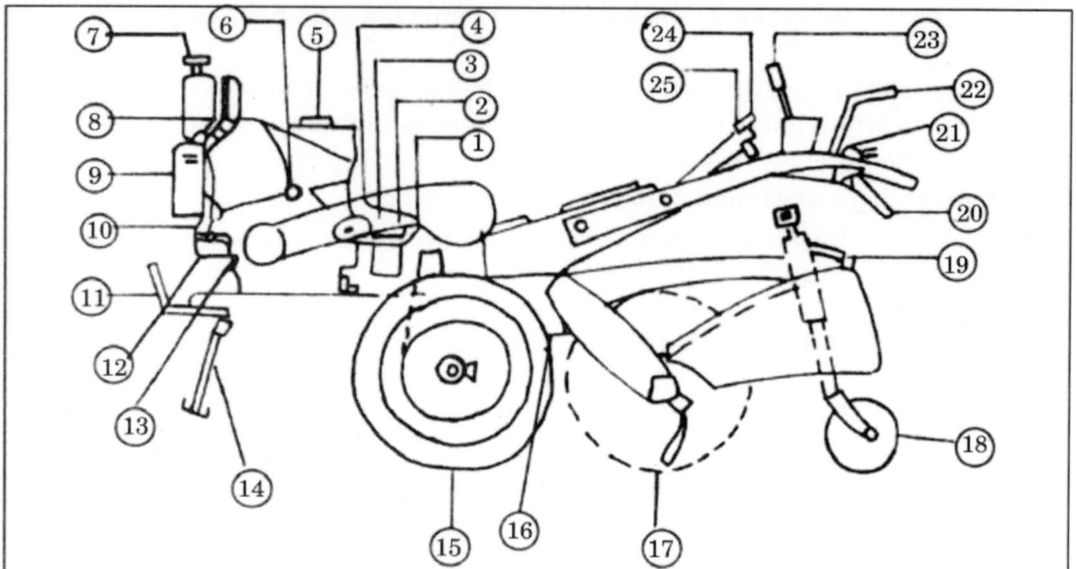

1. Oil filter
2. Tension pulley
3. Speed control lever (Engine)
4. V belt
5. Fuel tank
6. Fuel cock
7. Air cleaner
8. Nozzle
9. Muffler
10. Fuel injection pump
11. Bumper
12. Fuel filter
13. Dip stick
14. Stand
15. Tyre
16. Oil drain plug
17. Tynes
18. Tail wheel
19. Tail wheel adjust clamp
20. Side clutch lever
21. Speed control lever
22. Auxiliary handle
23. Main clutch lever
24. Rotary transmission lever
25. Main transmission lever

Fig. 24.1: Different components of power tiller.

The flow diagram for transmission of power is given below.

Chain sprocket

Engine (Belt pulley) → Main clutch →
- → Transmission gear → Steering clutch → wheels
- → Rotary tiller gear box → Rotary tiller shaft → Rotary tynes

Chain sprocket

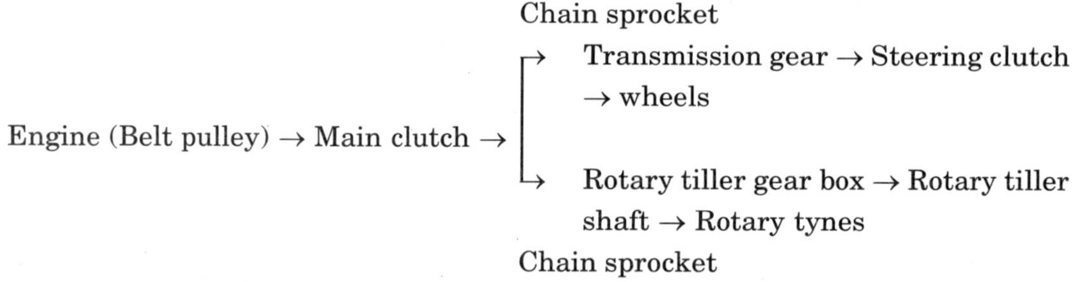

The main speed changing gear system provides three different forward speeds and one reverse speed. The arrangement of gears in sub speed changing system is such that one high and one low speed can be obtained. Thus six forward and two reverse speeds can be obtained by meshing different sets of gears in the speed changing gear box.

Power from the engine is transmitted through main shaft to the transmission line of the rotary tiller. A gear is mounted on the rotary tiller gearbox shaft, which receives the power from the gear mounted on the shaft. A sprocket is mounted at one end of the rotary tiller shaft which transmits the power to the another sprocket mounted on the rotary tyne shaft by means of chain. These two sprockets can be interchanged to obtain two different speeds of rotary tynes *i.e.* high and low.

Transmission box consists of gear, shafts and bearings. The speed change device may be of either gear type or belt type.

24.4.3 Clutch

The clutch is located between the engine and the gearbox. Its functions are:

(i) To temporarily cut off the power for starting the engine, gear, shifting and interim stopping of power tillers.
(ii) To engage power for smooth movement of power tiller and gradual taking up the load.
(iii) To protect components of transmission system from damage during overloading on power tillers, owing to the automatic slipping of clutch.

Types of Clutches: Power tillers may be provided with one of the following types of clutches.

(i) Two disc constant contact friction clutch
(ii) One disc constant contact friction clutch
(iii) Dog clutch
(iv) Belt tension pulley clutch.

Two disc constant contact friction clutch: The driving part of the clutch consists of driven belt pulley, pressure plates, driving plate and pulley cover. The driven part consists of two driven plates and clutch shaft. Two driven plates, one driving plate and pulley cover are pressed under the action of six clutch compressed springs held between the clutch housing and pressure plate. Power of the engine is transmitted from driven pulley to the clutch shaft due to its factional force between the driving and driven parts. This type of clutch is generally used for bigger power tiller or power tillers of high horse power.

One disc constant contact friction clutch: In some low horsepower power tillers, one disc constant contact friction clutch is used due to low torque of the engine. Its construction is similar to the two disc constant contact friction clutch excepting using only one driven plate.

Dog clutch: Dog clutches are provided in small size power tillers. The power from the engine is transmitted to the driven shaft through dog or jaw clutch. Jaw clutches do not have problem of slippage and generation of heat. These clutches are inexpensive and lighter, however, jaw clutches cannot be engaged at high speeds and when both the shafts are at rest. Engagement at any speed is accomplished by shock.

Belt tension pulley clutch: In few low horsepower power tillers, the belt tension pulleys are used as clutch. These clutches are not very effective in complete engagement and disengagement of power and are affected by weather conditions causing short service life.

24.4.4 Brakes

All power tillers have some braking arrangement for stopping the movement. The brake also helps in stopping immediately the power tiller in motion and prevent it from slopping downward on inclined surfaces. Power tillers are provided with simple disc type ring with inner expanding type, band type or shoe type braking system.

24.4.5 Steering System

The function of steering system is to enable turning of the power tillers at the field end during operation or for making a sharp turn on the road. In general, power tillers are equipped with jaw type starting clutch. Its function is to cut off the transmitted power to the left or right driving wheel for steering.

The jaw type steering clutch consists of a driven'gear of intermediate reduction with both ends of square jaw, steering gear with jaw, steering spring and steering fork. Both ends of driven gear have three square jaws. The side of steering gear facing driven gear of intermediate reduction also has three square jaws, which mesh with jaw of driven gear of intermediate reduction unit. The other side of steering gear is the seat for spring.

When right hand grip is pressed, right pull rod, steering lever and steering fork are actuated. The steering fork removes the right steering gear from jaw to rightward and the power tiller turns towards right due to retardation. In this situation, the right wheel is stopped while left wheel is still getting power for its movement. The power tiller can also be turned to left by pressing left hand grip. When power tiller moves in straight line, the power is transmitted to both the wheels.

24.4.6 Rotary Unit

Rotary tiller is the important component of the power tiller. It is an useful equipment to prepare land for wet and dry cultivation and interculture operation in garden and crops with large row spacing. It has also been proved an ideal device for uprooting, shredding, mixing of sugarcane stumps in soil and interculture operation in forest plantation.

The rotary tiller consists of a transverse shaft on which knives or tynes are mounted to cut the trash and soil. The effective length of this shaft may vary from 20 to 70 cm. Many types and shapes of tynes are available. The hook or pointed tynes are suitable for deep tillage in relatively clean ground. The L-shaped tynes are better for trashy conditions for weed control and where deep penetration is required. One of the major problems with rotary tiller is the breakage and bending of tynes in hard or stony ground. In order to overcome this problem, spring mounting tynes are used.

The tynes are spaced along the axis of rotor and are staggered around the periphery to take incremental bites of soil and to distribute the load on the machine more uniformly. As the rotor revolves, each tyne cuts a slice from the untilled soil and the cross-sectional area of the slice depends on the depth of tillage and the amount of forward travel per cut. Due to stow forward speed and high rotational velocity of tiller's tyne, the soil is cut into very fine pieces. The power tiller leaves very less untilled land due to its short longitudinal dimensions.

Mounting of rotary tiller: Depending upon the power transmission arrangement, the rotary tillers are classified as (i) Centre drive type, *(ii)* side drive type.

In centre drive type, the rotary tiller transmission box is located in the centre of the rotor shaft. The power is transmitted to the rotating shaft through chain and sprocket drive. In side drive type, the rotary tiller gearbox is located at the rear part of the speed changing gearbox. The left supporting arm and side transmission box of the rotary tiller are respectively fitted on left and right housing. The power in this type is transmitted from the rotary tiller shaft to the rotary tyne shaft through chain and sprocket arrangement.

Centre drive type has the following characteristics:

(i) Tilling width can be widened,
(ii) Rotary unit is light in weight,
(iii) Fixing of attachment is easy,
(iv) The tyne shaft can be detached easily,
(v) Mounting and dismounting of rotary unit is very easy,
(vi) It may leave some portion of the field untilled.

Side drive type has the following characteristics.

(i) Deeper tilling is possible,
(ii) The arrangement is useful for hard soil

Most of the power tiller available in south-East Asian countries are provided with side drive for rotary tillers.

24.4.7 Control Mechanism

The purpose of control mechanism of power tiller is mainly for cutting off the power from the engine, steering, changing speeds of movement of power tiller and rotary tiller, stopping the power tiller, changing speed of engine and adjusting depth of operation. The control mechanism consists of the levers, hand grip, pull rod or wire ropes. Lever and knobs are provided on or near the handle to facilitate easy access for the operator. Some of the components for control mechanism in power tiller have been shown in the following figures 24.2 *(a, b, c, d)*.

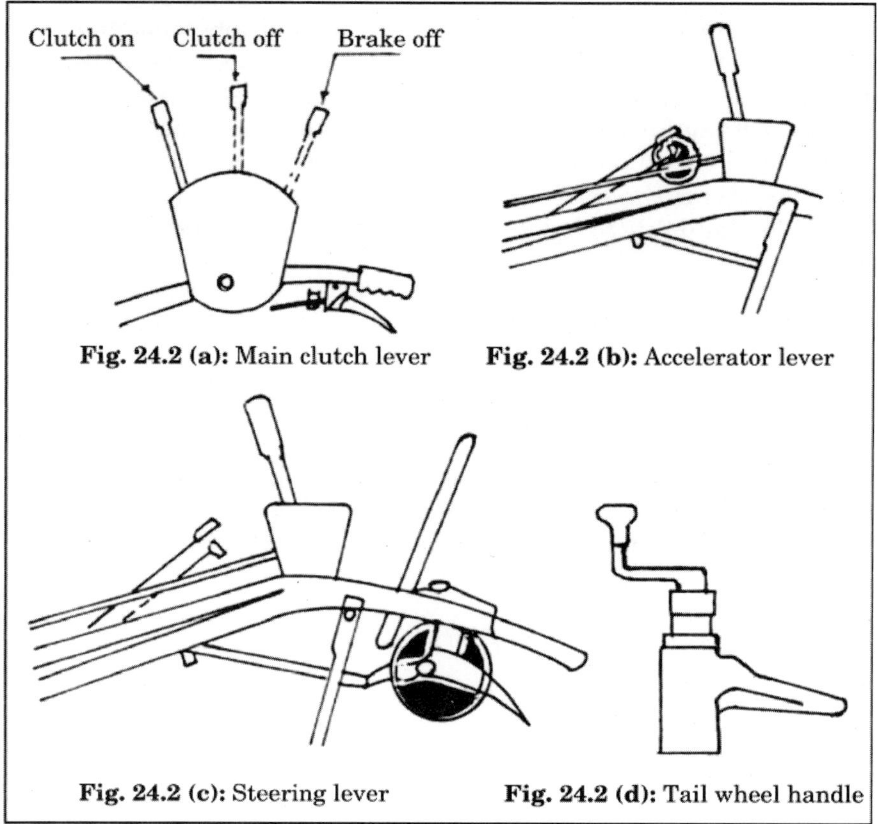

Fig. 24.2 (a): Main clutch lever **Fig. 24.2 (b):** Accelerator lever

Fig. 24.2 (c): Steering lever **Fig. 24.2 (d):** Tail wheel handle

24.4.8 Chassis

Chassis is used to support the engine and other parts of the power tiller. The position of the engine can be adjusted with the help of slotted holes according to requirements and for balancing the power tillers. On the front part of the chassis, a bumper is fixed. Stand is fixed beneath the front part of the frame. The power tiller rests on the stand when it is not in working position and is stationary for supplying power to stationary machines. When the power tiller is in working position, the stand is pressed closed to the frame with the help of lever provided near the handle.

24.4.9 Wheels

The wheels in power tiller include the ground drive wheels and tail wheel.

Drive wheel: The drive wheels carry most of the weight of power tiller, provide traction and propel the power tiller. The power tillers are provided with rubber tyre wheels. These wheels give good performance when the power tiller is used on the hard surface or on road or in dry fields".

The rubber tyre wheels offer the advantage of low vibrations, small rolling resistance and least damage to the road surface. The wheel size of 6" x 14" ; 4 ply rating is very common. Nevertheless, 5" x 14" wheel size are also used in power tillers. The steel wheels are used for wet cultivation. Advantages of the steel wheels or cage wheels are simple in construction, low initial cost and better traction but its disadvantages are greater vibration and higher rolling resistance.

Tail wheel: Power tillers are provided with one tail wheel which carries a small part of the weight of power tiller with the result, the power tiller has better stability during operations. It also helps in adjusting the depth of operations and in steering the power tiller during operation. The pressure of drive wheel generally ranges from 1.1 to 1.4 kg/cm^2.

24.5 MATCHING IMPLEMENTS WITH POWER TILLER

Power tillers were initially imported from Japan and they were conceived as a source of power for rotapuddling and to some extent for transportation. Mould board ploughs and ridgers were also available with power tillers. The limited use was due to the availability of only few matching implements. This did not favour the economical use of power tillers. In many cases, power tiller owners had to maintain bullocks and also power tillers remained only as a supplementary source of power. In order to enhance the utility of power tillers and to make them as an independent source of power, the work on development of matching implements was taken up at various research and development organization in our country. The popularization of power tillers has at present gained a momentum by the availability of a number of matching implements for performing various farming operations. Some of the power tiller matching implements have been shown in the following figures (Figs. 24.3 *a, b, c, d, e, f*).

24.6 TRENDS IN POWER TILLER POPULATION IN INDIA

In view of the potential of powers tillers for Indian Agriculture, field testing of

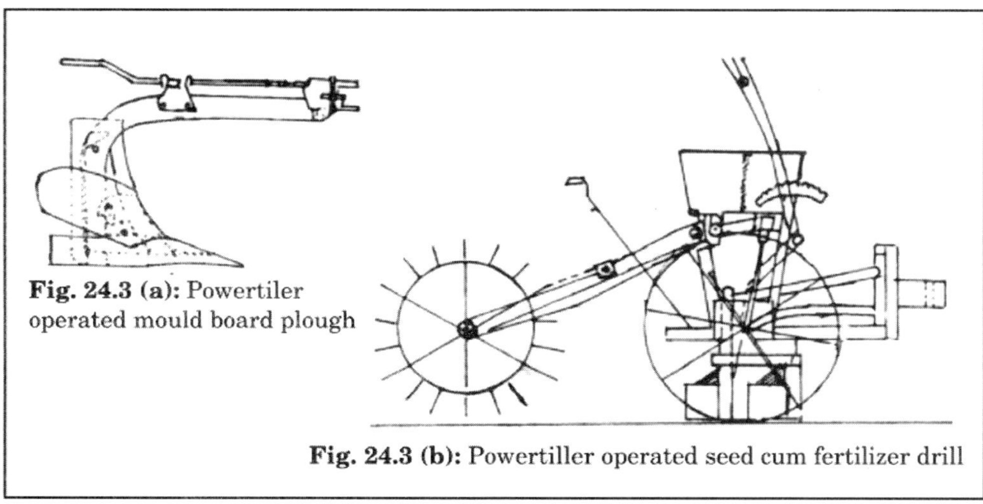

Fig. 24.3 (a): Powertiler operated mould board plough

Fig. 24.3 (b): Powertiller operated seed cum fertilizer drill

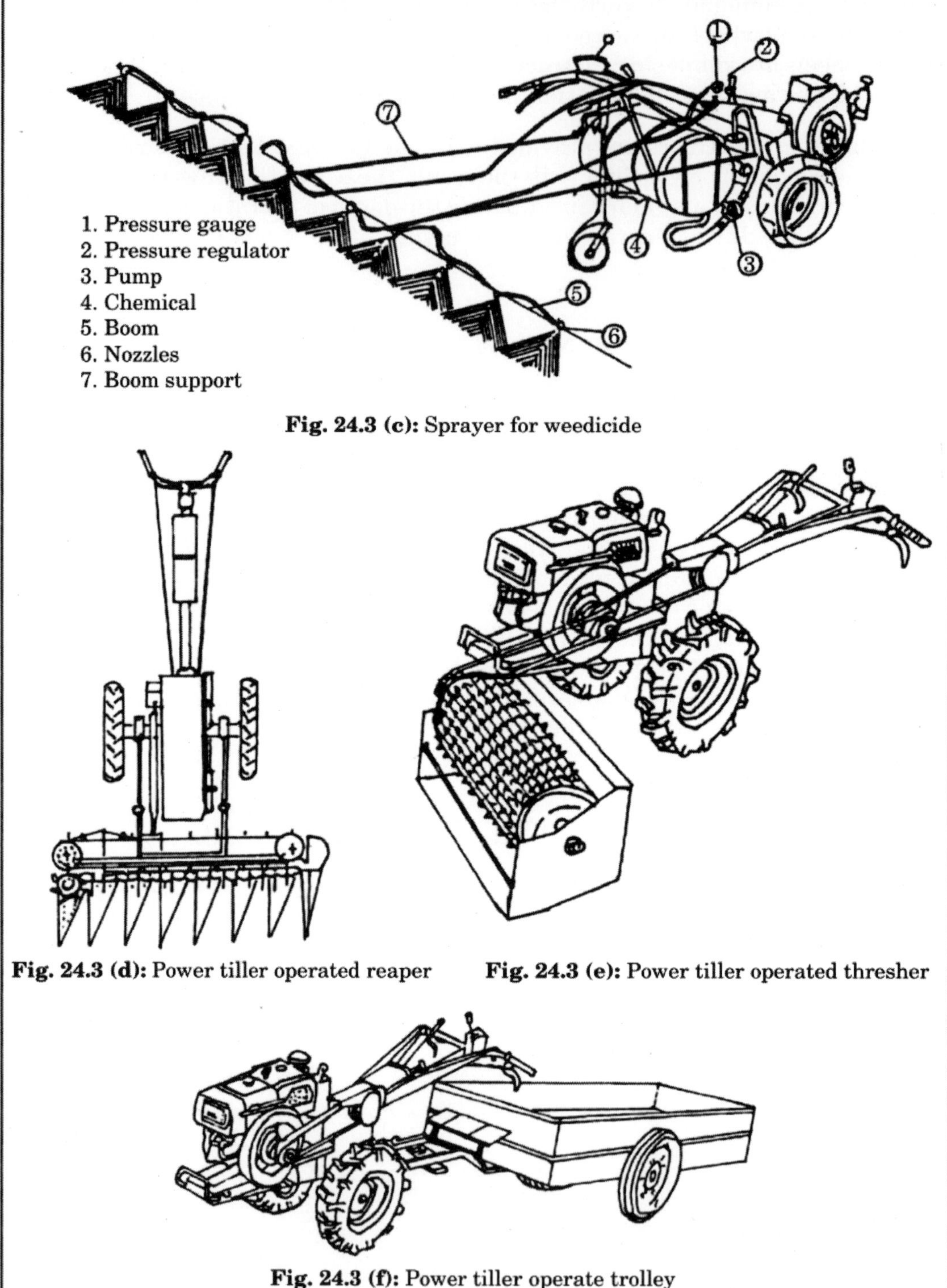

Fig. 24.3 (c): Sprayer for weedicide

Fig. 24.3 (d): Power tiller operated reaper

Fig. 24.3 (e): Power tiller operated thresher

Fig. 24.3 (f): Power tiller operate trolley

different types of power tillers was carried out at different research institutions of India during the period 1955 to 1960. Even though the indigenous production of power tiller started in the year 1961-65, the import of power tiller continued till 1973-74. During the period 1970-85, there has not been the significant increase in the production volume of power tillers. Major reasons for slow growth of power tiller industry in the country may be attributed to non-availability of low cost, light weight power tillers, poor after-sale service, non availability of well matched equipment and less favourable benefit-cost ratio.

Power tiller industry has been showing the signs of growth since 1984 after a dormant period of about two decades with rise of production from about 2500 to about 9000 power tillers per year. M/s VST Tillers Tractors Ltd., Bangalore (VST) and the Kerala Agro Machinery Corporation, Athani (KAMCO) are at present the two leading manufacturers of power tillers in our country. Presently the market demand for power tillers is rising. The production and sale of power tillers in India and different states along with the list of manufacturers have been mentioned in the following Tables.

Table 24.1: Production and sale of power tillers in India.

Year	*Production (Numbers)*	*Sale (Numbers)*
1987-88	3258	3258
1988-89	4798	4678
1989-90	5334	5442
1990-91	6228	6316
1991-92	7580	7520
1992-93	8648	8642
1993-94	9034	9449
1994-95	8334	8376
1995-96	10500	10045
1996-97	11210	11000
1997-98	13450	13100
1998-99	14488	14480
1999-2000	16891	16891
2000-01	17315	16018
2001-02	14837	13563
2002-03	14438	14613
2003-04	15849	15665

Table 24.2: Sale of power tillers in different states

Sl. no.	*States*	*1999-2000 (Numbers)*	*2000-01 (Numbers)*	*2001-02 (Numbers)*
1	West Bengal	5270	5161	4556
2	Tamil Nadu	2644	2125	1865
3	Assam	1506	1514	1103
4	Orissa	970	1150	990
5	Karnataka	1649	1623	979
6	Maharashtra	477	603	765
7	Kerala	1536	1194	584
8	Tripura	289	288	455
9	A.P	1142	469	429
10	Gujarat	290	460	277
11	Bihar	171	197	244
12	Other states U.T.	947	1234	1006
13	All India	16891	16018	13563

Table 24.3: Manufacturers, Production and Sale of power tillers during 2003-04

SI. no.	*Name of manufacturer*	*Production (Numbers)*	*Sale (Numbers)*
1	VST Tillers and Tractor Ltd., Bangalore	8070	8070
2	Kerala Agro Machinery Co. Ltd. (KAMCO) Athani, Kerala	6775	6582
3	Ganga Motors, Kolkata	181	181
4	Sunmoon Industires, Aurangabad, Maharashtra	8	8
5	Greaves Cotton Ltd., Kolkata	815	815
	Total	15849	15665

Chapter 25

Maintenance of Tractor and Power Tiller

25.1 MAINTENANCE AND TROUBLE SHOOTING

The term maintenance of a machine is usually meant for the upkeep of mechanical parts through systematic series of inspections and operations performed periodically with a view to improve its efficiency and performance. Tractor and power tiller maintenance comprises both daily and periodic measures, inspections and adjustments. Correctly carried out maintenance is probably the best guarantee of reliability, low repair costs and satisfactory life. It is in the interest of both the manufacturer and the customer that the tractor gives reliable service over a long period. The machine instruction manual for the customer is an absolute necessity. Spare part lists thoroughly and clearly illustrated can be of great value to the user. Proper and regular maintenance of the machine assists the user in up keeping the machine in good running condition causing reduced running costs and expensive repairs.

25.1.1 Objectives of Maintenance

Followings are the objectives of maintenance:

(a) It increases the efficiency of a machine
(b) It improves performance
(c) It increases the useful life of machine
(d) It reduces fuel cost
(e) It reduces maintenance cost
(f) It reduces the consumption of spare parts.

25.1.2 Element of Maintenance

Following are the important steps to be followed for complete maintenance of a machine.

(a) Routine external inspection,
(b) Periodic internal inspection

(c) Daily service
(d) Replacing faulty components before it spoils other parts
(e) Keeping record of oil-fuel consumption
(f) Schedule service
(g) Planning schedule overhaul.

25.1.3 Record Keeping

Trouble free running of a machine is as important as to keep the records that the machine runs economical, gives desired output, consumes less parts and is reliable. This can only be achieved if the record keeping is maintained up to date such as (a) Daily hours worked (b) Oil change register, (c) Oil and fuel consumption, (d) Parts replaced with date

25.1.4 Trouble Shooting

A machine is generally made of many different parts which need to be accurately adjusted and looked after for its efficient working. In running a machine, the operator comes across various troubles that affect its performance. Firstly, the trouble needs to be diagnosised, possible causes for it require to be detected, and finally the remedial measures to be taken immediately for its smooth functioning.-

In all trouble-shooting procedures, the following steps are adopted.

(a) When a part is apparently not Functioning, the part is not opened straight away, the connections leading to it are examined.
(b) Checking is done from simple to complex connections.
(c) Checking is done from less time consuming operations to more time consuming ones.
(d) For all these checking, safety for the operator and for engine is to be ensured.
(e) Before coming to any abrupt decision, the entire machine system is to be studied with the help of operators service manual.

25.2 GENERAL PRECAUTIONS FOR TRACTOR/POWER TILLER

The most important thing for preserving the long life of the tractor is to keep dirt out of its vital working parts, as tractor is subjected to run in most dusty conditions. Enclosed compartments, seals and filters need to be provided to keep the supply of air, fuel and lubricants clean. The effectiveness of these safeguards needs to be given utmost attentions. Filters should be replaced or cleaned regularly. Worn seals or broken gaskets should be quickly replaced. In addition to these, the following special precautions need to be adopted for the important components of the tractor for improving its durability and working capacity.

Lubrication system:

(i) Use clean oil of proper grade
(ii) Use crank case flushing oil for flushing the crank case.
(iii) Drain the crank case by agitating the oil only when the engine is hot.
(iv) Never check the oil level while the engine is in running condition.
(v) Never use cotton waste for cleaning the components.
(vi) Metal type oil filter element should be cleaned with a bristle brush along with petrol or diesel,

Cooling system:

(i) Never run the tractor without water
(ii) Always maintain the level of water as recommended by the manufacturer.
(iii) Never fill the water when the engine is hot and use only clean water.
(iv) Never remove the radiator cap abruptly when the engine is hot.
(v) Never lubricate the fan bearing when the engine is in running condition.

Air cleaner system:

(i) Never remove the oil cup when the engine is in running condition.
(ii) Refill only clean oil of the proper grade
(iii) Always clean the filter with a jet of compressed air
(iv) Never use a cracked rubber hose
(v) Check proper valve clearance of inlet and exhaust valve of engine and immediately replace the gasket in valve mechanism if worn out.

Fuel system:

(i) Use only clean fuel of proper grade
(ii) Handle the fuel filter very carefully as the mesh in extremely fine.

Transmission and wheel system:

(i) Avoid riding over the clutch pedal
(ii) Never overload the engine
(iii) Never over lubricate the bearings
(iv) Release the clutch slowly to avoid jerks
(v) Drain the transmission case only when the engine is warm
(vi) Always keep recommended inflation of the tyre.
(vii) The valve of rubber tyre of power tiller should be properly covered with the cap in order to prevent the entry of dust, mud, water into it.

Hydraulic system:

(i) Use only clean hydraulic oil of proper grade
(ii) Maintain the oil level as prescribed by the manufacturer.

Electrical system:

(i) Never touch the concentrated electrolyte
(ii) Never add concentrated electrolyte in the battery,
(iii) Never let the terminals of the battery to corrode
(iv) Never drive the tractor if dynamo is not functioning
(v) Never touch bare wirings

Before, during and after operation:

(i) Before staring the tractor, make sure that there is no leakage of the oil and fuel and the tyres are properly inflated.
(ii) After starting the tractor, make sure that all the instruments fitted in the panel are functioning properly.
(iii) While adjusting the hydraulic lift and cleaning the parts, do not stand on the implement.
(iv) Don't smoke or keep the flame near the fuel tank.
(v) Never ride or allow any body to ride on the drawbar or the implement during operation.
(vi) Always drive the tractor at a speed slow enough to ensure safety, especially on rough ground or near ditches
(vii) Reduce the speed before making a turn.
(viii) During night operation keep the front light on
(ix) Always keep the tractor in gear when going down steep slopes.
(x) After stopping the tractor, see that all the controls are on neutral or off position. Shut off the fuel tank valve and take out the contact key.

25.3 PERIODICAL MAINTENANCE OF TRACTOR

Periodic maintenance of tractor is an important task so as to get its trouble free service throughout the working season. But due to constant use, the tractor and its engine parts wear out, involving major overhauls such as replacing cylinder liners, connecting rod bearings, main bearings, replacing of piston rings, grinding of crankshaft, replacing of valves and valve seats etc. Hence there is the requirement of systematic periodical maintenance at the intervals of 8 to 10, 50 to 60, 100 to 120, 200 to 250, 480 to 500 and 960 to 1000 engine working hours. If periodical maintenance is performed properly, the premature wear and damage of parts can be avoided resulting in the long-term functioning of the mechanisms and units. Some of the important periodic attention which should be given to the tractor are as follows.

1. At 8 to 10 engine working hours

(i) Clean the tractor and implements,

(ii) Check the oil level in crankcase and hydraulic oil chamber. Top up if necessary when engine is cold

(iii) Remove sediments from the air pre-cleaner bowl. Clean the oil-bath of the air cleaner if the tractor operates in dusty conditions,

(iv) Top up the fuel tank, if necessary, preferably in the evening after day's work to avoid condensation,

(v) Clean the radiator. Remove dust and dirt accumulated in the core and top up the cooling system, using clean water.

(vi) Check the tension of V-belts driving dynamo and fan according to manufacturer's recommendations.

(vii) Check the tightness of fuel pipes, oil pipes, water pipes, hoses and drain plugs to avoid leakage.

(viii) Check the recommended air pressure of front and rear tyres.

(ix) Check the electrical units. The ends of cables should be properly connected to the terminals.

(x) Check the level of electrolyte in the battery and if necessary top up with distilled water.

(xi) Grease all points recommended by manufacturer.

(xii) Check all the ball joints of the steering linkage. Check the tightness of bolts and screws of the front axle, wheel hubs, wheel disks etc.

(xiii) Start the engine and ensure that

(a) The engine runs smoothly without knocking and abnormal noise.

(b) Oil pressure gauge is showing sufficient pressure

(c) The warning lights, if provided, are functioning properly.

(d) The dynamo is generating proper current.

2. At 50 to 60 engine working hours

(i) Carry out all the operations given in (l).

(ii) Clean the oil filter

(iii) Check the clearance between the clutch thrust bearing and disengaging levers.

(iv) Make sure that the brakes are in good working conditions.

3. At 100-120 engine working hours

(i) Carry out all the operations given in (2)

(ii) Disconnect the cables of the electrical equipment from terminals, apply grease to the terminal and reconnect.

(iii) Check the water pump and make sure that there are no leaking points

(iv) Lubricate the dynamo, by putting a few drops of engine oil in each of the oil caps.

4. At 200 to 250 engine working hours

(i) Carry out all the operations given in (3)

(ii) Drain oil from oil sump and flush with flushing oil. Refill with new engine oil up to mark.

(iii) Lubricate the joints of the throttle control linkage and other ball joints.

(iv) Check the clearance of the front wheel hub bearing.

(v) Check the toe-in of the front wheel.

5. At 480 to 500 engine working hours

(i) Carry out all the operations given in (4)

(ii) Add rust removing compound to the radiator and flush the cooling system.

(iii) Interchange the tyres of the front wheels to secure wearing

(iv) Flush the fuel tank

(v) Clean the self starter and dynamo of the tractor

(vi) Check the injector and adjust if necessary.

6. At 960 to 1000 engine working hours

(i) Carry out all the operations given in (5)

(ii) Check the oil in the gear box, steering housing, PTO case and hydraulic system of tractor.

(iii) Clean and adjust the brake lining.

(iv) Check the compression pressure of the engine and if necessary overhaul the engine.

(v) Check the valve clearance

(vi) Check the tension of valve spring.

25.4 PERIODICAL MAINTENANCE OF POWER TILLER

The periodical maintenance of power tiller are as follows:

1. After every 10 hours of use

(i) Clean the air cleaner

(ii) Check the level of water in the radiator and if necessary, pour clean water to it.

(iii) Grease all the points recommended by the manufacturer.

(iv) Check the level of oil in the crank case and gear box.
(v) Tighten the nut and bolt in the drive wheel, tynes of rotavator and in chassis.
(vi) Apply lubricating oil in all levers, clutch rod and the tail wheel.
(vii) Check the fuel in the fuel tank and top up if necessary.

2. At 50 to 60 engine working hours

(i) Carry out all the operations given in (1)
(ii) Check the tension of belt used in fan and clutch.
(iii) Clean the tyre and maintain the recommended air pressure inside it.
(iv) Apply lubricating oil in all control mechanism.
(v) Clean the oil filters
(vi) Make sure whether the brakes are in good working condition.

3.At 100 to 120 engine working hours

(i) Carry out all the operations given in (2)
(ii) Lubricate the chain and sprocket.
(iii) Check the valve clearance
(iv) Clean the filters
(v) Drain the oil from oil sump and flush with flushing oil.
(vi) Tighten the nut and bolts in cylinder head.

4.At 450 to 500 engine working hours

(i) Carry out all the operations given in (3)
(ii) Clean the radiator, crankcase, fuel tank, air cleaner, filters etc.
(iii) Change the oil in crankcase and gearbox.
(iv) Check the fuel pump and injector and adjust if necessary.
(v) Check the valve clearance
(vi) Check all fastening bolts and screws and make sure they are tight.

5.At 900 to 1000 engine working hours

(i) Carry out all the operations given in (4)
(ii) Check the proper adjustment of clutch system.
(iii) Check the sharpness of rotavator tynes.
(iv) Tighten the nut and bolts fitted in the chassis
(v) Clean the tyres and check their wearing.
(vi) Remove the chain case of rotavator and check the chain and sprocket for wear, tightness and lubricate them.

25.5 STORAGE OF TRACTOR

If the tractor is to be out of service for some time, it should be stored in a dry place. Leaving the tractor exposed to the outside environment shortens its life considerably. While placing the tractor in storage for more then 30 days, the following steps should be taken.

(i) Store the tractor in a dry, well protected place. If under-cover storage is not available, use a tarpaulin to cover the tractor.
(ii) Wash and clean the tractor thoroughly.
(iii) Clean all unpainted parts and where rust prevention is necessary, apply suitable grease on those parts.
(iv) Lubricate the chassis thoroughly.
(v) Drain the lubricant from the crankcase while engine is warm. Flush with crankcase cleaning oil. Refill fresh lubricant of the grade recommended by the manufacturer with 5 to 10 per cent rust-preventive engine oil. Run the tractor for a small period to splash the lubricant in all parts.
(vi) Remove the storage battery from the tractor and store in accordance with manufacturer's recommendation.
(vii) Drain the cooling system of the tractor.
(viii) Jack the tractor, so that the tyres are clear of the ground. If water is filled in tyres, drain the same.
(ix) Plug (fill the holes) all the orifices which expose the internal parts of the engine to atmosphere such as crankcase breather pipe and exhaust pipe if covers are not provided.
(x) Run the engine for 15 minutes and then switch off the engine .

25.5.1 Removal from Storage

While removing the tractor from storage, the following procedure should be followed:

(i) Remove the blocks placed under the tyre.
(ii) Check the electrolyte of the battery. Get the battery charged fully and fit it to the tractor.
(iii) Clean the tractor thoroughly.
(iv) Remove the plugs from the crankcase breather pipe and exhaust pipe.
(v) Lubricate the chassis.
(vi) Check oil level in air cleaner
(vii) Refill the clean water in cooling system
(viii) Fill the fuel tank with fuel and bleed the air from the fuel system.
(ix) Refill the crankcase with fresh oil of correct grade as recommended by the manufacturers.
(x) Start the engine and allow it to run slowly for several minutes in order to get the fresh oil distributed throughout the engine before putting load.

Similarly for storing of power tiller for more than 30 days the procedures adopted for tractor are also followed.

25.6 TROUBLE SHOOTING OF TRACTOR

Various troubles of tractor along with their possible causes and remedies have been discussed as follows.

A. Problem: Engine Fails to Start

Possible causes	*Remedies*
(i) Starting switch inoperative	(i) Inspect for loose terminals or faulty cables.
(ii) Battery discharged	(ii) Charge the battery
(iii) Air in fuel system	(iii) Bleed the fuel line by operating hand primer
(iv) No fuel in tank	(iv) Fill tank with diesel fuel
(v) One or several fuel pipes are closed	(v) Wash and blow with compressed air
(vi) Clogged fuel filter	(vi) Clean and replace the fuel filter
(vii) Air cleaner blocked	(vii) Service air cleaner
(viii) Defective fuel feed pump	(viii) Remove the fuel feed pump, inspect and eliminate fault
(ix) Defective fuel injection pump	(ix) Overhaul the injection pump
(x) Nozzle orifice chocked	(x) Clean it
(xi) Fuel pump timing not correct	(xi) Check up the timing

B. Problem: Engine Fails to Develop Full Power

Possible causes	*Remedies*
(i) Faulty fuel supply	(i) Check fuel supply
(ii) Air supply restricted	(ii) Check air cleaning system
(iii) Incorrect valve clearance	(iii) Adjust thc valve clearance properly
(iv) Fuel pump timing inaccurate	(iv) Adjust fuel pump
(v) Loss of compression at cylinder head gasket	(v) Fasten cylinder head and if necessary change the gasket

C. Trouble: Engine Gets Overheated Very Soon

Possible causes	*Remedies*
(i) Insufficient water in cooling system	(i) Add water into the radiator up to the required level
(ii) Weak tension of fan belt	(ii) Tighten it as required
(iii) Radiator clogged outside	(iii) Clean with compressed air and water
(iv) Thermostat valve opens immediately	(iv) Change the defective valve
(v) Incorrect engine lubrication, wrong grade oil, low level diluted oil, clogged strainer	(v) Check lubrications system
(vi) Engine overloaded	(vi) Reduce load

D. Engine Smoke

Engine generally generates black, blue and white smoke. Black smoke means excess fuel - it affects the combustion chamber on the daily run. Also it can lead to indigestion. This causes reduction in engine life. Hence black smoke is due to incomplete fuel combustion. Similarly blue smoke is due to excess burning of oil. It promotes heavy carbon deposits and shortens the life of your engine.

Black smoke

Possible causes	*Remedies*
(i) Air cleaner choked	(i) Clean it well
(ii) Engine overloaded	(ii) Lower engine load
(iii) Excess fuel injected	(iii) Check the injection pump and injector in authorized dealer
(iv) Insufficient air supply	(iv) Clean air cleaner
(v) Unsuitable fuel	(v) Use the correct fuel

Blue smoke

Possible causes	*Remedies*
(i) Excessive oil in crankcase	(i) Remove extra oil
(ii) Thin grade oil used	(ii) Change oil
(iii) Worn out oil rings	(iii) Replace the oil rings

White smoke

Possible causes	*Remedies*
(i) Engine too cold	(i) Warm up
(ii) Insufficient compression	(ii) Clean, regrind valves, fasten cylinder head, change worn out parts in the engine cylinder
(iii) Presence of water in fuel	(iii) Change fuel
(iv) Excessive oil in crankcase	(iv) Maintain correct level of oil in crankcase

E. Engine Knocks

Possible causes	*Remedies*
(i) One or more cylinders missing	(i) Locate and correct
(ii) Loose connecting rod	(ii) Tighten
(iii) Poor grade of fuel	(iii) Use recommended grade
(iv) Injection pump timed incorrectly	(iv) Time the injection pump correctly
(v) Incorrect engine temperature	(v) Keep temperature in working range as gauged by heat indicator.
(vi) Clogged air cleaner	(vi) Service air cleaner
(vii) Incorrect valve clearance	(vii) Adjust valve clearance
(viii) One or more fuel injectors inoperative	(viii) Check and if necessary, wash or replace defective injectors.

F. Trouble: Excessive Fuel Consumption

Possible causes	Remedies
(i) Fuel line leaks	(i) Check connections, pipes and gaskets of fuel supply system
(ii) Restricted air intake	(ii) Clean air cleaner
(iii) Faulty injectors	(iii) Change injectors
(iv) Improper quality fuel	(iv) Use recommended fuel
(v) Engine overloaded	(v) Reduce load

G. Excessive Oil Consumption

Possible causes	Remedies
(i) Wrong viscosity oil	(i) Change to specified grade of oil
(ii) Crankcase gasket leaking	(ii) Install new gasket
(iii) Oil level in crankcase too high	(iii) Maintain proper oil level
(iv) Engine overheating	(iv) Follow engine overheating remedy mentioned above
(v) Leaks in lubrication system	(v) Check connections and gaskets
(vi) Crankcase breather restricted	(vi) Service the breather
(vii) Piston rings not seating	(vii) Install new rings.

H. Engine Suddenly Stops

Possible causes	Remedies
(i) Vent hole of diesel tank clogged	(i) Clean
(ii) Fuel contains water	(ii) Check and replace
(iii) Piston seized in the cylinder	(iii) Check and replace
(iv) Connecting rod bearing or main bearing seized	(iv) Check and replace
(v) Camshaft bushes seized	(v) Check and replace
(vi) No fuel in tank	(vi) Check and fill
(vii) Air in fuel system	(vii) Bleed the air
(viii) Clogged fuel filter	(viii) Clean or replace this

I. Trouble: Engine Runs on Load But does not Run in Idle Condition

Possible causes	Remedies
(i) Valve clearance incorrect	(i) Adjust the clearance
(ii) Injection timing incorrect	(ii) Check the timing and set it
(iii) Governor adjustment faulty	(iii) Set the governor in authorized dealer
(iv) Injection pump faulty	(iv) Repair it in a specialized shop
(v) Cylinder worn	(v) Rebore and resleeve the cylinder

J. Troubles in Clutch

Clutch slips

Possible causes	Remedies
(i) Oil in the clutch lining because of leaks in engine seal or gear box.	(i) Replace the lining and correct leaks
(ii) Too much a clearance of clutch	(ii) Adjust the clearance
(iii) Worn out facing	(iii) Install a new friction plate
(iv) Weak clutch spring	(iv) Install a new set of springs

Clutch fails to disengage

Possible causes	Remedies
(i) Excessive clutch clearance	(i) Adjust clearance
(ii) Friction plate not moving	(ii) Clean the splined shaft and hub
(iii) Worn out facing of friction plate	(iii) Replace the facing on the friction plate

Clutch grabs

Possible causes	Remedies
(i) Wrapped friction plate	(i) Straighten or replace the friction plate
(ii) Release levers not evenly spaced	(ii) Adjust the lever properly
(iii) Damaged pressure plate	(iii) Replace

K. Troubles in Gear Box

Difficult gear shifting

Possible causes	Remedies
(i) Improper clutch adjustment	(i) Adjust clutch adjustment properly
(ii) Worn out clutch linging	(ii) Replace clutch plate
(iii) Shifter lock springs too strong	(iii) Adjust
(iv) Less oil in gear box	(iv) Top up oil
(v) Worn out splines of main shaft	(v) Replace main shaft

Gears slip out of mesh

Possible causes	Remedies
(i) Shifter lock spring too weak	(i) Install new spring
(ii) Worn out teeth of gear	(ii) Change the set
(iii) Mounting shaft misaligned to bent	(iii) Align or replace shaft

Excessive sound in differential

Possible causes	*Remedies*
(i) Less oil in gear box	(i) Top up oil with correct grade
(ii) Lubricant of poor quality	(ii) Replace it with correct grade of oil
(iii) Worn out gear	(iii) Replace with new ones
(iv) Worn-out bearings and shaft	(iv) Replace with new ones

L. Troubles in Differential

Sound in differential occasionally

Possible causes	*Remedies*
(i) Misadjustment of crown wheel and pinion	(i) Adjust it correctly
(ii) Loose differential bearings	(ii) Adjust the bearings

Excessive sound in differential

Possible causes	*Remedies*
(i) Worn out side bearings	(i) Replace new ones
(ii) Worn-out teeth of crown wheel and pinion	(ii) Replace with new one
(iii) Less or no oil in differential	(iii) Top up with correct grade oil
(iv) Improper adjustment of crown wheel and pinion	(iv) Adjust these correctly

M. Troubles on Brake

Possible causes	*Remedies*
(i) Excess play of the brake pedal	(i) Adjust the play
(ii) Oil leakage in brake chamber	(ii) Check the leakage
(iii) Broken brake shoe	(iii) Replace the shoe
(iv) Broken lining	(iv) Replace the lining
(v) Acting stroke of the right and left pedal different	(v) Stroke should be made same

Brake chatter

Possible causes	*Remedies*
(i) Too much brake lining	(i) Cut off excess material
(ii) Poor brake adjustment	(ii) Adjust brakes
(iii) Brake drum out of round	(iii) Rebore or replace the drum

Brake pedal does not return smoothly

Possible causes	*Remedies*
(i) Broken return spring	(i) Replace it
(ii) Grease runs short at each sliding section	(ii) Remove rust and apply grease

N. Troubles in Hydraulic System

Hydraulic pump not building up required pressure

Possible causes	Remedies
(i) Leakage in inlet pipe	(i) Check it
(ii) Clogged strainer fitted at inlet pipe	(ii) Clean it
(iii) Oil of poor viscosity	(iii) Change with recommended oil
(iv) Low oil leveling resources	(iv) Top up the reservoir

Hydraulic pressure dose not drop

Possible causes	Remedies
(i) Control valve out of order	(i) Check at the authorized shop
(ii) Broken cylinder	(ii) Check at the authorized shop and replace

Sound in hydraulic pump

Possible causes	Remedies
(i) Chockecd inlet pipe	(i) Clean it
(ii) Chocked inlet strainer	(ii) Clean it
(iii) Leakage in inlet pipe	(iii) Service it
(iv) Defective pump	(iv) Service it

Hydraulic system fails to lift the implement

Possible causes	Remedies
(i) Lack of hydraulic oil	(i) Top up the reservoir
(ii) Filter clogged	(ii) Clean the filter
(iii) Hydraulic pump defective	(iii) Repair at authorized service station
(iv) Broken control valve	(iv) Repair at authorized service station
(v) Broken cylinder	(v) Replace the cylinder
(vi) Pipe leakage	(vi) Remove leakage
(vii) Cracked pipes	(vii) Replace the pipes

25.7 TROUBLE SHOOTING OF POWER TILLER

The trouble shooting of power tiller are same as the tractor mentioned in section 25.6 excluding the hydraulic system. The problems encountered in the additional component *i.e.* rotavator used in power tiller need to be discussed in this section along with their possible causes and remedies. Hence the trouble shooting of rotavator are as follows.

A. Rotor won't run

Possible causes	Remedies
(i) Seizure of rotavator gear box	(i) Check and repair at authorized shop
(ii) Obstruction in rotor	(ii) Check rotor for stones or thrash

B. Blades do not hold correct depth

Possible causes	*Remedies*
(i) Blades bent or not arranged in staggered manner	(i) Check and fit accordingly
(ii) Rotor speed too low	(ii) Increase to the required speed

C. Severe blade breakage

Possible causes	*Remedies*
(i) Blade bolts not tight	(i) Tighten and use recommended bolts.
(ii) Working in soil with too many obstructions	(ii) Allow slower rotor speed to reduce blade impact on obstacles. Make first pass on shallow depth and then gradually increase depth in the later passes.

D. Excessive power requirement

Possible causes	*Remedies*
(i) Rotor speed too high	(i) Slow rotor speed
(ii) Blades bent, broken or incorrectly arranged	(ii) Straighten or replace damaged blades, check the blade arrangement pattern

E. Rotor plugs with soil

Possible causes	*Remedies*
(i) Soil too wet to work	(i) Wait till soil dries
(ii) Rotor speed too low for travel speed	(ii) Speed up rotor by changing gears
(iii) Blades bent or incorrectly arranged	(iii) Straighten or replace damaged blades and check the arrangement pattern.

Chapter 26

Some Important Formulae

1. Drawbar Horsepower (hp)= $\frac{Draft\,(kg)\,X\,Speed\,(m/s)}{75}$

2. Drawbar Horsepower (kW)= $\frac{Draft\,(kN)\,X\,Speed\,(km/hr)}{3.6}$

3. Power (Watt) = Draft (N) × Speed (m/s)
4. Frictional horsepower = I.H.P. – B.H.P or B.H.P. = I.H.P. – F.H.P.
5. Brake horsepower (kW) = $\frac{2\pi NT}{60000}$, where, T = Torque, N^{-m}, N = Engine speed, rpm
6. Brake horsepower (hp) = $\frac{2\pi NT}{4500}$, where, T = Torque, Kg^{-m}, N = Engine speed, rpm
7. Indicated horse power for 4 stroke cycle engine, I.H.P. = $\frac{P.L.A.N}{4500} \times \frac{n}{2}$
8. Indicated horse power for 2 stroke cycle engine, I.H.P. = $\frac{P.L.A.N}{4500} \times n$

   Where, P = Mean effective pressure, kg/cm^{-2}, L = Length of stroke, m; A = Area of the cylinder, cm^2, $A = \pi . r^2 = \frac{\pi}{4} . d^2$, Where, r = radius of cylinder, cm, D = diameter of cylinder, cm, N = Number of revolution of crankshaft, rpm, n = Number of cylinders

9. Thermal efficiency = $\frac{\text{Available useful work}}{\text{Heating value of supplied fuel}} \times 100$
10. Heat value = Quantity (kg) × calorific value (kcal/kg)

11. **Indicated thermal efficiency:** It is the ratio of the heat equivalent to one kW–hr to the heat in the fuel per indicated power hour, $\eta_{ith} = \frac{\text{indicated power (kW)} \times 3600}{m_f \times c}$ where m_f = fuel used in kg/h and c = calorific value kJ/kg
12. **Brake thermal efficiency:** It is the ratio of heat equivalent to one kW-hr to the heat in fuel per brake power hour, $\eta_{bth} = \frac{\text{brake power (kW)} \times 3600}{m_f \times c}$ where m_f = quantity of fuel kg/hr, c = calorific value of fuel = kJ/kg
13. Piston displacement (swept volume) = Cross-section area of piston × stroke length $P_d = \frac{\pi}{4} D^2 \times L$
14. Displacement volume (cm^3) = $\frac{\pi}{4} D^2 \times L \times N$; Where, D = diameter of piston, cm, L = length of stroke, cm, N = Number of power strokes per minute.
15. Total volume of cylinder (cm^3) = $P_d + C_v$ where, C_v = clearance volume [Note: C_v = total volume of cylinder- piston displacement]
16. Compression ratio = $\frac{Total\ volume}{Clearance\ volume} = \frac{p_d + c_v}{c_v}$
17. Piston speed (m/min) = $\frac{2 X L X N}{100}$ Where, L = length of stroke, cm, N = Number of revolution per minute, rpm
18. Rim speed or peripheral speed of flywheel (m/min) = π D N, Where, D = diameter of flywheel, m, N = Number of revolution per minute, rpm
19. Firing interval (4 stroke engine) = $\frac{720}{Number\ of\ cylinder}$
20. Firing interval (2 stroke engine) = $\frac{360}{number\ of\ cylinder}$
21. Efficiency of air cleaner = ($\frac{Wt.\ of\ dust\ collected\ in\ cleaner}{dust\ in\ cleaner\ and\ in\ absolute\ cleaner}$))
22. Dust entering in an engine (g/hr) = volume of air (m^3/hr) × dust (g/m^3)
23. Tension difference of belt is calculated as P (kW) = $\frac{(T_1 - T_2)V}{1000}$

Where, T_1 = tight side tension of belt, N, T_2 = slack side tension of belt, N, V = belt speed, m/s P = transmitted power, kW

24. Maximum allowable tension of belt is given as

$T_1/T_2 = e^{\mu\phi}$ Where, e = base of natural logarithm, μ = coefficient of friction between belt and pulley, ϕ = arc of contact of belt radians = $\frac{\text{Degree} \times \pi}{180}$

25. Length of belt can be calculated for open belt system as

$$L = \frac{n}{2}(D_1+D_2) + \frac{(D_1 - D_2)^2}{4C} + 2C$$

Where, D_1 = effective outside diameter of pulley 1, cm, D_2 = effective outside diameter of pulley 2, cm, C = distance between centre of the two pulleys, cm

26. API degree of oil = $\frac{141.5}{\textit{Specific gravity of oil at } 60^0\ F} - 131.5$

27. Governor regulation. R = Where, R = Regulation %, N_1 = speed of engine at no load, N_2 = sped of engine at load

28. Relationship between diameter and speed of pulley $N_1D_1 = N_2D_2$

Where, N_1 = revolution per minute of driving pulley. D_1 = diameter of driving pulley, N_2 = revolution of per minute of driven pulley, D_2 = diameter of driven pulley

29. Fuel injection time of a diesel engine

$t_i = \frac{\theta}{360} x \frac{1}{n} x 60$ where t_i = fuel injection time, minutes, θ = crank rotation angle, degree, N = engine rpm

30. Hydraulic/fluid power of pump (w_f)

$w_f\,(kW) = \frac{QxP}{60}$ where Q = actual pump discharge, l/m, p = pressure across the pump, MPa

31. Engine brake power

$W\,(kW) = \frac{2\pi N_e T_e}{6x10^4}$ where N_e = engine rpm; T_e = engine torque, N-m

32. Drawbar power of tractor

$W_d\,(kW) = \frac{P \times S}{3.6}$ where P = drawbar pull, kN, S = speed kmph

33. Shaft power of pump (w_S)

$w_S\,(kW) = \frac{2\pi NT}{60000}$ where T = input shaft torque N-m, N = shaft speed, rpm

Also $W_s = \frac{w_f}{\eta_m}$ where η_m = mechanical efficiency or overall efficiency of pump

34. Travel reduction or slippage, S (%) = $\frac{D_n - D_l}{D_n} \times 100$, where D_n = distance travelled at no load, m for one revolution, D_1 = distance travelled with load/ m for one revolution

    Alternatively, S= $(V_t - V_a)/V_t = (1 - \frac{V_a}{V_t})$; V_t = theoritically wheel speed = $r\times \omega,\ m.s^{-1}$, V_a actual wheel speed, m/s, r = rolling radius of wheel on hard surface, m, ω = angular velocity of wheel, rad. s^{-1}
35. Tractive force 'F' of tractor drive wheel, F = AC + w tanϕ or F = A (C + P tanϕ) Where A = contact area, cm^2, C = cohesion of soil N/cm^2, ϕ = Angel of internal friction of soil, degrees

    P = average normal soil pressure, N/cm^2, $P = \frac{w}{bl}$ for track type device and P = $(\frac{w}{0.78})\ (bl)$ for pneumatic tyres, w = weight on drive wheel, N, b = tyre width, cm, l = length of contact of tyre on soil surface, cm
36. Tractive efficiency (TE) = drawbar power/ axle power

    TE= $(P\times S)/(3.6\times w_a)$; where P = drawbar pull, kN; s = actual forward sped of tractor, kmph, w_a = axle power, kW
37. The maximum upgrade angle up to which tractor is stable

    $\tan\theta_{max} = {l_1}/{y}$ where θ_{max} is the maximum upgrade angle up to which tractor is stable, 'l_1' is the perpendicular distance from rear axle to c.g. of tractor and 'y' is the height of c.g. from the ground
38. The maximum downgrade slope $\tan\phi_{max} = \frac{l_2}{y}$ where l_2 is the perpendicular distance from front axle to c.g of tractor and 'y' is the height of c.g. from the ground
39. Sound pressure level, SPL = 20 log $\frac{p}{p_0}$, dB

    Where, P = R.M.S. sound pressure, N/m^2 P_0=reference sound pressure, 2×10^{-5} N/m^2
40. Vibration acceleration level, VAL = 20 log $\frac{a}{a_0}$, *dB*

    Where, a = measured RMS acceleration, m/s^2, a_0 = reference acceleration, 1 m/s^2

41. If sound pressure level is doubled, increase in resulting sound pressure level is 6 dB

42. Torque of steering kingpin (kingpin torque) = $W f \sqrt{\frac{b^2}{8} + e^2}$ where e= kingpin offset, b = width of wheel, W = vertical load on wheel, f = friction coefficient.

43. Optimum weight on rear wheel

    For cohesive soil, weight (kg) = $\frac{110 \times PTO\ (kW)}{Actual\ speed\ (kmph)}$ and for non-cohesive soil,

    weight (kg) = $\frac{123 \times PTO\ (kW)}{Actual\ speed\ (kmph)}$

SOLVED EXAMPLES

RENEWABLE ENERGY SOURCES

Example 1: A farmer constructed a 2 m^3 40 days HRT (hydraulic retention time) Deenbandhu model biogas plant. The gas will be solely used for cooking in a stove with a burner efficiency of 45 per cent. If the density of biogas is 0.94 kg m^{-3} with a heating value of 21 MJ kg^{-1}, find out the total effective energy available per day in MJ. ***(GATE-2010)***

Solution: Given, HRT = 40 days; Volume of biogas plant V = 2 m^3; Burner efficiency η = 45 per cent; Density of biogas ρ = 0.94 kg m^{-3}; Heating value H = 21 MJ kg^{-1}

Effective energy available per day E = $H\rho\eta V$ = 21 × 0.94 × 0.45 × 2 = 17.76 MJ

Answer: Total effective energy available per day is 17.76 MJ.

Example 2: What is the local Apparent time (LAT) corresponding to 14 h 30 min Indian Standard Time (IST) at a place in India (19° 07'N, 72° 51'E) in the month of April with a time correction of zero min. ***(GATE-2010)***

Solution: Given, IST = 14 h 30 min; Longitude of the location = 72° 51' = 72.85°

For India, standard longitude = 82.5°

So, local apparent time,

(LAT) = standard time – 4 (standard longitude – local longitude) + equation of time correction

= 14h 30' – 4 (82.5° – 72.85°) = 14h 30' – 38.6' = 13h 51.4'

Answer: The local apparent time is 13 h 51.4 min.

Example 3: A horizontal axis drag type wind mill with square blades and horizontal axis lift type wind mill with air foil section blades having same rotor size are installed at a height of 10 m above the ground. The average wind speed is 25 kmph. The maximum power coefficient for drag type and lift type wind mills is 0.148 and 0.593, respectively. If the maximum power extracted by drag type wind mill is 5 kW, find out the corresponding power extracted by lift type wind mill. ***(GATE-2012)***

Solution: Given, two wind mills a) horizontal axis drag type with square blades; Installation height = 10 m; Average wind speed v = 25 kmph = 6.95 m.sec^{-1}; Power coefficient for drag type C_{pd} = 0.148 and that for lift type C_{pl} = 0.593; Power extracted by drag type wind mill P_d = 5 kW

Input power to drag type wind mill $P_{input} = \frac{P_d}{C_{pd}} = \frac{5}{0.148} = 33.78\text{ kW}$

As both the wind mills are operated under same conditions, the input power is same for both type wind mills.

Answer: Power extracted by lift type wind mill is 26.787 kW.

Example 4: A propeller type wind turbine of 8 m diameter generates 4 kW electric power. If the overall efficiency of the generation system is 32 per cent and the density of air is 1.2 kg per m^3, then find the average wind speed in $m.s^{-1}$.

(GATE-2015)

Solution: Given, Wind turbine diameter D = 8m; Power output P = 4 kW; Overall efficiency η_0 = 32 percent = 0.32; Density of air ρ = 1.2 $kg.m^{-3}$.

Let the average wind velocity be v

$$\text{So, } v = \sqrt[3]{\frac{2P}{\eta_o \rho \frac{\pi}{4} D^2}} = \sqrt[3]{\frac{2 \times 4 \times 1000}{0.32 \times 1.2 \times \frac{\pi}{4} \times 8^2}} = 7.45 \text{ m. s}^{-1}$$

Answer: The average wind velocity is 7.45 $m.s^{-1}$.

Example 5: A 2 m^3 biogas plant is to be operated using cow-dung as feedstock. The cow-dung is mixed with water in proportion of 1:1 (by weight) to form a slurry of density 1090 $kg.m^{-3}$. The yield of biogas is 0.036 $m^3.kg^{-1}$ of cow-dung. If the hydraulic retention time is 40 days, find the volume of the digester of the biogas plant.

(GATE-2018)

Solution: Given, Biogas plant volume V = 2 m^{-3}; Cow dung mixed with water = 1:1; Slurry density ρ = 1090 $kg.m^{-3}$; Yield of biogas y = 0.036 $m^3.kg^{-1}$; Hydraulic retention time HR = 40 days

From 1 kg cow dung, 0.036 m^3 of biogas is generated. The quantity of cow dung required per day for generation of 2 m^3 of biogas = 55.55 kg

Total mass of slurry per day (1:1 ratio of cow dung and water) = 55.55 × 2 = 111.1 kg

Volume of slurry per day = 111.1 kg / 1090 kg/ m^3 = 0.1 m^3

In order to keep the materials for 40 days (HRT) in the digester, the volume of digester becomes 0.1 × 40 = 4 m^3

Hence the volume of the digester of the biogas plant is 4.00 m^3

Alternative, Volume of digester $V_d = \frac{2V\rho\, y}{HR} = \frac{2 \times 2 \times 1090 \times 0.036}{40} = 3.942 \text{ m}^3 \approx 4 \text{ m}^3$

Answer: Volume of digester of biogas plant is 4 m^3.

Example 6: A cylindrical parabolic collector is designed to heat fluid that enters the absorber at 140°C at a flow rate of 5 kg per minute. The specific heat capacity of fluid is 1.5 $kJ.kg^{-1}.°C^{-1}$ and its outlet temperature is 180°C. If the incident beam radiation on the plane of aperture is 3000 $kJ.h^{-1}.m^{-2}$ and useful projected area of the projector is 2m×10m, find the percentage efficiency of collector. ***(GATE-2017)***

Solution: Given, inlet temperature T_1 = 140°C; Mass flow rate = 5 kg. min^{-1}; Specific heat capacity of fluid = 1.5 $kJ.kg^{-1}.°C^{-1}$; Outlet temperature = 180°C; Incident beam radiation, I = 3000 $kJ.h^{-1}.m^{-2}$

Useful projected area A = 2m × 10m = 20 m^2

Useful work done per hour $= \dot{m}C_p(T_2 - T_1) \times 60 = 5 \times 1.5 \times (180 - 140) \times 60 = 18000 \text{ kJ}$

Input incident energy per hour $= I \times A = 3000 \times 20 = 60000 \text{ kJ}$

$$\text{So, efficiency} = \frac{\text{Useful work done per hour}}{\text{Incident energy per hour}} = \frac{18000}{60000} = 0.3 = 30 \text{ per cent}$$

Answer: Percentage efficiency of the collector is 30 per cent.

IC ENGINE AND ITS COMPONENTS

Example 7: An engine has four connecting rods having diameter 7 cm and length of 5 cm. It is rotating at a speed of 1800 rpm and uses SAE 20 oil with specific gravity 0.6 at mean temperature of 100°C. The kinematic viscosity of oil at 100°C is 7.5 centistokes. The journal and bearing to be concentric with a gap of 0.006 mm. Find the power lost in friction to run the engine.

Solution: Given, Journal and bearing are concentric; Gap between journal and bearing = 0.006 mm; Connecting rod having diameter = 7 cm; Connecting rod of length = 5 cm; Engine speed N = 1800 rpm; SAE 20 oil is used with specific gravity 0.6 *i.e.* density = 0.6 kg/m^3; Kinematic viscosity at 100°C = 7.5 centistokes

The friction force F by Newton's law is due to the viscosity of the lubricant and is given by

$$F = \mu A \frac{v}{h} \qquad (1)$$

Where μ = absolute viscosity of the oil; A = contact area with metal plate; v = peripheral velocity; h = clearance between the plates

$$\text{Peripheral velocity, v of the oil is given by } v = \frac{d \times N}{60 \times 100} = \frac{7 \times 1800}{60 \times 100} = 2.1 \text{ m. sec}^{-1}$$

Dynamic viscosity μ = kinemetic viscosity × mass density

$$= 7.5 \times 10^{-4} \times 0.6 = 4.5 \times 10^{-4} \text{kg.sec}^{-1}.\text{m}^{-2}$$

Hence frictional force from eqn.1 is calculated by

$$F = \mu A \frac{v}{h} = 4.5 \times 10^{-4} \times \frac{\pi \times 7 \times 5}{10000} \times \frac{2.1}{0.0006 \times 10^{-4}} = 173.180 \text{ kg} = 1.689 \text{ kN}$$

Hence frictional power lost = Frictional force × velocity = 1.689 × 2.1 = 3.567 kW

Answer: Frictional power lost is 3.567 kW

Example 8: An IC engine developing 330 kW indicated power and 255 kW brake power uses 100 kg oil per hour. The fuel has a calorific value of 10500 kCal.kg^{-1}. Jacket water was circulated at the rate of 232 kg per minute with temperature rise of 20°C. Draw the heat balance sheet.

Solution: Given, fuel consumption per hour = 100 kg; Indicated power = 330 kW; Brake power = 255 kW; Water circulation rate = 232 kg.min^{-1}=3.86 kg.sec^{-1} with temperature rise of 20°C; Specific heat of water = 1 kCal.kg^{-1}.°C^{-1}; calorific value of fuel C_f = 10500 kCal.kg^{-1}

$$\text{Brake specific fuel consumption} = \frac{100}{330} = 0.303 \text{kg.kW}^{-1}.\text{h}^{-1}$$

$$\text{Heat supplied per min} = \frac{100 \times 10500}{60} = 17500 \text{ kCal.min}^{-1}$$

Heat absorbed by circulating water = water flow rate × specific heat×temperature rise = 232 × 1 × 20 = 4640 kCal. min^{-1}

$$\text{Heat supplied for developing useful power} = 255 \times \frac{60}{4.2} = 3642.85 \text{ kCal.min}^{-1}$$

$$\text{Heat required to overcome friction loss} = (330 - 255) \times \frac{60}{4.2} = 1071.42 \text{ kCal.min}^{-1}$$

Heat lost by the exhaust gas = 17500 – 4640 – 3642.85 – 1071.42

= 8145.73 kCal.min^{-1}

$$\text{Percentage heat lost in developing useful power} = \frac{3642.85}{17500} \times 100 = 20.8 \text{ per cent}$$

$$\text{Percentage heat lost in cooling} = \frac{4640}{17500} \times 100 = 26.5 \text{ per cent}$$

Percentage heat lost in radiation and exhaust $= \frac{8145.73}{17500} \times 100 = 46.54$ per cent

Percentage heat lost in overcoming friction $= \frac{1071.42}{17500} \times 100 = 6.12$ per cent

Answer: Percentage of power lost in developing power, cooling, radiation and exhaust and overcoming friction is 20.8, 26.5, 46.54 and 6.12 per cent.

Example 9: A single cylinder four stroke engine is operated by petrol. It has a bore of 70 mm and stroke of 100 mm. The charge enters at a temperature of 25°C and pressure of 103 kPa. Power developed by the engine is 16.8 kW while operating at 2500 rpm. The engine has fuel consumption of 3 $kg.h^{-1}$. If the engine has a mechanical efficiency of 80 per cent and compression ratio of 6, find out work done, air standard efficiency, thermal efficiency, relative efficiency, indicated and brake mean effective pressure and air fuel ratio.

Solution: Given, single cylinder 4 stroke; Bore d = 70 mm; Stroke L = 100 mm; T_1 = 25°C and P_1 = 103 kPa; Power developed by the engine P = 16.8 kW; Engine speed N = 2500 rpm; Fuel consumption f = 3 $kg.h^{-1}$; Assuming Calorific value of fuel C_f = 10800 $kCal.kg^{-1}$; Mechanical efficiency η_m = 80 per cent; Compression ratio r = 6

Assuming $C_v = 0.169$ and $C_p = 0.238$

So, $k = \frac{C_p}{C_v} = \frac{0.238}{0.169} = 1.4$

Hence air standard efficiency $\eta_a = 1 - \frac{1}{r^{k-1}} = 1 - \frac{1}{6^{1.4-1}} = 51.16$ per cent

Density of air at inlet $\rho_a = \frac{P_1}{RT_1} = \frac{103 \times 10^3}{8.314 \times (273 + 25) \times 9.81} = 4.19\ kg.m^{-3}$

Stroke volume $V_s = \frac{\pi}{4} d^2 L = \frac{\pi}{4} \times 7^2 \times 10 = 384.85\ cc = 3.84 \times 10^{-4} m^3$

Let R_f = fuel air ratio, So, amount of fuel injected

$$= \frac{\rho_a V_s N R_f}{2} \times 60 = \frac{4.19 \times 3.84 \times 10^{-4}}{2} \times 2500 \times 60 \times R_f = 120.672 R_f\ kg.h^{-1}$$

But given f = 3 $kg.hr^{-1}$, So, $R_f = \frac{3}{120.672} = 0.024$

So, air fuel ratio = $\frac{1}{R_f} = \frac{1}{0.024}$ = 41.67, given r = $\frac{V_1}{V_2}$ = 6

And $V_1 - V_2$ = 384.85 cc. From this V_2 = 76.97 cc and V_1 = 461.82 cc

So volumetric efficiency = $\frac{V_1 - V_2}{V_1} = \frac{461.82 - 76.97}{461.82}$ = 83.34 per cent

Indicated power developed per cylinder is given by IP = $\frac{P_m V_s N}{2 \times 60}$

$$P_m = \frac{IP \times 2 \times 60}{V_s N} = \frac{21.0 \times 2 \times 60}{384.85 \times 10^{-6} \times 2500} = 2619.20 \text{ kPa}$$

So, brake mean effective pressure = $\eta_m \times P_m$ = 0.8 × 2619.20 = 2095.36 kPa

Indicated thermal efficiency

$$h_{ith} = \frac{\text{Indicated power}}{\text{Fuel power}} = \frac{IP}{f \times C_f} = \frac{21.0}{3 \times 10800 \times \frac{4.23}{3600}} = 55.16 \text{ per cent}$$

Relative efficiency = $\frac{\eta_{ith}}{\eta_a} = \frac{55.16}{51.16}$ × 100 = 107.8 per cent

Heat supplied

$$Q_1 = C_p(T_2 - T_1) = C_p T_1(r^{k-1} - 1) = 0.24 \times 298(6^{1.4-1} - 1) = 74.92 \text{ kCal. kg}^{-1}$$

So, work done W = $Q_1 \times \eta_a$ = 74.92 × 0.5116 = 38.34 kCal.kg^{-1}

Answer: Work done is 38.34 kCal.kg^{-1}, air standard efficiency is 51.16 per cent, thermal efficiency is 55.16 per cent, relative efficiency is 107.8 per cent, indicated mean effective pressure is 2619.20 kPa, brake mean effective pressure is 2095.36 kPa, air fuel ratio is 41.67.

Example 10 A 5 kW gas engine running at 200 rpm consumes 4.4 m^3 of gas per hour. It has swept volume of 0.0095 m^3, clearance volume of 0.0022 m^3, mechanical efficiency of 74 per cent and volumetric efficiency of 86 per cent. The gas has a calorific value of 4400 kCal.m^{-3}. Find out relative efficiency, compression ratio, air fuel ratio and air standard efficiency.

Solution: Given, Engine power P = 5 kW; Fuel consumption per hour f = 4.4m^{-3}; Engine speed N = 200 rpm; Clearance volume V_c = 0.0022 m^{-3}; Swept volume V_s = 0.0095 m^{-3}; Mechanical efficiency η = 74 per cent; Volumetric efficiency η_v = 86 per cent; Calorific value of gas C_v = 4400 kCal.m^{-3}

$$V_s = V_1 - V_2$$

Where V_1 = volume of cylinder at the end of suction

V_2 = volume of cylinder at the end of compression = V_c

So, $V_1 = V_s + V_c = 0.0095 + 0.0022 = 0.0117\ m^3$

$$\text{Compression ratio} = \frac{V_1}{V_2} = \frac{0.0117}{0.0022} = 5.32$$

Energy input to the engine per hour = 4.4 × 4400 = 19360 kCal = 81002 kJ

So, power input to engine = 22.5 kW

$$\eta_m = \frac{\text{brake power}}{\text{indicated power}} \times 100$$

$$\text{Or, indicated power} = \frac{\text{brake power}}{\eta_m} \times 100 = \frac{5}{74} \times 100 = 6.75 \text{ kW}$$

Hence indicated thermal efficiency

$$\eta_{th} = \frac{\text{indicated power}}{\text{power input}} \times 100 = \frac{6.75}{22.5} \times 100 = 30 \text{ per cent}$$

Air standard efficiency

$$\eta_{otto} = \left(1 - \frac{1}{r^{k-1}}\right) \times 100 = \left(1 - \frac{1}{5^{1.4-1}}\right) \times 100 = 47.46 \text{ per cent}$$

Relative efficiency =

$$\frac{\text{indicated thermal efficiency}}{\text{air standard efficiency}} \times 100 = \frac{30}{47.46} \times 100 = 63.21 \text{ per cent}$$

Volume of air fuel mixture sucked per stroke

Total volume of air sucked per minute = 0.0095 × 200 = 1.9 m^3

Hence, total volume of air sucked per hour = 1.9 × 60 = 114 m^3

$$\text{Air fuel ratio} = \frac{114 - 4.4}{4.4} = 24.9:1$$

Answer: Relative efficiency = 63.21 per cent; Air fuel ratio = 24.9:1; Compression ratio = 5.32; Air standard efficiency = 47.46 per cent

Example 11: A engine running on ethyl alcohol has an air fuel ratio of 15:1. At what partial vapour pressure that enable 75 per cent of the fuel to be vaporized at manifold at full throttle.

Solution: Given air fuel ratio = 15:1; 75 per cent of ethyl alcohol is vaporised at the manifold at full throttle

$$\text{By weight, air fuel ratio} = \frac{(P_{atm} - P_p)\text{ molecular weight of air}}{\text{molecular weight of fuel } \times P_p}$$

Where P_{atm} = atmospheric pressure; P_p = partial vapour pressure

The chemical formula of ethyl alcohol is C_2H_5OH and the molecular weight is 46 and that of air is 29.2.

$$\text{Again, air fuel ratio} = \frac{\text{Air}}{\text{Fuel } \times \text{perc entage of fuel vapoured}} \times 100 = \frac{15}{1} \times \frac{100}{75} = 20:1$$

$$\text{Hence } \frac{20}{1} = \frac{(1.03 - P_p)29.2}{46 \times P_p}$$

So, $P_p = 0.0316$ kg.cm^{-2}

Answer: Partial vapour pressure is 0.0316 kg.cm^{-2}

Example 12: A four-cylinder, four-stroke compression ignition engine has piston stroke of 10.5 cm and cylinder bore of 11 cm. At a mean piston speed of 7 m.sec^{-1}, the developed brake mean effective pressure is 650 kPa. Calculate the brake power in kW developed by the engine. ***(GATE-2013)***

Solution: Given, Four cylinder four stroke engine; Piston stroke L = 10.5 cm; Cylinder bore D = 11 cm; Piston speed v = 7 m.sec^{-1}; Mean effective pressure p_{mean} = 650 kPa

Here LN/2 = v (piston speed in m.sec^{-1}) where L = Piston stroke, m and N = RPM of engine

$$\text{Power} = p_m \times A \times \frac{v}{2} \times \text{no of cylinder} = 650 \times \frac{\pi}{4} \times 0.11^2 \times \frac{7}{2} \times 4 = 86.48 \text{ kW}$$

Answer: Power developed by the engine is 86.48 kW.

Example 13: The brake power of a four-cylinder engine is 30 kW with all cylinders firing and 20 kW with any one cylinder cut. Find mechanical efficiency of the engine. ***(GATE-2014)***

Solution: Given, 4-cylinder engine; Brake power BP = 30 kW; When any cylinder is cut, the power becomes 20 kW.

In one cylinder, the indicated power = (30 – 20) kW

Hence indicated power for all the cylinders in the engine, IP = 4(30 – 20) = 40 kW

$$\text{So mechanical efficiency} = \frac{BP}{IP} \times 100 = \frac{30}{40} \times 100 = 75 \text{ percent}$$

Answer: Mechanical efficiency of the engine is 75 percent.

Example 14: A four stroke four-cylinder diesel engine running at 2000 rpm develops brake power of 60 kW and the fuel consumption is 0.30 kg kW^{-1} h^{-1}. The engine has a bore of 120 mm and stroke of 100 mm. If the air-fuel ratio is 15:1 and air density is 1.15 kg m^{-3}, calculate the volumetric efficiency of the engine.

(GATE-2014)

Solution: Given, four-stroke four-cylinder engine; Engine speed N = 2000 rpm; Engine power P = 60 kW; Fuel consumption f = 0.30 kg kW^{-1} h^{-1}; Air-fuel ratio = 15:1; Air density is 1.15 kg m^{-3}; Bore D = 120 mm = 0.12 m; Stroke L = 100 mm = 0.1 m

$$\text{Amount of fuel consumed per stroke} = \frac{f \times P}{\frac{N}{2}} = \frac{\left(\frac{0.3}{60} \times 60\right)}{\left(\frac{2000}{2}\right)} = 3 \times 10^{-4} \text{ kg}$$

So, amount of air enters $15 \times 3 \times 10^{-4} = 45 \times 10^{-4}$ kg

$$\text{Volume per stroke} = \text{number of cylinder} \times \frac{\pi}{4} D^2 L = 4 \times \frac{\pi}{4} \times 0.12^2 \times 0.1 = 4.52 \times 10^{-3} \text{ m}^3$$

$$\text{So, density of air inside cylinder} = \frac{45 \times 10^{-4} \text{kg}}{4.52 \times 10^{-3} \text{m}^3} = 0.99 \text{ kg.m}^{-3}$$

$$\text{So volumetric efficiency} = \frac{\text{Density of air inside cylinder}}{\text{Density of air}} = \frac{0.99}{1.15} = 0.865 = 86.5 \text{ per cent}$$

Answer: Volumetric efficiency is 86.5 per cent.

Example 15: A single cylinder four stroke diesel engine of a tractor is developing a mean effective pressure of 15 bar. It develops a 35 kW power at a speed of 2200 rpm. Assuming the stroke bore ratio of 1.25, determine the stroke and bore.

Solution: Given, single cylinder four stroke diesel engine; Mean effective pressure p_{mean} = 15 bar = 1500 kPa; Power developed P = 35 kW; Speed = 2200 rpm; Stroke bore ratio = 1.25; Where L = stroke and d = bore

$$\text{So, } P = \frac{p_{mean} LAN}{60 \times 2} = \frac{1500 \times 1.25d \times \pi \times d^2 \times 2200}{4 \times 60 \times 2} = 26998.06 \, d^3$$

But given P = 35 kW, so on solving d = 0.109 m = 109 mm

So stroke = 1.25 × 109 = 136 mm

Answer: Bore is 109 mm and stroke is 136 mm.

Example 16: A single cylinder four stroke diesel engine of a tractor has 1050 cc stroke volume and is tested 30 revolutions per second. It has a braking torque of 70 N.m and a mean effective pressure of 1500 kPa. Calculate the brake power and mechanical efficiency.

Solution: Given, swept volume = 1050 cc; Engine speed = 30 rps; Brake torque = 70 N.m; Mean effective pressure p_{mean} = 1500 kPa

Brake power BP = $2\pi NT = 2\times\pi\times30\times70$ = 13194 W = 13.194 kW;

Indicated power = $p_{mean} V_s N = 1500 \times 1050 \times 10^{-6} \times 30 = 47.25$ kW

$$\text{So, mechanical efficiency} = \eta_m = \frac{BP}{IP} \times 100 = \frac{13.194}{47.25} \times 100 = 27.92 \text{ per cent}$$

Answer: Brake power is 13.194 kW and mechanical efficiency is 27.29 per cent.

THERMODYNAMIC PRINCIPLES OF IC ENGINES

Example 17: An gas engine with a compression ratio of 5:1 has pressure at the end of compression is 600 kPa above atmospheric. The pressure and temperature at 0.3 percent of the stroke were atmospheric and 87°C, respectively. Calculate the temperature at the end of compression and the index n for the compression stroke

Solution: Given P_2 = 600 kPa above atmospheric = 700 kPa(As atmospheric pressure is 100 kPa)

Let pressure, volume and temperature of gas at 0.3 percent of stroke be P_1, V_1 and T_1 respectively and that at the end of compression be P_2, V_2 and T_2, respectively.

Given compression ratio = 5:1

$$\frac{V_1}{V_2} = 5$$

If $V_2 = 1\text{m}^3$, then $V_1 = 5\text{ m}^3$

So, stroke volume = 5 – 1 = 4 m^3

Hence $V_1 = 5 - 0.3 \times 4 = 3.8\text{ m}^3$

And $T_1 = 273 + 87 = 360$ K

From gas law
$$\frac{P_1V_1}{T_1} = \frac{P_2V_2}{T_2}$$

Or,
$$T_2 = \frac{P_2V_2}{P_1V_1}T_1 = \frac{700 \times 1}{100 \times 3.8} \times 360 = 663.15 \text{ K}$$

Again
$$P_1V_1{}^n = P_2V_2{}^n$$

$$n = \frac{1}{\log\frac{P_1V_1}{P_2V_2}} = \frac{1}{\log\frac{100 \times 3.8}{700 \times 1}} = 1.292$$

Answer: Temperature is 663.15 K and index of compression is 1.292

Example 18: A certain volume of air is compressed from an initial condition of 1 atmosphere and 85°C through a volume ratio of 5. After heat addition at constant volume, the pressure is 30 atmospheres. True specific heat at constant volume is given by $0.18 + 12.5 \times 10^{-9}T^2$, where T is absolute temperature. Law of compression $PV^{1.3}= C$ is followed. Calculate heat energy supplied per kg of working fluid.

Solution:

If the gas obeys perfect gas laws PV = WRT (1)

Again $PV^n = C$ (2)

Hence from eqn. 1 and 2

$$\frac{WRT}{V}V^n = C; \text{ Or, } TV^{n-1} = K; \text{ Where } K = \frac{C}{RW}; \text{ So, } T_1V_1{}^{n-1} = T_2V_2{}^{n-1}$$

Or,
$$T_2 = T_1(\frac{V_1}{V_2})^{n-1}$$

$$\frac{V_1}{V_2} = \text{compression ratio} = 5 \text{ (given)}$$

$$T_1 = 273 + 85 = 358 \text{ K}$$

So,
$$T_2 = 358 \times 5^{1.3-1} = 580.19 \text{ K}$$

From eqn. 2 $P_2 = P_1\left(\frac{V_1}{V_2}\right)^n = 1 \times 5^{1.3} = 8.10 \text{ atm}$

From eqn. 1,
$$\frac{P_2V_2}{T_2} = \frac{P_3V_3}{T_3}$$

Or, $$T_3 = \frac{P_3V_3}{P_2V_2}T_2 \quad \ldots (3)$$

Since heat is added at constant volume $V_2 = V_3$

So, from eqn. 3 = $T_3 = \frac{P_3}{P_2}T_2 = \frac{30}{8.10} \times 580.19 = 2148.85$ K

Hence the heat supplied per kg of working fluid = $C_p dT$

$$= \int_{T_2}^{T_3} (0.17 + 12.3 \times 10^{-9}T^2)dT$$

$$= \left[0.17T + \frac{12.3 \times 10^{-9}T^3}{3}\right]_{580}^{2148} = 407.03 \text{ kCal. kg}^{-1}$$

Answer: Energy supplied per kg of fluid $= 407.03$ kCal. kg^{-1}

Example 19: Air having volume of 275 litres at a pressure of 150 kPa and temperature of 18°C is compressed to 90 litres. Find the temperature, pressure and work done during compression if the compression is adiabatic, isothermal and at constant pressure.

Solution: Given P_1 = 150 kPa; V_1 = 275 litres = 0.275 m^3; T_1 = 18°C = 273+18 =291 K; V_2 = 90 litres = 0.090 m^3

a) For isothermal process = T_2 = 291 K; and $P_1V_1 = P_2V_2$

Hence $P_2 = \frac{P_1V_1}{V_2} = \frac{1 \times 0.275}{0.09} = 3.05$ kg. cm^{-2}

$$\text{Work done} = \int_1^2 Pdv = RT_1 \times 2.303 \times \log_{10}\frac{V_2}{V_1} = 2.303 \times 8.314 \times 288 \log_{10}\frac{0.09}{0.275}$$

$$= -26747.97 \text{ N.m}$$

b) Adiabatic compression $K = \frac{C_p}{C_v} = \frac{0.238}{0.17} = 1.4$

For adiabatic process $P_1V_1^{1.4} = P_2V_2^{1.4}$

Or, $$P_2 = P_1(\frac{V_1}{V_2})^{1.4} = 1(\frac{0.275}{0.09})^{1.4} = 4.78\frac{\text{kg}}{\text{cm}^2}$$

And $$T_2 = T_1\left(\frac{P_2}{P_1}\right)^{\frac{n-1}{n}} = 291(4.78)^{\frac{1.4-1}{1.4}} = 454.91\ \text{K}$$

Work done $$= \frac{R(T_1 - T_2)}{n-1} = \frac{8.314(291 - 454.91)}{1.4-1} = -3406.86\ \text{N.m}$$

c) Constant pressure process

Here $P_1 = P_2 = 150$ kPa

$$\frac{V_1}{T_1} = \frac{V_2}{T_2}$$

Or, $$T_2 = T_1\frac{V_2}{V_1} = 291 \times \frac{0.09}{0.275} = 95.236\ \text{K}$$

Work done$$= \int_1^2 P\,dv = P(V_2 - V_1) = 15 \times 10^4(0.09 - 0.275) = -27750\ \text{N.m}$$

Example 20: A diesel tractor starts at –25°C by using a fuel having auto ignition temperature 345°C. Assuming brake specific fuel consumption to be 0.270 kg. $kW^{-1}hr^{-1}$, calculate minimum compression ratio, thermal efficiency and power developed per litre of fuel.

Solution: Assume the compression to be adiabatic, *i.e.* PV^k = constant

Where P is pressure, V is volume and k is specific heat ratio $= \frac{C_p}{C_v}$

For air, k can be assumed to be 1.4

Hence $$P_1V_1^{\,k} = P_2V_2^{\,k} \quad \text{... (1)}$$

From universal gas law $$PV = nRT \quad \text{... (2)}$$

Where R is gas constant and T is the temperature and n are no of moles(1mole)

Putting the value of V from eqn 1 in eqn 2 we get

$$P_1\left(\frac{RT_1}{P_1}\right)^k = P_2\left(\frac{RT_2}{P_2}\right)^k$$

Or, $$T_2 = T_1\left(\frac{P_2}{P_1}\right)^{\frac{k-1}{k}} \quad \text{... (3)}$$

In the present case, $T_2 = 345° + 273 = 618$ K

And $T_1 = 273° - 25° = 248$

Putting the values of T_1, T_2 and k in equation 3, we get $\frac{P_2}{P_1} = 24.42$

So, the compression ratio will be 24.42

Let the calorific value of fuel be 45 MJ.kg^{-1} and specific gravity be 0.78

Hence heat input for each kW-hr = 0.270 × 45 = 12.15 MJ

1kWh = 860 kCal = 3.6 MJ

Now heat output per kW-hr = 860 kCal = 3.54 MJ

$$\text{Hence thermal efficiency} = \frac{3.543}{12.15} \times 100 = 29.1 \text{ per cent.}$$

As specific gravity of fuel is 0.78 *i.e.* 1 lit of fuel is weighing 0.78 kg.

As 0.270 kg develop 1 kW-hr, from 1 lit fuel, energy developed will be 2.88kW.h

Answer: Minimum compression ratio = 24.42; Thermal efficiency = 29.1 per cent; Kilowatt-hr power developed per litre of fuel = 2.88 kW-hr

Example 21 Air at a pressure of 1.2 atm is compressed from 0.3 m^3 to 0.03 m^3 at a temperature of 18°C. During compression it obeys $PV^{1.4} = C$. Assuming $C_p = 0.24$ and $C_v = 0.17$, find the work done in compression, the change in internal energy and the heat exchanged to the walls per kg of air compressed.

Solution: For adiabatic compression, $PV^n = C$ ... (1)

And $PV = WRT$ ... (2)

Combining eq. 1 and 2

$$\frac{WRT}{V} V^n = C \quad \text{Or } T\,V^{n-1} = \frac{C}{WR} = \text{Say } C_1 \text{ Or } \frac{T_2}{T_1} = \left(\frac{V_1}{V_2}\right)^{n-1} \text{ Or}$$

$$T_2 = T_1 \left(\frac{V_1}{V_2}\right)^{n-1} = (273 + 18)\left(\frac{0.3}{0.03}\right)^{1.4-1} = 730.95 \text{ K}$$

$$\text{Again} P\,V^{1.4} = C \qquad \text{Or } P_1 {V_1}^{1.4} = P_2 {V_2}^{1.4}$$

Hence $$P_2 = P_1\left(\frac{V_1}{V_2}\right)^{1.4} = 1.2 \times \left(\frac{0.3}{0.03}\right)^{1.4} = 30.14 \text{ atm}$$

Now work done in compression, $W_1 = \int_{V_1}^{V_2} P\, dv$ ... (3)

From (1) $P = \frac{C}{V^n}$; Putting the value of P in eq. 3

$$W = \int_{V_1}^{V_2} \frac{C}{V^n} dv = \left[\frac{CV^{1-n}}{1-n}\right]_{V_1}^{V_2} = \frac{C}{1-n}[V_2^{\ 1-n} - V_1^{\ 1-n}]$$

$$= \frac{1}{1-n}[P_2V_2^{\ n}V_2^{\ 1-n} - P_1V_1^{\ n}V_1^{\ 1-n}] = \frac{1}{1-n}[P_2V_2 - P_1V_1] = \frac{P_1V_1 - P_2V_2}{n-1} \quad (4)$$

Work done in compression, W from eq. 4

$$= \frac{1.2 \text{ x } 10^4 \text{ x } 0.3 - 30.14 \text{ x } 0.03 \text{ x } 10^4}{1.4 - 1} = -13605 \text{ Kg.m} = -133.465 \text{ kJ}$$

Change in internal energy per kg air (du)

$$= C_v d_1 = 0.17\,(T_2 - T_1) = 0.17\,(730.95 - 291)\text{kCal} = 74.8 \text{ kCal.kg}^{-1}\text{K}^{-1}$$

$$= 312.66 \text{ kJ.kg}^{-1}\text{K}^{-1} \text{ air}$$

Heat exchange to the wall, dQ per kg of air

$$= du + \delta P\, dv = 312.66 - 133.465 = 179.19 \text{ kJ.kg}^{-1} \text{of air}$$

Answers: Work done = –133.465 kJ; Change in internal energy = 312.66 kJ.kg^{-1}; Heat exchanged = 179.19kJ.kg^{-1}

Example 22: The gas constant for hydrogen gas is 420 kg.m.moles^{-1}.K^{-1} and specific heat at constant pressure is 3.4. This gas of volume 90 litres at a pressure of 1.5 atm and temperature of 15°C is compressed to 15 atm adiabatically and then expanded isothermally to its original volume. Find the amount of heat required to be added during isothermal expansion and amount of heat to be removed from the gases after expansion to get back the gas to its original volume. Also find the final pressure of the gas.

Solution: Given R = 420 Kg.m.kgf^{-1}moles^{-1}K^{-1}; C_p = 3.4; V_1 = 0.09 m^{-3}; P_1 = 1.5 atm; P_2 = 15 atm

For adiabatic process the work done, W is given by

$$W = \frac{P_1V_1 - P_2V_2}{n-1} = \frac{P_1V_1}{n-1}(1 - \frac{P_2V_2}{P_1V_1})$$

For adiabatic process $P_1V_1{}^k = P_2V_2{}^k$; $\frac{V_2}{V_1} = (\frac{P_1}{P_2})^{\frac{1}{n}}$;

$$\text{Now}\quad W = \frac{P_1V_1}{n-1}\left(1 - \frac{P_2V_2}{P_1V_1}\right) = \frac{P_1V_1}{n-1}\left(1 - \frac{P_2}{P_1} \times \left(\frac{P_1}{P_2}\right)^{\frac{1}{n}}\right) = \frac{P_1V_1}{n-1}\left(1 - \left(\frac{P_1}{P_2}\right)^{\frac{n-1}{n}}\right)$$

$$\text{Again } C_p - C_V = \frac{R}{J};\ \text{Or, } 1 - \frac{C_V}{C_p} = \frac{R}{JC_p};\ \text{Or, } 1 - \frac{1}{K} = \frac{R}{JC_p} = \frac{420}{3.4 \times 427} = 0.289$$

Or, K = 1.407, Hence the law of adiabatic process is given by

So, $P_1V_1{}^{1.407} = P_2V_2{}^{1.407}$ Or, $1.5 \times 0.09^{1.407} = 15 \times V_2{}^{1.407}$

Or, $V_2 = 0.0175\ m^3$

Process 2 – 3 is isothermal process. Hence $P_2V_2 = P_3V_3$

$$\text{So, } P_3 = \frac{P_2V_2}{V_3} = \frac{15 \times 0.0175}{0.09} = 2.91\text{kg.cm}^{-2}$$

Heat to be added during process 2 – 3 is given by

$$Q_{2-3} = du + \int_2^3 Pdv = C_v dT + 2.303RT\log_{10}\frac{V_3}{V_2} = C_v(T_3 - T_2) + 2.303RT\log_{10}\frac{V_3}{V_2}$$

$$\text{Now } T_2 = T_1(\frac{V_1}{V_2})^{n-1} = (273 + 15)\left(\frac{0.09}{0.0175}\right)^{1.407-1} = 560.85\ \text{K}$$

Since the process is isothermal $T_2 = T_3 = 560.85K$

$$Q_{2-3} = C_v(T_3 - T_2) + 2.303RT\log_{10}\frac{V_3}{V_2}$$

$$= 2.303(0) + 2.303 \times \frac{420}{427}560.85\log_{10}\frac{0.09}{0.0175} = 903.55\ \text{kCal.kg}^{-1}$$

Hence heat to be extracted from the gas to bring to state 1,

$$Q_3 = C_v dT = 2.34(T_1 - T_3) = 2.34(288 - 560.85) = -638.47\ \text{kCal.kg}^{-1}\text{of air}$$

Answer: Final pressure of gas P_3 = 2.91 kg. cm^{-2}; Heat added = 903.55 $kCal.kg^{-1}$; Heat extracted = 638.47 $kCal.kg^{-1}$

Example 23: The compression curve for a tractor engine follows $PV^{1.3}$ = C where P = pressure and V = volume of the gas and C is a constant. At two points *i.e.* at 25 per cent of the stroke and 75 per cent of the stroke, pressure is 200 kPa and 500 kPa. Calculate compression ratio and air standard efficiency for this cycle.

Solution: Given, engine working on 4 stroke cycle principle $PV^{1.3}$ = C;

Two points on the curves are 25 per cent and 75 per cent on the stroke

Pressures are P_{25} = 200 kPa and P_{75} = 500 kPa

Compression ration $r = \frac{V_1}{V_2}$

Swept volume, $V_s = V_1 - V_2 = (r - 1)V_1 = (r - 1)V_c$

At 25 per cent of the compression stroke, the volume of gas in the cylinder

$$V_{25} = V_c + \frac{3}{4}V_c(r-1) = \frac{V_c}{4}(1+3r)$$

At 75 per cent of the compression stroke, the volume of gas in the cylinder

$$V_{75} = V_c + \frac{1}{4}V_c(r-1) = \frac{V_c}{4}(3+r)$$

Hence the adiabatic compression, $P_{25}(\frac{V_c}{4}(1+3r))^{1.3} = P_{75}\left(\frac{V_c}{4}(3+r)\right)^{1.3}$

Or, r = 5.7 and air standard $\eta = 1 - \frac{1}{r^{n-1}} = 1 - \frac{1}{5.7^{1.3-1}} = 0.406 = 40.6$ per cent

Answer: Compression ratio is 5.7; Air standard efficiency is 40.6 per cent

Example 24: In an engine cycle, the compression starts at 1.5 atm and 17°C. It has a compression ratio 15:1. The bore and stroke are 130 mm and 160 mm. If constant pressure heat addition ends at 5 per cent of the stroke, then find out the work done and air standard efficiency. Also find out the increase or decrease in efficiency if cut off ends at 20 per cent of the stroke.

Solution: Given P_1 = 1.5 atm T_1 = 17°C = 17 + 273 = 290 K; Compression ratio = 15:1 So, $r = \frac{V_1}{V_2}$; Bore r = 130 mm and stroke L = 160 mm

Swept volume $V_s = V_1 - V_2 = \frac{\pi}{4} r^2 L = \frac{\pi}{4} 13^2 \times 16 = 2123.71$ cc

So, by using the value of compression ratio and V_s, $V_2 = 151.69$ cc and $V_1 = 2275.41$cc

Again, given cut off takes place at 5 per cent of the swept volume. So,

$$V_3 - V_2 = 0.05V_s = 0.05 \times 2123.71 = 106.18 \text{ cc}$$

Or, $$V_3 = V_2 + 106.18 = 151.69 + 106.18 = 257.87 \text{ cc}$$

Cut off ratio $$\rho = \frac{V_3}{V_2} = \frac{257.87}{151.69} = 1.6$$

Compression and expansion takes place adiabatically

So $$T_2 {V_2}^{k-1} = T_1 {V_1}^{k-1}$$

Or, $$T_2 = T_1 (\frac{V_1}{V_2})^{k-1} = T_1 (r)^{k-1} = 290 \times 15^{1.4-1} = 856.72 \text{ K}$$

Similarly, $$T_3 = \rho \times T_2 = 1.75 \times 856.72 = 1499.24 \text{ K}$$

And $$T_4 = T_3 (\frac{V_3}{V_4})^{k-1} = T_3 (\frac{V_3}{V_1})^{k-1} = 1499.24 (\frac{257.87}{2275.41})^{1.4-1} = 627.49 \text{ K}$$

Heat supplied $Q_1 = C_p (T_3 - T_2) = 0.24 \times (1499.24 - 856.72) = 154.20 \text{ kCal. kg}^{-1}$

Heat rejected $Q_2 = C_v (T_4 - T_1) = 0.17 \times (627.49 - 290) = 57.37 \text{kCal. kg}^{-1}$

Net work done $W = Q_1 - Q_2 = 154.20 - 57.37 = 96.83 \text{ kCal. kg}^{-1}$

So, air standard efficiency if cut off takes place at 5 per cent of the stroke is given by

$$\eta = 1 - \frac{1}{r^{k-1}} \left(\frac{\rho^k - 1}{k(\rho - 1)} \right) = 1 - \frac{1}{15^{1.4-1}} \left(\frac{1.75^{1.4} - 1}{1.4(1.75 - 1)} \right) = 61.66 \text{ per cent}$$

If cut off takes place at 20 per cent of the stroke, then

$$V_3 = V_2 + 0.20V_s = 151.69 + 0.2 \times 2123.71 = 576.43 \text{ cc}$$

So, cut off ratio $$\rho = \frac{V_3}{V_2} = \frac{576.43}{151.69} = 3.8$$

So, air standard efficiency,

$$\eta = 1 - \frac{1}{r^{k-1}}\left(\frac{\rho^k - 1}{k(\rho - 1)}\right) = 1 - \frac{1}{15^{1.4-1}}\left(\frac{3.8^{1.4} - 1}{1.4(3.8 - 1)}\right) = 52.66 \text{ per cent}$$

Answers: Net work done is 96.83 kCal.kg^{-1}; Air standard efficiency for 5 per cent cut off is 61.66 per cent and 20 per cent cut off is 52.66 per cent

Example 25: The air fuel mixture enters the diesel engine at a temperature of 85°C and the maximum pressure developed by the engine is 3.6 MPa. The engine has a compression ratio of 14:1. Temperature at the beginning of the expansion is 1850°C. Calculate temperature at the end of compression and expansion, heat supplied and rejected per kg of fluid and thermal efficiency. Assume remaining necessary data.

Solution: Given T_1 = 85 + 273 = 358 K; Compression ratio = 14:1 *i.e.* $r = \frac{V_1}{V_2} = 14$; T_3 = 1850 + 273 = 2123 K; Assuming adiabatic constant K = 1.4, C_p = 0.238 and C_v = 0.17

$$T_2 = T_1\left(\frac{V_1}{V_2}\right)^{k-1} = 358 \times 14^{1.4-1} = 1028.80 \text{ K}$$

Cut off ratio $\rho = \frac{V_3}{V_2} = \frac{T_3}{T_2} = \frac{2123}{1028.80} = 2.07$

Hence $T_4 = T_1\left(\frac{V_3}{V_2}\right) = 358 \times \left(\frac{2123}{1028.80}\right) = 741.06 \text{ K}$

Heat supplied $= C_p(T_3 - T_2) = 0.238(2123 - 1028.80) = 260.41 \text{ kCal. kg}^{-1}$

Heat rejected $= C_v(T_4 - T_1) = 0.17(738.75 - 358) = 64.72 \text{ kCal. kg}^{-1}$

Air standard efficiency

$$\eta = 1 - \frac{1}{r^{k-1}}\frac{\rho^k - 1}{k(\rho - 1)} = 1 - \frac{1}{14^{1.4-1}}\left(\frac{2.07^{1.4} - 1}{1.4(2.07 - 1)}\right) = 58.90 \text{ per cent}$$

Answer: Temperature at the end compression is 1028.80 K and at the end of expansion 741.06 K; Heat supplied per kg 260.41 kCal; Hear rejected per kg 64.72 kCal; Thermal efficiency or air standard efficiency 58.90 per cent

Example 26: A mixed cycle engine has a swept volume of 0.013 m^3 and clearance volume of 0.0013 m^3. The mean effective pressure is 700 kPa. Inlet temperature of the mixture is around 85°C and pressure at the end of compression is around 2.5 MPa. The fuel injection ends at 6 per cent of the stroke. Maximum limiting pressure is around 4 MPa. Calculate the amount of heat supplied per cycle and its efficiency.

Solution: Given swept volume V_s = 0.013 m^3; Clearance volume V_c = 0.0013 m^3; Mean effective pressure P = 700 kPa; Inlet temperature T_1 = 273 + 85 = 358 K; Injection ends at 6 per cent of the stroke; P_1 = 4 MPa

Volume of injection $V_4 = 0.06V_s + V_c = 0.06 \times 0.013 + 0.0013 = 0.00208\ m^3$

$$V_1 = V_s + V_c = 0.013 + 0.0013 = 0.0143\ m^3 \text{ and } V_2 = V_c = 0.0013\ m^3$$

Compression ratio $r = \frac{V_1}{V_2} = \frac{0.0143}{0.0013} = 11.46$, Cut off ratio $\rho = \frac{V_4}{V_3} = \frac{0.00208}{0.0013} = 1.6$

The compression and expansion follow the adiabatic process.

Hence, $$T_2 = T_1(\frac{V_1}{V_2})^{k-1} = 358 \times 11.46^{1.4-1} = 949.63\ K$$

Again $$T_3 = T_2\frac{P_3}{P_2} = 949.63 \times \frac{4}{2.5} = 1519.41\ K$$

Again $$T_4 = T_3\frac{V_4}{V_3} = 1519.41 \times 1.6 = 2431.06\ K$$

Heat supplied per kg of air, $Q_1 = C_v(T_3 - T_2) + C_p(T_4 - T_3)$

$= 0.169(1519.41 - 949.63) + 0.238(2431.06 - 1519.41) = 313.26\ kCal.kg^{-1}$

$$T_5 = \frac{T_4}{\left(\frac{V_4}{V_3}\right)^{k-1}} = \frac{2431.06}{(1.6)^{1.4-1}} = 2014.40\ K$$

Heat rejected $Q_2 = C_v(T_5 - T_1) = 0.17 \times (2014.40 - 358) = 281.58\ kCal.kg^{-1}$

So, efficiency

$$= \frac{\text{work done}}{\text{heat supplied}} \times 100 = \frac{Q_2 - Q_1}{Q_1} \times 100 = \frac{313.26 - 281.58}{313.26} \times 100 = 10.10 \text{ per cent}$$

Answer: Heat supplied is 313.26 $kCal.kg^{-1}$; Thermal efficiency is 10.10 per cent

Example 27: Air enters into an engine cylinder at a pressure of 100 kPa and temperature of 300 K in the suction stroke of an ideal Otto cycle. The cylinder clearance volume is 600 cm^3 and compression ratio is 8:1. The expansion stroke of the cycle is polytropic in nature. The pressure at the beginning of the expansion stroke is 10 MPa and the temperature at the end of the expansion stroke is 1800 K. Find the polytropic exponent of the expansion stroke. ***(GATE-2018)***

Solution:

Given, ideal Otto cycle; Pressure at the start of suction P_1 = 100 kPa; Temperature at the start of suction T_1 = 300 K; Clearance volume V_c = 600 cc; Compression ratio CR = 8:1; Expansion stroke is polytropic; Pressure at the start of expansion P_3 = 10 MPa = 10000 kPa; Temperature at the end of expansion T_4 = 1800 K

Let volume at the start and at the end of suction is V_1 and V_2 and swept volume is V_s

So, $V_2 = V_c = 600$ cc, So, $\frac{V_1}{V_2} = \frac{V_2 + V_s}{V_2} = 8$

On solving $V_s = 7 \times V_2 = 7 \times 600 = 4200$ cc

So, $V_1 = 8 \times V_2 = 8 \times 600 = 4800$ cc

Let, volume at the start of expansion is V_3 and temperature at the end of expansion is T_4

For ideal otto cycle $V_2 = V_2 = 600$ cc and $V_4 = V_1 = 4800$ cc

And $P_4 = P_1\frac{T_4}{T_1} = 100 \times \frac{1800}{300} = 600$ kPa

So, the polytropic exponent for expansion process $n = \frac{\log\frac{P_3}{P_4}}{\log\frac{V_4}{V_3}} = \frac{\log\frac{10000}{600}}{\log\frac{4800}{600}} = 1.35$

Answer: Polytropic exponent for expansion process is 1.35

Example 28: The intake pressure of a diesel engine is 1 bar and pressure at the end of the compression is 34 bar. The adiabatic exponent is 1.3 and the expansion ratio is 7. Find the diesel cycle efficiency. ***(GATE-2016)***

Solution: Given, intake pressure P_1 = 1 bar; Pressure at the end of compression P_2 = 34 bar; Adiabatic exponent k = 1.3; Expansion ratio $\rho = 7$

Compression ratio $r = \left(\frac{P_2}{P_1}\right)^{\frac{1}{k}} = \left(\frac{34}{1}\right)^{\frac{1}{1.3}} = 15.06$

So, efficiency of diesel engine

$$\eta = 1 - \frac{1}{r^{k-1}} \times \frac{\rho^k - 1}{k(\rho - 1)} = 1 - \frac{1}{15.06^{1.3-1}} \times \frac{7^{1.3} - 1}{1.3(7-1)} = 34.36 \text{ per cent}$$

Answer: Efficiency of diesel engine is 34.36 per cent.

Example 29: Air temperature at the beginning of compression of an internal compression engine is 27°C and the engine compression ratio is 16:1. Calculate the temperature of air at the end of compression. ***(GATE-2017)***

Solution: Given, temperature at the beginning of compression T_1 = 27°C = 273 + 27 = 300 K; Compression ratio r = 16

So, temperature at the end of compression

$$T_2 = T_1 r^{\frac{1-k}{k}} - 273 = \left(300 \times 16^{\frac{1.4-1}{1.4}}\right) - 273 = 662.45^0\text{C}$$

Answer: The temperature at the end of compression is 632.2°C

ENGINE BALANCING

Example 30: A single cylinder engine has a reciprocating part of 40 kg and revolving part of 30 kg at crank radius. The speed of the engine is 150 revolutions per minute and stroke is 350 mm. If 60 per cent of the reciprocating part and all the revolving parts are to be balanced, find the balance mass required at a radius of 320 mm and the unbalanced force when the crank has turned 45° from top dead centre.

Solution: Given, mass of reciprocating part m_1 = 40 kg; Mass of revolving part m_2 = 30 kg; Engine speed N = 150 rpm; Stroke L = 350 mm; Balanced portion of reciprocating mass c = 60 per cent = 0.6

Angular speed $$\omega = \frac{2\pi N}{60} = \frac{2\pi \times 150}{60} = 15.7 \text{ rad. sec}^{-1}$$

Crank radius $$r = \frac{L}{2} = \frac{350}{2} = 175 \text{ mm} = 0.175 \text{ m}$$

Mass to be balanced at crank pin m = $m_1c + m_2$ = 40 × 0.6 + 30 = 54 kg

Let the balance mass be m_b which is at a radius of r_b = 320 nm

$$\text{So, } m_b = \frac{mr}{r_b} = \frac{54 \times 175}{320} = 29.53 \text{ kg}$$

Unbalanced force at θ = 45° from top dead centre is

$$\sqrt{[(1-c)m_1r\omega^2\cos\theta]^2+(cm_1r\omega^2\sin\theta)^2}$$

$$=\sqrt{[(1-0.6)\times 40\times 0.175\times 15.7^2\times\cos 45]^2+(0.6\times 40\times 0.175\times 15.7^2\times\sin 45)^2}$$

$$= 880.7 \text{ N}$$

Answer: Balance mass required is 29.53 kg and unbalanced force is 880.7 N.

Example 31: Find out (i) the lag in terms of degrees of crank shaft rotation between injection and starting of the combustion (ii) the percentage of fuel injected during the period of ignition delay (iii) percentage fuel supplied during injection delay and rapid combustion if rapid combustion takes place 5° of crank rotation. When the injection pump of an I.C. engine running at 2000 rpm delivers fuel once for an interval of 20° of crank shaft rotation. The engine has an ignition delay of 0.001 seconds.

Solution: Given engine speed N = 2000 rpm; Fuel delivery interval = 20°

a) Time required by the crank shaft to rotate one degree

$$\frac{60}{\frac{N}{2}\times 720}=\frac{60}{\frac{2000}{2}\times 720}=8.34\times 10^{-5}\text{sec}$$

Hence the crank rotation during the period of ignition delay

$$=\frac{1}{8.34\times 10^{-5}}\times 0.001 = 12°$$

b) Time of injection

$$=\frac{20\times 60}{720\times\frac{2000}{2}}=1.67\times 10^{-3}\text{sec}$$

So, percentage of fuel injected during delay

$$=\frac{0.001}{1.67\times 10^{-3}}\times 100 = 60 \text{ per cent}$$

c) Percentage of fuel injected during delay and rapid combustion

$$=60+\left(\frac{5}{20}\times 100\right)=85 \text{ per cent}$$

Example 32: There are two engines, one is having a bore of 12 cm running at 1800 rpm and other is having a bore of 20 cm running at 1200 rpm. Fuels of different grades are used in the engines such that when combustion starts, the pressure is 3.0 MPa in both cases. In both engines the injections are so set that the oil is injected for duration of 30° of crank rotation. Determine the pressure in the smaller engine at the end of rapid combustion if the corresponding pressure is 5.64 MPa in large engine. Assuming the fuels used in both engines with ignition lag of 0.0014 seconds and rapid combustion period is 4°.

Solution: Given, bore d_1= 12 cm and d_2 = 20 cm; Engine speed N_1 = 1800 rpm and N_2 = 1200 rpm; For both engines the fuel injection duration = 30°of crank rotation; Pressure at the starting of fuel combustion = 3.0 MPa; Ignition lag = 0.0014 sec; Rapid combustion period = 4°; Pressure at the end of combustion = 5.64 MPa

Injection time for small engine $T_1 = \frac{30}{360} \times \frac{60}{1800} = 2.78 \times 10^{-3}$ sec

So, percentage of fuel injected during delay and rapid combustion

$$= \left(\frac{0.0014}{2.78 \times 10^{-3}} + \frac{4}{30}\right) \times 100 = 63.69 \text{ per cent}$$

Injection time for the large engine $T_2 = \frac{30}{360} \times \frac{60}{1200} = 4.17 \times 10^{-3}$ sec

So, percentage of fuel injected during delay and rapid combustion

$$= \left(\frac{0.0014}{4.17 \times 10^{-3}} + \frac{4}{30}\right) \times 100 = 46.90 \text{ per cent}$$

Hence the pressure in smaller engine is $= \frac{46.90}{63.69} \times 5.64 = 4.153$ MPa

Answer: The pressure at the end of compression is 4.153 MPa.

Example 33: Figure below shows the valve timing to tractor. Engine has 3 cylinders with stroke 8.9 cm, bore 7 cm. If the piston reciprocates 2000 times/minute and the length of the connecting rod is 30 cm, find out

(a) location of the valves (cm)

(b) at what percentage of displacement, the exhaust valve remains open.

(c) what is the time in second, the intake valve and the exhaust valves will remain open.

(d) if the cam gear has 48 teeth, find the number of crank shaft gear teeth that will be shifted from I.V.O. to I. V. C. and E. V. O to E. V. C

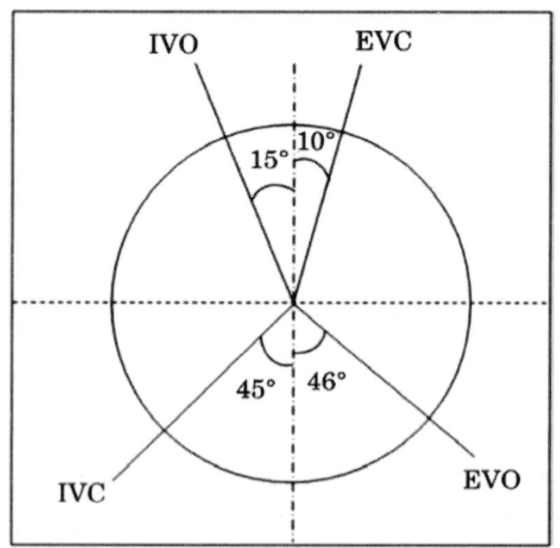

Solution: Given stroke L = 8.9 cm; Bore d = 7 cm; Piston reciprocates 2000 times/min; Length of connecting rod l = 29 cm; Inlet valve opens at θ_{io} = 13° and closes at θ_{ic} = 43°; Exhaust valve opens at θ_{eo} = 46° and closes at θ_{ec} = 10°

a) Location of inlet valve open from TDC = 4.45 – 4.45 cos13° = 0.114 cm
   Location of inlet valve open from BDC = 4.45 – 4.45 cos43° = 1.19 cm
   Location of outlet valve open from BDC = 4.45 – 4.45 cos46° = 1.35 cm
   Location of outlet valve closes from TDC = 4.45 – 4.45 cos10° = 0.067 cm

b) It is seen from fig that exhaust valve remain open for 46° + 180° + 10° = 236°

   For a complete cycle the total revolution = 720°

   So, the percentage of displacement for which the exhaust valve remains open

$$= \frac{236}{720} \times 100 = 32.8 \; per \; cent$$

c) Exhaust valve remains open for 236°

   inlet valve remains open for 15° + 180° + 43° = 238°

   So, time duration for which exhaust valve remain open

$$= \frac{236 \times 60}{720 \times 2000} = 9.83 \text{ milisecond}$$

   And that for inlet valve is $= \frac{238 \times 60}{720 \times 2000} = 9.91$ milisecond

d) given no of teeth in crank is 24 *i.e.* for 720° revolutions no of teeth covered is 24

   So, for exhaust valve $= \frac{236}{720} \times 24 = 7.86$ no of teeth is covered

   And for inlet valve $= \frac{238}{720} \times 24 = 7.93$ no of teeth is covered

FUEL AND COMBUSTION

Example 34: The analysis of hydrocarbon fuel gives 84 per cent carbon and 16 per cent hydrogen by weight. What would be the respective molecular formula?

Solution: Given, 1 kg of hydrocarbon contain 0.84 kg of carbon and 0.16 kg of hydrogen

Molecular weight of carbon is 12 and that of hydrogen is 1

So no. of moles of hydrogen $= \frac{0.84}{12} = 0.07$

No of moles of hydrogen $= \frac{0.16}{1} = 0.16$ mole

Hence no of C = 0.16/0.16 = 1 and H = 0.16/0.07 = 2.28

Hence molecular formula is $C_1H_{2.28}$or $C_{55}H_{114}$

Example 35: The carbon and hydrogen content of the fuel are 84 per cent and 4 per cent respectively by weight. The remaining portion is ash. Find the weight of necessary air for complete combustion of one kg of fuel if calorific value of a fuel is found by experiment to be 18000 Kcal.kg^{-1}.

Solution: After complete combustion of fuel, carbon dioxide, water vapour and sulphur dioxide are formed according to the chemical equation

$$C + O_2 = CO_2 \qquad (1)$$

$$2H_2 + O_2 = 2H_2O \qquad (2)$$

$$S + O_2 = SO_2 \qquad (3)$$

In 1 kg of fuel, carbon is 0.84 kg, hydrogen is 0.04 kg and ash is 0.12 kg. Usually 40 per cent of ash is formed due to sulphur present in the fuel. Hence sulphur content is = 0.4 × 0.12 = 0.048 kg

Eqn. 1 indicates that 12 kg of carbon requires 32 kg of oxygen. So, 0.84 kg carbon requires $\frac{32}{12} \times 0.84 = 2.24$ kg of oxygen

Similarly, from eqn. 2, 0.04 kg of hydrogen requires $\frac{32}{4} \times 0.04 = 0.32$ kg of oxygen

Similarly, 0.048 kg of sulphur requires $\frac{32}{32} \times 0.048 = 0.048$ kg of oxygen

So total amount of oxygen requires = 2.32 + 0.32 + 0.048 = 2.688 kg

Assuming air contains 23 per cent of oxygen by weight, then amount required is calculated by 2.688/0.23 = 11.686 kg.

Answer: Weight of air required for complete combustion is 11.686 kg.

Example 36: A fuel contains 85 per cent carbon and 15 percent hydrogen by weight. Ninety percent of the theoretical air required is supplied while engine is running. If all the hydrogen is burnt and carbon burns to carbon monoxide and carbon dioxide and assuming there is no free carbon left, then analyse the dry exhaust gas by volume.

Solution: Quantity of oxygen required for complete combustion can be calculated from the chemical equations

$$C + O_2 = CO_2 \quad \text{... (1)}$$

$$2H_2 + O_2 = 2H_2O \quad \text{... (2)}$$

Given, fuel contains 85 per cent carbon and 15 per cent hydrogen *i.e.* 100 gm of fuel contain 85 gm carbon and 15 gm hydrogen.

Amount of oxygen required for complete combustion of carbon $\frac{32}{12} \times 85 = 226.67$ kg

Amount of oxygen required for complete combustion of hydrogen $\frac{32}{4} \times 15 = 120$ kg

Hence total amount of oxygen required for complete combustion = 226.67 + 120 = 346.67 kg

Air supplied is 90 per cent of the theoretical

So, oxygen supplied = 346.67 × 0.9 = 312 gm

Since all the hydrogen is burnt, oxygen used to form carbon dioxide and carbon monoxide is given by 312 – 120 = 192 gm

Let x gm of carbon is converted to carbon dioxide. So, 85 – x gm is converted to carbon monoxide.

Oxygen required for x gm of carbon $\frac{32x}{12}$

Oxygen required for 85 – x gm of carbon to convert into carbon monoxide is given by

$$\frac{32(85-x)}{24} \text{ gm, So, } \frac{32x}{12} + \frac{32(85-x)}{24} = 200, \text{ Or, } x = 65 \text{ gms}$$

Hence carbon burns completely are 65 gm and incompletely is 20 gm

Eqn. 1, 2 and 3 shows that the products of combustion are H_2O, CO_2, CO and N_2. Hence dry exhaust gas has CO_2, CO and N_2.

By burning 100 gm of fuel $\frac{85}{12}$ moles of carbon give $\frac{65}{12} = 5.416$ moles of carbon dioxide and $\frac{20}{12} = 1.67$ moles of carbon monoxide.

Air contain 23 per cent of oxygen and 77 per cent of nitrogen by weight. So, for 312 gm of oxygen there will be $\frac{77}{23} \times 312 = 1044.52$ gm of nitrogen *i.e* 37.304 moles

So, in dry exhaust total no of moles of gases = 5.416 + 1.67 + 37.304 = 44.39 moles

Therefore, percentage composition of exhaust gases is

Carbon dioxide $\frac{5.416}{44.39} \times 100 = 12.30$ per cent.

Carbon monoxide $\frac{1.67}{44.39} \times 100 = 3.7$ per cent

Nitrogen $\frac{37.304}{44.39} \times 100 = 84.03$ per cent

Answer: Percentage of carbon dioxide is 12.20 per cent, carbon monoxide is 3.7 per cent and nitrogen is 84.03 per cent.

Example 37: A diesel fuel contains 84 per cent carbon and 16 percent hydrogen. It has an API gravity of 38. While using the fuel in a tractor, 7 litres fuel is consumed in 1.5 hrs. Find out the air fuel ratio, air admitted in the cylinder per minute and fuel water ratio assuming complete combustion of the fuel. The pressure of the mixture at 17°C is atmospheric.

Solution: For complete combustion it is assumed that the whole carbon is converted to carbon dioxide and the hydrogen is converted to water vapour. *i.e.*

$$C + O_2 = CO_2 \qquad (1)$$

$$2H_2 + O_2 = 2H_2O \qquad (2)$$

Given fuel contains 84 per cent carbon and 16 per cent hydrogen *i.e.* 1 kg of fuel contains 0.84 kg of carbon and 0.16 kg of hydrogen.

From eqn. 1, 0.84 kg of carbon for complete combustion requires $\frac{32 \times 0.84}{12} = 2.24$ kg of oxygen

Similarly, from eqn. 2, 0.16 kg of hydrogen requires $\frac{32 \times 0.16}{4} = 1.28$ kg of oxygen

Total oxygen required = 2.24 + 1.28 = 3.52 kg

As air contains 23 per cent of oxygen by weight, the amount of air required

$$= \frac{3.52}{0.23} = 15.30 \text{ kg}$$

Let the density of air is 1.225 kg.m^{-3}. So, for complete combustion of 1 kg of fuel requires

$$\frac{15.30}{1.225} = 12.48\,\text{m}^3\,\text{of air}$$

Let this air is admitted into the cylinder at NTP *i.e.* temperature is 27 °C and pressure is atmospheric.

At NTP the volume of air = $\frac{\text{WRT}}{\text{P}}$

Where, W_2 = amount of air, R = gas constant, T_2 = temperature (273+27=300K), P_2 = pressure (1.03×10^4 kg.cm^{-2})

Given API gravity = 38 = $\frac{141.5}{\text{specific gravity of fuel at } 15.5^0\text{C}} - 131.5 = 38$ Or, specific gravity = 0.834, So, density of fuel is 834 kg.m^{-3}(ρ_1)

From gas law $\frac{P_1}{\rho_1 T_1} = \frac{P_2 V_2}{\rho_2 T_2}$, Where, P_1 and P_2 = pressure, T_1 and T_2 = temperature, ρ_1 and ρ_2 = density

T_1 = 273+15.5 =288.5 K, So, $\rho_2 = \frac{P_2 T_1}{P_1 T_2}\rho_1 = \frac{1.03 \times 288.5}{1.02 \times 300} \times 834 = 809.89\ \text{kg. m}^{-3}$

Hence specific gravity = 0.809

As given, 7 litres of fuel is burnt in 1.5 h *i.e.* 7 litres of fuel is burnt in 90 minutes

Weight of fuel burnt per minute $= \frac{7}{90} \times 0.809 = 0.063\ \text{kg. min}^{-1}$

1 kg of fuel needs 12.48 m^3 of air, So, 0.063 kg of fuel requires 12.48 × 0.063 = 0.786 m^3 per min

For complete combustion, 4 kg of hydrogen forms 36 kg of water vapour. Hence 0.16 kg of hydrogen will form $\frac{36 \times 0.16}{4} = 1.44$ kg of water vapour

Therefore, water ratio by weight = 1:1.44

Answer: Air fuel ratio = 12.48:1; Air admitted per minute = 0.786 m^3.min^{-1}; Fuel water ratio = 1:1.44

Example 38: A diesel fuel has same ignition delay as that of blend of two reference fuel of 56 per cent n-cetane and 44 per cent heptamethylnonane. Find the cetane number of the diesel fuel. ***(GATE-2017)***

Solution: Given, percentage of n-cetane; Percentage of heptamethylnonane = 44 per cent

The reference fuel n-cetane has cetane number 100 per cent and heptamethylnonane has cetane number 15 per cent. Cetane number = 56 × 1 + 44 × 0.15 = 62.6

Answer: Cetane number of the fuel is 62.6.

Example 39: A four-cylinder diesel operated tractor has a cylinder of running at a rpm of 2000. It has an air fuel ratio of 15:1. Assuming coefficient of air admission in the engine is 0.85 and density of air is 1.209 $kg.m^{-3}$, find air and fuel supplied to the engine.

Solution: Given, number of cylinder n = 4; Engine speed N = 2000 rpm; Bore D = 10 cm = 0.1 m; Stroke L = 10 cm = 0.1 m; Air fuel ratio = 15:1; Coefficient of air admission C_a = 0.85; Density of air ρ_{air} = 1.209 $kg.m^{-3}$

Swept volume $$V_s = \frac{\pi}{4}d^2L = \frac{\pi}{4} \times 0.1^2 \times 0.1 = 7.85 \times 10^{-4}\ m^3$$

Amount of air admission in the combustion chamber

$$M_a = \frac{C_a n V_s N}{2 \times \rho_{air} \times 60} = \frac{0.85 \times 4 \times 7.85 \times 10^{-4} \times 2000}{2 \times 1.209 \times 60} = 0.036\ kg.sec^{-1}$$

$$= 2.16 kg.min^{-1}$$

So, mass of fuel consumed $$M_f = \frac{M_a}{\text{air fuel ratio}} = \frac{2.16}{15} = 0.144\ kg.min^{-1}$$

Answer: Mass of air and fuel consumed are 2.16 $kg.min^{-1}$ and 0.144 $kg.min^{-1}$, respectively.

CARBURETOR

Example 40: Calculate indicated power, brake power and mechanical efficiency of the engine. A 4 – stroke, 4-cylinder petrol engine is coupled to a brake dynamometer having an effective radius of 50 cm when running at 1500 rpm. When spark plug of each cylinder was short circuited in turn, the brake loads adjusted to bring the same value of speed, were found to be 84.0 N, 86.5 N, 85 N and 89 N respectively. The brake load was 120 N, at that time the carburettor was adjusted to air fuel ratio of 13:1 to evaporate 75 per cent of fuel.

Solution: Given, 4-cylinder petrol engine coupled to brake dynamometer of radius r = 60 cm running at N = 1500 rpm; Brake load W = 120 N; When spark plug of each cylinder was short circuited in turn, the brake loads are 84.0N, 86.5N, 85 N and 89 N; Air fuel ratio = 13:1; Fuel evaporation percentage = 75 per cent

Brake power of the engine $P = \frac{2\pi NT}{60000} = \frac{2\pi \times 1500 \times 0.5 \times 120}{60000} = 9.42 \text{ kW}$

When 1st cylinder is short circuited, brake power

$$P_1 = \frac{2\pi NT_1}{60000} = \frac{2\pi \times 1500 \times 0.5 \times 84.0}{60000} = 6.59 \text{ kW}$$

When 2nd cylinder is short circuited, brake power

$$P_2 = \frac{2\pi NT_2}{60000} = \frac{2\pi \times 1500 \times 0.5 \times 86.5}{60000} = 6.79 \text{ kW}$$

When 3rd cylinder is short circuited, brake power

$$P_3 = \frac{2\pi NT_3}{60000} = \frac{2\pi \times 1500 \times 0.5 \times 85}{60000} = 6.67 \text{ kW}$$

When 4th cylinder is short circuited, brake power

$$P_4 = \frac{2\pi NT_4}{60000} = \frac{2\pi \times 1500 \times 0.5 \times 89}{60000} = 6.99 \text{ kW}$$

Let frictional power of all cylinder remain constant, then indicated power of

1st cylinder $I_1 = 9.42 - 6.59 = 2.83$ kW

2nd cylinder $I_2 = 9.42 - 6.79 = 2.63$ kW

3rd cylinder $I_3 = 9.42 - 6.67 = 2.75$ kW

4th cylinder $I_4 = 9.42 - 6.99 = 2.43$ kW

Hence indicated power of the engine = 2.83 + 2.63 + 2.75 + 2.43 = 10.64 kW

So frictional power = indicated power – brake power = 10.64 – 9.42 = 1.22 kW

So, mechanical efficiency $= \frac{\text{brake power}}{\text{indicated power}} \times 100 = \frac{9.42}{10.64} \times 100 = 88.53$ per cent

Air fuel ratio $= \frac{\text{air fuel ratio} \times 100}{\text{per cent of fuel evaporated}} = \frac{13 \times 100}{75} = 17.33$

Again air fuel ratio $= \frac{(P_{atm} - P_p)\text{Molecular weight of air}}{P_p \times \text{molecular weight of fuel}}$

Where P_{atm} = atmospheric pressure and P_p = partial vapour pressure

Let the molecular weight of air 29.2. The petrol has a molecular formula C_8H_{18} which has molecular weight 114. Let the atmospheric pressure be 103 kPa.

Putting these in the above equation and on solving P_p was found to be 1.537 kPa

Answer: Indicated power is 10.64 kW; Brake power is 9.42 kW; Mechanical efficiency is 88.53 per cent; Partial vapour pressure is 1.537 kPa

Example 41:

(i) Determine the diameter of the fuel jet to get an air-fuel ratio of 11:1 if the petrol has specific gravity 0.80 and coefficient of discharge through the jet is 0.80 in a 4 cylinder four stroke petrol engine with a 3.20 cm diameter venturi and a coefficient of air flow 0.85. The density of air may be taken as constant throughout the operation having the value of 1.06 kg/m³.

(ii) Compute the air-fuel ratio if the depression head across fuel jet changes to 55 cm of water for the carburettor and 4 cm head of water is sufficient to cause the fuel to begin to flow from the jet.

(iii) If the air flow ratio has to come back to 11:1 then what should be the size of the fuel jet

Solution: Given, 4 stroke 4-cylinder petrol engine; For air flow venturi diameter = 3.20 cm; Discharge coefficient through ventui C_a = 0.85; Density of air ρ_a = 1.06 kg/m³; Air fuel ratio = 11:1; Coefficient of discharge of carburation jet C_f = 0.80; Specific gravity S_g = 0.80; Depression jet across fuel jet 55 cm of water for carburettor; Head of water sufficient to cause the fuel to flow = 4 cm

In general case if no pressure head is required to flow the fuel from jet, then the air fuel ratio is given by $= \frac{A_a}{A_f} \times \frac{C_a}{C_f} \sqrt{\frac{\rho_a}{\rho_f}}$

Where A_a = area of venturi, A_f = area of fuel jet, ρ_a = density of air, ρ_f = density of fuel

Putting the known values, $11 = \frac{3.20^2}{d_f^{\,2}} \times \frac{0.85}{0.80} \times \sqrt{\frac{1.06}{0.80 \times 10^3}}$

So, d_f = 0.189 cm = 1.89 mm

When the air begins to flow through the venturi, the fuel begins to rise because of the difference in air pressures on the fuel in the float chamber and that on the fuel in the jet at the venturi. Effective depression across the venturi affecting the fuel flow

$= h_w - h_w'$, where h_w = head of water across the venturi, h_w = head of water required to cause the fluid to flow, hence in actual case, the fuel flow is proportional to

$$A_fC_f\sqrt{(h_w - h_w{'})\rho_f}$$

So, actual air fuel ratio

$$= \frac{A_a}{A_f} \times \frac{C_a}{C_f}\sqrt{\frac{\rho_a}{\rho_f} \times \frac{h_w}{h_w - h_w}} = \frac{3.20^2}{0.189^2} \times \frac{0.85}{0.80}\sqrt{\frac{1.06}{0.80 \times 10^3} \times \frac{55}{55-4}} = 11.51:1$$

c) So, to get the air fuel ratio 11:1 the diameter of the venturi has to be changed

$$\text{So, } 11 = \frac{3.20^2}{d_f{}^2} \times \frac{0.85}{0.80}\sqrt{\frac{1.06}{0.80 \times 10^3} \times \frac{55}{55-4}}$$

Hence $d_f = 0.1933$ cm = 1.933 mm

Answer: Diameter of fuel jet is 1.89 mm; Air fuel ratio is 11.51:1; Diameter of fuel jet is 1.933 mm

FUEL SUPPLY SYSTEM OF A DIESEL ENGINE

Example 42: Design a cylinder for 12.30 kW diesel engine running at 1200 rpm, using high speed diesel fuel with calorific value of 45980 kJ/kg. For complete combustion, 34 per cent extra air is to be supplied. Take the thermal efficiency as 30 per cent, mechanical efficiency as 85 per cent and over load as 115. The fuel contains 12 per cent hydrogen and 85 per cent carbon. Assume other necessary relevant values.

Solution: Given, engine power = 12.30 kW; Engine speed = 1200 rpm; High speed diesel fuel has calorific value C_f = 45980 kJ.kg^{-1}; Fuel has 85 per cent carbon and 12 per cent hydrogen; For complete combustion 34 per cent extra air is supplied; Mechanical efficiency η_m = 85 per cent; Thermal efficiency η_{ith} = 30 per cent; Overload = 115 per cent

1 kg of fuel contain 0.85 kg of carbon and 0.12 kg of hydrogen

For theoretically complete combustion, the chemical equations are

$$C + O_2 = CO_2 \qquad (1)$$

$$2H_2 + O_2 = 2H_2O \qquad (2)$$

Amount of oxygen required for complete combustion of 0.85 kg carbon

$$\frac{32}{12} \times 0.85 = 2.27 \text{ kg}$$

Amount of oxygen required for complete combustion of 0.12 kg hydrogen $\frac{32}{4} \times 0.12$ = 0.96 kg, hence total amount of oxygen needed 2.27 + 0.96 = 3.23 kg

Assuming air contains 23 per cent of oxygen by weight, weight of air required $= \frac{3.23}{0.23}$, Hence air fuel ratio 14.04:1

Let the air enters the cylinder at a pressure of 100 kPa(P) and at 27° C *i.e.* 300K (T), then the volume of air per kg of air $V = \frac{mRT}{MP} = \frac{1000 \times 8.314 \times 300}{28.97 \times 100 \times 1000} = 0.860\ m^3$

As given 34 per cent extra air is to be supplied, the amount of air at 100 kPa and 27° C is =0.860 × 1.34 = 1.152 m^3

$$\text{Indicated power} = \frac{\text{Brake power}}{\eta_m} = \frac{12.30}{0.85} = 14.47\ kW$$

$$\text{Thermal efficiency} = \frac{\text{indicated power}}{\text{fuel power}} \times 100 = \frac{\text{indicated power} \times 3600}{f \times C_f} \times 100$$

Where f = mass of fuel consumed, kg/hr

$$\text{So, } f = \frac{\text{indicated power} \times 3600}{\eta_{ith} \times C_f} = \frac{14.47 \times 3600}{0.30 \times 45980} = 3.77 kg.h^{-1} = 0.063\ kg.min^{-1}$$

$$\text{So, fuel consumption per stroke} = \frac{0.063}{\frac{1200}{2}} = 10.5 \times 10^{-4} kg$$

As volume of 1 kg of air is 1.152 m^3

$$\text{So, volume of fuel injected per stroke} = \frac{10.5 \times 10^{-4}}{\frac{1}{1.152}} = 1.21 \times 10^{-4} m^3 = 121\ cc$$

Assuming the length to stroke 1.25:1 *i.e.* L = 1.25 d

$$\text{So, stroke volume } V_s = \frac{\pi}{4} d^2 L = \frac{\pi}{4} d^3 \times 1.25 = 121\ cc$$

On solving d = 4.97 cm, L = 1.25 × 4.97 = 6.21 cm

Let P_m be the mean effective pressure then

$$P_m = \frac{\text{indicated power} \times 60 \times 2}{V_s \times N} = \frac{14.47 \times 60 \times 2}{121 \times 10^{-4} \times 1200} = 11.95\ \text{kN.m}^{-2} = 11.95\ \text{kPa}$$

The air goes to the cylinder through the inlet manifold at a pressure less than atmospheric. Let the pressure be P_1 92 kPa. If P_2 is the pressure at the end of compression for adiabatic process

$$P_2 = P_1 \left(\frac{V_1}{V_2}\right)^k = 92 \times (14.5)^{1.4} = 3887.79\ \text{kPa}$$

Maximum force acting on the piston $= \frac{\pi}{4} \times d^2 P_2 = \frac{\pi}{4} \times \left(\frac{4.97}{100}\right)^2 \times 3887.79 = 7.54\ \text{kN}$

The piston of the engine is assumed to be a circular container with one end closed. This maximum pressure is acting uniformly all over the surface. Hence this P_2 is the designed pressure.

Let S_b be the bending stress, So, $S_b = \frac{3Pr^2}{4t^2}$, Where P = maximum pressure acting on the plate r = inner radius of piston, t = thickness of piston

Let the material of construction of material is aluminium alloy whose allowable bending stress S_b is around 90 MPa *i.e.* 90 N.mm^{-2}

And $r_i = r_o - t$, Where r_i and r_o = inner and outer radius respectively

So $$t = \sqrt{\frac{3Pr^2}{4S_b}} = \sqrt{\frac{3 \times 3887.79 \times r^2}{4 \times 90}} = 0.180r = 0.180(r_o - t)$$

On solving t = 3.5 mm

Answer: Diameter of cylinder is 4.97 cm; Stroke length is 6.21 cm; Diameter of the piston is 4.97 cm; Thickness of piston is 3.5 mm

Example 43: If the specific gravity of fuel as 0.70 and absolute viscosity as 26 centipoises then find out the Reynolds Number for flow through an orifice of diameter 0.012 cm when the pressure difference is 25 MPa.

Solution: Given, orifice diameter d = 0.012 cm; Pressure difference ΔP = 25 MPa = 250 kg.cm^{-2}; Specific gravity of fuel S_g = 0.70 *i.e.*; Absolute viscosity μ = 26 centipoise = 0.26 gm.cm^{-1}.sec^{-1}, Pressure head $h = \frac{\Delta P}{\rho}$

Velocity of flow $v = C_d\sqrt{2gh} = C_d\sqrt{2g\frac{\Delta P}{\rho}} = 0.93\sqrt{\frac{2 \times 981 \times 250}{0.70 \times 10^{-3}}} = 24618\ \text{cm.sec}^{-1}$

Where C_d = discharge coefficient, its value is assumed to be 0.93

Reynolds number $R = \frac{\rho vd}{\mu} = \frac{0.70 \times 0.012 \times 24618}{0.26} = 795.35$ where d is diameter of orifice, cm

Answer: Reynolds's number is 795.35

Example 44: At an injection pressure of 12 MPa, spray of an injector penetrates 13.5 cm in 13.90 millisecond. Find out the time required for the spray to penetrate this distance when the injector is set at a pressure of 18.5 MPa. Combustion chamber pressure is 2 MPa. It can be assumed that orifice and combustion chamber air density remain constant.

Solution: Given depth of penetration S_1 = 13.5 cm; Duration t_1 = 13.9 millisecond; Injection pressure P_1 = 12 MPa; P_2 = 18.5 MPa; Combustion chamber pressure P_c = 2 MPa

The depth of penetration, S, from a spray can be calculated from the following relationship.

S is a function of time of penetration and difference between the injection pressure and combustion pressure.

$S = f(t\sqrt{\Delta P})$ where t = time of penetration and ΔP = difference between the injection pressure and combustion pressure

Let t_2 is the time required for the spray to penetrate S_1distance when the injector is set at a pressure of P_2

In the present case, $S_1 = S_2$, Or, $t_1\sqrt{\Delta P_1} = t_2\sqrt{\Delta P_2}$ Or, $t_1\sqrt{P_1 - P_c} = t_2\sqrt{P_2 - P_c}$

Or, $3.9 \times \sqrt{12-2} = t_2\sqrt{18.5-2}$ Or,$t_2 = 10.82$ millisecond

Answer: Time required for the spray to penetrate is 10.82 millisecond.

Example 45: A single jet nozzle in an injector delivers 1.2×10^{-4} kg of fuel during a 28° period of crank travel of an engine running at 2500 revolution per minute. Assuming the injection pressure to be 15 MPa and pressure at the end of compression be 3.5 MPa, find the orifice size. Specific weight of the fuel be 0.78 and coefficient of discharge be 0.85.

Solution: Given mass of fuel injected in the injector = 1.2×10^{-4} kg during crank rotation θ = 28°; Speed n = 2500 rpm; Injection pressure p_i = 15 MPa; Pressure at the end of compression p_{ec} = 3.5 MPa; Specific gravity of fuel S_g = 0.78; Coefficient of discharge C_d = 0.85

Let A = Area of the orifice

Amount of fuel supplied = flow area × velcity × time to supply

$$= A \times C_d \times \sqrt{2g(p_i - p_{ec})S_g} \times \frac{\theta \times 60}{n \times 360}$$

$$= A \times 0.85 \times \sqrt{9.81 \times (15 - 3.5) \times 10^6 \times 780} \times \frac{28 \times 60}{2500 \times 360}$$

But given amount of fuel supplied is 1.2×10^{-4} kg

So, on solving we get $A = 3.01 \times 10^{-4}$ m^2 = 301 mm^2, which results in diameter of orifice is 19.58 mm.

Answer: Diameter of orifice is 19.58 mm.

Example 46: A four stroke engine has 6 cylinders, developing 224 kW power at a speed of 2000 revolution per minute. While using API 32 fuel, the fuel consumption is 0.25 kg.kW^{-1}.h^{-1}. The fuel injection pressure is 125 atm. The fuel injection duration is 35° of crank travel. The fuel has specific gravity of 0.88. The orifice has discharge coefficient of 0.58. Determine the weight and volume of fuel injected per cylinder per cycle. Also determine the diameter of fuel orifice for a single hole nozzle, for a five-hole nozzle. Find out the power loss in injecting the fuel.

Solution:

Given, 6-cylinder four stroke engine develop BP = 224 kW at N = 2000 rpm; API degree of fuel = 32°; Brake specific fuel consumption BSFC = 0.25 kg. kW^{-1}hr^{-1}; Injection pressure P_i = 125 atm; Injection duration θ = 35°; Specific gravity S_g = 0.88 *i.e.* density = 0.88 gm.cc^{-1}; Coefficient of discharge C_d = 0.58

$$\text{Fuel consumption per cylinder} = \frac{\text{BP} \times \text{BSFC}}{\text{no of cyinder}} = \frac{224 \times .25}{6} = 9.33 \text{ kg}$$

Fuel consumed per cycle per cylinder

$$= \frac{\text{fuel consumption } \times 2}{\text{N} \times 60} = \frac{9.33 \times 2}{2000 \times 60} = 1.56 \times 10^{-4}\text{kg}$$

$$= .156 \text{ gm} = \frac{0.156}{0.88} = .177 \text{ cc}$$

$$\text{Injection pressure } P_i = 125 \text{ atm} = 125 \times 10^5 \text{ Pa}$$

$$\text{Head causing flow } h = \frac{P_i}{\rho \times g} = \frac{125 \times 10^5}{880 \times 9.81} = 1447.96 \text{ m}$$

Velocity of flow

$$v = \sqrt{2gh} = \sqrt{2 \times 9.81 \times 1447.96} = 168.54 \text{ m.sec}^{-1} = 16854 \text{ cm.sec}^{-1}$$

Volume of fuel injected per cycle per cylinder

$$= C_d Avt = 0.58 \times A \times 16854 \times \frac{35 \times 60}{2000 \times 720}$$

but it was found that amount of fuel injected is 0.177 cc. So, on solving A= 2.28 cm^2

For single nozzle let diameter is d_1

So $A = \frac{\pi}{4} {d_1}^2 = 2.28 \text{ cm}^2 = d_1 = 1.7 \text{ cm} = 17 \text{ mm}$

For five number of nozzles let diameter of each is d_2

So $A = \frac{\pi}{4} {d_2}^2 \times 5 = 0.0228 \text{ cm}^2, d_2 = 0.0763 \text{ cm} = 0.763 \text{ mm}$

Amount of fuel supplied per sec

$$= \frac{\text{BSFC} \times \text{BP}}{3600 \times \text{density of fuel}} = \frac{224 \times .25}{3600 \times 880} = 1.76 \times 10^{-5} \text{m}^3.\text{sec}^{-1}$$

Pressure required for injecting that amount of fuel = 125×10^5 Pa

So, power lost in injection = 125×10^5 Pa $\times$ 1.76×10^{-5} $m^3.sec^{-1}$ = 220 W = 0.220 kW

So, percentage power lost $= \frac{0.220}{224} \times 100 = 0.1$ per cent

Answer: Fuel injected per cylinder per cycle = 0.177 cc; Diameter of single hole orifice = 1.7 mm; Diameter of five-hole orifice = 0.763 mm; Percentage of loss of power in injection = 0.1 per cent

Example 47: In Example 46, if the fuel injection pump has full stroke displacement of 4 times the rated engine fuel consumption and if the injection pump plunger has a stroke twice of its diameter, what are the pump plunger bore and stroke dimensions.

Solution:

According to question, full stroke displacement = 4 × rated fuel consumption

The fuel consumption per sec $= 0.25 \times 224 \times \frac{1}{880} \times \frac{1}{3600} \times 10^6 = 17.67\text{cc}$

So, fuel stroke displacement $= 4 \times 17.67 = 70.70 \text{ cc.sec}^{-1}$

Hence, fuel supplied per cylinder $= \frac{70.70}{6} = 11.78 \text{ cc.sec}^{-1}$

Let L = stroke length and d = bore dia

So, then fuel supplied per stroke $= \frac{\pi}{4} d^2 L$

Hence, fuel supplied per second $= \frac{\pi}{4} d^2 L \times \frac{n}{60}$

Where, n = speed of crank shaft

So, fuel supplied per second $= \frac{\pi}{4} \times d^2 \times 2d \times \frac{n}{60} = 15.7d^3 \text{ cc. sec}^{-1}$ or $11.78 = 15.7d^3$

Hence, d = 0.908 cm = 9.80 mm

So, L = 2 × d = 18.17 mm

Answer: Plunger has a bore of 9.80 mm and stroke of 18.17 mm.

Example 48: A two-cylinder four stroke diesel engine develops 15 KW power at 2400 rpm. Its brake specific fuel consumption is 0.268 kg kW^{-1} h^{-1}. If the specific gravity of fuel is 0.85, find out quantity of fuel injected per cylinder per cycle? ***(GATE-2011)***

Solution: Given, two-cylinder four stroke engine; Power developed P = 15 kW; Speed of engine N = 2400 rpm; Brake specific fuel consumption bsfc = 0.268 kg. $kW^{-1}.h^{-1}$; Specific gravity of fuel S = 0.85

$$\text{Amount of fuel consumed per hour} = \frac{\text{bsfc} \times P}{S \times 1000} = \frac{0.268 \times 15}{0.85 \times 1000} = 4.72 \times 10^{-3} m^3 h^{-1}$$

$$\text{Amount of fuel consumed per stroke} = \frac{\text{Amount of fuel consumed per hour}}{60 \times \frac{N}{2}}$$

$$= \frac{4.72 \times 10^{-3}}{60 \times \frac{2400}{2}} = 6.56 \times 10^{-8} m^3$$

$$\text{Amount of fuel consumed per cylinder} = \frac{\text{Amount of fuel consumed per stroke}}{\text{Number of cylinder}}$$

$$= \frac{6.56 \times 10^{-8}}{2} = 3.28 \times 10^{-8} m^3$$

Answer: Quantity of fuel consumed per cylinder per stroke is 3.28×10^{-8} m^3

Example 49: A diesel engine running in dual fuel mode with diesel as pilot fuel and producer gas as primary fuel produces 3.5 kW at rated engine speed and is coupled directly to a generator for producing electricity. The amount of diesel and producer gas consumed per hour is 460 ml and 12.5 m^3, respectively

a) Assuming calorific value of diesel and producer gas as 35280 and 3.97MJ m^{-3}, respectively, Find out the brake thermal efficiency.

b) If generator efficiency is 90 per cent, what is the maximum electricity produced in kW? ***(GATE-2012)***

Solution: Given, diesel engine running on dual fuel mode diesel as pilot fuel and producer gas as primary fuel; Power produced (brake power) BP = 3.5 kW at rated engine speed; Amount of diesel consumed f_d = 460 ml = 460 × 10^{-6} m^3; Amount of producer gas consumed f_p = 12.5 m^3; Calorific value of diesel CF_d = 35280 MJ.m^{-3} = 35280000 kJ.m^{-3}; Calorific value of producer gas CF_p = 3.97 MJ.m^{-3} = 3970 kJ.m^{-3};

Generator efficiency η_g = 90 per cent = 0.9

Brake thermal efficiency

$$\eta_{bth} = \frac{BP \times 3600}{f_d CF_d + f_p CF_p} = \frac{3.5 \times 3600}{(460 \times 10^{-6} \times 35280000) + (12.5 \times 3970)} \times 100 = 19.13 \text{ per cent}$$

Maximum electricity produced = BP × η_g = 3.5 × 0.9 = 3.15 kW

Answer:

a) Brake thermal efficiency is 19.13 per cent

b) Maximum electricity produced is 3.15 kW

Example 50: The brake power of a six cylinder four-stroke diesel engine at 3000 rpm is 125 kW. It has a brake specific fuel consumption of 200 gm. kW^{-1}.h^{-1}. Specific gravity of fuel is 0.85. Find the amount of fuel injected per cycle per cylinder in ml. ***(GATE-2015)***

Solution: Given, six cylinder 4-stroke diesel engine; Engine speed N = 3000 rpm; Engine power P = 125 kW; Brake specific fuel consumption BSFC = 200 gm. kW^{-1}.h^{-1}; Specific gravity of fuel SG = 0.85

Fuel consumption per hour FC_h = BSFC × P = 200 × 125 = 25000 gm.h^{-1}

Fuel consumption per cycle $FC_{cycle} = \dfrac{FC_h}{60 \times \frac{N}{2}} = \dfrac{25000}{60 \times \frac{3000}{2}} = 0.27$ gm

Fuel consumption per cylinder $FC = \dfrac{FC_{cycle}}{6} = \dfrac{0.27}{6} = 0.046$ gm

Fuel consumption per cylinder in ml $= \dfrac{FC}{SG} = \dfrac{0.046}{0.85} = 0.054$ ml

Answer: Fuel consumption per cycle per cylinder is 0.054 ml.

Example 51 A 3-cylinder tractor engine operated on diesel has a bore of 120 mm and stroke of 120 mm. It is running at 1500 revolution per minute. The air fuel ratio is 15:1. Assuming an air density of 1.25 $kg.m^{-3}$, find air and fuel supplied to the engine per minute.

Solution: Given, 3-cylinder tractor engine; Cylinder diameter d = 120 mm = 0.12 m; Stroke L = 120 mm = 0.12 m; Engine speed N = 1500 rpm; Air density ρ_a = 1.25 $kg.m^{-3}$; Air fuel ratio = 15:1

$$\text{Swept volume } V_s = \frac{\pi}{4}d^2L = \frac{\pi}{4} \times 0.12^2 \times 0.12 = 1.35 \times 10^{-3} \text{ m}^3$$

Total swept volume V_s' = number of cylinder × V_s = 3 × 1.35 × 10^{-3} = 4.07 × 10^{-3} m^3

$$\text{So volume of air injected per minute } F_{air} = \frac{V_s' \times N}{2} = \frac{4.07 \times 10^{-3} \times 1500}{2} = 3.05 \text{ m}^3$$

$$\text{Mass of air injected per minute } M_{air} = F_{air} \times \rho_a = 3.05 \times 1.25 = 3.81 \text{ kg}$$

$$\text{Mass of fuel injected per minute } M_{fuel} = \frac{M_{air}}{15} = \frac{3.81}{15} = 0.254 \text{ kg}$$

Answer: Mass of air and fuel consumed per minute are 3.81 kg and 0.254 kg, respectively.

Example 52: A 3-cylinder tractor has a cylinder of 10 × 10 cm running at 2000 revolution per minute develops 24 kW power. It consumes 7.8 Lh^{-1} at full load. Brake mean effective pressure is 0.8 MPa and density of diesel is 825 $kg.m^{-3}$. Calculate the fuel delivery rate of fuel injection pump of tractor engine.

Solution: Given, No. of cylinders n = 3; Power developed (brake power) BP = 24 kW; Cylinder diameter d = 10 cm; Stroke length L = 10 cm; Break Power P = 24 kW; Engine speed N = 2500 rpm; Fuel consumption m_{fuel} = 7.8 $L.h^{-1}$; Break mean effective pressure p_{mean} = 0.8 MPa; Density of diesel ρ_{diesel} = 825 $kg.m^{-3}$ = 0.825 $kg.L^{-1}$

$$\text{Swept volume } V_s = \frac{\pi}{4}d^2L = \frac{\pi}{4} \times 0.1^2 \times 0.1 = 7.853 \times 10^{-4} \text{ m}^3$$

Brake specific fuel consumption

$$\text{BSFC} = \frac{m_{fuel} \times \rho_{diesel}}{\text{BP}} = \frac{7.8 \times 0.825}{24} = 0.268 \text{ kg. kW}^{-1}\text{h}^{-1}$$

So specific fuel delivered

$$= \frac{p_{mean} \times \text{BSFC} \times V_s}{\rho_{diesel}} = \frac{0.8 \times 10^3 \times 0.268 \times 7.85 \times 10^{-4}}{830} \times 10^6 = 202.78 \text{ cc}$$

Answer: Fuel delivery rate of fuel injection pump is 202.78 cc.

Example 53: A four stroke diesel engine developing 35 kW brake power at a rpm of 2500 has three cylinders. The brake specific fuel consumption is 0.2 $kg.kW^{-1}.h^{-1}$. Assuming fuel injection pump has full stroke displacement of 4 times the rated engine fuel consumption and if the injector pump plunger has stroke length twice its diameter and density of diesel is 830 $kg.m^{-3}$, find total fuel delivery rate of the pump, pump plunger diameter and plunger stroke length.

Solution: Given, four stroke diesel engine; Number of cylinder n = 3; Brake power of engine BP = 35 kW; Rated speed of engine N= 2500 rpm; Specific fuel consumption BSFC = 0.20 $kg.kW^{-1}.h^{-1}$; Plunger full stroke displacement = 4 times rated engine fuel consumption; Plunger stroke length = twice plunger diameter *i.e.* L = 2d; Density of diesel ρ_{diesel} = 830 $kg.m^{-3}$

Total fuel supplied to the engine per second

$$= \frac{BSFC \times BP}{\rho_{diesel}} = \frac{0.2 \times 35}{830 \times 3600} = 2.34 \times 10^{-6} m^3 sec^{-1}$$

Plunger full stroke displacement

$$= 4 \times \text{fuel supplied} = 4 \times 2.34 \times 10^{-6} = 9.37 \times 10^{-6}\ m^3\ sec^{-1}$$

Fuel supplied per cylinder

$$= \frac{\text{full stroke displacement}}{n} = \frac{9.37 \times 10^{-6}}{3} = 3.12 \times 10^{-6} m^3 sec^{-1}$$

Fuel supplied per stroke $V_s = \frac{\pi}{4} d^2 L = \frac{\pi}{4} \times d^2 \times 2d = 1.57\ d^3\ m^3$

Fuel supplied per second $= \frac{V_s \times N}{2 \times 60} = \frac{1.57 \times d^3 \times 2500}{2 \times 60} = 32.70\ d^3 m^3 sec^{-1}$

But as calculated fuel supplied per cylinder is 3.12 × 10^{-6} m^3 sec^{-1}. So, on solving d = 4.52 mm and stroke L = 2d = 2 × 4.52 = 9.04 mm

Answer: Bore is 4.52 mm and stroke is 9.04 mm.

Example 54: A direct ignition tractor engine has an injector having four holes and ignition delay of 18°. Calculate the swirl ratio of engine needed to fan one spray plume half way to the next plume during ignition delay, if the engine speed is 2000 rpm.

Solution: As the injector has four holes, therefore spray plumes of nozzles are $\frac{360}{4}$ = 90° apart. Theoretically the air swirl must move $\frac{90}{4}$ = 45° during the delay period.

So, time available for air to move during delay period is

$$t = \frac{\text{delay degree}}{360 \times \text{engine speed}} = \frac{18}{360 \times 2000} = 2.5 \times 10^{-5} \text{ min}$$

So, swirl speed $N_s = \frac{45}{360 \times t} = \frac{45}{360 \times 2.5 \times 10^{-5}} = 5000 \text{ rpm}$

Swirl ratio $SR = \frac{N_s}{N} = \frac{5000}{2000} = 2.5$

Answer: Swirl ratio is 2.5.

ENGINE GOVERNING SYSTEM

Example 55: An engine governor is operated by hit and miss principle. The measured fuel consumption is 3.6 m^3/hr at 18°C and 760 mm mercury. The engine is operated at a speed of 300 rpm and number of explosion is 60 per minute. For fuel consumption measurement, an orifice of diameter 23 mm is provided. The pressure across the orifice is 8 cm of water column at a temperature of 13°C and a pressure of 770 mm of mercury. The discharge coefficient for the orifice is 0.58. Find amount of air supplied per minute and air fuel ratio by volume.

Solution: Given T_1 = 18°C = 291 K; T_2 = 13°C = 286 K; P_1 = 760 mm of mercury = $\frac{760}{760} \times 10^4 = 10^4$ kg.cm^{-2}; P_2 = 770 mm of mercury = $\frac{770}{760} \times 10^4 = 1.013 \times 10^4$ kg.cm^{-2}; Discharge coefficient C_d = 0.58; Orifice diameter d = 23 mm = 0.023 m; Area of orifice $A = \frac{\pi}{4}(d)^2 = \frac{\pi}{4}(0.023)^2$; No of explosion = 60 per min; Engine speed = 300 rpm

Perfect gas law for 1 kg of gas is $P_1V_1 = RT_1$

Or, $V_1 = \frac{RT_1}{P_1} = \frac{29.29 \times 291}{10^4} = 0.852 \text{ m}^3$

So, density of air $w_a = \frac{1}{V_1} = \frac{1}{0.852} = 1.173 \text{ kg.m}^{-3}$

Now $w_a h_a = wh$, Where, w = density of water (1000 kg.m^{-3}), h_a = head of air, h = head of water (8 cm = 0.08 m, given). So, $h_a = \frac{wh}{w_a} = \frac{1000 \times 0.08}{1.173} = 68.20$ m of air

Discharge through orifice $Q = C_d A\sqrt{2gh_a} = 0.58 \times \frac{\pi}{4}(0.023)^2 \times \sqrt{2 \times 9.81 \times 68.20}$

$$= 0.0088 \text{ m}^3.\text{sec}^{-1} = 0.528 \text{ m}^3.\text{min}^{-1}$$

Hence the weight of air = 1.173 × 0.528 = 0.62 kg.min^{-1}

b) Volume of gas supplied per working stroke

$$= \frac{3.6}{60 \times 60} = 0.001\ m^3 \text{ at } 18°C \text{ and } 760 \text{ mm of mercury}$$

$$\text{Hence at NTP total gas supplied} = 0.001 \times \frac{273}{291} \times \frac{76}{76} = 0.000938\ m^3$$

$$\text{Volume of air per suction stroke} = \frac{0.62}{\frac{300}{2}} = 0.0042\ m^3$$

$$\text{Hence volume of air per suction stroke at NTP} = 0.0042 \times \frac{273 \times 76}{291 \times 76} = 0.004\ m^3$$

$$\text{Air fuel ratio} = \frac{0.004}{0.000938} = 4.48$$

Answers: The weight of air supplied is 0.62 kg.min^{-1} and air fuel ratio is 4.48:1

Example 56: An engine rotating at a speed of 2000 rpm. It has a centrifugal governor which is running at half of the engine speed. Angle of arm with axis of rotation is 0° when the weights are in closed position and 28° when in open position. Effective lever arm, a = 3.5 cm. Given b = 4.2 cm, L = 6 cm, governor makes 32° with the axis of rotation. It has a spring constant of 7500 N/m. Assuming the weights are spherical and arms to be weightless, find out the weight W to be attached at the ends of arm.

Solution: Given engine speed = 2000 rpm; Governor speed = half of engine speed = 1000 rpm The centrifugal force due to rotating mass $F = \frac{W}{g} \times \omega^2 r$

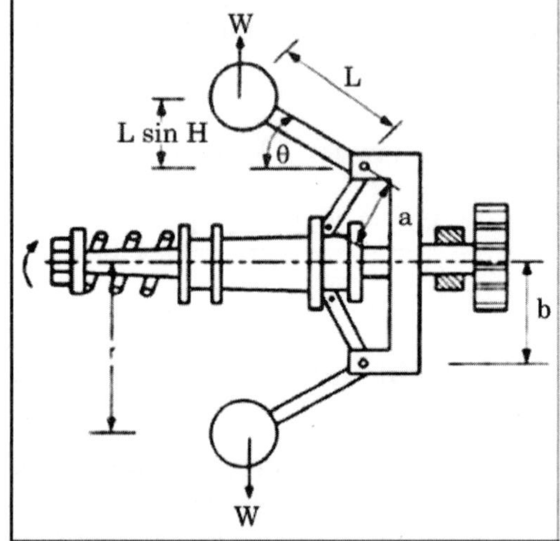

Where, W = weight of rotating mass, ω = angular velocity, rad.sec^{-1}, r = radius of rotating mass from the axis of rotation

At a certain speed the above centrifugal force is balanced by spring tension. From fig, the governor rotates at 1000 rpm while rotating at an angle 32° with the axis of rotation.

So, from fig r = b + L sin β = 4.2 + 6 × sin 32° = 7.37 cm = 0.0737 m

$$\text{So, centrifugal force } F = \frac{w}{9.81} \times \left(2\pi \times \frac{1000}{60}\right)^2 \times 0.0737 = 82.49w \text{ N}$$

Total displacement of the spring = 3 × sin 32° = 1.58 cm = 0.0158 m

So, spring force = 7500 × 0.0158 = 119.231 N

Now taking moment about the hinge, 119.231 × 3 × cos 28° = 82.38 w × 6 × cos 28°

On solving w = 0.723N = 73.76 gm

Answer: Weight to be attached at the end of each arm = 73.76 gm

Example 57: Calculate the governor regulation of a tractor engine having no load speed of 1400 rpm and full load speed of 1600 rpm.

Solution: Given, no-load speed N_l = 1400 rpm; speed at full load N_h = 1600 rpm

$$\text{Governor regulation} = \frac{2(N_h - N_l)}{(N_h + N_l)} = \frac{2(1600 - 1400)}{1600 + 1400} = 0.134 = 13.4 \text{ per cent}$$

Answer: governor regulation is 13.4 per cent.

Example 58: A porter governor has equal arms each of 250 mm long and pivoted on the axis of rotation. The mass of each ball is 4 kg and mass of sleeve is 25 kg. The radius of rotation is 150 mm and height of governor is 180 mm when it begins to lift. The radius of rotation is 200 mm when it is at maximum speed with height of governor 160 mm. Find effort and power of governor. Assume per cent increase in speed from no load to high load is 14 per cent.

Solution: Given, arm length l = 250 mm; Mass of ball m = 4 kg; Mass of sleeve M = 25 kg; Radius of governor when sleeve begins to lift r_1 = 150 mm; Height of governor when sleeve begins to lift h_1 = 180 mm; Radius of governor at maximum speed r_2 = 200 mm; Height of governor at maximum speed h_2 = 160 mm; per cent increase in speed N_p = 14 per cent = 0.14

Sleeve lift L = 2(h_1 – h_2) = 2(180 – 160) = 40 mm

Governor effort F = N_{pg}(M + m) = 0.14 × 9.81 × (25 + 4) = 39.82 N

Power of governor P = F × L = 39.82 × 40 = 1593.15 N.mm = 1.593 N.m

Answer: Piston effort is 39.82 N and power is 1.593 N.m.

ENGINE LUBRICATION SYSTEM

Example 59 An engine operating at a speed of 2500 rpm. It has a gear type pump in lubricating system which develop a pressure of 15 MPa and requires an input power of 12 kW. If the mechanical efficiency is 0.80, calculate the oil flow rate and displacement volume per revolution. Assume the volumetric efficiency is 0.7.

Solution: Given, speed of engine N = 2500 rpm; Shaft power P = 12 kW; Pressure developed p = 15 MPa; Mechanical efficiency ρ_m = 0.8

Fluid power $P_{fluid} = P_m = 12 \times 0.8 = 9.6$ kW

$$\text{Fuel flow rate } Q = \frac{P_{fluid} \times 60}{p} = \frac{9.6 \times 60}{15} = 38.4 \text{ L.min}^{-1}$$

$$\text{Pump displacement volume per revolution} = \frac{Q}{N} = \frac{38.4}{2500} \times 1000 = 15.36 \text{ cc}$$

Answer: Oil flow rate is 38.4 $L.min^{-1}$ and pump displacement is 15.36 cc.

Example 60: A capillary tube viscometer is 2.5 mm in diameter and 120 mm long. 200 cc of oil is passed through the tube in 15 seconds which has a specific gravity of 0.9. A head of 6 mm of oil above inlet of viscometer tube maintains a pressure difference through the tube. Calculate the pressure difference across the tube and dynamic viscosity of oil.

Solution: Given, diameter of tube d = 2.5 mm; Length of tube L = 120 mm; Head of oil above inlet h = 6 mm; Specific gravity of oil SG = 0.92; Volume of oil flow V = 200 cc; Time required to flow the liquid t = 15 sec

Pressure head of water column $h_{water} = SG \times h = 0.92 \times 6 = 5.52$ mm

So, pressure difference across the tube

$$\frac{h_{water} \times 100 \text{ Pa}}{1000 \text{ mm}} = \frac{5.52 \text{ mm} \times 100 \text{ Pa}}{1000 \text{ mm}} = 0.552 \text{ Pa} = 5.52 \text{ dyne.cm}^{-2}$$

According to Poiseville's law of viscosity

$$\mu = \frac{\pi p d^4 t}{16 \times 8VL} = \frac{\pi \times 5.52 \times 0.25^4 \times 15}{16 \times 8 \times 200 \times 12} = 3.30 \times 10^{-6} \text{ poise}$$

Answer: The pressure difference is 0.552 Pa and dynamic viscosity is 3.30×10^{-6} poise.

ENGINE COOLING SYSTEM

Example 61: Find the water circulation rate of a 35-kW tractor engine which maintain a temperature difference of 6°C of coolant. Assume that for cooling 0.15 $litre.sec^{-1}$ per kW water circulation required and it rejects 0.55 kW per kW of engine power.

Solution: Given tractor engine power is 35 kW; Coolant temperature difference is 6°C;Power rejected by engine is 0.55 kW per kW of engine power

So, total power rejected = 0.55 × 35 = 19.25 kW

Water circulation required is 0.15 $L.sec^{-1}$ per kW

So, total water circulation required = 0.15 × 19.25 = 2.88 litre.sec^{-1}

Answer: So, water circulation required is 2.88 litre.sec^{-1}

ENGINE INTAKE AND EXHAUST SYSTEM

Example 62 A four stroke diesel engine of 30 kW brake power has mechanical efficiency of 80 percent. The fuel consumption is 125 gm per minute. Mass of water circulated through cooling system is 450 kg per hour. Inlet and outlet water temperature are 30°C and 80°C, respectively. Air fuel ratio is 25:1. Mean specific heat of exhaust gas is 1.05 kJ.kg^{-1}.K^{-1} and calorific value of fuel is 45000 kJ.kg^{-1}. Room temperature is 30°C and exhaust gas temperature is 450°C. Make an energy balance sheet based on energy and percentage.

Solution: Given, brake power BP = 30 kW; Mechanical efficiency η_m = 0.8; Fuel consumption m_f = 125 gm.min^{-1}; Water circulation rate m_w = 450 kg.h^{-1}; Water inlet temperature T_{iw} = 30°C; Water outlet temperature T_{ow} = 80°C; Air fuel ratio = 25:1; Mean specific heat of exhaust gas C_{ve} = 1.05 kJ.kg^{-1}.K^{-1}; Calorific value of fuel, CF = 45000 kJ.kg^{-1}; Room temperature T_a = 30°C, Exhaust gas temperature $T_{exhaust}$ = 450°C

a) Energy input to engine by fuel H_f

$$H_f = m_f \times CF = \frac{125 \times 45000}{60 \times 1000} = 93.75 \text{ kJ.sec}^{-1}$$

b) Heat converted to useful work equivalent to bp (H_w)

$$H_w = BP = 30 \text{ kW} = 30 \text{ kJ.sec}^{-1}$$

c) Heat loss due to incomplete combustion (H_{ic})

$$H_{ic} = \frac{m_f \times CF}{\text{Air fuel ratio}} = \frac{125 \times 45000}{25 \times 60} = 3.75 \text{ kJ.sec}^{-1}$$

d) Heat loss in exhaust gas (H_g)

$$H_g = (m_f + 25 \times m_f)C_{ve}(T_{exhaust} - T_a)$$

$$= \frac{(125 + 25 \times 125)}{1000 \times 60} \times 1.05 \times (450 - 30) = 23.88 \text{ kJ.sec}^{-1}$$

e) Heat loss in cooling system (H_c)

$$H_c = m_w C_w (T_{ow} - T_{iw}) = \frac{450}{3600} \times 4.2 \times (80 - 30) = 26.25 \text{ kJ.sec}^{-1}$$

f) Unaccounted losses (H_u)

$$H_u = H_f - (H_w + H_{ic} + H_g + H_c)$$

$$= 93.75 - (30 + 3.75 + 23.88 + 26.25) = 9.87 \text{ kJ.sec}^{-1}$$

CLUTCH

Example 63: A tractor having single plate clutch with both side effectives, is operating on the principle of uniform wear. It transmits power of 25 kW at a rpm of 2500. Intensity of maximum pressure is 85 kN.m^{-2}. The maximum speed at mean radius should not exceed 4 m/sec. Assuming the coefficient of friction as 0.35, calculate the inner and outer radius. The outer radius is 1.5 times of the inner radius. The axial thrust is taken up by 6 springs of 30 mm coil diameter. Allowable modulus of rigidity is MPa and shear stress is 600 MPa. Calculate the diameter of the spring wire.

Solution: Given, Power P = 25 kW; Speed N = 2500 rpm; Maximum intensity of pressure p = 85 kN/m^2; Coefficient of friction μ = 0.35; Outside radius = 1.5 times of inner radius; 6 no of springs of diameter(D) 30 mm; Allowable modulus of rigidity MPa; Shear stress S = 600 MPa; Principle of wear is followed

Total power transmitted $P = \dfrac{2\pi NT}{1000 \times 60}$

Or $T = \dfrac{1000 \times 60 \times P}{2\pi N} = \dfrac{1000 \times 60 \times 25}{2\pi \times 2500} = 95.49$ Nm

Hence, taking factor of safety = 1.3, designed torque T = 95.49 × 1.3 = 124.14 Nm

Total frictional force = 2 × 0.35 × F = 0.7F Where F is the axial thrust

Let r is the effective radius, Torque T = 0.7Fr

$$Fr = 177.34\text{Nm} \qquad \ldots (1)$$

Let r_i = inner radius and r_o = outer radius

As given $r_o = 1.5 \times r_i$ and $r = \dfrac{r_o + r_i}{2} = 1.25\ r_i$

Now total axial thrust $F = p \times A = p \times \pi(r_o^2 - r_i^2)$

$$= 85 \times \pi((1.5r_i)^2 - r_i^2) = 333.79r_i^2 = 213.62\ r^2\text{kN}$$

From eqn. 1, $F_r = 177.34$ Or $213.62\ r^3 \times 1000 = 177.34$, Or r = 0.093 m

$$r_i = \frac{r}{1.25} = \frac{0.0902}{1.25} = 0.075 \text{ m} = 7.5 \text{ cm}$$

$$r_o = 1.5 \times r_i = 1.5 \times 0.075 = 0.112 \text{ m} = 11.2 \text{ cm}$$

Axial thrust $F = 213.62\ r^2 = 213.62 \times 0.093^2 = 1.84$ kN

Load taken by each spring $\dfrac{F}{6} = \dfrac{1.84}{6} = 0.307$

The diameter of the spring wire is given by

$$d = \sqrt[3]{\frac{8FD}{S}} = \sqrt[3]{\frac{8 \times 0.307 \times 1000 \times 0.030}{600 \times 10^6}} = 4.97 \times 10^{-3}\ \text{m} = 4.97\ \text{mm}$$

Answers: Outside diameter of clutch 22.4 cm; Inside diameter of clutch 15 cm; Diameter of spring wire 4.97 mm

Example 64: In a disc clutch, the inside and outside radii of the clutch plate are 50 and 100 mm, respectively. If the axial force exerted on the disc is 4 kN, what is the maximum pressure experienced by clutch plate operating under uniform wear conditions? ***(GATE-2012)***

Solution: Given, disc clutch inside radius r_i = 50 mm and outside radius r_o = 100 mm; Axial force exerted W = 4 kN; Clutch is operating under uniform wear principle

$$\text{Hence constant } C = \frac{W}{2\pi(r_o - r_i)} = \frac{4}{2\pi(100 - 50)} = 0.0127\ \text{kN.mm}^{-1}$$

Let p_{max} is the maximum pressure experienced by the clutch plate

$$\text{So, } p_{max} = \frac{C}{r_i} = \frac{0.0127}{50} = 0.254\ \text{N.mm}^{-2}$$

Answer: Maximum pressure experienced by the clutch plate is 0.254 $N.mm^{-2}$

Example 65: A multi-disc clutch has 4 steel discs and 3 bronze discs. The outside and inside diameters of contact surfaces are 250 mm and 180 mm, respectively. The coefficient of friction is 0.3 and axial force is 400 N. Assuming uniform wear, find the power in kW that the clutch can transmit at 1000 rpm. ***(GATE-2014)***

Solution: Given, multiple disc clutch has 4 steel discs and 3 bronze discs; Outside diameter D_o = 250 mm = 0.25 m; Inside diameter D_i = 180 mm = 0.18 m; Coefficient of friction μ = 0.3; Axial force W = 400 N; Speed of rotation N = 1000 rpm; Clutch is operated under uniform wear

No. of pair of contact surfaces n = 4 + 3 – 1 = 6

$$\text{Mean radius } r_m = \frac{D_o + D_i}{2} = \frac{0.25 + 0.18}{2} = 0.215\ \text{m}$$

Torque transmitted T = nμWrm = 6 × 0.3 × 400 × 0.215 = 154.8 Nm

$$\text{Power transmitted } P = \frac{2\pi NT}{60 \times 1000} = \frac{2\pi \times 1000 \times 154.8}{60 \times 1000} = 16.21\ \text{kW}$$

Answer: The power transmitted by the clutch is 16.21 kW.

Example 66: A multiple disc clutch is to transmit 15 kW at 750 rpm. The inner and outer radii of the friction surfaces are 60 mm and 100 mm, respectively. The coefficient of friction is 0.1 and maximum allowable pressure is 350 kPa. Assuming uniform wear, find the no of pair of contact surfaces required. ***(GATE-2015)***

Solution: Given, multiple disc clutch; Power to be transmitted P = 15 kW; Speed N = 750 rpm; Inner radius r_i = 60 mm = 0.06 m; Outer radius r_o = 100 mm = 0.1 m; Coefficient of friction μ = 0.1; Maximum allowable pressure P = 350 kPa

Clutch operated under uniform wear

$$\text{Mean radius } r_m = \frac{r_i + r_o}{2} = \frac{0.06 + 0.1}{2} = 0.08 \text{ m}$$

Axial force $W = P \times \pi(r_o^2 - r_i^2) = 350 \times \pi(0.1^2 - 0.06^2) = 7.03$ kN

$$\text{Torque transmitted } T = \frac{P \times 60}{2\pi N} = \frac{15 \times 60}{2 \times \pi \times 750} = 0.19 \text{ kN.m}$$

So, number of contact surface required

$$n = \frac{T}{\mu \times W \times r_m} = \frac{0.19}{0.1 \times 7.03 \times 0.08} = 3.37 \approx 3$$

Answer: Number of contact surface required is 3.

Example 67: A single plate dry clutch transmits 15 kW power at 1200 rpm. The clutch sustains a maximum axial load of 2.65 kN. The ratio of outer to inner diameter of the frictional surface is 1.25:1. Considering uniform wear with a coefficient of friction of 0.3 on both the frictional surfaces, calculate the outer diameter of the clutch plate. ***(GATE-2018)***

Solution: Given, single plate dry clutch *i.e.* number of contact surface n = 1; Power transmitted P = 15 kW; Speed N = 1200 rpm; Maximum axial load W = 2.65 kN; Ratio of outer diameter to inner diameter = 1.25; Clutch operated on uniform wear principle; Coefficient of friction μ = 0.3

Let, outer and inner diameter and radius are d_o and d_i and r_o and r_i respectively

$$\text{As given } \frac{d_o}{d_i} = \frac{r_o}{r_i} = 1.25$$

$$\text{So, mean radius} \quad r_m = \left(\frac{r_o + r_i}{2}\right) = \left(\frac{1.25 r_i + r_i}{2}\right) = 1.125\, r_i \quad \text{...(1)}$$

$$\text{Torque transmitted } T = \frac{P \times 60}{2\pi N} = \frac{15 \times 60}{2 \times \pi \times 1200} = 0.119 \text{ kN.m}$$

So, $$r_m = \frac{T}{\mu Wn} = \frac{0.119}{0.3 \times 2.65 \times 1} = 0.149 \text{ m} \qquad ...(2)$$

On solving equation 1 and 2,

So, $$d_i = 2 \times r_i = 2 \times 0.133 = 0.266 \text{ m}$$

And $$d_o = 1.25 \times d_i = 1.25 \times 0.266 = 0.3326 \text{ m} = 332.6 \text{ mm}$$

Answer: Outer diameter is 332.6 mm.

GEAR BOX

Example 68: A two-wheel drive tractor has a PTO speed of 540 rpm and it produces 35 kW net engine power. Calculate the corresponding torque available at PTO.
(GATE-2010)

Solution: Given, PTO speed N = 540 rpm; Engine power P = 35 kW

So, torque available at PTO $$T = \frac{P \times 1000}{2\pi N} = \frac{35 \times 1000}{2\pi \times 540} = 618 \text{ Nm}$$

Example 69: A centrifugal pump is driven by tractor PTO through a belt and pulley system. The thickness of the belt is 6 mm permissible stress of 2.8 MN/m^2. Calculate the width of belt based on the following data.

	Tractor P.T.O. Pully	Pump Pully
Diameter, d	0.45m	1.65 m
Angle of wrap, θ	2.3 radians	3.7 radians
Coefficient of friction, K	0.32	0.28
RPM, N	1000	...
Power transmitted, P	25 kW	...

Solution: Since angle of wrap and coefficient of friction are different for different pulleys, the design will be based on the lesser value of $e^{K\theta}$.

So, for tractor PTO pulley $e^{K\theta} = e^{0.32 \times 2.3} = 2.08$

and for pump pulley $e^{K\theta} = e^{0.28 \times 3.7} = 2.81$

So, the design will be based on the tractor pulley.

Now $$\frac{T_1}{T_2} = 2.08 \qquad ... (1)$$

Where T_1 and T_2 = tension on tight side and slack side respectively

Velocity $v = \frac{\pi dN}{60} = \frac{\pi \times 0.45 \times 1000}{60} = 23.56 \text{ m. sec}^{-1}$

Hence, $\frac{(T_1 - T_2)v}{1000} = 25$ Or, $\frac{(T_1 - T_2)23.56}{1000} = 25$

On solving $T_1 - T_2 = 1061.12$ N ... (2)

On solving eqn. 1 and 2, $T_2 = 982.51$ N and $T_1 = 2043.63$ N

Let b = width of belt, given thickness of belt t = 6 mm = 0.006 m and permissible stress = 2.8 MN/m^2, So, $b \times t \times 2.8 \times 10^6 = T_1$

Or, $b \times 0.006 \times 2.5 \times 10^6 = 2043.63$

On solving b = 0.136 m = 13.6 cm

Answer: Width of belt is 13.6 cm

Example 70 A tractor power take-off (PTO) driven stationary peg tooth type wheat thresher operating at a cylinder speed of 540 rpm requires a torque of 250 Nm at PTO. Find the minimum net engine power required. ***(GATE-2012)***

Solution: Given speed N = 540 rpm; Torque T = 250 Nm

Minimum net engine power required $P = \frac{2\pi NT}{60 \times 1000} = \frac{2 \times \pi \times 540 \times 250}{60 \times 1000} = 14.13$ kW

Answer: Minimum net engine power required is 14.13 kW.

Example 71: A tractor gear box has 8 forward speeds. The speed ratios (number of engine revolutions for one revolution of driving wheel) vary in exact geometrical progression. If the speed ratios in highest and lowest gear are 14.9 and 108.8, respectively, then find the geometric constant. ***(GATE-2014)***

Solution: Given, tractor gear box has 8 forward speed which have gear ratio in geometric progression; Speed ratio lowest *i.e.* speed ratio in highest gear $T_n = 14.9$; Speed ratio highest *i.e.*speed ratio in lowestgear a = 108.8

Let, the geometric constant = r, nth term in geometric progression is $T_n = ar^{n-1}$

So, $14.9 = 108.8 \times r^{8-1}$, on solving r = 0.752

Answer: The geometric constant is 0.752

Example 72: The speed reduction in 1st low gear of a tractor gear box and differential with final drive are 5:1 and 40:1, respectively. For the tractor developing 24 kW power at an engine rpm of 2000 with an overall power transmission efficiency of 80 per cent, calculate the total torque in kN.m available at the wheel axle. ***(GATE-2015)***

Solution: Given, speed reduction of 1st low gear = 5:1; Speed reduction in differential with final drive = 40:1; Tractor power P = 24 kW; Engine speed N = 2000 rpm; Power transmission efficiency η = 80 per cent = 0.80

$$\text{Torque at engine} = \frac{P \times 60 \times 1000 \times \eta}{2\pi N} = \frac{24 \times 60 \times 1000 \times 0.8}{2 \times \pi \times 2000} = 91.67 \text{ Nm}$$

$$\text{So, torque available at wheel} = \frac{91.67 \times 5 \times 40}{1000} = 18.33 \text{ kN. m}$$

Answer: Torque available at wheel axle is 18.33 kN.m.

Example 73: An epicyclical gear train consists of a sun gear, four planet gears pinned on a planet carrier (which is mounted on the sun gear) and a fixed ring gear (in which the entire sun-planet gear assembly is inserted). If the ring gear has 56 teeth and each planet gear has 18 teeth, find the number of teeth on the sun gear. ***(GATE-2018)***

Solution: Given, teeth on ring gear T_r = 56; Teeth on each planet gear T_p = 18

Let teeth on sun gear is T_s, So, Ts = $T_r - 2T_p$ = 56 – 2 × 18 = 20

Answer: The number of teeth on sun gear is 20.

Example 74: In a tractor power transmission, the pinion (24 teeth) is in mesh with a gear (46 teeth) on counter shaft and another gear (20 teeth) of counter shaft is in mesh with main shaft gear (50 teeth). The engine is running at 1800 rpm with differential gear ratio 3.5:1 and planetary gear ratio 4:1. If the tractor is fitted with 1.2 m diameter rear wheel, find the forward speed of tractor. ***(GATE-2017)***

Solution: Given, teeth of pinion T_{pinion} = 24; Teeth of first gear on counter shaft T_{g1} = 46; Teeth of second gear on counter shaft T_{g2} = 20; Teeth of gear on main shaft T_g = 50; Engine speed N = 1800 rpm; Differential gear ratio 3.5:1; Planetary gear ratio 4:1; Tractor rear tyre diameter d = 1.2 m

$$\text{Total speed reduction VR} = \frac{T_g \times T_{g1}}{T_{g2} \times T_{pinion}} \times \text{Differential gear ratio} \times \text{Planetary gear ratio} = \frac{50 \times 46}{20 \times 24} \times 3.5 \times 4 = 67.08$$

$$\text{So, wheel revolution } N_{wheel} = \frac{N}{VR} = \frac{1800}{67.08} = 26.83 \text{ rpm}$$

$$\text{So, wheel speed} = \frac{\pi d N_{wheel}}{60} = \frac{\pi \times 1.2 \times 26.83}{60} = 1.68 \text{ m. sec}^{-1} = 6.06 \text{ kmph}$$

Answer: forward speed of tractor is 6.06 kmph.

DIFFERENTIAL AND FINAL DRIVE

Example 75 In tractor differential, the pinion on the propeller shaft has 12 teeth and the crown gear has 60 teeth. The propeller shaft rotates at 1000 rpm and the right rear axle rotate at 210 rpm while taking a left turn. What is the rpm of rear left axle? ***(GATE-2010)***

Solution: Given, teeth on propeller shaft $T_p = 12$; Teeth on crown gear $T_c = 60$; Speed of propeller shaft $N_p = 1000$ rpm; Speed of right rear axle $N_r = 210$ rpm; The above speeds occur when the vehicle is taking left turn.

Let, speed of left rear axle = N_1

Crown gear speed $N_c = \frac{T_p}{T_c} N_p = \frac{12}{60} \times 1000 = 200$ rpm

As, $N_c = N_r = N_l$ during straight run. But during left turn number of revolution added to the right is equal to number of revolution deducted from the left. Thus, left rear axle revolution is

$$N_l = 200 - (210 - 200) = 190 \text{ rpm}$$

Answer: Left rear axle revolution is 190 rpm.

STEERING SYSTEM AND FRONT AXLE

Example 76: A tractors is equipped with the Ackermann steering mechanism. The inner front wheel makes an angel 43° whereas the outer front wheel makes 33°. The distance between kingpins at the axles is 1200 m. assuming correct steering, calculate the wheel base of the vehicle.

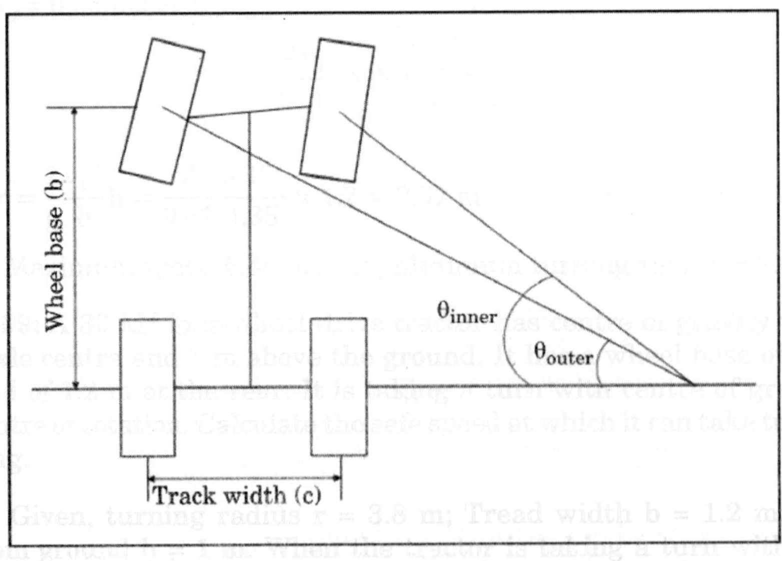

Solution: Given, Angle made by inner front wheel θ_i = 43°; Angle made by outer front wheel θ_O = 33°; distance between kingpins at the axle c = 1200 mm

For correct steering cot θ_O – cot θ_i = where b = wheel base

So, $\cot 33 - \cot 43 = \frac{1200}{b}$

On solving $b = \frac{1200}{\cot 33 - \cot 43} = \frac{1200}{1.54 - 1.07}$ = 2553.19 mm

Answer: The wheel base is 2553.19 mm

WHEELS AND TYRES

Example 77: A load of 3 kN is acting on a tyre having 150 mm nominal width. The effective coefficient of friction of ground and tyre is 0.6 and kingpin offset is 10 mm. Assuming tyre impression on ground is circular having diameter equal to nominal width, find the kingpin torque of the tyre. ***(GATE-2017)***

Solution: Given, load W = 3 kN; Nominal width b = 150 mm = 0.15 m; Effective coefficient of friction between ground and tyre μ = 0.6; Kingpin offset e = 10 mm = 0.01 m

Tyre impression on ground is circular having diameter is equal to nominal width

$$\text{So, kingpin torque } T = W\mu\sqrt{\frac{b^2}{8} + e} = 3 \times 0.6\sqrt{\frac{0.15^2}{8} + 0.01} = 0.2037 \text{ kN.m}$$

$$= 203.7 \text{ Nm}$$

Answer: Kingpin torque is 203.7 N.m

Example 78: A tractor is operated at a speed of 5 kmph in cohesive soil and 6 kmph on non-cohesive soil. If it has a PTO power of 35 kW, find the optimum weight on the rear wheels.

Solution: Given, tractor speed at cohesive soil V_1 = 5 kmph and in non-cohesive soil V_2 = 6kmph; PTO power of tractor P = 35 kW

So, optimum weight on rear wheels during operation in

i) Cohesive soil $= \frac{110 \times P}{V_1} = \frac{110 \times 35}{5} = 770 \text{ kN}$

ii) Non-cohesive soil $= \frac{123 \times P}{V_2} = \frac{123 \times 35}{6} = 717.5 \text{ kN}$

Answer: Optimum weight on rear wheels during operation in cohesive soil is 770 kN and in non-cohesive soil is 717.5 kN.

BRAKES

Example 79 A four-wheel drive tractor weighing 15 kN is travelling down a slope of 12° at a speed of 18 kmph. The brake drum is attached to the final drive pinion which has 12 teeth. The bull gear has 60 teeth. The outer diameter of the rear wheel is 1.25 m. If the tractor is stopped at a distance of 6 m, then calculate the energy to be dissipated. The kinetic energy of the rotating parts to be 20 per cent of the kinetic energy of the tractor load and change in potential energy of tractor.

Solution: Given, slope $\theta = 12°$; Speed of operation $v = 18$ kmph $= 5$ m.sec^{-1}; Tractor weight $W = 15$ kN; Distance stopped $= H_1 - H_2 = 6$ m; Rotational energy = 20 per cent kinetic and potential energy. Total energy absorbed by the drum E_t is given by $E_t = E_k + E_p + E_r$

Where, E_k = kinetic energy, E_p = potential energy, E_r = rotational energy

$$E_k = \frac{Wv^2}{2g} = \frac{15 \times 5^2}{2 \times 9.81} = 19.113 \text{ kN.m}$$

$$E_p = W(H_1 - H_2) = Ws\sin\theta = 15 \times 6 \times \sin 12^0 = 18.71 \text{ kN.m}$$

$$E_r = 0.2(E_k + E_p) = 0.2(19.113 + 18.71) = 7.56 \text{ kN.m}$$

So, total energy absorbed $E_t = E_k + E_p + E_r = 19.113 + 18.71 + 7.56 = 45.388$ kN.m

Answer: The total energy absorbed is 45.388 kN.m.

HYDRAULIC SYSTEM

Example 80: A double acting hydraulic cylinder has a rod diameter equal to one-half the piston diameter. If the system pressure is maintained constant, Find the ratio of load carrying capacity of extension stroke to that of retraction stroke.

(GATE-2013)

Solution: Let D_1 = rod diameter and D_2= piston diameter

Given $D_1 = 0.5 \times D_2$

Let W_1 = load carrying capacity during extension and W_2 = load carrying capacity during retraction

For this system pressure is constant *i.e.* pressure on piston is equal to pressure on rod

So $\frac{W_1}{A_1} = \frac{W_2}{A_2}$ Or, $\frac{W_1}{W_2} = \frac{A_1}{A_2} = \frac{{D_1}^2}{{D_1}^2 - {D_2}^2} = \frac{{D_1}^2}{{D_1}^2 - \frac{1}{4}{D_1}^2} = \frac{1}{1 - \frac{1}{4}} = 1.33$

Answer: Ratio of load carrying capacity of extension stroke to that of retraction stroke is 1.33.

Example 81: A piston pump is driven by a 5 m diameter horizontal axis wind turbine for supplying water from a borehole with a total pump head of 10m. The mean velocity of air is 18 km h^{-1} and the density of air is 1.29 kg m^{-3}. The actual power coefficient of the wind turbine is 0.30 and the overall pump efficiency is 60 per cent. Neglecting the transmission losses, calculate the expected pump discharge in lps. ***(GATE-2013)***

Solution: Given, diameter of horizontal axis wind turbine d = 5 m; Total pump head H = 10 m; Mean velocity of air v = 18 kmph = 5 m.sec^{-1}; Density of air ρ_{air} = 1.29 kg.m^{-3}; Power coefficient C_p = 0.3

Overall pump efficiency η_{pump} = 60 per cent = 0.60

Power available from wind turbine $P = \frac{1}{2}\rho_{air} A v^3 C_D$

$$= \frac{1}{2} \times 1.29 \times \frac{\pi}{4} \times 5^2 \times 5^3 \times 0.3 = 474.92 \text{ W}$$

Power available at the pump $P_{pump} = P \times \eta_{pump}$ = 474.92 × 0.6 = 284.952 W

Let Q = discharge of pump in $m^3 sec^{-1}$

But pump power = η_{water} gHQ = 1000 × 9.81 × 10 × Q

As the power available at pump is 284.952 W, on solving Q = 2.904×10^{-3} m^3 sec^{-1} = 2.904 lps

Answer: Expected pump discharge is 2.904 lps.

Example 82: A hydraulic circuit uses a pump having a fixed displacement volume of 12.5 cm^3 rev^{-1} driven at 1500 rpm. The pump has a volumetric efficiency of 85 per cent and an overall efficiency of 75 per cent. If the system pressure is set at 15 MPa by the relief valve, calculate the power required to drive the pump. ***(GATE-2014)***

Solution: Given, volume of fixed displacement pump V_d = 12.5 cm^3 rev^{-1}; Speed of pump N = 1500 rpm; Volumetric efficiency η_v = 85 per cent = 0.85; Overall efficiency η_o = 75 per cent = 0.75; System pressure P = 15 MPa = 15×10^6 Pa

Theoretical flow rate

$$Q_{th} = V_d \times N = 12.5 \times 1500 = 18750 \ cm^3 min^{-1} = 3.125 \times 10^{-4} m^3 sec^{-1}$$

Actual flow rate $Q_a = Q_{th} \times \eta_v = 3.125 \times 10^{-4} \times 0.85 = 2.66 \times 10^{-4} m^3 sec^{-1}$

So, power required to drive the pump

$$= Q_a \times \frac{P}{\eta_o} = 2.66 \times 10^{-4} \times 15 \times \frac{10^6}{0.75} = 5.32 \text{ kW}$$

Answer: Power required to drive the pump is 5.32 kW.

Example 83: A double acting hydraulic cylinder has a bore of 200 mm with a piston rod of 140 mm diameter. The extending speed of the piston is 80 mm s^{-1}. If the flow of the oil is same as that of extending, find the retract speed of the piston in mm.s^{-1}. ***(GATE-2015)***

Solution: Given, double acting cylinder; Piston diameter D_p = 200 mm; Piston rod diameter D_r = 140 mm; Extend speed of piston V_p = 80 mm.sec^{-1}

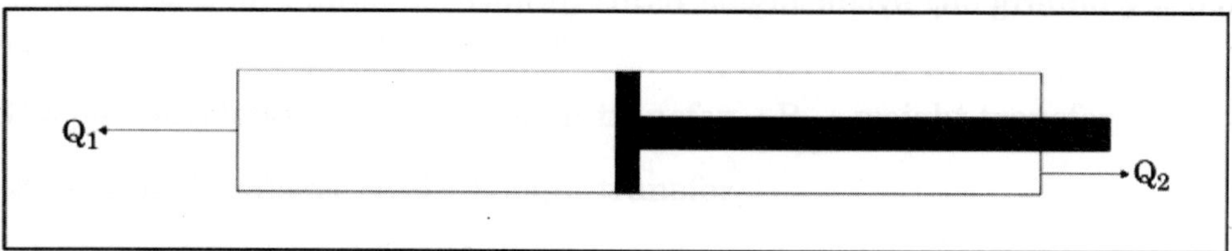

Let the retraction speed be V_r

Since piston is double acting, the discharge at both the end is same

So, $$Q_1 = Q_2$$

Or, $$\frac{\pi}{4} D_p^2 \times V_p = \frac{\pi}{4}(D_p^2 - D_r^2) \times V_r$$

Or, $$\frac{\pi}{4} 200^2 \times 80 = \frac{\pi}{4}(200^2 - 140^2) \times V_r$$

On solving $$V_r = 156.86 \text{ mm.sec}^{-1}$$

Answer: The retract speed of piston is 156.86 mm.sec^{-1}

Example 84: A gear pump gives a discharge of 100 L.min^{-1} against a system pressure of 15 MPa. If the overall efficiency of the pump is 0.75. Calculate the input power to run the pump in kW. ***(GATE-2016)***

Solution:

Given, pump discharge, Q = 100 $L.min^{-1}$ = 0.1 m^3 min^{-1}; System pressure P = 15 MPa = 15000 kPa; Overall efficiency η = 0.75

$$\text{So, input power to the pump} = \frac{Q \times P}{\eta \times 60} = \frac{0.1 \times 15000}{0.75 \times 60} = 33.34\ \text{kW}$$

Answer: Input power to the pump is 33.34 kW.

Example 85: A hydraulic system comprising of a pump and a single acting cylinder lifts 11 kN load. The pump flow rate is 5 $L.min^{-1}$ and overall system efficiency is 80 per cent. The cylinder diameter is 80 mm and its efficiency is 90 per cent. If pressure drop in the circuit is 500 kPa, then find the power required to drive the pump. ***(GATE-2017)***

Solution: Given, load lifted by the system W = 11 kN;

Pump flow rate Q = 5 $L.min^{-1}$ = 8.34 × 10^{-5} m^3 sec^{-1}; Overall system efficiency η = 80 per cent = 0.8; Cylinder diameter d = 80 mm = 0.08 m; Cylinder efficiency $\eta_{cylinder}$ = 90 per cent = 0.9; Pressure drop in the circuit P = 500 kPa

$$\text{Pressure required to lift 11 kN, } P = \frac{W}{\frac{\pi}{4}d^2} = \frac{11}{\frac{\pi}{4}0.08^2} = 2188.38\ \text{kPa}$$

So, theoretical power required to lift the load

$$= PQ = 2188.38 \times 8.34 \times 10^{-5} = 0.182\ \text{kW}$$

$$\text{Actual power required} = \frac{0.182}{0.9} = 0.202\ \text{kW}$$

Power loss due to pressure drop = ΔP × Q = 500 × 8.34 × 10^{-5} = 0.0417 kW

$$\text{So, power required to drive the pump} = \frac{0.202 + 0.0417}{0.8} = 0.304\ \text{kW}$$

Answer: Power required to drive the pump is 0.304 kW.

Example 86: In a tractor hydraulic system, the plunger barrel type pump gives a discharge of 25 $L.min^{-1}$ at a speed of 1500 rpm. Assuming the pump efficiency is 0.8 and length to diameter ratio of the pump is 1.25, find out the designed diameter and stroke of the pump. Find out the power required to operate the pump. The pressure developed in the pump is 25 MPa.

Solution: Given, pump discharge Q = 25 $L.min^{-1}$; speed = 1500 rpm; pump efficiency η = 0.8; pressure developed in the pump p = 25 MPa; length to diameter ratio L:D = 1.25:1

Stroke volume $V_s = \frac{\pi}{4}D^2L = \frac{\pi}{4}D^2(1.25D) = 0.981\ D^3$

So, pump discharge $Q = V_sN\eta = 0.981\ D^3 \times \frac{1500}{60} \times 0.8 = 19.634\ D^3$

But given Q = 25 $l.min^{-1} = 4.167 \times 10^{-4}\ m^3.sec^{-1}$

So, on solving D = 27.68 mm and L = 1.25 × D = 1.25 × 27.68 = 34.60 mm

Power developed by the pump $P_{out} = \frac{Q \times p}{60} = \frac{25 \times 25}{60} = 10.42\,kW$

So, power required to drive the pump is $P_{in} = \frac{P_{out}}{\eta} = \frac{10.42}{0.8} = 13.025\ kW$

Answer: Pump diameter is 17.68 mm, stroke is 34.60 mm and power required to drive is 13.025 kW.

Example 87: In a hydraulic system, the pump delivers fluid at a rate of 20 L. min^{-1} at a speed of 1000 rpm. It develops a pressure of 15 MPa. The power required to operate the pump is 8 kW. Find out the efficiency of the pump and the torque requirement of the pump.

Solution: Given, pump discharge Q = 20 $L.min^{-1}$; Speed = 1000 rpm; Pressure developed in the pump p = 15 MPa; Input power to pump P_{in} = 8 kW

Output power of the pump $P_{out} = \frac{Q \times p}{60} = \frac{20 \times 15}{60} = 5\ kW$

So, efficiency of the pump $\eta = \frac{P_{out}}{P_{in}} = \frac{5}{8} \times 100 = 62.5$ per cent

So, torque required $T = \frac{pQ}{2\pi N\eta} = \frac{15 \times 10^6 \times 20 \times 10^{-3}}{2 \times \pi \times 1500 \times 0.625} = 50.92$ N.m

Answer: Efficiency of the pump is 62.5 per cent and torque required is 50.92N.m

Example 88: A pump requires 12 kW power to develop a pressure of 12 MPa which is operated at a speed of 2000 rpm. Assuming the mechanical efficiency be 0.85, find out the pump flow rate, displacement volume per revolution and theoretical discharge of the pump. The volumetric efficiency of the pump is 0.8.

Solution: Given, input power P_{in} = 12 kW; Developed pressure p = 12 MPa; Speed N = 2000 rpm; Mechanical efficiency η_m = 0.85; Volumetric efficiency η_v = 0.8

Output power $P_{out} = P_{in} \times \eta_m = 12 \times 0.85 = 10.2$ kW

$$\text{Pump flow rate } Q = P_{out} \times \frac{60}{p} = 10.2 \times \frac{60}{12} = 51 \text{ L.min}^{-1}$$

$$\text{Volume of fluid discharged per revolution } D_p = \frac{Q}{N} = \frac{51}{2000} = 0.025 \text{ L}$$

$$\text{Theoretical discharge of the pump } Q_{th} = \frac{Q}{\eta_v} = \frac{51}{0.8} = 63.75 \text{ L.min}^{-1}$$

Answer: Pump flow rate is 51 L.min^{-1}, volume of fluid discharged per revolution is 0.025 L and theoretical discharge of the pump is 63.75 L.min^{-1}

Example 89: A double acting hydraulic cylinder has a piston diameter of 80 mm and rod diameter of 30 mm. The maximum pressure and discharge is 20 MPa and 50 L.min^{-1}, respectively. The inlet port is connected to the pump while the outlet port is to the reservoir. Calculate the maximum load on cylinder while the piston is extending and retracting and the piston speed when extending.

Solution: Given, piston diameter, d_p = 80 mm = 8 cm; rod diameter d_r = 30 mm = 3 cm; Developed pressure p = 20 MPa; Pump discharge Q = 50 L.min^{-1}; Inlet port is connected to the pump; Outlet port is connected to the reservoir

$$\text{Piston area } A_p = \frac{\pi}{4} d_p^{\,2} = \frac{\pi}{4} \times 8^2 = 50.26 \text{ cm}^2$$

$$\text{Rod area } A_r = \frac{\pi}{4} d_r^{\,2} = \frac{\pi}{4} \times 3^2 = 7.06 \text{ cm}^2$$

Let P_1 is the pressure while extending and P_2 is the pressure while retracting

While extending P_2 = 0 and during retracting P_1 = 0 and the maximum value of P_1 and P_2 is 20 MPa

$$\text{Maximum force while extending } F_e = \frac{P_1 \times A_p}{10} = \frac{20 \times 50.26}{10} = 100.52 \text{ kN}$$

$$\text{Maximum force while retracting } F_r = \frac{P_2 \times (A_p - A_r)}{10} = \frac{20 \times (50.26 - 7.06)}{10} = 86.40 \text{ kN}$$

$$\text{Piston speed while extending } V_p = \frac{Q}{6A_p} = \frac{50}{6 \times 50.26} = 0.16 \text{ m.sec}^{-1}$$

Answer: Maximum force while extending is 100.52 kN and while retracting is 86.40 kN and piston speed while extending 0.16 m.sec^{-1}.

Example 90: A pump has a flow rate of 40 $L.min^{-1}$ and pressure rise in the pump is 15 MPa. The volumetric efficiency is 0.9 and the torque efficiency is 0.85. Find out the theoretical fluid discharge and the power required to drive the pump.

Solution: Given, pump flow rate Q = 40 $L.min^{-1}$; pressure rise p = 15 MPa; Volumetric efficiency η_v = 0.9; torque efficiency η_m = 0.85

$$\text{Output power } P_{out} = \frac{Q \times P}{60} = \frac{40 \times 15}{60} = 10 \text{ kW}$$

$$\text{So, input power required } P_{in} = \frac{P_{out}}{\eta_m \eta_v} = \frac{10}{0.85 \times 0.9} = 13.07 \text{ kW}$$

$$\text{Theoretical pump discharge } Q_{th} = \frac{Q}{\eta_v} = \frac{40}{0.9} = 44.45 \text{ L.min}^{-1}$$

Answer: Input power required is 13.07 kW and theoretical pump discharge is 44.45 $L.min^{-1}$.

ELECTRICAL SYSTEM

Example 91: Calculate the power of battery and amount of energy stored in a 12 V battery of capacity 500 Ah. Assume battery discharge duration is 10 hours.

Solution: Given, battery terminal voltage (V) = 12 V; Battery capacity (C) = 500 Ah. Since, the battery is discharged in 10 hours, the current drawn from the battery is

Discharge current = capacity (Ah) / discharge duration (h) = 500/10 = 50 A

Power = voltage (V) x Current (A) = 12 x 50 = 600 watt. Thus, the power of battery is 600 watt.

The energy stored in a battery is given by the expression,

Energy = voltage (V) x capacity (C) = 12 V x 500 Ah = 6000 wh = 6.0 kWh.

Therefore, energy stored in the battery is 6.0 kWh.

TRACTION AND MECHANICS OF TRACTOR CHASSIS

Example 92: A tractor weighing 25kN has a drawbar power of 25 kW. It is pulling an implement at a speed of 5 kmph on a level ground. The centre of gravity of tractor is 0.92 m ahead of rear axle and 1.2 m above the ground and wheel base is 2.3 m. The total contact area of rear wheel is 0.102 m^2 while the point of hitch is located 0.5 m above the ground surface and 0.45 m behind the rear axle. Assuming the angle of internal frication as 18° and cohesion coefficient as 110 kN/m^2 for soil.

Calculate i) Angle of inclination of line of pull ii) Soil reaction at rear and front wheel iii) Weight transfer and iv) maximum drawbar pull that the tractor can exert without overturning, if angle of inclination remains unchanged.

Solution: Given, weight of tractor, W_t = 25kN; Drawbar power, P_d = 25 kW; Speed of operation, $v = 5\ \frac{km}{hr} = 5 \times \frac{5}{18} m.sec^{-1} = 1.38\ m.sec^{-1}$; CG location ahead of rear axle, l_1 = 0.92 m and above the ground, y = 1.2 m; Wheel base, l = 2.3 m; Rear wheel contact area, A = 0.102 m^2; Hitch point location behind the rear wheel, l_3 = 0.5 m and above ground y_1 = 0.45 m; Angle of internal friction ϕ = 18°; Cohesion coefficient, C = 110 kN.m^{-2}

i. Drawbar pull, $H = \frac{P_d}{v} = \frac{25}{1.38} = 18.11\ kN$

Tractive force, $F = AC + R_r \tan \emptyset$

Where, R_r = dynamic soil reaction forces on rear wheels

So, $F = (0.102 \times 110) + R_r \tan 18° = 11.22 + 0.324\ R_r$ ....(1)

Fig. 1 shows the forces acting on a tractor while moving on a level ground with line of pull, P inclined at an angle α with the horizontal. Applying Newton's law of motion, summation of vertical forces give

$$R_r + R_f = W_t + H \sin \alpha$$

Or $$R_r + R_f = 25 + 18.11 \sin \alpha \quad ...(2)$$

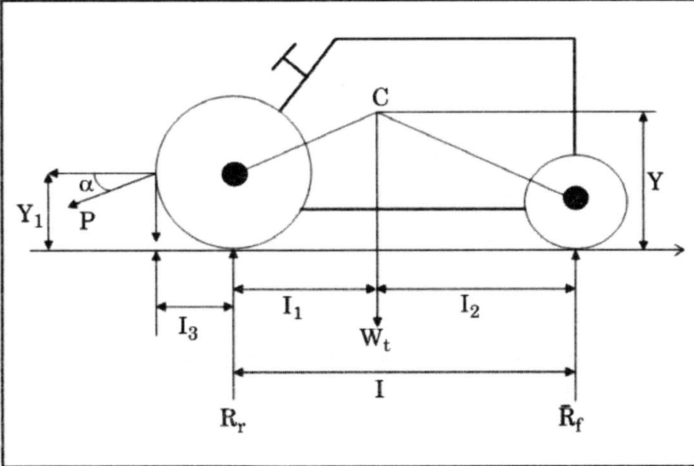

Fig. 1: Tractor free body diagram while moving on level ground with pull

From summation of horizontal forces

$$F = H \cos \alpha \text{ Or } F = 18.11 \cos \alpha$$

Equating eqn. 1 and 3

$$11.2 + 0.324R_r = 18.11 \cos \alpha \quad ... (3)$$

Taking moment about point A,

$$R_f \times l + P \times y_1 \times \cos\alpha + P \times l_3 \times \sin\alpha = W_t \times l_1$$

$$\text{Or, } R_f \times 2.3 + 18.11 \times 0.45 \times \cos\alpha + 18.11 \times 0.5 \times \sin\alpha = 30 \times 0.92 \quad (4)$$

On solving eqn. 2, 3 and 4 to eliminate R_f and R_r, we get

$$\sin\alpha - 19.798\cos\alpha + 3.35 = 0$$

On solving we get $\alpha = 23.41°$

ii. Substituting the value of α in eqn. 3

$11.22 + 0.324\ R_r = 18.11\cos\alpha$ Or, $11.22 + 0.324\ R_r = 18.11\cos\alpha\ 23.41°$

Or, $R_r = 16.72$ kN

Substituting the value in eqn. 2

$R_r + R_f = 30 + 18.11\sin\alpha$ Or, $16.72 + R_f = 25 + 18.11\sin\alpha\ 23.41°$

Or, $R_f = 15.47$ kN

iii. Weight transfer $= \dfrac{H(y_1\cos\alpha + l_3\sin\alpha)}{l}$

The maximum permissible drawbar pull will be when reaction on front wheel reduces to zero. For that condition taking moment about point A

$P_{max} \times y_1 \times \cos\alpha + P_{max} \times l_3 \times \sin\alpha = W_t \times l_1$

Or, $P_{max} \times 0.45 \times \cos\alpha\ 23.41° + P_{max} \times 0.5 \times \sin\alpha\ 23.41° = 25 \times 0.92$

Or, $P_{max} = 37.60$ kN

Answers: $\alpha = 23.41°$; $R_r = 16.72$ kN; $R_f = 15.47$ kN; Weight transfer = 4.81 kN; $P_{max} = 37.60$ kN

Example 93: Determine the location of the centre of gravity for the tractor having wheel base 2.2 m, radius of rear wheel 0.9 m and front wheel 0.45 m and width of rear wheel 0.3 m. Weight of the tractor is 30 kN. Reaction of front wheel when it is on level ground is 7.5 kN and that when it is lifted to 0.5 from ground is 5.6 kN.

Solution: Given, wheel base, l = 2.2 m; Radius of rear wheel, $r_r = 0.9$ m; Radius of front wheel, $r_f = 0.45$ m; Width of rear wheel, b = 0.3 m; Total weight of the tractor, $W_t = 30$ kN; Weight carried by front wheels on level ground, $R_f = 7.5$ kN; Weight carried by front wheels when lifted 0.5 m(n) from ground, $R_f' = 5.6$ kN

Taking moment about point A

$$W_t \times l_1 = R_r \times l \text{ Or, } 30 \times l_1 = 7.5 \times 2.2 \text{ Or, } l_1 = 0.55 \text{ m}$$

Difference in radius of both the wheels, $\Delta r = r_r - r_f = 0.9 - 0.45 = 0.45$

After lifting the front wheel, the wheel base becomes l' and is given by

$$l' = \sqrt{l^2 - (n - r_r)^2 + \Delta r^2} = \sqrt{2.2^2 - (0.95 - 0.9)^2 + 0.45^2} = 2.24\text{m}$$

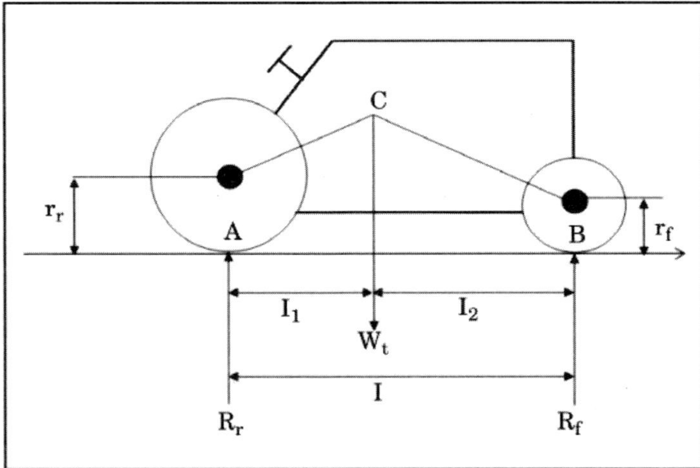

Fig. 1: Tractor free body diagram while moving on level ground

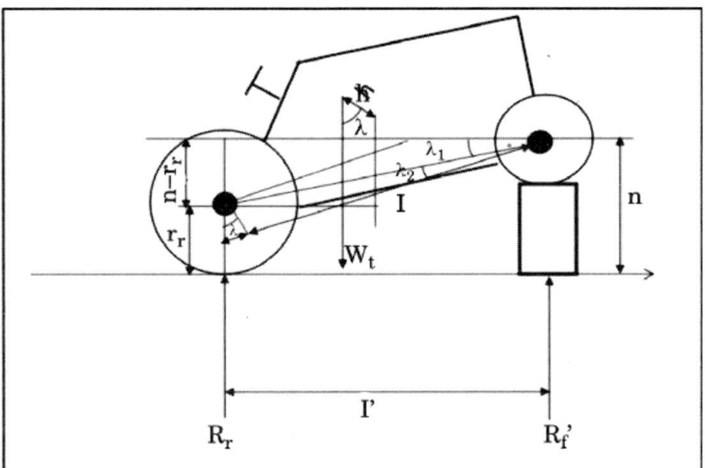

Fig. 2: Tractor free body diagram in front wheel lifted position

Now the angles $\tan \lambda_1 = \dfrac{n - r_r}{l'} = \dfrac{0.95 - 0.9}{2.24} = 0.022$

$$\tan \lambda_2 = \frac{\Delta r}{l} = \frac{0.45}{2.2} = 0.204$$

So $\lambda_1 = 1.27°$ and $\lambda_2 = 11.56°$

And $\lambda = \lambda_1 + \lambda_2 = 1.27° + 11.56° = 12.83°$

Now the shifting of centre of gravity h is given by $h = \dfrac{W_t l_1 - R_f' l}{W_t \tan \lambda} - \dfrac{R_f' \Delta r}{W_t}$

$$\text{or, } h = \frac{30 \times 0.55 - 5.4 \times 2.2}{30 \times tan12.08} - \frac{5.6 \times 0.45}{30} = 0.63$$

So, y = 0.9 + 0.63 = 1.53 m

Answer: Location of centre of gravity will be 0.55 m ahead of rear axle and 1.53 m above the ground level.

Example 94: A track type tractor weighing 40 kN with tracks 0.4 m wide and 1.8 m long. Soil parameters are: Cohesion, C = 12 kN/m^2, Angle of internal friction, φ = 28°, Cohesive modulus of determination K_c = 2.8, Friction modulus of deformation, K_φ = 0.7, Coefficient of wheel shrinkage, n = 0.3. Assume the lugs of the tractor to be such that the soil is sheared off in a plane area at the ends of the lugs. Determine the maximum pull that the tractor can exert.

Solution: Given weight of tractor W_t = 40 kN; Width of track = 0.4 m and length of track = 1.8 m, pull is given by P = F – R, Where, P = pull, F = thrust, R = rolling resistance, Soil thrust F is related with angle of internal friction and cohesion is given by

F = AC + W tanϕ Where, A = contact area of wheel, W = weight on traction wheel. So, area is given by A = width × length = 0.4 × 1.8 = 0.72 m^2

In a track type tractor, the centre of gravity is at the centre line of base. So

$$W = \frac{W_t}{2} = \frac{40}{2} = 20 \text{ kN}$$

So F = AC + W tanϕ = (0.72 × 12) + (20 × tan 28°) = 19.27 kN

The rolling resistance(R) of the wheel is given by

$$R = \frac{2}{(n+1)(K_c + bK_\emptyset)^{1/n}} \left(\frac{W}{2L}\right)^{\frac{n+1}{n}}$$

$$= \frac{2}{(0.3+1)(2.8 + 40 \times 0.7)^{\frac{1}{0.3}}} \left(\frac{20 \times 1000}{2 \times 180 \times 9.81}\right)^{\frac{0.3+1}{0.3}} = 1.36 \times 10^{-3}\text{kg}$$

So, for pull P = 19.27 – 1.34 × 10^{-5} = 19.26 kN

For both the tracks, pull = 19.26 kN × 2 = 38.53 kN

Answer: The maximum pull that the tractor can exert 38.53 kN

Example 95: A four-wheel drive tractor weighing 35 kN is climbing up a 20 per cent slope. The tractor has a wheel base of 2.3 m and drawbar height of 0.35 m.

The centre of gravity is located 1.2 m front of the rear axle and 0.6 m above the ground. The area of contact of each wheel with soil is 0.3 m². The soil parameters are: Cohesion, C = 4.5 kN/m², Angle of internal friction, φ = 30°. If the pull is assumed to be parallel to the soil surface and engine horse power is not limiting, determine the maximum drawbar pull. The rolling resistance can be ignored.

Solution:

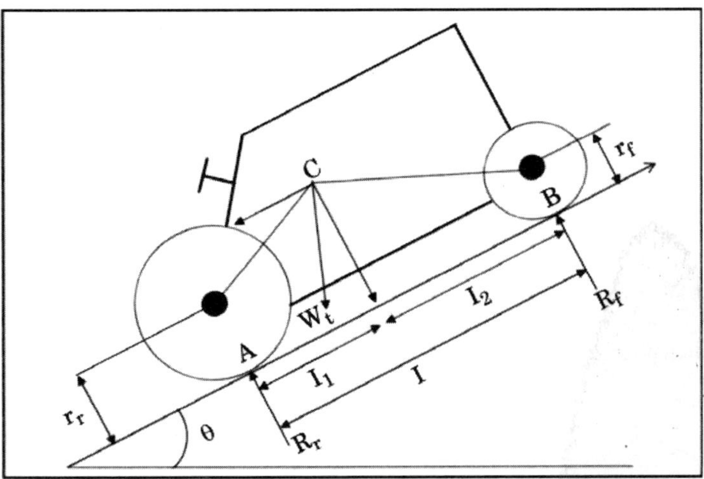

Fig. 3: Tractor on a slope of θ

Fig. 3 shows the tractor climbing up the slope of θ and the reaction at wheels. Given, weight of the tractor W_t = 35 kN; Wheel base l = 2.3 m; Drawbar height y_1 = 0.35 m; Location of centre of gravity from rear axle centre l_1 = 1.2 m and from ground level y = 0.6 m; Area of contact of each wheel A = 0.3 m²; Cohesion C = 4.5 kN.m^{-2}; Angle of internal friction φ = 30°; Pull is parallel to the ground; Slope sin θ

$= 20 \text{ per cent} = \frac{20}{100} = \frac{1}{5}$

So θ = 11.53° and cos θ = 0.979

If R_f and R_r are the soil reaction forces on front and rear wheel, then

$$R_f + R_r = W_t \cos\theta \text{ Or, } R_f + R_r = 35 \times \cos\theta\ 11.53° = 34.29 \text{ kN} \quad (1)$$

Taking moment about point A,

$$W_t \cos\theta \times l_1 = R_f \times l + P \times y_1 + W_t \sin\theta \times y$$

$$\text{Or, } 35 \cos\theta\ 11.53° \times 1.2 = R_f \times 2.3 + P \times 0.35 + 35 \sin\theta\ 11.53° \times 0.6$$

$$\text{Or, } 2.3\ R_f - 0.35\ P = 36.954 \quad \ldots (2)$$

Given F – P = 0

Where $F = AC + W \tan\theta = 0.3 \times 4.5 + R_r \tan\theta\ 30° = 1.35 + 0.58\ R_r$ ... (3)

On solving eqn. 1, 2 and 3 we get maximum drawbar pull P = 11.17 kN

Answer: The maximum drawbar pull is 11.17 kN

Example 96: A tractor with rubber tyres weighs 22 kN. It is pulling a plough at 6 kmph on a level ground and comes to a sudden stop after striking an immovable boulder. Assuming the coefficient of traction as 0.61, the maximum possible drawbar power as 20 kW and 60 per cent of the tractor weight on the rear wheel, calculate the kinetic energy to be abstracted to stop the tractor. Find the maximum drawbar pull and the percentage excess pull above the maximum tractive ability assuming the tractor comes to a stop (not declutched) by a drawbar spring in 20 cm.

Solution: Given, weight of the tractor W_t = 22 kN; Speed of operation v = 6 kmph =1.67 $m.sec^{-1}$; Coefficient of friction $\mu = 0.61$; Maximum possible drawbar power P_1 = 20 kW; Tractor weight on rear wheel = 60 per cent of weight of tractor

So, $R_r = 0.6 \times W_t = 0.6 \times 22 = 13.2$ kN, $y_2 - y_1 = 0.2$ m

Increase in potential energy in the drawbar spring is given by

$$(y_2 - y_1)\left(\frac{P_{max} + T_1}{2}\right)$$

Where P_{max} = maximum drawbar pull, T_1 = drawbar pull

$$T_1 = \frac{P_1}{v} = \frac{20}{1.67} = 11.97 \text{ kN}$$

So, increase in potential energy $0.2\left(\frac{P_{max} + 11.97}{2}\right) \text{kN.m}$

$$\text{Kinetic energy of the tractor } = \frac{1.1}{2}\frac{W_t}{g}v^2 = \frac{1.1}{2}\frac{22}{9.81}1.67^2 = 3.439 \text{ kN.m}$$

Work done by the drive wheel when the tractor is stopping, is given by

$$\mu R_r (y_2 - y_1) = 0.61 \times 13.2 \times 0.2 = 1.6104 \text{ kNm}$$

Since the tractor is equipped with a drawbar and spring to absorb energy and the clutch is being kept engaged after the implement is stopped, the increase in potential energy in drawbar spring is equal to the kinetic energy and the work done by the drive wheels while the tractor is stopping. This forms the law of conservation of energy.

i.e. $0.2\left(\frac{P_{max} + 11.97}{2}\right) = \frac{1.1}{2}\frac{W_t}{g}v^2 + \mu R_r(y_2 - y_1)$

or, $0.2\left(\frac{P_{max}+11.97}{2}\right) = 3.439 + 1.6104$ or, $P_{max} = 18.524$ kN

percentage increase in pull is given by

$$\frac{P_{max} - T_1}{P_{max}} \times 100 = \frac{18.524 - 11.97}{38.524} = 35.38 \text{ per cent}$$

Answers: Kinetic energy to be extracted = 3.439 kJ; Maximum drawbar pull = 18.524 kN; Percentage increase in pull = 35.38 per cent

Example 97: A four-wheel tractor weighing 25 kN is travelling at 6 kmph speed on a slope of 12°. The centre of gravity is located 0.72 m from the centre of rear axle. It is pulling a 3-bottom mould board plough having width 0.35 m. The line of pull makes an angle 25° with the horizontal. The hitch point is located at a distance of 0.4 m from the rear wheel contact point. Reaction force of front wheels when it is on the level ground is 6.3 kN and 5.4 kN when it is lifted to about 0.46 m. The wheel base is 2.3 m. Determine drawbar pull and drawbar power. when the tractor is about to topple on rear wheels. Taking the average soil pressure of 75 $kN.m^{-2}$. Calculate the maximum depth of penetration that the plough can go.

Solution: Given a four-wheel drive tractor; Total weight of the tractor, W_t = 25 kN; Location of centre of gravity from rear axle centre l_1 = 0.72 m; Speed of operation = 6 kmph = 1.67 $m.sec^{-1}$; Slope θ = 12°; 3 bottom mb plough having width 0.35 m, so total width = 3 × 0.35 = 1.05 m; Line of pull makes an angle α = 25° with the ground; Hitch point distance from rear wheel centre l_3 = 0.4 m; Average soil pressure = 75 kN/m^2; Weight carried by front wheels on level ground, R_f = 6.3 kN; Weight carried by front wheels when lifted 0.46m(n) from ground, R_f = 5.4 kN; Wheel base, l = 2.3 m

Now summing the forces along the direction of motion

$$F - P\cos\alpha - W_t \sin\theta = 0 \qquad \text{... (1)}$$

Again, summing the forces perpendicular to the direction of motion

$$R_f + R_r - P\sin\alpha - W_t\cos\theta = 0 \qquad \text{... (2)}$$

For the limiting case, when the tractor is toppling about the rear wheel, $R_f = 0$

So, from eqn. 2, $R_r = P\sin\alpha + W_t\cos\theta$ ... (3)

Taking moments about point A,

$$W_t l_1 - R_f l - P l_3 = 0 \qquad \text{... (4)}$$

For the limiting condition, eqn. 4 results in

$$W_t l_1 = P l_3 \text{ Or, } P = \frac{W_t l_1}{l_3} = \frac{25 \times 0.72}{0.4} = 45 \text{ kN}$$

From eqn. 3, $R_r = P\sin\alpha + W_t \cos\theta = 45 \times \sin\alpha\ 25° + 25 \times \cos\alpha\ 25° = 41.67$ kN

Drawbar power $P \times v = 45 \times 1.67 = 75.15$ kW

Assuming the soil pressure to be constant, soil pressure × depth × width = total pull

$$\text{or, } 75 \times \text{depth} \times 1.05 = 45, \text{ or, depth} = \frac{45}{75 \times 1.05} = 0.5714 \text{ m} = 57.14 \text{ cm}$$

Answer: Maximum depth of penetration is 57.14 cm.

Example 98: A four-wheel drive tractor has centre of gravity located at 1.2 m from the level ground. It has a turning radius of 4.5 m and a tread width of 1.35 m. Assuming no other lateral forces, determine the turning speed in m/sec. Also calculate the minimum stable turning radius when it is operated at 23 kmph.

Solution: Given, turning radius r = 4.5 m; Tread width b = 1.35 m; Speed V_2 = 23 kmph = 6.38 m.sec^{-1}; Centre of gravity from ground h = 1.2 m

a) The centrifugal force F due to turning is given by

$$F = \frac{W}{g}\frac{v_1^2}{r}$$

Taking moment about the contact point of the rear wheel,

$$F \times h = W \times \frac{b}{2} \text{ Or, } \frac{W}{g}\frac{v_1^2}{r} \times h = W \times \frac{b}{2}$$

$$\text{Or, Or, } v_1 = \sqrt{\frac{b}{2} \times g \times r} = \sqrt{\frac{1.35}{2} \times 9.81 \times 4.5} = 5.45 \text{ m.sec}^{-1}$$

b) For v_2 = 6.38 m.sec^{-1},

$$\frac{W}{g}\frac{v_1^2}{r} \times h = W \times \frac{b}{2}$$

$$\text{Or, } r = \frac{2}{g}\frac{v_1^2}{b}h = \frac{2}{9.81}\frac{6.38^2}{1.35} \times 1.2 = 7.37 \text{ m}$$

Answers: Maximum speed 5.45 m.sec^{-1}; Minimum turning radius = 7.37 m

Example 99: A 30 kN four-wheel drive tractor has centre of gravity 0.8 m from the rear axle centre and 1 m above the ground. It has a wheel base of 2.2 m and wheel tread of 1.2 m at the rear. It is taking a turn with centre of gravity 3.8 m from its centre of rotation. Calculate the safe speed at which it can take turn without overturning.

Solution: Given, turning radius r = 3.8 m; Tread width b = 1.2 m; Centre of gravity from ground h = 1 m. When the tractor is taking a turn with a turning radius there is a centrifugal force $\frac{W}{g}\frac{v^2}{r}$ where W is the weight of the tractor and v is the peripheral velocity.

For turning, the moment causing the tipping should be more or equal to the restoring moment. Mathematically $\frac{W}{g}\frac{v^2}{r}h > W\frac{b}{2}$

For limiting case, $\frac{W}{g}\frac{v^2}{r}h = W\frac{b}{2}$ Or, $v = \sqrt{\frac{g\,r\,b}{2h}} = \sqrt{\frac{9.81\times 3.8\times 1.2}{2\times 1}} = 4.72\ \text{m.sec}^{-1}$

Answer: Turning speed = 4.72 m.sec^{-1}

Example 100: A 30 kN three-wheel drive tractor has centre of gravity 0.9 m ahead of rear axle centre and 1 m above the ground. The horizontal distance of centre of gravity of the other rear wheel is 0.5 m. It is taking a turn at a speed of 6 kmph. Calculate the safe turning radius at which it can take turn without overturning.

Solution: Given weight of the tractor W_t = 30 kN; Speed = 6 kmph = 1.67 m/sec; Location of centre of gravity ahead of the rear axle l_1 = 0.9 m and above the ground = h = 1.0 m; Distance between centre of gravity and rear wheel $\frac{b}{2}$ = 0.5 m

Let r = turning radius, the centrifugal force resulting from turning passes through centre of gravity is given by $F = \frac{W}{g}\frac{v^2}{r}$

For stability taking, moment about A, the stability moment Wx1 must be less than or equal to overturning moment $\frac{W}{g}\frac{v^2}{r}h$ *i.e.* $W\frac{b}{2} \le \frac{W}{g}\frac{v^2}{r}h$. For limiting case

$$\frac{b}{2} = \frac{1}{g}\frac{v^2}{r}h$$

$$\text{Or, } r = \frac{1}{g}\frac{v^2}{\frac{b}{2}}h = \frac{1}{9.81}\frac{1.67^2}{0.5}\times 1.0 = 0.56\ \text{m}$$

Answer: The turning radius is 0.56 m

Example 101: A 37 kW two-wheel drive tractor weighing 20 kN with a wheel base of 2.1 m is having the option to be fitted with either 12.4-28 12PR or 13.6-28 12PR at the rear axle. The ratio of section height and section width for all tyres is 0.75. On a level ground, the weight distribution on the front and rear axles is 35 and 65 per cent of the total tractor weight, respectively. Cone index of soil is 1200 kPa.

a) Calculate the motion resistance ratio of both the above rear wheels at recommended inflation pressure and operating on a level ground.

b) Find out the net traction developed by 13.6-28 12 PR tyre operating at recommended inflation pressure and on a level ground with 15 per cent slip.

(GATE-2010)

Solution: Given, 37 kW two-wheel drive tractor; weight of tractor W = 20 kN; wheel base l = 2.1 m; rear wheel size i) 12.4–28 12 PR ii) 13.6–28 12 PR; Ratio of section height and section width = 0.75; Weight distribution on front W_1 = 35 per cent = 0.35 × 20 = 7 kN; Weight distribution on rear W_2 = 65 per cent = 0.65 × 20 = 13 kN; Cone index of soil C_I = 1200 kPa

a) For 12.4–28 12 PR tyre

b_1 = 12.4" = 0.314 m

h_1 = 0.75 × 0.314 = 0.2355 m

Rim diameter = 28" = 0.711 m

d_1 = 1.06 × 0.711 + 2 × 0.2355 = 1.22 m (generally, diameter of wheel is taken 6 per cent more than the rim diameter)

$$W_1 = \frac{7}{2} = 3.5 \text{ kN}$$

$$C_{n1} = \frac{CIb_1d_1}{W_1} = \frac{1200 \times 0.314 \times 1.22}{3.5} = 130.37$$

So, motion resistance ratio $\rho_1 = \frac{1.2}{C_{n1}} + 0.04 = \frac{1.2}{130.37} + 0.04 = 0.05$

b_2 = 13.6" = 0.34 m

h_2 = 0.75 × 0.34 = 0.255 m

Rim diameter = 28" = 0.711 m

d_2 = 1.06 × 0.711 + 2 × 0.255 = 1.26 m

$$W_2 = \frac{13}{2} = 6.5 \text{ kN}$$

$$C_{n2} = \frac{CIb_1d_1}{W_2} = \frac{1200 \times 0.34 \times 1.26}{6.5} = 79.08$$

So, motion resistance ratio $\rho_1 = \frac{1.2}{C_{n2}} + 0.04 = \frac{1.2}{79.08} + 0.04 = 0.055$

b) For 13.6-28 12 PR tyre

C_n = 79.08

Given slip s = 15 per cent = 0.15

$$\text{So, } \frac{\text{Net traction}}{\text{Load}} = 0.75(1 - e^{-0.3C_n s}) = 0.75(1 - e^{-0.3 \times 79.08 \times 0.15}) = 0.729$$

Net traction = 0.729 × 13 = 9.48 kN

Example 102: A tow wheel drive tractor, weighing 15.84 kN with a wheel base of 21600 mm. has the static weight divided between the front and rear axles in the ratio of 30:70 on a horizontal level surface. The hitch pint is at a height of 700 mm from the ground and at a horizontal distance of 120 mm to the rear side from the centre of the rear axle. Pull acts at an angle of 12° downwards from the horizontal. Calculate the maximum pull in kN, when the front wheels would just start rising from the ground. ***(GATE-2012)***

Solution: Given, weight of tractor W = 15.84 kN; Wheel basel = 2160 mm = 2.160 m; Weight distribution on front and rear wheel = 30:70; Location of hitching above the ground y = 700 mm = 0.7 m; Behind rear wheel centre l_3 = 120 mm = 0.12 m; Pull makes angle with horizontal α = 12°

So, reaction force on front wheel R_f = 0.3 × 15.84 = 4.752kN

Distance between centre of gravity and rear axle centre

$$l_1 = \frac{R_f \times l}{W} = \frac{4.752 \times 2.16}{15.84} = 0.648 \text{ m}$$

$$\text{Weight transfer} = \frac{P(y\cos\alpha + l_3\sin\alpha)}{l}$$

In dynamic condition, weights on front wheel $R_f = \frac{W \times l_1}{l} - \frac{P(y\cos\alpha + l_3\sin\alpha)}{l}$

For limiting condition $R_f = 0$

$$\textit{i.e.} \Rightarrow \frac{W \times l_1}{l} = \frac{P(y\cos\alpha + l_3\sin\alpha)}{l} \Rightarrow P = \frac{W \times l_1}{y\cos\alpha + l_3\sin\alpha} = \frac{15.84 \times 0.648}{0.7\cos 12 + 0.12\sin 12} = 14.46 \text{ kN}$$

Answer: Maximum pull is 14.46 kN.

Example 103: A two-wheel drive tractor pulls an implement that has a draft force of 11.5 kN. The total motion resistance of the tractor is 2.5 kN. Under these circumstances, the slip of the drive wheels is 20 per cent. If the power loss in transmission is 20 per cent, find the percentage of power lost in converting engine power into drawbar power. ***(GATE-2013)***

Solution: Given, draft force F_d = 11.5 kN; Motion resistance force F_{mr} = 2.5 kN; Slip s = 20 per cent = 0.2; Power loss in transmission = 20 per cent

So, transmission efficiency η = 100 – 20 = 80 per cent

Net force available at engine $F_e = (F_d + F_{mr})(1 + s)\frac{1}{\eta} = (11.5+2.5)\times 1.2\times \frac{1}{0.80} = 21$ kN

So, power lost in converting engine power to drawbar power

$$= \frac{F_e - F_d}{F_e} \times 100 = \frac{21 - 11.5}{21} \times 100 = 45.23 \text{ per cent}$$

Answer: Power loss in converting engine power to drawbar power is 45.23 per cent.

Example 104: A four-wheel-drive tractor has a static weight of 50 kN with 40 per cent weight on rear axle and 60 per cent weight on front axle. The wheel base is 2 m. The tractor is pulling a disc harrow that exerts a level drawbar pull at a hitch height of 0.5 m from the ground. During operation, when the dynamic reaction on each axle is same, find out the dynamic reaction ratio developed by the tractor. ***(GATE-2014)***

Solution: Given, tractor static weight W = 50 kN; Weight on rear R_r = 40 per cent of W = 0.4 × 50 = 20 kN; Weight on front R_f = 60 per cent of W = 0.6 × 50 = 30 kN; Wheel base l = 2 m; Pull is horizontal; Hitch height above the ground y = 0.5 m; Dynamic reaction on each axle is same.

Let P = pull developed, and R_f – weight transfer = R_r + weight transfer

Or, 30 – weight transfer = 20 + weight transfer

On solving, weight transfer = 5 kN

As weight transfer $= \frac{Py}{l} = \frac{P \times 0.5}{2} = 0.25\ P = 5$ kN

On solving P = 20 kN, so, dynamic reaction ratio $= \frac{P}{W} = \frac{20}{50} = 0.4$

Answer: Dynamic reaction ratio is 0.4

Example 105: A two-wheel drive tractor weighing 18 kN has a wheel base 1.8 m. Its centre of gravity is located 600 mm ahead of the rear axle centre, under static condition, on a level ground. When this tractor is used to pull a disc, plough hitched at a height of 390 mm from the ground, the draft observed is 6 kN. Calculate change in reaction on rear wheels of the tractor due to pull in kN. ***(GATE-2018)***

Solution: Given, weight of tractor W = 18 kN; Wheel base l = 1.8 m; CG location ahead of rear axle centre l_1 = 600 mm = 0.6 m; Hitch height y = 390 mm = 0.39 m; Draft observed L = 6 kN

Weight on rear axle without application of pull

$$R_r = \frac{W(l - l_1)}{l} = \frac{18(1.8 - 0.6)}{1.8} = 12 \text{ kN}$$

Weight on rear axle with application of pull

$$R_r' = \frac{W(l - l_1)}{l} - \frac{Ly}{l} = \frac{18(1.8 - 0.6)}{1.8} - \frac{6 \times 0.39}{1.8} = 10.7 \text{ kN.}$$

So, change in reaction on rear wheel = $R_r - R_r'$= 12 – 10.7 = 1.3 kN

Answer: Change in reaction of rear wheel on tractor due to application of pull is 1.3kN.

Example 106: A two-wheel drive tractor with rear wheel rolling radius of 600 mm develops a tractive force of 18.5 kN at a wheel slip of 12 per cent. The engine speed is 2400 rpm and the transmission ratio from the engine to the rear wheels is 120:1. If the tractor experiences a motion resistance of 1.75 kN, calculate the drawbar power developed by the two-wheel drive tractor. ***(GATE-2018)***

Solution: Given, rear wheel rolling radius r_r = 600 mm; Developed tractive force H = 18.5 kN; Slip s = 12 per cent = 0.12; Engine speed N = 2400 rpm; Engine to wheel speed reduction 120:1; Motion resistance R = 1.75 kN

Torque available at the wheel T = $(H - R)r_r$ = (18.5 – 1.75) × 0.6 = 10.05 kN.m

Wheel speed $N_w = \frac{N}{120} = \frac{2400}{120}$ =20 rpm

Drawbar power developed = $\frac{2\pi N_w T}{60}(1 - s) = \frac{2 \times \pi \times 20 \times 10.05}{60}$ = 18.52 kW

Answer: Drawbar power developed is 18.52 kW.

Example 107: A tractor pulls 8 kN drawbar pull against 4 kN rolling resistance. If tractor develops 57 per cent tractive efficiency, find the slip experienced by tractor in percentage. ***(GATE-2017)***

Solution: Given, drawbar pull P = 8 kN; Rolling resistance TF = 4 kN; Tractive efficiency η = 57 per cent = 0.57

As $\eta = \frac{P}{P + TF}$ (100 – s) where s is the slip experienced in percentage

So, $s = 100 - \frac{\eta(P + TF)}{P} = 100 - \frac{57(8 + 4)}{8}$ = 14.5 per cent

Answer: Slip experienced by the tractor is 14.5 per cent.

Example 108: A tractor of 19.5 kN weight and 1.8 m wheel base has 70 per cent static weight on rear axle. It pulls 8 kN drawbar pull on a level ground through a hitch point located 450 mm above the ground. Calculate the dynamic weight on the rear wheel. ***(GATE-2017)***

Solution: Given, tractor weight W = 19.5 kN; Wheel basel = 1.8 m; Load on rear wheel R_r = 70 per cent of W = 13.65 kN; Pull P = 8 kN; Hitch height above the ground y = 450 mm = 0.45 m

$$\text{Due to pull, weight transfer } W_{tr} = \frac{Py}{l} = \frac{8 \times 0.45}{1.8} = 2 \text{ kN}$$

So, dynamic weight on rear wheels = static weight + W_{tr} = 13.65 + 2 = 15.65 kN

Answer: Dynamic weight on rear wheel is 15.65 kN

Example 109: A two-wheel drive tractor weighing 25 kN has a wheel base of 2.2 m and front and rear wheel reaction is 40 and 60 per cent of the total weight. It pulls a drawbar load of 15kN in such a way that distance of drawbar pull line from front wheel contact point is 0.8 m and that from rear wheel contact point is 0.5 m. Find the weight transfer occurs in the both the wheels due to the pull.

Solution: Given, tractor weight W_t = 25 kN; Wheel base l = 2.2 m; Reaction on front wheel R_f = 40 per cent of W_t = 0.4 × 25 = 10 kN; Reaction on rear wheel R_r = 60 per cent of W_t = 0.6 × 25 = 15 kN; Distance of drawbar pull line from front wheel contact point h_f = 0.8 m; Distance of drawbar pull line from rear wheel contact point h_r = 0.5 m

$$\text{Weight transfer on front} = \frac{R_f h_f}{l} = \frac{10 \times 0.8}{2.2} = 3.63 \text{ kN}$$

$$\text{Weight transfer on rear} = \frac{R_r h_r}{l} = \frac{15 \times 0.5}{2.2} = 3.40 \text{ kN}$$

Answer: Weight transfer on front and rear is 3.63 and 3.40 kN.

Example 110: A pneumatic tyre weighing of 300 N is 30 cm wide and 15 cm contact length. The soil cohesion coefficient is 1.3 $N.cm^{-2}$ and angle of internal friction of the soil is 28°. Find the tractive force developed by the wheel.

Solution: Given, weight of the wheel W = 300 N; Tyre width b = 30 cm; Contact length l = 15 cm; Soil cohesion coefficient C = 1.3 $N.cm^{-2}$; Angle of internal friction ϕ = 28°

Ground contact area A = 0.78bl = 0.78 × 30 × 15 = 351 cm^2

So, tractive force developed by the wheel

$$F = AC + W\tan\phi = 351 \times 1.3 + 300 \tan 28° = 615.81 \text{ N}$$

Answer: Tractive force developed by the wheel is 615.81 N.

Example 111: A pneumatic wheel weighing 300 N has a width of 30 cm and 45 cm contact length. Coefficient of wheel sinkage is 0.3. The cohesion and friction modulus of soil deformation are 3 and 0.5 respectively. The soil parameters are C = 1.3 $N.cm^{-2}$ and $\phi = 28°$. Find the drawbar pull developed by the wheel.

Solution: Given, weight of the pneumatic wheel W = 300 N; Tyre width b = 30 cm; Contact length l = 45 cm; Cohesion modulus of soil deformation $K_c = 3$; Friction modulus of soil deformation $K_\phi = 0.5$; Coefficient of wheel sinkage n = 0.3; Soil cohesion coefficient C = 1.3 $N.cm^{-2}$; Tractive effort F = AC + W tanϕ = 0.78blC + Wtanϕ

$$= 0.78 \times 30 \times 45 \times 1.3 + 300 \tan 28 = 1528.41 \text{ N}$$

Rolling resistance

$$R = \frac{2}{\left[(n+1)(K_c + bK_\phi)^{\frac{1}{n}}\right]}\left[\frac{W}{2l}\right]^{\frac{n+1}{n}} = \frac{2}{\left[(0.3+1)(3+30\times 0.5)^{\frac{1}{0.3}}\right]}\left[\frac{300}{2\times 45}\right]^{\frac{0.3+1}{0.3}}$$

$$= 0.018 \text{ N}$$

So, maximum drawbar pull developed by the wheel

$$P = F - R = 1528.41 - 0.018 = 1528.39 \text{ kN}$$

Answer: Maximum drawbar pull developed by the wheel is 1528.39 kN.

Example 112: A two-wheel drive tractor moving on a hill has a weight of 30 kN and wheel base of 2.1 m. The centre of gravity location is 0.9 m ahead of the rear axle centre and 0.6 m above the ground. Find out the maximum slope when the tractor is moving up hill and downhill for which there will be no overturning.

Solution: Given, weight of tractor W = 30 kN; Wheel base l = 2.1 m; Centre of gravity location in front of rear axle centre l_1 = 0.9 m and above the ground y = 0.6 m

For the tractor moving up hill, slope $\theta_{up} = \tan^{-1}\frac{l_1}{y} = \tan^{-1}\frac{0.9}{0.6} = 56.30^0$

For the tractor moving downhill, slope $\theta_{down} = \tan^{-1}\frac{l - l_1}{y} = \tan^{-1}\frac{2.1 - 0.9}{0.6} = 63.43^0$

Answer: Limiting slope for tractor moving up hill is 56.30° and for downhill is 63.43°

Example 113: A tractor weighing 30 kN is taking turn at a radius of 6 m on concrete road without application of brakes. The centre of gravity is 0.85 m above the ground and at a distance of 0.65 m from the tipping axis. Find out the critical turning speed at which the tipping will begin.

Solution: Given, weight of tractor W = 30 kN; Turning radius r = 6 m; Centre of gravity location above the ground h = 0.6 m; Centre of gravity locationfrom the tipping axis x = 0.65 m. So, critical turning speed

So, critical turning speed $V = \left(\frac{grx}{h}\right)^{0.5} = \left(\frac{9.81 \times 6 \times 0.65}{0.6}\right)^{0.5}$

$$= 7.98 \text{ m.sec}^{-1} = 28.74 \text{ kmph}$$

Answer: Critical speed is 28.74 kmph.

TRACTOR TESTS AND PERFORMANCE

Example 114: The tractor seat vibrates with a frequency of 1 Hz. When no damping is provided, the frequency of damped vibration is reduced by 10 per cent. What is the damping factor? ***(GATE-2010)***

Solution: Given, tractor vibration frequency, f = 1 Hz; Reduction in vibration = 10 per cent

So, damped vibration, $f_d = 1 - 0.1 \times 1 = 0.9$ Hz

Natural frequency of un-damped vibration $\omega_n = 2\pi f = 2\pi \times 1 = 6.28$ rad.sec^{-1}

Natural frequency of damped vibration $\omega_d = 2\pi f_d = 2\pi \times 0.9 = 5.66$ rad.sec^{-1}

So, reduction in frequency $a = \sqrt{\omega_n{}^2 - \omega_d{}^2} = \sqrt{6.28^2 - 5.66^2} = 2.72$ rad.sec^{-1}

Again $a = \frac{c}{2m}$, Where c = damping coefficient, m = mass of the system

Again $a = \frac{c}{2(m)}$, where c = damping coefficient, m = mass of the system

So, $c = a \times 2(m) = 2.72 \times 2(m) = 5.44(m)$

Critical damping coefficient $c_c = 2(m)\omega_n = 2(m) \times 6.28 = 12.56(m)$

Damping factor $= \frac{c}{c_c} = \frac{5.44(m)}{12.56(m)} = 0.436$

Answer: Damping factor is 0.436.

Example 115: During a test, sound level was measured as 90 dB in the operator's cabin on a tractor. Taking reference sound pressure as 2×10^{-5} Nm^{-2}, calculate the measured Root Mean Square (RMS) sound pressure. ***(GATE-2013)***

Solution: Given, measured sound pressure level SPL = 90 dB; Reference sound pressure $P_0 = 2 \times 10^{-5}$ Nm^{-2}

Let, the measured RMS sound pressure be P, so, SPL = $20 \log \frac{P}{P_0}$

Or, $90 = 20 \log \frac{P}{2 \times 10^{-5}}$, on solving P = 0.632 Nm^{-2}

Answer: Measured RMS sound pressure is 0.632 Nm^{-2}.

Example 116: For a reference pressure of 2×10^{-5} Pa, the sound pressure level at operator workspace of a tractor was found 80 dB. If RMS sound pressure is increased by 8 times, calculate the resulting sound pressure in dB. ***(GATE-2015)***

Solution: Given, reference sound pressure $P_0 = 2 \times 10^{-5}$ Pa; Sound pressure level at operator workspace SPL_1 = 80 dB

RMS sound pressure is increased by 8 times

Let, for SPL_1 the sound pressure be P_1

So, $P_1 = P_0 \left(10^{\frac{SPL_1}{20}}\right) = 2 \times 10^{-5} (10^{(80/20)}) = 0.2$ Pa

The increased RMS sound pressure $P_2 = 8 \times P_1 = 8 \times 0.2 = 1.6$ Pa

So, sound pressure level $SPL_2 = 20 \log \frac{P_2}{P_0} = 20 \log \frac{1.6}{2 \times 10^{-5}} = 98.06$ dB

Answer: The RMS sound pressure level is 98.06 dB.

Example 117: Natural frequencies of an undamped operator seat is 5 Hz, and combined weight of the seat and the operator is 880 N. If there are four springs fitted in parallel belowthe operator seat, calculate the spring rate (or stiffness) of each spring in kN.m^{-1}. ***(GATE-2016)***

Solution: Given, undamped natural frequency of an operator seat ω_n = 5 Hz; Weight of operator and seat W = 880 N = 89.70 kg; Number of spring n = 4

Springs are connected in parallel

So effective spring rate $K_{effctive} = \omega_n^2 \times W = 5^2 \times 89.7 = 2242.60$ Nm^{-1}

Spring rate of each spring $K = \frac{K_{effctive}}{4} = \frac{2242.60}{4}$ = 560.65 N.m = 0.560 kN.m^{-1}

Answer: Spring rate of each spring is 0.560 kN.m^{-1}

Example 118: In a farm, there are 5 tractors, all are operating at the same time. The noise level from them are 85 dB, 86 dB, 87 dB, 88 dB and 90 dB, respectively. Find the combined noise level.

Solution: Given, noise level from 5 different tractors are $L_1 = 85$ dB, $L_2 = 86$ dB, $L_3 = 87$ dB, $L_4 = 88$ dB, $L_5 = 90$ dB

So, combined noise level

$$SPL_C = 10\log[10^{\frac{L1}{10}} + 10^{\frac{L2}{10}} + 10^{\frac{L3}{10}} + 10^{\frac{L4}{10}} + 10^{\frac{L5}{10}}]$$

$$= 10\ \log\left[10^{\frac{85}{10}} + 10^{\frac{86}{10}} + 10^{\frac{87}{10}} + 10^{\frac{88}{10}} + 10^{\frac{90}{10}}\right] = 94.54 \text{ dB}$$

Answer: The combined noise level is 94.54 dB.

POWER TILLER

Example 119: A power tiller with diesel engine produces 12 kW of brake power with a brake thermal efficiency of 30 per cent. If the air-fuel ratio in the combustion process is 14:1 and heating value of the fuel is 45 MJ kg^{-1}, calculate air consumption rate of the engine in hour. ***(GATE-2018)***

Solution: Given, brake power BP = 12 kW; Brake thermal efficiency η = 30 per cent = 0.3; Air fuel ratio = 14:1; Heating value of fuel H = 45 MJ.kg^{-1}

Fuel consumption rate per hour

$$F_{fuel} = \frac{BP}{H \times \eta} \times 3600 = \frac{12}{45 \times 1000 \times 0.3} \times 3600 = 3.2 \text{ kg}$$

So, air consumption rate per hour $F_{air} = F_{fuel} \times 14 = 3.2 \times 14 = 44.8$ kg

Answer: Air consumption rate is 44.8 kg.h^{-1}

Example 120: A flat belt pulley transmits 7.5 kW power at a speed of 250 revolutions per minute. The diameter of the pulley is 1.2 m. The tension ratio for the belt is 2.5. If maximum tension in belt does not exceed 200 N/cm, calculate width of a suitable leather belt. Find the dimension of various parts of pulley for 6 numbers of arms. Taking shear stress to be 75 MN/m^2, find diameter of the shaft. The cross section of the arm is ellipsoid having major axis twice the minor axis.

Solution: Given, flat pulley diameter d = 1.2 m; Speed N = 250 rpm; Power transmitted P = 7.5 kW; Maximum tension in the belt T = 2 Nm; Tension ratio $\frac{T_1}{T_2}$ = 2.5, Where, T_1 and T_2 = tension on tight side and slack side respectively; No of arms in pulley = 6; Shear stress S = 75 MN.m^{-2}; Cross section of arm is ellipsoid

Peripheral speed of pulley $v = \frac{\pi dN}{60} = \frac{\pi \times 1.2 \times 250}{60} = 15.7\ m.sec^{-1}$

Now power $P = \frac{(T_1 - T_2)v}{1000}$ Or, $7.5 = \frac{(T_1 - T_2) \times 15.7}{1000}$,

on solving $T_1 - T_2 = 477.46$ N

But given $\frac{T_1}{T_2} = 2.5$, so, on solving $T_2 = 318.30$ N and $T_1 = 795.77$ N

So, width of belt $b = \frac{T_1}{T} = \frac{795.77}{200} = 3.98$ cm

Generally, width of pulley is 25 per cent larger than the width of belt.

So, width of pulley = 3.98 × 1.25 = 4.97 cm

From the consideration of tensile stress in the rim due to centrifugal force, the thickness of rim for C.I. pulley is obtained by empirical formula.

Thickness of rim $= \frac{d}{80} + 0.003 = \frac{1.2}{80} + 0.003 = 0.018$ m = 1.8 cm

Net effective pull in the belt $= T_1 - T_2 = 477.7$ N

So, total bending moment offered by each arm $= \left(318.47 \times \frac{1}{2}\right)\frac{1}{6} = 26.53$ Nm

For ellipsoidal section having major axis twice the minor axis, the bending moment will be $M = \frac{\pi}{32} b^3 af$ where, b = length of major axis, m, a = length of minor axis, m,

f = allowable stress in CI $= \frac{\text{maximum permissible stress}}{\text{factor of safety}} = \frac{75}{5} = 15$ MPa

Hence, $26.53 = \frac{\pi}{32} \times b^3 \times \frac{1}{2} \times 15 \times 10^6$, on solving b = 0.033m = 3.3 cm

If d = diameter of shaft, then torque T is given by $T = \frac{\pi}{16} f_s d^3$

Where, f_s = maximum permissible shear stress for steel, T = 318.47 × 1.2/2 = 191.08 Nm

So, $T = \frac{\pi}{16} \times 75 \times 10^6 d^3 = 191.08$, On solving d = 0.0234 m = 2.3 cm

Diameter of pulley hub = 2.5 d = 2.5 × 2.3 = 5.75 cm

Length of hub L = $\frac{2}{3}$ × width of pulley = $\frac{2}{3}$ × 4.97 = 3.31 cm

Answer: Width of belt = 3.98 cm; Width of pulley = 4.97 cm; Thickness of rim = 1.8 cm; Minor axis of ellipsoidal arm = 0.5 cm; Major axis of ellipsoidal arm = 3.3 cm; Diameter of shaft = 2.3 cm; Diameter of hub = 5.75 cm; Length of hub = 3.31 cm

Example 121: 200 litres of hot liquid is cooled in a refrigerator per hour. The liquid has a density of 1150 $kg.m^{-3}$ and specific heat of 0.72 $kCal.kg^{-10}C^{-1}$. The inlet temperature is 130°C and outlet temperature is 60°. Water available for cooling purpose is 1000 litres per hour at a temperature of 10°C. Overall heat transfer coefficient is given as 1000 kCal per sq. m per hr per °C. Calculate the required heating area for parallel flow and counter flow.

Solution: Given, hourly liquid flow = 200 L; Density of liquid ρ = 1150 kgm^{-3} = 1.15 $kg.L^{-1}$; Specific heat Cp = 0.72 $kCal.kg^{-1}.°C^{-1}$; T_1 = 130°C; T_2 = 60°C; Water available for cooling purpose = 1500 lit per hour at 10°C

Let T_s = final temperature of cooling water

Based on heat balance equation, Heat loss = Heat gain

$200 \times 1.15 \times 0.72 \times (130 - 60) = 1000 \times 1 \times (T_s - 10)$

On solving, T_s = 21.59°C = 22°C

a) Parallel flow: T_{11} = 130 – 10 = 120°C, T_{12} = 60 – 22 = 38°C
Now log mean temperature difference

$$\Delta T_m = \frac{T_{11} - T_{12}}{2.303 \log \frac{T_{11}}{T_{12}}} = \frac{120-38}{2.303 \log \frac{120}{38}} = 71.30°C$$

$$\text{Hence, } Q = UA\Delta T_m = 200 \times \frac{1150}{1000} \times 0.72 \times (130 - 60) = 1000 \times A \times 71.29$$

On solving A = 0.162 m^2

b) For counter flow: T_{11} = 130 – 22 = 108°C, T_{12} = 60 – 10 = 50°C
Now log mean temperature difference

$$\Delta T_m = \frac{T_{11} - T_{12}}{2.303 \log \frac{T_{11}}{T_{12}}} = \frac{108-50}{2.303 \log \frac{108}{50}} = 75.3°C$$

$$\text{Hence, } Q = UA\Delta T_m = 200 \times \frac{1150}{1000} \times 0.72 \times (130 - 60) = 1000 \times A \times 75.3$$

On solving A = 0.153 m^2

Answer: Surface area for parallel flow is 0.162 m^2 and that for counter flow is 0.153 m^2.

Air Compressor

Example 122: A single stage reciprocating air compressor is required to compress 1 kg of air from 1 bar to 4 bar. The initial temperature is 27° C. compare the work requirement in the following cases

1. Isothermal compression
2. Compression with $pv^{1.2}$= constant and
3. Isentropic compression

Solution: Given: m = 1 kg; p_1 = 1 bar; p_2 =4 bar; T_1 = 27°C = 27 + 273 = 300K; n = 1.2

1. Work required for isothermal compression

   We know that the work required by the compressor $W = 2.3\ p_1v_1 \log\left(\frac{p_2}{p_1}\right) = 2.3$ $m\ RT_1 \log\left(\frac{p_2}{p_1}\right)$ $(\because p_1v_1 = mRT_1) = 2.3 \times 1 \times 287 \times 300 \log\left(\frac{4}{1}\right) = 119\ 230$ J = 119.23 kJ ($\because$R for air = 287J/kg K)

2. Work required for polytrophic compression (*i.e.* $pv^{1.2}$ = constant)

   We know that work required by the compressor, $W = \frac{n}{n-1} \times m\ RT_1$

   $$\left[\left(\frac{p_2}{p_1}\right)^{\frac{n-1}{n}} - 1\right] = \frac{1.2}{1.2-1} \times 1 \times 287 \times 300 \left[\left(\frac{4}{1}\right)^{\frac{1.2-1}{1.2}} - 1\right] = 134\ 320 \text{ J} = 134.32 \text{ kJ}$$

3. Work required for isentropic compression

   We know that work required by the compressor, $W = \frac{\gamma}{\gamma-1} \times mRT_1$

   $$\left[\left(\frac{p_2}{p_1}\right)^{\frac{\gamma-1}{\gamma}} - 1\right] = \frac{1.4}{1.4-1} \times 1\times 287 \times 300 \left[\left(\frac{4}{1}\right)^{\frac{1.4-1}{1.4}} - 1\right] = 146\ 630\text{J} = 146.63 \text{ kJ}$$

Example 123: Determine the size of the cylindrical double acting air compressor of 40 kW indicated power, in which air is drawn in at 1 bar and 15° C and compressed according to the law $pv^{1.2}$=constant, to 6 bar.The compressor runs at 100 r.p.m with average piston speed of 152.5 $m.min^{-1}$. Neglect clearance

Solution: Given: I.P = 40 kW = 40 × 10^3 W; p_1 = 1 bar =1 × 10^5 N/m^2; T_1 =15°C = 15 + 273 = 288 K; n = 1.2; p_2 = 6 bar; N = 100 r.p.m; average piston speed = 152.5 m/min. Let D = diameter of the cylindrical in meters and L = length of the stroke in meters. We know that average piston speed 2LN = 152.5; ∴ L = 152.5 /2N = 152.5/ 2 × 100 = 0.7625 m

Volume of air before compressor, $v_1 = \frac{\pi}{4} \times D^2 L = \frac{\pi}{4} \times D^2 \times 0.7625 = 0.6\ D^2\ m^3$ and work done by the compressor, $W = \frac{n}{n-1} \times \left[\left(\frac{p_2}{p_1}\right)^{\frac{n-1}{n}} - 1\right] = \frac{1.2}{1.2-1} \times 1 \times 10^5 \times 0.6\, D^2$ $\left[\left(\frac{6}{1}\right)^{\frac{1.2-1}{1.2}} - 1\right]$ N – m = 125310 D^2N-m. Since the compressor is double acting, therefore number of working strokes per minute, N_w = 2N × 100 = 200

We know that indicated power (I.P), $40 \times 10^3 = \frac{W \times N_W}{60} = \frac{125\,310 D^2 \times 200}{60}$ = 417.7 × 10^3 D^2; D^2 = 00.096 or D = 0.31 m or 310 mm.

Example 124: A single acting reciprocating air compressor has cylinder diameter and stroke of 200 mm and 300 mm respectively. The compressor sucks air at 1 bar and 27°C and delivers at 8 bar while running at 100 r.p.m. Find: 1. Indicated power of the compressor; 2. Mass of air delivered by the compressor per minute; and 3. Temperature of air delivered by the compressor. The compression follows the law $pv^{1.25}$ = C. Take R as 287J/kg K.

Solution: Given, D = 200 mm = 0.2 m; L = 300 mm = 0.3 m; p_1 = 1 bar = 1 × 10^5 N/m²; T_1 = 27°C = 27 + 273 = 300 K; P_2 = 8 bar; N = 100 rpm; n = 1.25; R = 287 J/kg K

We know that volume of air before compression, $v_1 = \frac{\pi}{4} \times D^2 \times L = \frac{\pi}{4}(0.2)^2(0.3)$ = 0.0094 m³

1. Indicated power of the compressor

   We know that work done by the compressor for polytropic compression of air,

   $$W = \frac{n}{n-1} \times \left[\left(\frac{p_2}{p_1}\right)^{\frac{n-1}{n}} - 1\right] = \frac{1.25}{1.25-1} \times 1 \times 10^5 \times 0.0094 \left[\left(\frac{8}{1}\right)^{\frac{1.25-1}{1.25}} - 1\right] \text{N-m} =$$

   4700 (1.516 – 1) = 2425 N-m. Since the compressor is single acting, therefore number of working strokes per minute, N_w = N = 100, ∴ indicated power of the compressor $= \frac{W \times N_W}{60} = \frac{2425 \times 100}{60}$ = 4042 W = 4.042 kW

2. Mass of air delivered by the compressor per minute

   Let m = mass of air delivered by the compressor per stroke

   We know that $p_1v_1 = mRT_1$ ∴ $m = \frac{p_1 v_1}{R\,T_1} = \frac{1 \times 10^{5} \times 0.0094}{287 \times 300}$ = 0.0109 kg per stroke and mass delivered minute = m × N_w = 0.0109 × 100 = 1.09 kg

3. Temperature of air delivered by the compressor

   Let T_2 =temperature of air delivered by the compressor. We know that

$\frac{T_2}{T_1} = \left(\frac{p_2}{p_1}\right)^{\frac{n-1}{n}} = \left(\frac{8}{1}\right)^{\frac{1.25-1}{1.25}}$ $= 8^{0.2} = 1.516\ T_2 = 1.516 \times T_1 = 1.516 \times 300 = 454.8\ K =$ 181.8°C

Mechanical Power Transmission

Example 125: Find out the speed of the centrifugal pump having diameters of pump pulley and tractor driving pulley are respectively 40 cm and 44 cm. the rpm of tractor is 1000.

Solution: Power transmission between two pulleys *i.e.* tractor and centrifugal pump is given by $N_1 \times D_1 = N_2 \times D_2 \rightarrow N_2 = \frac{N_1 \times D_1}{D_2} = \frac{1000 \times 40}{44} = 909$ rpm

Example 126: If a 10 cm spur gear is driving a bigger gear of 50 cm at a torque of 20 N-m, find out the velocity ratio and torque developed at bigger gear.

Solution: Given, D_1 = 10 cm; D_2 = 50 cm; T_1 = 20 N.m.

$$\text{Velocity ratio} = \frac{D_1}{D_2} = \frac{T_1}{T_2} = \frac{N_2}{N_1} = \frac{10}{50} = \frac{1}{5}$$

$$\text{Torque } T_2 = T_1\left(\frac{D_2}{D_1}\right) = 20\left(\frac{50}{10}\right) = 100 \text{ N.m}$$

Example 127: An electric motor drives a counter shaft with the help of a canvas belt and pulley arrangement, such that the speed of the belt is 1000 m/min and transmits 80 kW power. Find the difference between two tensions on either side of the belt. If tension on the tight side of the belt be twice to that on the slack side, calculate both the tensions of the belt.

Solution: Given, speed = 1000 $m.min^{-1}$ = 16.66 $m.s^{-1}$; power transmitted = 80 kW; $T_1 = 2\ T_2$

Power (W) = $\frac{(T_1 - T_2)\ V}{60}$ where T_1 and T_2 are in N

velocity, m/min $\rightarrow 80 \times 1000 = (T_1 - T_2) \times \frac{1000}{60}$

$(T_1 - T_2) = 4800$ N $\rightarrow (2T_1 - T_2) = 4800$ N, so, $T_2 = 4800$ N and $T_1 = 2T_2 = 2 \times 4800 = 9600$ N

Example 128: A stationary pulley of 20 cm diameter is driving a countershaft with two pulleys of 30 cm and 25 cm diameter. The 20 cm. pulley of the engine is directly connected to the 30 cm pulley by means of a belt while 25 cm pulley is connected to a power paddy thresher pulley of 35 cm diameter. Calculate the revolutions of the thresher when the engine is rotating at 1000 rpm.

Solution: Let the rpm of counter shaft is N_1. $30 \times N_1 = 20 \times 1000 \rightarrow N_1$

$$= \frac{20{,}000}{30} = 666.66$$

Since both the pulleys of 30 cm and 25 cm diameters are on the same counter shaft, the speed of both the pulleys would be same.

Let rpm of thresher be N_2, so, $35 \times N_2 = 25 \times 666.66 \rightarrow N_2 = 476$ rpm. Hence, rpm of thresher is 476.

Example 129: Calculate the width of belt to transmit 10 kW power to a pulley of 40 cm diameter. Pulley makes 2000 rpm, coefficient of friction between belt and pulley is 0.2 and angle of contact is 120°. Maximum tension in belt not to exceed 80 N per cm. width.

Solution: The ratio between the belt tension on the tight and slack side is given by $\frac{T_1}{T_2} = e^{\mu\theta}$ where μ is the coefficient of friction, θ is the angle of lap of the belt over the pulley in radians.

$$\theta = \frac{\text{Degree} \times \pi}{180} = \frac{120 \times \pi}{180} = 2.09 \text{ radians. } \mu = 0.20, \frac{T_1}{T_2} = e^{(0.20)\times(2.09)} = 1.52$$

$$10 \times 1000 \text{ (W)} = \pi \times \frac{40}{100} \times 2000 \times \frac{(T_1 - T_2)}{60} \rightarrow T_1 = 349 \text{ N and } T_2 = 229.53 \text{ N}$$

$$\text{Width of belt} = \frac{T_1}{\text{maximum tension, N/cm}} = \frac{349}{80} = 4.36 \text{ cm}$$

Example 130: There is 50 cm. pulley on the main shaft which drives a 40 cm. pulley on the counter shaft. From the counter shaft, a 35 cm. pulley drives 20 cm. pulley on the machine. If the main shaft speed is 400 rpm, calculate the speed of the machine (i) without belt slip (ii) with 10% belt slip.

Solution: The drive is shown above, where A, B, C and D are pulleys. Speed of pulley B= (400 × 50)/40 = 500 rpm.

Being on the same shaft, speed of pulley B and C would be same. Speed of C = 500 rpm. Pulley C drives pulley D; so, speed of D is 500 × (35/20) = 875

(i) Speed of machine without slip = 875 rpm

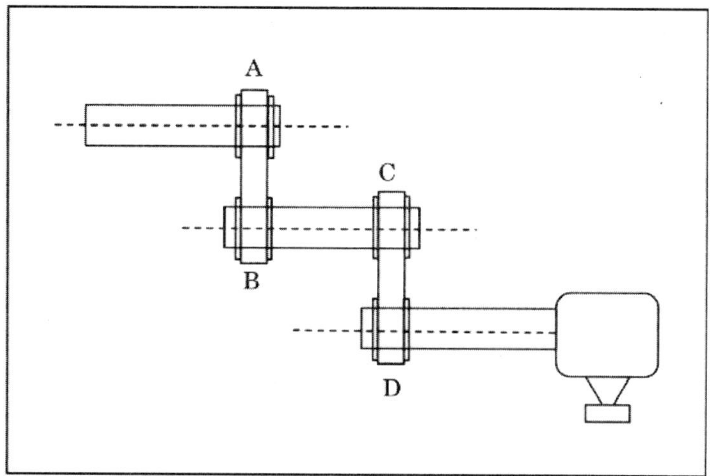

(ii) Speed of B and C with 10% slip = 500 × (90/100) = 450

So, speed of D = 450 × (35/20) = 787.5 rpm.

Example 131: An open belt drive system transmits 5 kW power using a flat belt of width and thickness as 80 mm and 5 mm, respectively. The driving shaft speed is 1500 rpm and the driven shaft speed is 500 rpm. The coefficient of friction between the belt and pulley is 0.20, and the wrap angle of the belt is 168°. If diameter of the smaller pulley is 200 mm, find the maximum stress induced in the belt. ***(GATE-2018)***

Solution: Given, open belt drive system; Power transmitted P = 5 kW; Belt width b = 80 mm = 0.08 m; Belt thickness t = 5 mm = 0.005 m; Driving shaft speed $N_{driving}$ = 1500 rpm; Driven shaft speed N_{driven} = 500 rpm; Coefficient of friction $\mu = 0.2$

$$\text{Wrap angle } \theta = 168° = 168 \times \frac{\pi}{180} = 2.93 \text{ rad}$$

Diameter of smaller pulley $D_{driving}$ = 200 mm = 0.2 m

$$\text{Pulley speed } v = \frac{\pi(D_{driving} + 2t)N_{driving}}{60} = \frac{\pi \times (0.2 + 2 \times 0.005) \times 1500}{60}$$

$$= 16.49 \text{ m.sec}^{-1}$$

Let tight and slack side tension are T_1 and T_2, respectively

$$\text{So, } T_1 - T_2 = \frac{P}{v} = \frac{5 \times 1000}{16.49} = 303.15 \text{ N.m} \quad ..... (1)$$

$$\text{Again } \frac{T_1}{T_2} = e^{\mu\theta} = e^{0.2 \times 2.93} = 1.8 \quad ..... (2)$$

On solving equation 1 and 2 T_2 = 378.93 N and T_1 = 682.08 N

So, maximum stress induced in the belt $\frac{T_1}{bt} = \frac{682.08}{0.08 \times 0.005} \times 10^{-6} = 1.7$ MPa

Answer: Maximum stress induced in the belt is 1.7 MPa.

Example 132: A belt drive has two v-belts connected in parallel. The groove angle is 40° and cross-sectional area is 800 mm². The density of the belt material is 1080 kg.m⁻³. The coefficient of friction between belt and pulley is 0.3. Each pulley of the drive is 320 mm in diameter and speed of pulley is 1500 revolutions per minute. If allowable stress in the belt is 8 MPa, find the power transmitted by the belt and the pulley speed at which it transmits maximum power.

Solution: Given, groove angle $2\beta = 40°$; Cross sectional area $A = 800$ mm²; Density of material $\rho = 1080$ kg.m⁻³; Coefficient of friction between belt and pulley $\mu = 0.3$; Pulley diameter $d = 320$ mm $= 0.32$ m; Pulley speed $N = 1500$ rpm; Allowable stress $\sigma = 8$ MPa $= 8$ N.mm⁻²; Assuming angle of wrap $\theta = 180° = \pi$ radian

$$\text{Linear velocity } v = \frac{\pi d N}{60} = \frac{\pi \times 0.32 \times 1500}{60} = 25.13 \text{ m.sec}^{-1}$$

$$\text{Mass of belt per unit length } m = \rho \times A = 1080 \times 800 \times 10^{-6} = 0.864 \text{ kg}$$

$$\text{Centrifugal tension } T_c = mv^2 = 0.864 \times 25.13^2 = 545.63 \text{ N}$$

$$\text{Maximum tension in the belt } T = \sigma \times A = 8 \times 800 = 6400 \text{ N}$$

$$\text{So, tension on tight side } T_1 = T - T_c = 6400 - 545.63 = 5854.36 \text{ N}$$

$$\text{Tension on slack side } T_2 = \frac{T_1}{e^{\frac{\mu\theta}{\sin\beta}}} = \frac{5854.36}{e^{\frac{0.3 \times \pi}{\sin\frac{40}{2}}}} = 372.34 \text{ N}$$

Power transmitted in the belt is

$$P = 2(T_1 - T_2)v = 2(5854.36 - 372.34) \times \frac{25.13}{1000} = 275.52 \text{ kW}$$

Speed at which maximum power is transmitted

$$v_{max} = \sqrt{\frac{T}{3m}} = \sqrt{\frac{6400}{3 \times 0.864}} = 49.7 \text{ m.sec}^{-1}$$

Answer: Power transmitted is 221.56 kW and at a speed of 49.7 m.sec⁻¹ maximum power is transmitted.

Example 133: The centre to centre distance between two sprockets of a chain drive is 600 mm. It is used to reduce the speed from 200 rpm to 100 rpm. The driving sprocket has a pitch circle diameter of 480 mm and the number of teeth is 18. Find the number of teeth on the driven sprocket and pitch and length of chain.

Solution: Given, centre to centre distance C = 600 mm; Speed of driving sprocket N_1 = 200 rpm; Speed of driven sprocket N_2 = 100 rpm; Number of teeth on driving sprocket T_1 = 18; Pitch circle diameter of driving sprocket d_1 = 480 mm

$$\text{Number of teeth on driven shaft } T_2 = T_1 \frac{N_1}{N_2} = 18 \times \frac{200}{100} = 36$$

$$\text{Pitch of the chain } p = 2 \times \frac{d_1}{2} \times \sin\frac{180}{T_2} = 2 \times \frac{480}{2} \times \sin\frac{180}{36} = 41.8 \text{ mm}$$

$$\text{Constant } k = \frac{C}{p} = \frac{0.6}{0.0418} = 14.342$$

$$\text{Length of chain } L = p\left[\frac{T_1 + T_2}{2} + \frac{\csc\frac{180}{T_1} - \csc\frac{180}{T_2}}{4k} + 2k\right]$$

$$(\text{As csc} = \text{cosec}) = 0.0418\left[\frac{18 + 36}{2} + \frac{\csc\frac{180}{18} - \csc\frac{180}{36}}{4 \times 14.342} + 2 \times 14.342\right] = 2.351 \text{ m}$$

Answer: Number of teeth on driven sprocket is 36, Pitch and length of chain are 41.8 mm and 2.351 m.

Example 134: The number of teeth on a spur gear is 30 which rotates at a speed of 200 rpm. If it has a module of 2 mm, find the circular pitch and pitch line velocity.

Solution: Given, teeth on spur gear T = 30; Speed N = 200 rpm; Module m = 2 mm

Circular pitch $p = \pi m = \pi \times 2 = 6.28$ mm

$$\text{Pitch line velocity } v_p = \frac{\pi N m T}{60 \times 1000} = \frac{\pi \times 200 \times 2 \times 30}{60 \times 1000} = 0.628 \text{ m.sec}^{-1}$$

Answer: Circular pitch is 6.28 mm and pitch line velocity is 0.628 m.sec^{-1}.

Multiple Choice Objective Type Questions

1. A man can develop
   (A) 0.1 hp (B) 0.5 hp (C) 0.75 hp (D) 1.0 hp
2. Medium size bullock can develop
   (A) 0.50 hp (B) 1 hp (C) 0.1 hp (D) 1.25 hp
3. The average force that a bullock can exert
   (A) 1/5th of its body weight
   (B) l/10th of its body weight
   (C) l/20th of its body weight
   (D) 1/15th of its body weight.
4. The velocity required to operate windmill is more than
   (A) 5 kmph (B) 10 kmph
   (C) 20 kmph (D) 5 miles per hour
5. The average capacity of a windmill is around
   (A) 0.50 hp (B) 1.0 hp (C) 1.5 hp (D) 2.0 hp
6. Renewable sources of energy are
   (A) Exhaustible (B) Inexhaustible
   (C) Finite (D) Nuclear based only
7. Solar energy is a
   (A) Inexhaustible source of energy (B) Non-renewal source of energy
   (C) Finite source of energy (D) None of above.
8. The fossil fuel is
   (A) Coal (B) Oil (C) Natural gas (D) all above.
9. Non commercial source of energy is
   (A) Wind energy (B) Fossil fuel
   (C) Electrical energy (D) All the above.
10. Natural gas is the
    (A) Fossil fuel (B) Renewable source of energy
    (C) Biogas (D) None of above.
11. Biogas contains
    (A) Methane gas (B) CO_2 gas
    (C) Ethane and CO_2 (D) Both (A) and (B)

12. Percent of Methane gas in biogas varies from
    (A) 45 -to 70 % (B) 15 to 30 %
    (C) 80 to 85 % (D) 5 to 10 %
13. A device which converts sunlight directly into electrical energy, is called
    (A) Solar heater (B) Solar cell
    (C) Solar furnace (D) Solar power plant.
14. The single cylinder engine is generally used in
    (A) Tractor (B) Motor car
    (C) Stationary engine (D) Power tiller engine
15. The diesel engine used on tractors are
    (A) One stroke engine (B) Two stroke engine
    (C) Four stroke engine (D) None of these
16. The carburetor is the main part of
    (A) Diesel engine (B) Steam engine
    (C) Gas engine (D) Petrol engine
17. The compression ratio of diesel engine is
    (A) 4 to 8:1 (B) 14 to 20:1 (C) 4 to 15:1 (D) 14 to 22:1
18. The compression ratio of petrol engine is
    (A) 4 to 8:1 (B) 4 to 10:1 (C) 8 to 15:1 (D) 14 to 20:1
19. The air fuel ratio of petrol engine is
    (A) 10:1 (B) 15:l (C) 20:1 (D) None of these
20. The IHP of an engine indicates
    (A) Power on piston (B) Power on flywheel
    (C) Power on drawbar (D) Loss of power
21. The BHP of an engine indicates
    (A) Power on piston (B) Power on flywheel
    (C) Power on drawbar (D) Power on PTO.
22. In diesel cycle engine, heat is added at
    (A) Constant pressure (B) Constant temperature
    (C) Constant volume (D) None of the above.
23. In Otto cycle, heat is added at
    (A) Constant pressure (B) constant temperature
    (C) Constant volume (D) None of these
24. The fluctuation of engine speed during a cycle depends upon
    (A) Mass of flywheel
    (B) Mass of crankshaft
    (C) Speed of flywheel
    (D) Governor speed

25. The material used for construction of cylinder block is
    (A) Cast iron and steel (B) Cast iron and aluminum alloy
    (C) Brass and Steel (D) Steel and aluminum alloy.
26. The compression rings mounted on piston are generally made of
    (A) Low carbon steel (B) Aluminum
    (C) High carbon steel (D) Chromium
27. The camshaft in an engine is always mounted
    (A) Perpendicular to the crankshaft (B) Inclined to the crankshaft
    (C) Parallel to the crankshaft (D) None of these
28. The oil pan of an engine is made of
    (A) Steel or aluminum (B) Cast iron or brass
    (C) Steel or cast iron (D) Cast iron or zinc.
29. The camshaft controls
    (A) Valve closing (B) Valve opening
    (C) Valve timing (D) All are correct.
30. The specific fuel consumption of diesel engine is about
    (A) 180 gm/bhp/hr (B) 200 gm/bhp/hr
    (C) 150 gm/bhp/hr (D) 290 gm/bhp/hr.
31. The specific fuel consumption of petrol engine is about
    (A) 180 gm/bhp/hr (B) 200 gm/bhp/hr
    (C) 150 gm/bhp/hr (D) 290 gm/bhp/hr
32. Oil pump is driven by
    (A) Camshaft (B) Timing gear
    (C) Crankshaft (D) All are correct.
33. Flywheel is made of
    (A) Cast iron (B) Aluminum alloy
    (C) High carbon steel (D) All of above.
34. The stroke: bore ratio of tractor engines varies from
    (A) 1.0 to 1.5 (B) 1.0 to 1.25 (C) 1.0 to 1.35 (D) 1.0 to 1.45
35. White smoke indicates
    (A) Presence of water in fuel (B) Burning of lubricants in cylinder
    (C) Presence of water in lubricant (D) Rich air and fuel mixture.
36. Blue smoke indicates
    (A) Presence of water in fuel (B) Burning of lubricants in cylinder
    (B) Engine is overloaded (D) Rich air and fuel mixture.
37. The compression pressure of diesel engine inside the cylinder varies from
    (A) 30–40 kg/cm^2 (B) 35–45 kg/cm^2
    (C) 40– 50 kg/cm^2 (D) 50–60 kg/cm^2

38. The compression pressure of petrol engine inside the cylinder varies from
    (A) 6–10 kg/cm^2 (B) 10–15 kg/cm^2
    (C) 8–12 kg/cm^2 (D) 12–15 kg/cm^2.
39. The rate of doing work (power) is equal to
    (A) 75 kg m/sec. (B) 4500 kg-m/min
    (C) 27,000 kg-m/hr (D) All pf the above.
40. Thermal efficiency of CI engine is
    (A) Higher than SI engine (B) Lower than SI engine
    (C) Equal to SI engine (D) None of these.
41. The weight of diesel engine per horsepower is
    (A) Lighter than petrol engine (B) Equal to petrol engine
    (C) Heavier than petrol engine (D) None of these
42. Gudgeon pin of engine piston is made of
    (A) Case hardened steel (B) Cast Iron
    (C) Aluminum alloy (D) All of the above.
43. Piston of an engine is made of
    (A) Cast iron (B) Aluminum alloy
    (C) Nickel iron alloy (D) All of the above.
44. The common face and seat angle of valve is
    (A) 45° (B) 60° (C) 90° (D) 120°
45. The gear used in timing gear is
    (A) Spur gear (B) Bevel gear (C) Helical gear (D) Worm gear
46. Gasket is made of
    (A) Copper (B) Asbestos (C) Fibre (D) Both (A) and (B)
47. Constant speed engine is used in
    (A) Tractor (B) Power tiller
    (C) Motor car (D) Electric generator
48. Variable speed engines are used in
    (A) Tractor (B) Four wheel vehicle
    (C) Two wheel vehicle (D) All of the above.
49. Constant speed governor is used in
    (A) Petrol engine (B) Stationary engine
    (C) Tractor engine (D) All of the above.
50. The violent noise heard in engine during the process of combustion is
    (A) Scavenging (B) Detonation
    (C) Wearing (D) All of the above.
51. The process of removal of burnt gases from the engine cylinder is
    (A) Scavenging (B) Detonation (C) Knocking (D) Burning.

52. If engines of different horsepower, rpm, cylinder dimensions are to be compared for their performance, the common parameter would be
    (A) Unit speed (B) Fuel consumption
    (C) Mean effective pressure (D) Exhaust temperature.
53. The process that follows the equation PVn = Constant is called
    (A) Constant pressure process (B) constant volume process
    (C) Adiabatic process (D) Polytropic process
54. In the polytropic process, equation PV^n = constant. If the value of n is infinitely large, the process is termed as
    (A) Constant volume (B) Adiabatic
    (C) Constant temperature (D) Constant pressure.
55. In the polytropic process, equation PV^n = constant. If n = 1, the process is termed as
    (A) Constant volume (B) Adiabatic
    (C) Constant temperature (D) Constant pressure.
56. An isentropic process is always
    (A) Irreversible and adiabatic
    (B) Reversible and isothermal
    (C) Reversible and adiabatic
    (D) None of the above.
57. Octane number of gasoline generally available in India is in the range of
    (A) 10 to 20 (B) 40 to 60 (C) 80 to 90 (D) 90 to 95
58. Cetane number of diesel fuel generally available in India is in the range of
    (A) 10 to 15 (B) 15 to 20 (C) 20 to 30 (D) 40 to 55
59. The use of diesel fuel with high sulphur contents may result in an increased.
    (A) Wear of cylinder face
    (B) Wear of piston rings
    (C) Delay period
    (D) Rate of pressure rise
60. The tetra-ethyl lead added in small quantities to most SI engine fuels is
    (A) To bring their octane number to standard values
    (B) To improve the brake thermal efficiency
    (C) To improve the specific output
    (D) All of the above.
61. Pre-ignition is defined as
    (A) Ignition of the charge after passage of the spark
    (B) Ignition of charge prior to passage of the spark
    (C) Ignition of end charge
    (D) Ignition of the charge near the inlet valve.

62. Tetra-ethyl lead may cause
    (A) Absorbed into the lungs in vapor form
    (B) Promote spark plug and exhaust valve corrosion
    (C) Cause troublesome deposit on Cylinder walls
    (D) All of the above.
63. In a diesel engine, the duration between the time of injection and time of ignition is called
    (A) Period of ignition (B) Explosion period
    (C) Delay period (D) Burning period
64. If petrol is used in a diesel engine, then
    (A) Engine output will be reduced (B) Engine efficiency be reduced
    (C) Engine may knock severely (D) There would not be any effect.
65. The size of the intake valve in S.I. engine is
    (A) Equal to the exhaust valve
    (B) Larger than exhaust valve
    (C) Smaller than exhaust valve
    (D) Does not deped on the type of engine
66. Blowby losses in diesel engines as compared to petrol engines are
    (A) Higher (B) Lower (C) Same (D) Fluctuating
67. The most popular firing order in case of four stroke four cylinder in line engine is
    (A) 1-2-3-4 (B) 1-3-2-4 (C) 1-3-4-2 (D) 1-2-4-3
68. The correct sequence of the decreasing order of brake thermal efficiency of the three given basic types of 1C engine is
    (A) 4-stroke CI engine, 4-stroke SI engine, 2-stroke SI engine
    (B) 4-stroke SI engine, 4 stroke CI engine, 2-stroke SI engine
    (C) 4-stroke CI engine, 2 stroke SI engine, 4-stroke SI engine
    (D) 2-stroke SI engine, 4-stroke SI engine, 4-stroke CI engine
69. The method of determination of indicated horse power of a multi cylinder SI engine
    (A) Morse test (B) Prony brake test
    (C) Motoring test (D) Heat balance test.
70. Which one of the following engine will have heaviest fly wheel.
    (A) 40 HP four stroke petrol engine (B) 40 HP 2 stroke petrol engine
    (C) 40 HP two stroke diesel engine (D) 40 HP four stroke diesel engine.
71. Alcohol fuels have the disadvantage as compression ignition fuel because
    (A) They have low octane number (B) High cetane number
    (C) Low catene number (D) Air-fuel ratio required is higher.

72. Thermal efficiency of a two stroke petrol engine as compared to that of four stroke petrol engine for same output, same speed and same compression ratio is
    (A) More
    (B) Less
    (C) Same as long as compression ratio is same
    (D) Same as long as output is same.
73. The CI engine fuels as compared to gasoline used in SI engine has
    (A) Longer ignition lag (B) Lower calorific value
    (C) Heavier and more viscous (D) All of the above.
74. The locations of the cranks on the crank shaft in the various radial directions depends on the
    (A) Firing order (B) Engine size
    (C) Engine output (D) Number of cylinders
75. Diesel engine knock occurs at the
    (A) End of combustion (B) Mid-point of combustion
    (C) Start of combustion (D) Any time during combustion.
76. Petrol engine knock occurs at the
    (A) End of combustion (B) Mid-point of combustion
    (C) Start of combustion (D) Any time during combustion.
77. The main drawback of diesel engine over petrol engine is that
    (A) It requires more cranking power (B) It is noisier
    (C) It is heavier (D) All of the above.
78. The reason for increased emphasis on diesel engine is that
    (A) It is relatively fuel efficient engine
    (B) It has no electrical ignition system
    (C) It has better torque speed characteristics
    (D) All of the above.
79. For the same power output and same compression ratio, as compared to two stroke engine, four stroke SI, engines have
    (A) Higher fuel consumption (B) Lower thermal efficiency
    (C) Higher exhaust temperatures (D) Higher thermal efficiency
80. For the same power output and engine operating conditions, as compared to four stroke engine, two stroke engines have
    (A) Higher un-burnt hydrocarbons (B) Lowest un-burnt hydrocarbons
    (C) Higher oxides of nitrogen (D) Lower exhaust temperature.
81. The two stroke cycle engine has ports in the
    (A) Cylinder head (B) Piston
    (C) Cylinder wall (D) Crankcase

82. Tappet clearance is provided between tappet and
    (A) Spring (B) Stem
    (C) Push rod (D) Valve lifter
83. The purpose of the condenser in spark ignition circuit is to
    (A) Reduce the arcing at contact breaker
    (B) Reduces secondary spark
    (C) Increase build-up voltage time in primary circuit
    (D) Increase arcing at contact breaker
84. For the same power output, a multi cylinder engine as compared to single cylinder engine is
    (A) Heavier (B) Lighter
    (C) Same weight (D) None of the above.
85. The size of an engine cylinder is referred to in terms of its
    (A) Diameter and bore (B) Displacement and efficiency
    (C) Bore and stroke (D) Bore and length.
86. The topmost piston ring near the piston crown is known as
    (A) Oil ring (B) Compression ring
    (C) Groove ring (D) Scrapper ring.
87. Which of the following component is not needed in case of two stroke engine
    (A) Condenser (B) Ignition coil (C) Distributor (D) Battery.
88. Normally the engines used in two wheelers are
    (A) Four stroke air cooled (B) Four stroke water cooled
    (C) Two stroke water cooled (D) Two stroke air cooled.
89. Which of the following chemical is used in radiator to avoid the freezing of the coolant
    (A) Glycol (B) Ethyl alcohol
    (C) Methyl, alcohol (D) Naptha
90. Freezing point of petrol is around
    (A) 10°C (B) 0°C (C) –30°C (D) –180°C
91. If diesel fuel is supplied to petrol engine
    (A) The engine will explode
    (B) The engine will not run
    (C) The engine will run with very high knocking
    (D) The engine will run at very high speed.
92. If petrol fuel is supplied to diesel engine
    (A) The engine will explode
    (B) The engine will not run
    (C) The engine will run with very high knocking
    (D) The engine will run at very high speed.

93. The longer the time required to ignite the diesel fuel is an indication of
    (A) Lower octane number (B) Lower self ignition temperature
    (C) Higher cetane number (D) Lower cetane number
94. If a petrol engine is to be converted to run on natural gas , the spark timing has to be
    (A) Advanced (B) Retarded
    (C) Maintained same (D) None of the above.
95. In an engine cylinder, heat is transferred by
    (A) Conduction
    (B) Convection
    (C) Radiation
    (D) Conduction, convection and radiation.
96. The approximate exhaust temperatures of the diesel engines are of the order of
    (A) 100–400°C (B) 150–450°C (C) 200–500°C (D) 200–600°C
97. The approximate exhaust temperatures of the petrol engines are of the order of
    (A) 100–500°C (B) 150–650°C (C) 300–800°C (D) 300–100CPC
98. Engine cylinder head gasket is placed between the
    (A) Cylinder block and the valves
    (B) Cylinder head and cylinder block
    (C) Cylinder head and intake manifold
    (D) Intake manifold and carburetor.
99. The mechanism used to convert the reciprocating motion of the piston to rotary motion of the crank shaft is known as
    (A) Slider crank mechanism (B) Four bar mechanism
    (C) Geneva mechanism (D) Quick return mechanism.
100. The purpose of an ignition coil in an IC engine is to
    (A) Step up the voltage supplied to the spark plug
    (B) Step down the voltage supplied to the spark plug
    (C) Step up the current supplied to the spark plug
    (D) Step down the current supplied to the spark plug.
101. Connecting rods are generally of the following form
    (A) Forged I section (B) Forged round section
    (C) Cast iron triangular section (D) Forged square section steel.
102. As the number of cylinders on multi cylinder engines increases, the power to weight ratio
    (A) Remains same (B) Decreases
    (C) Increases (D) Becomes zero.

103. The device for smoothing out the power impulses from the engine is called
   (A) Clutch (B) Differential (C) Flywheel (D) Gearbox
104. Ignition advance is expressed in terms of
   (A) Crank angle
   (B) Millimeters of piston travel before TDC
   (C) Time in milli seconds
   (D) Any of the above.
105. Too rich mixture for a SI engine means air: fuel ratio of about
   (A) 17:1 (B) 15:1 (C) 10:1 (D) 8:1
106. In an I.C. engine, the approximate percentage of the combustion heat that passes to the cylinder wall is
   (A) 5 % (B) 10 % (C) 30 % (D) 60 %
107. At the first sight, a petrol engine is identified by
   (A) Cylinder size (B) Power output
   (C) Size of air cleaner (D) Spark plug
108. Which part is not common between the petrol and diesel engines
   (A) Air cleaner (B) Exhaust silencer
   (C) Battery (D) Fuel injector.
109. The heating value of petrol is
   (A) 7000 kcal/kg (B) 8000 kcal/kg
   (C) 10,500 kcal/kg (D) 12,500 kcal/kg
110. The heating value of diesel is approximately
   (A) 8000 kcal/kg (B) 9000 kcal/kg
   (C) 10,000 kcal/kg (D) 13,000 kcal/kg
111. In a scooter engine, the cylinder is lubricated by
   (A) Wet lubrication (B) Charge lubrication
   (C) Dry sump lubrication (D) Pressure lubrication.
112. Water in lubricating oil aids in
   (A) Decomposition (B) Oxidation
   (C) Formation of sludge (D) Dilution.
113. The centrifugal advance mechanism is located on
   (A) Camshaft (B) Crankshaft
   (C) Distributor (D) Drive gear
114. The purpose of crankcase ventilation is to
   (A) Cool the oil
   (B) Remove vaporized water and fuel
   (C) Supply oxygen to crankcase
   (D) Assist in maintaining the viscosity of oil.

115. The escape of burnt gases from the combustion chamber past the pistons into the crankcase is called.
(A) Gas loss (B) Blow-by
(C) By-pass (D) Crank case explosion

116. The function of oil scraper rings is to
(A) Retain compression (B) Lubricate cylinder wall
(C) Reduce piston wear (D) Increase compression pressure

117. Excess oil consumption in an engine may be due to
(A) Leakage of oil through oil pan gasket
(B) Poor quality or improper viscosity of engine oil
(C) Badly worn piston rings
(D) Any of the above.

118. The percentage of heat released from fuel-air mixture, in an I.C. engine which is converted into useful work is about
(A) 10 per cent (B) 15–20 per cent
(C) 25–30 per cent (D) 35–45 per cent.

119. Brake lining is mounted on
(A) Brake shoe (B) Brake drum
(C) Master cylinder (D) Wheel cylinder

120. When diameter of a tyre is specified as 1000 mm, it means
(A) Outer diameter of tyre is 1000 mm
(B) Inner diameter of tyre is 1000 mm
(C) Tube diameter is 1000 mm
(D) None of the above.

121. Which tractor has an air cooled engine
(A) Ford (B) Eicher (C) HMT (D) Hindustan

122. A four stroke diesel engine is operating at 1800 rpm. The duration of fuel injection is 20°. The time in seconds during which fuel is injected would be
(A) 1/540 sec (B) 1/270 sec (C) 1/1080 sec (D) 0.02 sec.

123. Engine starter motor is
(A) Single phase AC motor (B) B.C. series motor
(C) D.C. shunt motor (D) Synchronous motor

124. For balancing single cylinder engine, a counter weight is added to
(A) Piston (B) Connecting rod
(C) Crankshaft (D) Gudgeon pin

125. Engine piston is generally made of aluminum alloy because
(A) It is lighter (B) It is stronger
(C) It has less wear (D) It absorbs shock.

126. Radiator tubes are generally made of
   (A) Steel (B) Brass (C) Cast iron (D) Plastics
127. In a diesel engine, the pipe carrying fuel from fuel pump to nozzle is made of
   (A) Plastic (B) PVC (C) Steel (D) Copper
128. If water level in the engine is maintained low, the likely consequence would be
   (A) Piston seizure (B) Engine knocking
   (C) Bearing deterioration (D) All of the above.
129. If the oil level in the oil pan of the engine is maintained above the gauge mark, the likely damage, it may cause is
   (A) Engine fuming (B) Carbon deposit
   (C) Blue colour in exhaust (D) All of the above.
130. The minimum cranking speed in petrol engines is
   (A) Equal to the normal operating speed
   (B) Half of the operating speed
   (C) One fourth of operating speed
   (D) 60–80 rpm.
131. While starting an engine, the starter pinion meshes with
   (A) Clutch (B) Propeller shaft
   (C) Engine flywheel ring gear (D) Gears in gearbox,
132. The level of electrolyte in tractor battery should be
   (A) 15 mm below the top of plates
   (B) 5–10 mm below the top of plates
   (C) Exactly at the level of plates
   (D) 10–15 mm above top of plates.
133. Which oil is more viscous
   (A) SAE30 (B) SAE40 (C) SAE 50 (D) SAE 60
134. Engine oil in tractor is usually renewed after
   (A) 100 km (B) 500 km (C) 600 km (D) 800 km
135. Which transmission unit disengages the drive and provides a smooth take up of the drive.
   (A) Gear box (B) Clutch (C) Final drive (D) Differential
136. Which transmission unit allows the inner drive wheel to rotate slower than the outer wheel but still maintains a drive to both wheels.
   (A) Clutch (B) Gearbox (C) Differential (D) Final drive.
137. A single cylinder, four stroke engine is rotating at 2000 rpm. The number of power stroke occurring in one minute is
   (A) 500 (B) 2000 (C) 1000 (D) 4000

138. The stroke is increased when the
    (A) Piston is shortened
    (B) Connecting rod is lengthened
    (C) Piston pin is moved nearer to the crankshaft
    (D) Crankshaft throw is lengthened.
139. Ignition in an engine cylinder should occur at
    (A) BDC at the start of compression stroke
    (B) TDC at the end of compression stroke
    (C) TDC at the start of compression stroke
    (D) BDC at the end of suction stroke.
140. The purpose of piston pin is to
    (A) Prevent the valve from rotating
    (B) Link the connecting rod to the crankshaft
    (C) Ensure the piston ring to the piston
    (D) Connect the piston to the connecting rod.
141. How is the inlet valve opened and closed ? It is opened by
    (A) Cam and closed by spring
    (B) Spring and closed by cam
    (C) Gas pressure and closed by Cam
    (D) Cylinder vacuum and closed by a spring.
142. A four cylinder engine has a capacity of 2.4 litres. The swept volume of one cylinder is
    (A) 400 cm^3 (B) 600 cm^3 (C) 1200 cm^3 (D) 2400 cm^3
143. On modern four stroke engines, the exhaust valve opens just
    (A) Before TDC (B) After TDC
    (C) Before BDC (D) After BDC
144. On modern four stroke engines, the inlet valve opens just
    (A) Before TDC (B) After TDC
    (C) Before BDC (D) After BDC
145. The reason why petrol flows from the float chamber to the venturi is because of
    (A) Difference in pressure
    (B) Difference in level
    (C) Float level is higher
    (D) The air sucks out the petrol
146. The pressure in an engine cylinder is less than atmospheric pressure when the engine is performing the stroke called
    (A) Suction (B) Compression (C) Power (D) Exhaust

147. The advantage of a multi cylinder engine over a single cylinder engine is that the former is
    (A) Easier to start (B) Simple to service
    (C) Smoother in operation (D) Shortest in length.
148. The two firing orders used on four-cylinder in-line engines are
    (A) 1-3-4-2 and 1-4-2-3 (B) 1-4-3-2 and 1-3-2-4
    (C) 1-3-2-4 and 1-2-4-3 (D) 1-2-4-3 and 1-3-4-2
149. What is happening below the piston of a two-stroke engine at the instant when the spark occurs.
    (A) New gas is being compressed
    (B) Transfer port has just opened
    (C) New charge flowing in through the inlet port
    (D) Inlet port is closed and depression is being formed.
150. How does the petrol-air mixture enter the cylinder of a two stroke engine
    (A) The charge is pumped up the transfer port
    (B) The depression draws in the gas through the inlet port
    (C) The exhaust gas in the crank case drives out to the new gas
    (D) The upward moving piston pumps in the gas.
151. Water circulation in a thermo-siphon cooling system is caused by
    (A) Conduction currents
    (B) A belt driven water impeller
    (C) A gear driven water pump
    (D) The change in density of water.
152. The direction of flow of water through the radiator of a thermo siphon cooling system is from the
    (A) Top to bottom (B) Bottom to top
    (C) Front to back (D) Back to front.
153. Modern engines use a pump operated cooling system instead of the thermo-siphon system because the later
    (A) Could not be pressurized
    (B) Overcooled the cylinder head and valves
    (C) Required a large quantity of water
    (D) Does not allow a thermostat to be fitted.
154. Extra care must be taken when removing a modern radiator cap from a hot engine because
    (A) The seal can be damaged
    (B) Of the risk of scalding
    (C) The cooling system pressure is lower than atmospheric
    (D) The sudden increase in pressure can damage the radiator.

155. The purpose of adding antifreeze solution to the coolant is to
   (A) Prevent the coolant from freezing
   (B) Lower the freezing point of the coolant
   (C) Stop the formation of ice in the radiator
   (D) Avoid piston seizure due to ice in the water jacket.

156. The thermostat is normally positioned in the cooling system between the
   (A) Header hose and radiator
   (B) Radiator and bottom hose
   (C) Bottom hose and engine water jacket
   (D) Engine water jacket and header house.

157. The purpose of large spring loaded valves in a radiator cap is to
   (A) Lower the temperature at which the coolant boils
   (B) Prevent the coolant escaping when it boils
   (C) Reduce the risk of the rubber hoses collapsing when the pressure is low
   (D) Pressurize the system which raises boiling point of coolant

158. A relief valve is fitted to the main oil gallery of an engine. The purpose of this valve is to
   (A) Limit the maximum oil pressure
   (B) Open when the oil is hot
   (C) Maintain the supply if the gallery becomes blocked
   (D) Stop the oil flow to the bearings when the pressure is low

159. One reason for richening the petrol-air mixture for cold starting is
   (A) Fuel particles are smaller
   (B) Quantity of air is smaller
   (C) Cold engine does not vaporize fuel
   (D) Cold petrol will not flow through jet.

160. The part of an ignition system which transforms the voltage from 12V to more than 9000 V is the
   (A) Coil (B) Distributor (C) Capacitor (D) Contact breaker

161. The purpose of a capacitor in a coil ignition system is to
   (A) Transform the voltage
   (B) Act as a mechanical switch
   (C) Prevent arcing at the contact breaker
   (D) Direct the current to the appropriate plug.

162. A spark occurs at the sparking plug when the contact-breaker of a coil ignition system
   (A) Just opens (B) Just closes
   (C) Is fully opened (D) Is fully closed.

163. The action which takes place in the clutch when the pedal is depressed
    - (A) Pressure plate comes to rest
    - (B) Pressure plate moves away from the flywheel
    - (C) Driven plate moves towards the fly wheel
    - (D) Driven plate slows down to flywheel speed.
164. The gears in a constant mesh gear box have teeth which are inclined to the shaft axis. This type of gear is called
    - (A) Spur (B) Worm (C) Bevel (D) Helical.
165. Which one of the following gear box types uses friction to equalize the speed of the members prior to gear engagement
    - (A) Progressive
    - (B) Sliding mesh
    - (C) Constant mesh
    - (D) Synchromesh
166. Lubrication of a manual type gear box is achieved by
    - (A) Submerging all gears in oil
    - (B) Immersing the lay shaft in oil
    - (C) Connecting the oil ways to the engine pump
    - (D) A pressure pump driven from the input shaft.
167. Direct drive in a gear box is generally obtained by connecting
    - (A) Together two gears of equal size
    - (B) The smallest main shaft gear to the main shaft
    - (C) The smallest lay shaft pinion with largest main shaft wheel
    - (D) The clutch shaft to the main shaft by a dog clutch.
168. When a vehicle is cornering, the crown wheel is rotating at 500 rpm and the outer wheel is turning at 520 rpm. The speed of the inner wheel is
    - (A) 20 rpm (B) 480 rpm
    - (C) 500 rpm (D) 540 rpm
169. The purpose of a brake is to
    - (A) Store energy
    - (B) Change friction to heat
    - (C) Convert energy to kinetic energy
    - (D) Convert kinetic energy to heat energy.
170. A thick cable is needed to supply the starter motor, because
    - (A) Motor requires a very large current
    - (B) Voltage needed is higher than 200 V
    - (C) Thick insulation is needed to prevent a short circuit
    - (D) Extra strength is needed to resist the vibration.

171. If the battery polarity is reversed on a vehicle fitted with an alternator, the effect will be
    (A) Fan belt will slip
    (B) Cut out will not operate
    (C) The light will be dimmer than normal
    (D) The semi-conductor devices will be damaged.
172. One reason why an alternator produces a higher output than a dynamo is because the alternator
    (A) Is driven faster (B) Has a commutator
    (C) Generates direct current (D) Always uses a negative earth.
173. The relative density of the electrolyte in a lead acid battery is 1.280. This value indicates that the battery is
    (A) Fully discharged (B) Half charged
    (C) Three quarters charged (D) Fully charged.
174. The electrolyte used in a lead acid battery is
    (A) Sulphuric acid and diluted lead (B) Diluted lead and pure water
    (C) Pure water and distilled water (D) Distilled water and sulphuric acid.
175. A fuse in an electrical circuit blows as a result of a high
    (A) Voltage caused by a short circuit
    (B) Current flows caused by a short circuit
    (C) Voltage caused by an open circuit
    (D) Current flow caused by an open circuit.
176. A crankshaft is made by
    (A) Casting (B) Forging (C) Pressing (D) Turning.
177. In an automobile, choke is applied for
    (A) Increasing speed (B) Fuel economy
    (C) Starting in cold weather (D) Starting in hot weather.
178. Mixture formation in carburetor is based on the principle of
    (A) Law of vapors (B) Pascal's law
    (C) Venturi principle (D) Newton's laws of motion.
179. The function of a fuel pump in petrol engine is
    (A) To improve thermal efficiency
    (B) To pump fuel to carburetor
    (C) To prepare mixture of air and fuel
    (D) None of the above.
180. The function of a distributor in tractor is
    (A) To distribute power (B) To intensify spark
    (C) To distribute spark (D) To time the spark

181. The gap between spark plug electrodes is generally in the range
    (A) 0.375 to 1 mm (B) 0.375 to 1 cm
    (C) 1 to 5 mm (D) 1 to 5 cm.
182. Firing order of a six cylinder engine
    (A) 1-6-2-4-3-5 (B) 1-5-3-6-2-4
    (C) 1-3-2-6-5-4 (D) 1-4-3-2-1-5.
183. Insulating material generally used in spark plug is
    (A) Wood (B) Bakelite
    (C) Semi conductor (D) Porcelain
184. Higher proportion of residual gases in the cylinder would mean
    (A) Higher volumetric efficiency
    (B) Lower volumetric efficiency
    (C) No change in volumetric efficiency
    (D) None of the above.
185. Volumetric efficiency of a well adjusted engine at full load may be
    (A) 75–95 per cent (B) 50–60 per cent
    (C) 40–50 per cent (D) 25–40 per cent.
186. A petrol engine is easily identified by the sight of
    (A) Air cleaner (B) Oil cleaner (C) Fly wheel (D) Spark plugs.
187. The charge in a motor cycle engine consists of
    (A) Air+ petrol + Lubricating oil (B) Air + Petrol
    (C) Air + Lubricating oil (D) Air.
188. If the ignition of a charge inside the engine cylinder occurs before the passage of spark, it is known as
    (A) Detonation (B) Ping
    (C) Pre-ignition (D) Secondary ignition.
189. Pre-ignition in an engine may be detected by
    (A) Increase in speed (B) Sudden loss of power
    (C) Typical sound (D) None of the above.
190. An engine can be easily identified as petrol or diesel engine by looking at
    (A) Piston (B) Crankshaft
    (C) Camshaft (D) Fuel supply system
191. Fuel injection pressure in diesel engine is usually of the order of
    (A) 10 kg/cm^2 (B) 20–30 kg/cm^2
    (C) 60–80 kg/cm^2 (D) 90–130 kg/cm^2
192. A 5 bhp engine will consume fuel per hour at the rate of about
    (A) 1 kg (B) 2 kg
    (C) 3 kg (D) 4 kg

193. Detonation or pinging noise causes due to
(A) Early time of fuel injection
(B) Late timing of fuel injection
(C) Piston rings badly worn
(D) Valve spring weak

194. Fins are provided on engines is to
(A) Improve heat transfer rate
(B) Improve engine performance
(C) Reduce fuel consumption
(D) Assist in engine starting

195. Self-ignition temperature of diesel oil as compared to petrol is
(A) Higher
(B) Lower
(C) Same
(D) None of the above.

196. The level of oil in engine cylinder should be checked when the engine is
(A) Running
(B) Not running
(C) During cranking
(D) During idling

197. In a four stroke cycle, minimum temperature inside the engine cylinder occurs
(A) At the beginning of suction stroke
(B) At the end of suction stroke
(C) At the end of compression stroke
(D) At the end of exhaust stroke.

198. For heat transfer to occur in cooling system, essential requirement is
(A) Medium
(B) Source of heat
(C) Fluid flow
(D) Temperature difference.

199. The heat from combustion zone in engine cylinder to cooling medium is by
(A) Conduction
(B) Convection
(C) Radiation
(D) Conduction, Convection and radiation.

200. Motor cycles generally have
(A) Air cooling
(B) Water cooling
(C) Liquid cooling
(D) Evaporative cooling.

201. In order to know whether an engine is air-cooled or water cooled, one should observe
(A) Cooling fan and radiator
(B) Piston
(C) Lubricating system
(D) Ignition system.

202. A mechanical indicator is used to determine
(A) IHP
(B) BHP
(C) FHP
(D) Pumping horse power

203. The instrument used for the measurement of area of PV diagram is known as
(A) Area meter
(B) Clino meter
(C) Graph meter
(D) Planimeter.

204. Prony brake is used for testing
   (A) Small engines (B) Large capacity engines
   (C) Slow speed engines (D) High speed engines.
205. A rope brake type dynamometer falls under the category
   (A) Mechanical friction type dynamometer
   (B) Hydraulic type dynamometer
   (C) Transmission type dynamometer
   (D) Torsion type dynamometer
206. A hydrometer is used to determine
   (A) Relative humidity (B) Buoyancy force
   (C) Specific gravity of liquid (D) Viscosity of liquids.
207. If clutch fails to disengage, the possible cause may be
   (A) Excessive clutch clearance
   (B) Friction plate not moving freely on the spline shaft
   (C) Facing on the friction plate has worn down to the rivet heads
   (D) All of the above.
208. Excessive clutch clearance generally results in
   (A) Clutch failure to disengage (B) Clutch slip
   (C) Clutch plate overheating (D) Improper clutch engagement.
209. An oil immersed multiple disc is usually used for
   (A) Increasing the torque
   (B) Decreasing the diametric dimensions
   (C) Prolong wear life
   (D) All of the above.
210. The multiple discs are usually oil immersed because
   (A) Oil acts as a cushion to provide a smooth engagement
   (B) Energy released as heat is carried away by the oil
   (C) Oil prolongs the wear life
   (D) All of the above.
211. In a cone clutch, the cone angle commonly used are
   (A) 1 to 2 degrees (B) 12 to 16 degrees
   (C) 20 to 30 degrees (D) 40 to 60 degrees
212. A too small a cone angle may cause
   (A) An excessive wedge action (B) Easy clutch disengagement
   (C) Reduced torque capacities (D) 40 to 60 degrees.
213. Cone clutches are widely used with
   (A) Constant mesh gear boxes (B) Sliding mesh gear boxes
   (C) Synchromesh gear box (D) Epicyclic gear trains.

214. To make clutch engagement and disengagement as smooth as possible, the clutch plate has a series of
   (A) Cushion pads (B) Cushion springs
   (C) Torsion springs (D) Friction discs.
215. The term cc stands for
   (A) Combustion chamber (B) Engine capacity in cm^3
   (C) Clearance volume in cm^3 (D) Combustion chamber volume in cm^3.
216. In the case of vehicle having four wheel drive
   (A) Clutch linkages are simple (B) The road adhesion increases
   (C) The road adhesion decreases (D) Fuel consumption decreases
217. The main advantage of four wheel drive vehicle is
   (A) Lower initial cost
   (B) Lower running cost
   (C) When front wheels and all into a ditch, they can be driven out easily
   (D) None of the above.
218. The cooling of an engine is simple when the engine is mounted at the
   (A) Centre of the rear axle
   (B) Centre of the vehicle
   (C) Left side of the rear axle
   (D) Front of the vehicle.
219. Wheel base is
   (A) The distance between the centers of the front wheel and rear wheel
   (B) Distance between the front end of the vehicle to the rear most end including all the projections
   (C) Maximum width of the vehicle with all projections.
   (D) The distance between the two wheels on the same axle.
220. Overall length is
   (A) The distance between the centers of the front wheel and rear wheel
   (B) Distance between the front end of the vehicle to the rear most end including all the projections
   (C) The maximum width of the vehicle with all projections
   (D) The distance between the two wheels on the same axle.
221. Wheel track is
   (A) The distance between the centers of the front wheel and rear wheel
   (B) Distance between the front end of the vehicle to the rear most end including all the projections
   (C) The maximum width of the vehicle with all projections
   (D) The distance between the two wheels on the same axle.

222. Semi-centrifugal clutch is used in high power engines so that
    (A) Strain to the driver while disengaging is reduced
    (B) Strain on clutch plate may be reduced
    (C) Coil spring can be avoided
    (D) Life of the pressure plate can be increased.
223. Centrifugal clutches are used these days and these clutches
    (A) Obviates the need of clutch pedal
    (B) Helps in stopping the car in gear without stalling the engine
    (C) Make the driving operation very easy
    (D) All of the above.
224. Clutch linings are usually attached to the plate by
    (A) Brass rivets (B) Aluminum rivets
    (C) Steel rivets (D) Steel screws
225. Clutch chattering or grabbing is noticeable
    (A) During idle (B) At low speed
    (C) During acceleration (D) At very high speed.
226. It is strongly recommended that the asbestos powder should not be blown out from the clutch housing with compressed air because.
    (A) It is costly (B) It should be reused
    (C) It can cause lungs cancer (D) All of the above.
227. For an Indian vehicle, the gear ratio specified for first and reverse gear is
    (A) 1:1 (B) 1.5:1 (C) 2.25: 1 (D) 3.8:1
228. The meshing gear has 3:1 gear ratio. If the smaller gear has 20 teeth, the bigger gear will have
    (A) 20 teeths (B) 40 teeth (C) 60 teeths (D) 80 teeths.
229. Synchronizing devices are normally used when shifting into
    (A) Neutral to first gear (B) First to second gear
    (C) Second to third gear (D) Third to top gear.
230. The typical gear oil used in most of the vehicles has a classification of
    (A) SAE 20 (B) SAE 30
    (C) SAE 40 (D) Multipurpose SAE 85 W-90.
231. The gear box oil has to do the following job
    (A) To lubricate all moving parts and prevent wear
    (B) To reduce friction and power loss
    (C) To cool the gear box
    (D) All of the above.
232. Usually transmission grade oil in two stroke engine is
    (A) SAE 10 (B) SAE 30 (C) SAE 20W 40 (D) SAE 80W 90.

233. Diesel vehicles are more efficient than petrol vehicles because of
(A) Higher compression ratio
(B) Improved combustion because it operates on excess air
(C) Avoids energy losses at part load linked with butterfly back pressure on inlet gases
(D) All of the above.

234. Gear box oil leaks could be caused by
(A) Foaming due to incorrect lubricant
(B) Excessive lubricant
(C) Damaged or missing oil seals
(D) All of the above

235. The shaft which transmit the drive from the transmission to the bevel pinion is known as
(A) Clutch shaft (B) Propeller shaft
(C) Lay shaft (D) Main shaft.

236. The basic purpose of the universal joint in drive line is
(A) Drive angle to change
(B) Drive shaft to be supported at centre
(C) Drive shaft length to change
(D) Drive shaft to be removed and installed easily.

237. Ratio of lubricating oil and petrol in two stroke engine varies from
(A) 1:1 to 1:5 (B) 1:5 to 1:10
(C) 1:10 to 1:20 (D) 1:25 to 1:40.

238. Sulphur content in the diesel fuel is limited to
(A) 0.05 % (B) 1 %
(C) 2 % (D) 5 %

239. In a progressive type gear box used in motor cycles
(A) There is only one neutral position between all gears
(B) The gears pass through the intervening speeds while shifting from one speed to another
(C) All the gears are synchromesh type
(D) All the gears are sliding mesh type.

240. In a selective type gear box
(A) Any speed may be selected from the neutral position
(B) No need to obtain neutral position before selecting any forward or reverse position
(C) There is a neutral position between all gears
(D) The gears pass through the intervening speeds while shifting from one speed to another.

241. The forces which oppose the movement of the vehicle are
   (A) Air or wind resistance (B) Gradient resistance
   (C) Rolling resistance (D) All of the above.
242. The ratio of oil to be added to a litre of petrol in two stroke engine is normally.
   (A) 25 ml to a litre (B) 10 ml to a litre
   (C) 5 ml to a litre (D) 2 ml to a litre.
243. Ignition timing light is used to adjust the
   (A) Ignition timing of diesel engines
   (B) Ignition timing of petrol engines
   (C) Injection timing of diesel engines
   (D) Start of combustion in petrol engines.
244. The ignition timing light, working principle is
   (A) It emits an intense camera flesh light exactly at the same moment of spark occurrence
   (B) It counts the number of spark/min
   (C) It emits an ordinary light on engine flywheel or pulley with the fixed reference mark.
   (D) None of the above.
245. Aluminum metal and alloy is widely used in automobile engines because of
   (A) Higher thermal conductivity and better heat distribution
   (B) Increased compression ratio without detonation due to high thermal conductivity
   (C) Reduced engine weight
   (D) All of the above.
246. The gears in gear box are generally made of
   (A) Brass (B) Cast iron
   (C) Stainless steel (D) Alloy steel.
247. Normally starting motors used in passenger cars draw the current when cranking in amperes
   (A) 6 (B) 60
   (C) 600 (D) 6000
248. Starting motors are linked to the engine flywheel and turning of flywheel and the crankshaft is in between
   (A) 150 to 180 rpm (B) 500 to 550 rpm
   (C) 1000 to 1500 rpm (D) 2500 to 3000 rpm.
249. Starting motors revolve in the range of
   (A) 150 and 180 rpm (B) 500 and 550 rpm
   (C) 1000 and 1500 rpm (D) 2500 and 3000 rpm.

250. 6-volt battery rated at 120 ampere/hour should be able to furnish 6 amperes for

(A) 6 hours (B) 10 hours (C) 20 hours (D) 120 hours.

251. Piston rings are to be replaced if

(A) Excessive oil consumption

(B) Loss of compression

(C) High speed fluctuation

(D) Both A and B.

252. Rapid wear of the piston rings is caused by the

(A) Oils not reaching the cylinder walls

(B) Gap clearance is too small

(C) A bent connecting rod

(D) All of the above.

253. The purpose of the dynamo in an automobile is to

(A) Generate electric power

(B) Act as a reservoir of electrical energy

(C) Convert a portion of engine power into electrical energy

(D) Continuously recharge the battery

254. Kick is used in a vehicle to

(A) Changing the gears of motorcycle

(B) Changing the speed of motorcycle

(C) Starting the motor cycle

(D) Stopping the motorcycle.

255. The braking system provided in railway train is

(A) Hydraulic

(B) Air brakes

(C) Vacuum brakes

(D) Mechanical

256. The tyre specification 4.00 x 16 indicates

(A) Rim width is 4 inches, rim dia 16 inch

(B) Rim width 16 inch rim dia 4 inch

(C) Tyre width 4 inch, rim diameter 16 inch

(D) Rim width 4 inch, tyre diameter 16 inch.

257. The purpose of the caster angle on an automobile/tractor is to

(A) Prevent tyre width

(B) Bring the road contact of the tyre under the point of load

(C) Compensate for wear in the steering linkage

(D) Maintain direction control.

258. Camber angle plus steering axis angle is called the
   (A) Caster (B) Included angle
   (C) Point of intersection (D) Toe-out
259. The camber angle provided in the vehicle is in range of
   (A) 0.5–1.5° (B) 2.0 – 5.0° (C) 5.0 – 10.0° (D) 10–20.0°
260. During braking, the brake shoe is moved to force the lining against the
   (A) Wheel piston on the cylinder (B) Anchor pin
   (C) Brake drum (D) Wheel rim or axle
261. If the tyre size is 8.0 x 14.0, the diameter of tyre will be
   (A) 16 inches (B) 24 inches
   (C) 30 inches (D) 42 inches.
262. If the pressure in tyre is too low.
   (A) There will be excessive flexing in the local band around the wall
   (B) There will be fatigue and failure of the cords
   (C) The tread wear will be rapid
   (D) All of the above.
263. If the tyre size is 9-20-12, it indicates
   (A) A tire of 9 inches width, 20 inches rim diameter and 12 inches tyre diameter
   (B) A tyre of 9 inches width, 20 inches rim diameter and 12 ply rating
   (C) A tyre of 9 inches diameter, 20 inches rim diameter and 12 ply rating
   (D) A tyre of 9 inches width, 20 inches rim width and 12 inches tyre diameter
264. The number of plies in a heavy duty vehicle (tractor) is usually
   (A) 2–3 (B) 5–8
   (C) 8–10 (D) 12–16
265. Effect of under inflation may be
   (A) More wear of tyre tread on one side on than centre
   (B) Separation of piles
   (C) Cracking of tyre side walls
   (D) All of the above
266. Effect of over inflation may be
   (A) More wear in centre
   (B) Poor cushioning effect
   (C) Due to less contact area with road, less resistance between road and tyre
   (D) All of above.
267. The water pump used in cooling system usually is of
   (A) Centrifugal type (B) Reciprocating type
   (C) Vane type (D) None of the above.

268. A cylinder head gasket is used to
   (A) Seal in fuel, oil or coolant
   (B) Seal out dirt, water and air
   (C) Close contact between cylinder head and block
   (D) All of the above.

269. In an engine, the self starter works, but the engine is cranked very slowly, possible cause may be
   (A) Very low battery charge
   (B) Defective starter
   (C) Starter pinion is jammed in the fly wheel teeth
   (D) All of the above.

270. The basic difference between the spark-ignition engine and the diesel engine is
   (A) The diesel engine compresses air alone instead of an air-fuel mixture
   (B) Air temperature ignites the fuel in the diesel engine
   (C) The fuel is sprayed into the combustion chamber in the diesel engine as the piston nears TDC on the compression stroke
   (D) All of the above.

271. The main purpose of the engine fly wheel is to
   (A) Smooth out the flow of power
   (B) Serves as a part of clutch
   (C) Has teeth that mesh with the starting-motor drive pinion
   (D) All of the above.

272. The firing order is the
   (A) Order in which the cylinders are numbered
   (B) Order in which the cylinders deliver their power
   (C) Standard arrangement that can be changed by changing the crankshaft
   (D) Order in which the pistons are arranged.

273. The gear ratio between ring gear on the flywheel and gear on starter motor is of the order of
   (A) 1:1 to 5:1 (B) 5:1 to 8:1
   (C) 8:1 to 16:1 (D) 16:1 to 32:1

274. The term OHV related to engine represents
   (A) Overhead valve engine (B) Overhead vehicle
   (C) Over high velocity (D) None of the above.

275. The name that describes the escape of unburned air-fuel mixture and burned gases from the combustion chamber, past the piston rings and into the crankcase is
   (A) Blow-up (B) Blow past (C) Blow-by (D) Blow-off

276. The meaning of 1.3 litre engine is
(A) 1300 cc displacement volume (B) 1300 cc clearance volume
(C) 1300 cc total Volume (D) None of the above.
277. The main components of a generator are
(A) Frame (B) Armature
(C) Field coils (D) All of the above.
278. The main advantage of disc type of brake as compared to drum type of brake is
(A) Cooling is better (B) Heavier than drum brakes
(C) Constant friction Coefficient (D) All of the above.
279. The force of adhesion between the wheels and the road depends on
(A) Vehicle weight (B) Tyre inflation pressure
(C) Tread pattern (D) All of the above.
280. The bleeding process in braking system removes
(A) Air (B) Dirt
(C) Excessive braking fluid (D) Excessive fluid pressure
281. The maximum temperature which brake lining make with asbestos can sustain
(A) 50°C (B) 100°C (C) 350°C (D) 1000°C
282. Most commonly used lubricating oil system in tractor is
(A) Wet sump pressure system (B) Dry sump pressure system
(C) Pre lubrication System (D) Charge lubrication (Petroil system)
283. The percentage of fuel energy lost from the cylinder wall in an engine is approximately
(A) 1 % (B) 10 % (C) 30 % (D) 80 %
284. Most commonly used lubrication system in two stroke engines is
(A) Splash system
(B) Pressure system
(C) Charge lubrication (Petroil system)
(D) Dry sump system.
285. A spark plug sometimes misfires due to
(A) Presence of carbon deposits on spark plug
(B) Large air gap between plug electrodes
(C) Lower voltage than the required
(D) All of the above.
286. Richest fuel air mixture is required for
(A) Starting (B) Idling
(C) Cruising (D) Acceleration

287. Lean air fuel mixture is required for
    (A) Starting (B) Idling (C) Cruising (D) Acceleration
288. The major components of primary ignition circuit does not include
    (A) Spark plug (B) Contact breaker
    (C) Ignition switch (D) condenser
289. The major components of secondary ignition circuit include the secondary winding of ignition coil, distributor rotor, distributor cap and
    (A) Transistor (B) Spark plug
    (C) Ignition Switch (D) contact breaker
290. The viscosity rating SAE 10W/30 means
    (A) The oil has same viscosity as SAE 10W at –18°C
    (B) The oil has same viscosity as SAE 30 at 99°C
    (C) The oil has same viscosity as SAE 10 W at –18°C and same viscosity as SAE 30 at 99°C.
    (D) None of the above.
291. The term SAE is used for
    (A) Society of automotive Engineers (B) Society of American Engineers
    (C) Society of American Environment (D) None of the above
292. Aspect ratio of a tyre section is given as
    (A) Width of tyre section/height of tyre section
    (B) Height of tyre section/width of tyre section
    (C) Width of tyre X height of tyre
    (D) None of the above.
293. A tyre having aspect ratio 100 is termed as
    (A) High profile tyre (B) Low profile tyre
    (C) Square tyre (D) None of the above.
294. A tyre having aspect ratio 70 is termed as
    (A) High profile tyre (B) Low profile tyre
    (C) Square tyre (D) None of the' above
295. Air brake is preferred over hydraulic brakes because
    (A) Air brakes are more reliable
    (B) In air brakes, stopping distance is less
    (C) Air brakes need less amount of pedal effort
    (D) Air brakes require less maintenance.
296. Volatility of gasoline is important as it decides
    (A) Lubricating properties of the fuel
    (B) Starting and warming up characteristics of engine
    (C) Idling and high speed characteristics of the engine
    (D) Calorific value of fuel.

297. Easy starting of an engine means
  (A) It starts in one revolution of crankshaft
  (B) It starts in five revolutions of crankshaft
  (C) It starts in ten revolutions of crankshaft
  (D) It starts in one hundred revolutions of crankshaft
298. Vapour lock results in
  (A) Loss in power or stopping of the engine
  (B) Engine Knock
  (C) Back fire
  (D) Uncontrolled engine speed.
299. Vapour locking can be eliminated by
  (A) Fuel handling with minimum temperature rise
  (B) Proper cooling of fuel supply system
  (C) Minimum pumping of fuel
  (D) All of the above
300. Fuel saving in two wheeler driving can be achieved by
  (A) Steady driving
  (B) Avoiding frequent braking
  (C) Having speed range 40–50 kmph and reducing gear changes
  (D) All of the above.
301. External combustion engine is
  (A) Steam engine (B) Petrol engine
  (C) Diesel engine (D) Both (B) and (C)
302. Internal combustion engine is
  (A) Steam engine (B) Petrol engine
  (C) Diesel engine (D) Both (B) and (C)
303. The combustion, in which air-fuel mixture is ignited within cylinder by spark, is called.
  (A) Constant volume combustion (B) Variable volume combustion
  (C) Adiabatic combustion (D) None of the above.
304. The combustion, in which fuel is injected into highly compressed heated air in the cylinder, is called
  (A) Constant-volume combustion (B) Constant-pressure combustion
  (C) Adiabatic combustion (D) None of the above.
305. The fine spray of diesel oil is injected into the cylinder, during
  (A) Start of intake stroke
  (B) End of compression stroke
  (C) End of power stroke
  (D) Start of exhaust stroke

306. Engine weight per horse power (HP) is more in case of
   (A) Petrol engines (B) Diesel engines
   (C) Air-cooled engines (D) Motor cycle
307. A piston pin is also called
   (A) Wrist pin (B) Gudgeon pin
   (C) Needle (D) Both (A) and (B)
308. The part of engine, which supports and encloses the crank shaft and camshaft, is called.
   (A) Crank case (B) Crank journal
   (C) Main journal (D) Sleeve
309. The lower part of the crank case, is commonly called
   (A) Oil pan (B) Cam (C) Crank pin (D) Main journal
310. Camshaft gear is also called
   (A) Half time gear (B) Timing gear
   (C) Bevel-gear (D) Spiral gear
311. In 6-cylinder engines, the most commonly used firing order is
   (A) 1-4-2-6-3-5 (B) 1-5-3-6-2-4
   (C) 5-4-2-3-1-1 (D) Both (A) and (B).
312. In multi cylinder engines, pistons are connected to the same crankshaft, so that the
   (A) Power strokes can occur more often
   (B) Motion will be uniform
   (C) Efficiency will be more
   (D) Both (A) and (B).
313. A device by means of which, the heat is converted into work, is called
   (A) Heater (B) Motor (C) Heat engine (D) Rotor
314. In engine, the 'back fire' takes place in
   (A) Cylinder (B) Air chamber
   (C) Manifold (D) Exhaust chamber
315. In single cylinder engines, the purpose of counter balance, is to
   (A) Carry the load through idle stroke
   (B) Overcome friction between moving parts
   (C) Control the engine speed
   (D) Both (A) and (B).
316. In single cylinder engines, the purpose of large fly wheel, is to
   (A) Carry the load through idle stroke
   (B) Control piston speed
   (C) Overcome friction between moving parts
   (D) Both (A) and (C).

317. The use of engines in tractor is generally of
   (A) 4 cylinder 4 stroke cycle engine (B) 4 cylinder 2 stroke cycle engine
   (C) 6 cylinder 4 stroke cycle engine (D) 6 cylinder 2 stroke cycle engine
318. In 4-cylinder engines, which number of pistons move together in one direction
   (A) 1 and 4 (B) 2 and 4
   (C) 1 and 3 (D) 2 and 3
319. In tractor engines, the cylinders are numbered from
   (A) Front side of tractor (B) Rear side of tractor
   (C) Central Cylinder (D) Left most cylinder,
320. In tractor's engine, the first cylinder is counted from the
   (A) Close to radiator (B) Close to carburetor
   (C) Opposite to radiator (D) Central cylinder
321. In diesel engines, to vary the loads and speeds, the
   (A) Quantity of fuel injection, is changed
   (B) Amount of lubricants is increased
   (C) Air-percentage in fuel is increased
   (D) All above.
322. In carburetor type engines, to vary the loads and speeds, the
   (A) Quantity of fuel injection is changed
   (B) Lubricant amount is increased
   (C) Quantity of air-fuel mixture is changed
   (D) All above.
323. For the same size of engine, the 2 stroke engine gives
   (A) Twice the power of 4-stroke engine
   (B) Half the power of 4 stroke engine
   (C) 4 times the power of 4 stroke engine
   (D) None of the above.
324. For the same size of engines, the four stroke engine gives
   (A) Twice the power of 2 stroke engine
   (B) Half the power of 2 stroke engine
   (C) The power equal to 2 stroke engine
   (D) None of the above.
325. Quality of the fuel is judged from its
   (A) Volatility (B) Ignition quality
   (C) Consumption rate (D) Both (A) and (B)
326. 'Vapour lock' is associated to the
   (A) Fuel supply system of engine (B) Cooling system of engine
   (C) Ignition system of engine (D) All above.

327. Volatility of fuel affects the
   (A) Engine speed
   (B) Ease of start of engine
   (C) Fuel calorific value
   (D) Both (A) and (B).
328. Which of the following has highest vaporizing temperature
   (A) HSD oil
   (B) Petrol
   (C) Water
   (D) Kerosene
329. Which of the following has lowest vaporizing temperature
   (A) HSD oil
   (B) Petrol
   (C) Kerosene
   (D) Water.
330. Ignition quality of a fuel refers to
   (A) Less consumption of fuel
   (B) Ease of burning in cylinder
   (C) Higher consumption of fuel
   (D) Higher calorific value
331. Diesel fuels are rated by
   (A) Octane number
   (B) Cetane number
   (C) Calorific value
   (D) Engine efficiency
332. Petrol is rated by
   (A) Octane number
   (B) Cetane number
   (C) Calorific value
   (D) Engine efficiency
333. API gravity of pure water, is
   (A) 10
   (B) 5
   (C) 20
   (D) 30
334. Distillation test measures
   (A) Calorific value of fuel
   (B) Volatility of fuel
   (C) Fuel specific gravity
   (D) Fuel purity
335. Reid vapour pressure test, is associated with
   (A) Specific gravity of fuel
   (B) Fuel purity
   (C) Measurement of vapour pressure of fuel
   (D) None of the above.
336. High speed diesel (HSD) is
   (A) Lighter than the light diesel oil
   (B) Heavier than light diesel oil
   (C) Used in low speed diesel engines
   (D) Both (A) and (C).
337. Low speed diesel oil is
   (A) Heavier than HSD
   (B) Used in low speed diesel engines
   (C) Both (A) and (B)
   (D) Used in 2- stroke engines
338. The function of sediment bowl in fuel supply system, is to
   (A) Hold the dust and dirt of the fuel
   (B) Maintain a good pressure
   (C) Hold the air in the cylinder
   (D) Clean the air, entering into combustion chamber.

339. Regulation of air-fuel ratio at different speeds of engine, is performed by
   (A) Carburetor (B) Radiator
   (C) Cylinder (D) All above.
340. The function of choke, is to
   (A) Restrict the air supply in cylinder
   (B) Control the fuel in cylinder
   (C) Restrict the air supply in carburetor
   (D) None of the above.
341. Turbo charger is a
   (A) Turbo-compressor
   (B) Fuel controller
   (C) Lock in fuel supply line
   (D) Timber cell used for ignition.
342. In IC engines, the turbo charger is driven by
   (A) Intake air (B) Exhaust gas
   (C) Flywheel of engine (D) Gasoline
343. In IC engines, the function of turbo charger, is to
   (A) Supply the air under pressure to cylinder
   (B) Mix the air and fuel
   (C) Control fuel (D) All above.
344. Device used to control engine speed within a specified limit is called
   (A) Choke (B) Governor
   (C) Turbo-charger (D) Carburetor
345. In tractor engines, the type of governor used is
   (A) Constant speed governor (B) Variable speed governor
   (C) Hit and miss type governor (D) None of the above.
346. In stationary engines, the type of governor used, is
   (A) Constant speed governor (B) Variable speed governor
   (C) Hit and miss type governor (D) None of the above.
347. In gas engines, the governor system is
   (A) Hit and miss system (B) Constant speed governor
   (C) Variable speed Governor (D) None of the above.
348. Pneumatic governor is the type of
   (A) Throttle governor (B) Constant speed governor
   (C) Hit and miss governor (D) None of the above.
349. Centrifugal governors are very common in
   (A) Tractor engines (B) Stationary engines
   (C) Scooters (D) Both (A) and (B)

350. In engines, the governor weights are mountedon
   (A) Crank shaft
   (B) Shaft, positively driven by crankshaft
   (C) Camshaft
   (D) Both (A) and (B)
351. Source of mineral lubricant, is
   (A) Crude petroleum (B) Petrol
   (C) Diesel (D) Coal mixed with petroleum.
352. Mineral lubricants are most suitable for
   (A) Engines (B) Machines
   (C) Scooter and motor cycles (D) All above.
353. In forced feed system of lubrication, the pump used, is
   (A) Reciprocating type (B) Centrifugal pump
   (C) Positive displacement pump (D) All above.
354. In high speed multi-cylinder engines, the lubrication system, used is
   (A) Forced feed system (B) Splash system
   (C) Siphon system (D) None of the above.
355. In tractors, the lubrication system used is
   (A) Forced feed system (B) Splash system
   (C) Manual system (D) All above
356. An example of a semi-solid lubricant is
   (A) Fluid oil (B) Gear oil (C) Grease (D) Mica
357. Solid lubricant is
   (A) Graphite (B) Mica (C) Grease (D) Both (A) & (B)
358. Removing of water vapour from crankcase, is done by
   (A) Breather in crankcase (B) Vacuum meter
   (C) Allowing warm air (D) Sucking with pump.
359. API stands for
   (A) Asian Petroleum Institute (B) Accurate Petroleum Index
   (C) Ancient Petroleum Indicator (D) American Petroleum Institute
360. The crank case breather is mostly made of
   (A) Oil wetted screen (B) Cloth
   (C) Felt (D) Thin metal plate
361. The main function of distributor of ignition system of engine, is to
   (A) Close the primary electrical circuit
   (B) Open the primary circuit
   (C) Supply fuel into carburetor
   (D) Both (A) and (B).

362. The function of condenser in battery ignition system, is to
    (A) Produce a quick collapse of magnetic field in the coil
    (B) Produce a high voltage
    (C) Close the primary circuit
    (D) Both (A) and (B).
363. The device, used for starting/stopping the engine is
    (A) Spark plug (B) Ignition switch
    (C) Combustion chamber (D) Radiator
364. Capacity of a battery is indicated in terms of
    (A) Ampere-hour (B) Voltage developed
    (C) Number of hours consumed (D) Life of battery.
365. In magnetic ignition system, the primary current is produced by
    (A) Battery (B) Magnet
    (C) Electrical current (D) All above.
366. Of the total heat generated during operation of I.C. engine, the quantity of heat utilized by engine for running, is about
    (A) 30% (B) 20%
    (C) 70% (D) 60%
367. Function of 'thermostat' is to
    (A) Close the water flow into radiator
    (B) Allow the air-fuel mixture into combustion chamber
    (C) Cool the engine
    (D) Control the engine speed.
368. In IC engines, the thermostat valve is opened at the temperature from
    (A) 10 to 15°C (B) 70 to 75°C
    (C) 20 to 50°C (D) None of the above.
369. Scale formation inside the water passage of cooling system, can be prevented by using
    (A) Lime free water (B) Lime water
    (C) H_2SO_4 solution (D) HCL solution
370. Velocity ratio of drive and driven pulleys, can be expressed as
    (A) $N_1/N_2 = D_2/D_1$ (B) $N_2/N_1 = D_1/D_2$
    (C) $N_1/N_2 = D_1/D_2$ (D) Both (A) and (B).
371. If two pulleys (driving and driven) are connected by a belt, so that the rotation of one causes rotation to other, than relation between two pulleys (under no slippage) is
    (A) $\text{II}\ N_1 D_1 = \text{II}\ N_2 D_2$ (B) $\text{II}\ D_1 = \text{II}\ D_2$
    (C) $\text{II}\ N_1/D_1 = \text{II}\ N_2/D_2$ (D) $D^\wedge = D_2/N_2$

372. Chain and sprocket drive is used, where
(A) Two shafts are at short distance
(B) Power is to transmit at 90° angle
(C) Two shafts are at long distance
(D) Power is to transmit radially.

373. Gear drive is recommended, when two shafts are
(A) At long gap (B) Very close to each other
(C) 90° angle (D) At 45° angles

374. Gear drive is used to change the
(A) Speed of rotation (B) Direction of rotation
(C) Direction of shafting (D) All above.

375. Spur gear transmits the power through, its
(A) Teeth (B) Tension (C) Weight (D) All above.

376. Spur gears are used for power transmission, when two shafts are
(A) Parallel to each other (B) At some inclination
(C) Perpendicular to each other (D) Crossed to each other

377. Worm gear is used, when two shafts are at
(A) Right angle to each other (B) Some distance apart
(C) Radial direction (D) 45 cm apart.

378. Worm gear transmits the power at
(A) 90° angles (B) 30° angles (C) 45° angles (D) 180° angles

379. The point, where the axis of two shafts are at right angles and intersect each other, the suitable gear to use for power transmission, is
(A) Bevel gear (B) Helical gear (C) Worm gear (D) Spur gear

380. In helical gears, the teeth are
(A) Inclined to the axis of wheel (B) Cut at 90° angles
(C) Cut at 45° angle (D) cut parallel to the shaft.

381. The Oldham's couplings are used on two shafts, whose centers are at
(A) Right angles to each other (B) Short distance apart
(C) Out of alignment (D) None of the above.

382. Universal couplings are used, when two shafts are
(A) At right angles (B) At out of alignment
(C) Appreciably out of line (D) None of the above.

383. Flanged couplings are used for
(A) Joining two shafts in line
(B) Transmitting power at 30° inclination
(C) Jointing two different diameters of the shaft
(D) None of the above.

384. Splines are used for
   (A) Joining two shafts in line
   (B) Joining two different kinds of shafts and Pulleys
   (C) Transmitting power at 90° inclination
   (D) None of the above.
385. BHP of engine is measured by
   (A) Dynamometer (B) Tachometer
   (C) Voltammeter (D) Wattmeter
386. The power of a tractor measured at the end of drawbar, is called
   (A) DBHP (B) IHP (C) BHP (D) FHP
387. The power required to run the engine, at a given speed without producing any work, is called
   (A) BHP (B) IHP (C) FHP (D) DBHP
388. The pressure, that forces the piston down, during the power stroke is called
   (A) Mean effective pressure (B) IHP
   (C) DBHP (D) None of the above.
389. Any force, applied at a point to cause a turning effect, is called
   (A) Torque (B) Mean effective force
   (C) IHP (D) None of the above.
390. The power available to pull the loads or draw machines by tractor, is known as
   (A) Draw bar power (B) Draw bar pull
   (C) Draw bar draft (D) None of the above.
391. Average daily working hours of a man is counted as
   (A) 6 hours (B) 10 hours
   (C) 24 hours (D) 12 hours
392. In automobiles/tractors, the 'power train' stands for
   (A) Complete power transmission path
   (B) Hydraulic system
   (C) Ignition system
   (D) PTO.
393. Dog clutch is generally used in
   (A) Farm tractors (B) Scooters
   (C) Power tillers (D) None of above.
394. In tractors, the final drive is a gear reduction unit of
   (A) Power train
   (B) PTO
   (C) Hydraulic system
   (D) Ignition system.

395. A device, which transmits the power of tractor engine to the trailing implement is called
(A) PTO (B) Drawbar
(C) Hydraulic (D) None of the above.

396. The function of decompression lever, is to
(A) Release compression pressure from combustion chamber
(B) Control ignition timing
(C) Check fuel consumption
(D) Control hydraulic system.

397. The ply ratings of tyres, used generally in tractors is as
(A) 4, 6 or 8 (B) 2, 4, or 6 (C) 6, 8 or 12 (D) 8, 10 or 12.

398. The comparative strength of tyre, is indicated by
(A) Thickness and materials of construction
(B) Size and tyre width
(C) Ply ratings
(D) All above.

399. Which of the following tyres of rated ply, is stronger in use
(A) 4 (B) 6 (C) 2 (D) 8

400. In tyre size of 12–38", the number 12" represents
(A) Tyre ply (B) Sectional diameter of tyre
(C) Tyre strength as 12 kg/cm^2 (D) None of the above.

401. In tyre size of 12–38", the number 38" represents the
(A) Sectional diameter of tyre
(B) Ply rating
(C) Sectional diameter of rim on which tyre to be mounted
(D) None of the above.

402. The horizontal distance between the front and rear wheels of a tractor is called
(A) Ground clearance (B) Wheel base
(C) Toe-in (D) None of the above.

403. Ground clearance of tractor is measured under
(A) Maximum permissible load condition
(B) Without loading conditions
(C) Normal loading condition
(D) None of the above.

404. The diameter of smallest circle, described by the outermost point of the tractor, at the speed not exceeding 2 kmph, is called
(A) Track (B) Toe -in
(C) Ground clearance (D) Turning space.

405. The traction of tractor in wet field, can be improved by using
   (A) Cage wheel (B) Bigger size of tyre
   (C) Additional weight (D) None of the above.
406. The loss of value of the machine with lapse of time, is called
   (A) Salvage value (B) Lost value
   (C) Depreciation (D) Resale value
407. Before starting a diesel tractor, the gear shift lever is put into
   (A) Upper gear position (B) Lower gear position
   (C) Neutral position (D) None of the above.
408. Before starting a tractor, the PTO lever is kept into
   (A) Zero position (B) Neutral position
   (C) Lower position (D) None of the above.
409. Before starting the tractor, the hydraulic control lever is kept into
   (A) Upper position (B) Neutral position
   (C) Lowered position (D) None of the above.
410. In tractor, the traction can be increased by using
   (A) Rubber tires with grooves (B) Chain tire
   (C) Lugs, cleats or grousers (D) All above.
411. A device, used for propelling the vehicle by using the traction forces from the supporting surface, is called
   (A) Cage wheel (B) Fly wheel
   (C) Traction device (D) All the above
412. Ratio of total force output of the traction device in the direction of travel to the dynamic weight on the traction device, is called
   (A) Coefficient of friction (B) Rolling resistance
   (C) Viscosity (D) Coefficient of traction.
413. Tractive efficiency is affected by
   (A) Tyre inflation pressure (B) Soil condition
   (C) Wheel size and speed (D) All the above.
414. The force required in the direction of travel to overcome the resistance of motion, is called
   (A) Tractive force (B) Coefficient of friction
   (C) Rolling resistance (D) Tractive efficiency
415. A brake is considered to be cold, when its temperature is
   (A) Less than 80° C (B) 80°C
   (C) 10 to 15°C (D) 50°C
416. A brake is considered to be hot, when its temperature is
   (A) 80°C or more (B) 10 to 15°C
   (C) 75°C or less (D) None of the above.

417. Which is of the following country is the maximum user of power tiller
    (A) India (B) Russia (C) Japan (D) America
418. In India, the introduction of power tiller was in the year
    (A) 1963 (B) 1940 (C) 1950 (D) 1980
419. A gear reduction unit between differential and drive wheels of the tractor is referred to
    (A) Final drive (B) Gear unit
    (C) Ultimate Power system (D) None of the above.
420. In power tiller, the steering clutch lever is provided
    (A) On right handle
    (B) On left handle
    (C) On the grip of right and left handles
    (D) In front of the driver seat.
421. In power tillers, the most commonly used brake is,
    (A) Shoe type (B) Friction type
    (C) Rubber plate type (D) None of the above.
422. In power tillers, the plies of pneumatic tyres, used generally vary from
    (A) 2 to 4 (B) 4 to 8 (C) 4 to 6 (D) 8 to 12
423. Useful life of tractor is
    (A) 10,000 working hours (B) 10 years
    (C) 15 years (D) Both (A) and (B)
424. Useful life of a power tiller, is
    (A) 8,000 working hours (B) 10 years
    (C) 15 years (D) Both (A) and (B)
425. The tractors, used for earth moving works, are the
    (A) Drawbar type (B) PTO type
    (C) Crawler type (D) Trailed type.
426. Crawler tractors are not suitable for
    (A) Road transport (B) Ploughing
    (C) Harrowing and seeding (D) All above.
427. The tractors, to be used for cultivation of row crops, should have the property of
    (A) A large axle (B) Provision for tread adjustment
    (C) Speed adjustable (D) Both (A) and (B).
428. In tractor, if one wheel is locked, the speed of the other one is increased by
    (A) 4 times (B) 2 times (C) 3 times (D) 1.5 times
429. Final drives on tractors, provide additional gear reduction between
    (A) Engine and rear axle (B) Connecting rod and cam
    (C) Piston and crank shaft (D) Crank shaft and cam gear

430. The camber angle is associated to the
   (A) Steering mechanism (B) Transmission system
   (C) Hydraulic system (D) Differential unit.
431. In tractors, the radiator should be drained/cleaned
   (A) After 600 working hours (B) Once in a year
   (C) Twice in a year (D) At every 120 working hours,
432. In tractors, the complete air cleaner unit should be cleaned
   (A) Once in a year (B) At every 600 working hours
   (C) After 6 months (D) Twice in a year
433. The manufacturer of Swaraj make tractor, is
   (A) HMT Pinjore (B) Hindustan Tractors Ltd. Boroda
   (C) TAFE Madras (D) Punjab Tractor Ltd., Chandigarh.
434. The manufacturer of Massey Fergusson Tractor, is
   (A) PTL, Chandigarh (B) TAFE Madras
   (C) HMT Pinjore (D) Hindustan Tractor Ltd., Baroda.
435. The manufacturer of International Tractor is
   (A) TAFE Madras
   (B) HMT, Pinjore
   (C) International Tractor Co. of India, Bombay
   (D) PTL, Chandigarh.
436. The manufacturer of HMT (Zetor) tractor is
   (A) HMT Bangalore (B) HMT Shrinagar
   (C) HMT Pinjore (D) None of the above.
437. The manufacturer of Hindustan tractor, is
   (A) Hindustan Tractor Ltd., Baroda
   (B) HMT Pinjore
   (C) PTL Chandigarh
   (D) TAFE Madras
438. The manufacturer of Ford tractor, is
   (A) PTL, Chandigarh (B) HMT Pinjore
   (C) Escort Ltd., Faridabad (D) None of the above.
439. The manufacturer of Eicher tractor is
   (A) Eicher Tractors India Ltd., Faridabad
   (B) HMT Pinjore
   (C) TAFE Madras (D) PTL, Chandigarh.
440. Valve timing diagram is a function of
   (A) Engine compression (B) Engine speed
   (C) Engine torque (D) Mean effective Pressure

441. When stroke: bore ratio of an engine increases, the volumetric efficiency of the engine
(A) Decreases (B) Increases
(C) Remains unchanged (D) None of the above.

442. Working temperature of diesel engine ranges from
(A) 71–82°C (B) 82–88°C
(C) 75–80°C (D) None of the above,

443. Working temperature of petrol engine ranges from
(A) 71–82°C (B) 82–88°C
(C) 75–80°C (D) None of the above.

444. The compression pressure of diesel engine inside the cylinder varies from
(A) 30–40 kg/cm^2 (B) 35–45 kg/cm^2
(C) 40–50 kg/cm^2 (D) None of the above.

445. The compression pressure of petrol engine inside the cylinder varies from
(A) 6–10 kg/cm^2 (B) 10–15 kg/cm^2
(C) 8–12 kg/cm^2 (D) None of the above.

446. Thermal efficiency of CI engine is
(A) Higher than SI engine (B) Lower than SI engine
(C) Equal to SI engine (D) None of the above.

447. I-head type valves are generally used in
(A) Tractor engines (B) Stationary engines
(C) Steam engines (D) Both (A) and (B).

448. The teeth on the crankshaft gear is
(A) Half to camshaft gear (B) Equal to camshaft gear
(C) Double of Camshaft gear (D) Four times camshaft gear.

449. As 'cc' of engine increases, the power of engine will
(A) Decrease (B) Increase
(C) Remains same (D) None of the above.

450. A gear used in timing gear is
(A) Spur gear (B) Bevel gear (C) Helical gear (D) Worm gear

451. The gaskets are used to prevent
(A) Leakage (B) Dust
(C) Wear (D) All of the above.

452. The total number of rings on piston may vary from
(A) 2–4 (B) 3–5 (C) 3–7 (D) 2–7.

453. The firing order of eight-cylinder engine is
(A) 1-5-4-2-6-3-7-8 (B) 1-2-6-3-4-7-5-8
(C) 1-5-3-4-2-7-8 (D) 1-2-4-6-7-3-5-8

454. The temperature at which fuel ceases to flow is
(A) Flash point (B) Pour point
(C) Cloud point (D) None of the above.

455. The temperature at which fuel catches fire is
(A) Pour point (B) Cloud point
(C) Flash point (D) None of the above.

456. The lowest temperature at which fuel begins to crystallize is
(A) Cloud point (B) Pour point
(C) Flash point (D) None of the above.

457. Bomb calorimeter is used for determining
(A) Calorific value of fuel (B) Specific gravity of fuel
(C) Cetane number of fuel (D) SAE number of fuel

458. The boiling temperature of petrol ranges from
(A) 30 to 180°C (B) 30 to 200°C (C) 30 to 230° (D) 30 to 250°C

459. The boiling temperature of kerosene ranges from
(A) 200 to 300°C (B) 100 to 200°C
(C) 300 to 400°C (D) 300 to 500°C

460. If the volatility of fuel is high, it ignites
(A) Shortly (B) Late
(C) Will not ignite (D) None of the above.

461. If the fuel has higher octane number, its ignition quality will be
(A) Poor (B) Better
(C) Not affected (D) None of the above.

462. Venturi tube is provided in carburetor for
(A) Producing high pressure (B) Producing low pressure
(C) Pressure remains unchanged (D) None of the above.

463. Fuel pump used in carburetor engine is
(A) Fuel injection pump (B) Centrifugal pump
(C) Gasoline pump (D) None of these

464. A pump that transfers fuel from the fuel line to the fuel injection pump is
(A) Fuel lift pump (B) Feed pump
(C) Transfer pump (D) All of the above.

465. A pump that delivers the metered quantity of fuel to each cylinder at appropriate time under high pressure is
(A) Transfer pump (B) Fuel injection pump
(C) Centrifugal pump (D) Oil pump

466. The primary fuel filter used in fuel supply system of diesel engine is made of
(A) Cloth (B) Paper
(C) Both (A) and (B) (D) None of these

467. Secondary fuel filter is made of
(A) Thin wire (B) Cloth (C) Paper (D) All the above.

468. The carburetor employed on most of tractor engine are
(A) Down draft (B) Up draft (C) Middle draft (D) None of these

469. A good lubricant should have
(A) Sufficient viscosity
(B) Remains stable under changing temperature
(C) It should not corrode metallic surface
(D) All of the above.

470. Saybolt viscometer is used for
(A) Flash point test (B) Viscosity test
(C) Pour point test (D) None of the above.

471. Splash system of lubrication is generally used in
(A) Single cylinder and stationary engine
(B) Multi cylinder engines
(C) Tractor engines
(D) All of the above.

472. Oil filters are made of
(A) Cloth (B) Paper
(C) Felt or wire screen (D) All of the above.

473. Viscosity of oil is expressed at two temperatures *i.e.*,
(A) –18°C and 99°C (B) –20°C and 100°C
(C) –20°C and 120°C (D) –18°C and 150°C

474. Petroil lubrication system is generally used in
(A) Two stroke engines (B) Four stroke engines
(C) Tractor and vehicles (D) Aircrafts

475. A disadvantage of air-cooling system is
(A) Uneven cooling of engine parts (B) Engine temperature is high
(C) Both (A) and (B) (D) None of these.

476. Radiator is generally fabricated as
(A) Tubular type (B) Cellular type
(C) Both (A) and (B) (D) None of these

477. The pressure of radiator cap in cooling system of engine is about
(A) 0.34 to 0.4 kg/cm^2 (B) 0.4 to 0.5 kg/cm^2
(C) 0.4 to 0.6 kg/cm^2 (D) 0.4 to 1.0 kg/cm^2.

478. The boiling temperature of water in the radiator is raised to about
(A) 100°C (B) 110°C
(C) 90°C (D) 120°C

479. In case of air cooled engines, the reduction of weight of engines is about
(A) 10 per cent (B) 20 per cent (C) 25 per cent (D) 30 per cent.

480. A device is used for the high voltage current to jump and ignite the charge is known as
(A) Carburetor (B) Spark plug
(C) Injector (D) Battery ignition.

481. Lead acid battery is generally used in
(A) Tractor (B) Air crafts
(C) Light motors (D) None of these

482. Forward speed of crawler tractor is restricted to about
(A) 5 kmph (B) 6 kmph
(C) 10 kmph (D) Both (B) and (C).

483. The average body surface area of a man will be
(A) $1m^2$ (B) $1.5m^2$
(C) $2.0m^2$ (D) $2.5m^2$

484. The noise is expressed in
(A) Kg per sq.cm (B) Vibration
(C) Decibel (D) None of the above.

485. Maximum noise level from a tractor near the operator's ear should not exceed
(A) 85 dB (B) 90 dB (C) 95 dB (D) 100 dB

486. A noise level for eight hours of working is
(A) 85 dB (B) 90 dB (C) 95 dB (D) 100 dB

487. A noise level for six hours of working is
(A) 80 dB (B) 85 dB (C) 90 dB (D) 92 dB

488. Tractor seat suspension should have its natural frequency
(A) 0.5 to 1.5 cycle per second (B) 0.5 to 2.0 cycle per second
(C) 1 to 2.5 cycle per second (C) All the above.

489. In a tractor, centre of gravity is located at
(A) 1/3 of wheel base ahead of rear axle
(B) 2/3 of wheel base ahead of rear axle
(C) 1/4 of wheel base ahead of rear axle
(D) All of the above.

490. Supercharger of an engine is driven by
(A) Camshaft (B) Crankshaft
(C) Exhaust gases (D) Fly wheel

491. The function of drawbar in tractors is to
(A) Lift the load (B) Hitch the implement
(C) Pull load (D) All the above

492. Ballasting is done by
 (A) Water (B) Oil
 (C) Calcium chloride (D) All are correct

493. Ballasting in front tyre of tractor is done to
 (A) Increase traction (B) Decrease load
 (C) Increase stability (D) All of the above.

494. Ballasting in rear tyre of tractor is done to
 (A) Increase traction (B) Reduce tractor vibration
 (C) Increase weight of tractor (D) Prevent over turning of tractor.

495. Spark occurs when spark plug points
 (A) Open (B) Close
 (C) Both (A) and (B) (D) None of the above.

496. In 4 WD tractors power occurs in
 (A) Both front wheels (B) Both rear wheels
 (C) One front and one rear wheel (D) All four wheels.

497. Slogging or lugging ability is associated with
 (A) Engine speed (B) Engine torque
 (C) Engine load (D) All of the above.

498. Weight transfer of tractor is directly affected by
 (A) Pull (B) Hitch height
 (C) Both (A) and (B) (D) Wheel base

499. Brakes work on the principle of
 (A) Friction (B) Vibration
 (C) Sliding (D) Suction

500. When brake is applied, the kinetic energy of body is converted into
 (A) Electrical energy (B) Heat energy
 (C) Chemical energy (D) Mechanical energy.

501. Most commonly used method of determination of centre of gravity of tractor is
 (A) Suspension method (B) Balancing method
 (C) Weighing method (D) All of the above.

502. The rear part of a tractor is heavier than the front part of tractor to get
 (A) Higher tractive efficiency (B) Lower tractive efficiency
 (C) Less wheel slip (D) All of the above.

KEY ANSWERS FOR MULTIPLE CHOICE OBJECTIVE TYPE QUESTIONS

1. A
2. A
3. B
4. B
5. A
6. B
7. A
8. D
9. A
10. A
11. D
12. A
13. B
14. C
15. C
16. D
17. D
18. A
19. B
20. A
21. B
22. A
23. C
24. A
25. B
26. D
27. C
28. A
29. D
30. B
31. D
32. B
33. A
34. D
35. A
36. B
37. B
38. A
39. D
40. A
41. C
42. A
43. B
44. A
45. A
46. D
47. D
48. D
49. B
50. B
51. A
52. D
53. D
54. A
55. C
56. C
57. C
58. D
59. A&B
60. D
61. B
62. D
63. C
64. C
65. B
66. B
67. C
68. A
69. A
70. D
71. C
72. B
73. D
74. A
75. C
76. A
77. D
78. D
79. D
80. A
81. C
82. B
83. A
84. B
85. C
86. B
87. D
88. D
89. A
90. C
91. C
92. C
93. D
94. A
95. D
96. D
97. D
98. B
99. A
100. A
101. A
102. B
103. C
104. D
105. D
106. C
107. D
108. D
109. C
110. C
111. B
112. C
113. C
114. B
115. C
116. B
117. D
118. C
119. A
120. C
121. B
122. A
123. B
124. C
125. A
126. B
127. C
128. D
129. D
130. D
131. C
132. D
133. D
134. D
135. B
136. C
137. C
138. D
139. B
140. D
141. A
142. B
143. C
144. A
145. A
146. A
147. C
148. D
149. C
150. A
151. D
152. A
153. C
154. B
155. B
156. D
157. D
158. A
159. C
160. A
161. C
162. A
163. B
164. D
165. D
166. B
167. D
168. B
169. D
170. A
171. D
172. A
173. D
174. D
175. B
176. B
177. C
178. C
179. B
180. D
181. A
182. B
183. D
184. B
185. A
186. D
187. A
188. C
189. B
190. D
191. D
192. A
193. B
194. A
195. A
196. B
197. B
198. D
199. D
200. A
201. A^
202. A
203. D
204. A
205. A
206. C
207. D
208. A
209. D
210. D
211. B
212. A
213. D
214. C
215. B
216. B

217. C
218. D
219. A
220. B
221. D
222. A
223. D
224. A
225. A
226. C
227. D
228. C
229. D
230. D
231. D
232. C
233. D
234. D
235. B
236. A
237. D
238. A
239. B
240. A
241. D
242. A
243. B
244. A
245. D
246. D
247. C
248. A
249. D
250. C
251. D
252. D
253. D
254. C
255. C
256. C
257. D
258. B
259. A
260. C
261. C
262. D
263. B
264. D
265. D
266. D
267. A
268. D
269. C
270. D
271. D
272. B
273. C
274. A
275. C
276. A
277. D
278. A
279. D
280. A
281. C
282. A
283. C
284. C
285. D
286. A
287. C
288. A
289. B
290. C
291. A
292. B
293. A
294. B
295. C
296. B
297. C
298. A
299. D
300. D
301. A
302. D
303. A
304. B
305. B
306. B
307. D
308. A
309. A
310. A
311. D
312. D
313. C
314. C
315. D
316. D
317. A
318. A
319. A
320. A
321. A
322. C
323. A
324. B
325. D
326. A
327. D
328. A
329. B
330. B
331. B
332. A
333. A
334. B
335. C
336. D
337. C
338. A
339. A
340. C
341. A
342. B
343. A
344. B
345. B
346. A
347. A
348. A
349. D
350. D
351. A
352. D
353. C
354. A
355. A
356. C
357. D
358. A
359. D
360. A
361. A
362. D
363. B
364. A
365. B
366. A
367. B
368. A
369. B
370. A
371. D
372. A
373. A
374. B
375. D
376. A
377. A
378. A
379. A
380. A
381. A
382. B
383. C
384. A
385. A
386. A
387. A
388. C
389. A
390. A
391. A
392. A
393. A
394. C
395. A
396. B
397. A
398. A
399. C
400. D
401. B
402. C
403. B
404. A
405. D
406. A
407. C
408. C
409. B
410. C
411. D
412. C
413. D
414. D
415. C
416. A
417. A
418. C
419. A
420. A
421. C
422. B
423. A
424. D
425. D
426. C
427. D
428. D
429. B
430. A
431. A
432. B
433. A
434. D
435. B
436. C
437. C
438. A
439. C
440. A
441. B
442. A
443. A
444. B

445. B	**455.** B	**465.** D	**475.** A	**485.** C	**495.** A
446. A	**456.** C	**466.** B	**476.** C	**486.** B	**496.** A
447. A	**457.** A	**467.** A	**477.** C	**487.** B	**497.** D
448. D	**458.** A	**468.** C	**478.** A	**488.** D	**498.** B
449. A	**459.** B	**469.** B	**479.** B	**489.** B	**499.** C
450. B	**460.** A	**470.** D	**480.** B	**490.** A	**500.** A
451. A	**461.** A	**471.** B	**481.** B	**491.** B	**501.** B
452. D	**462.** B	**472.** A	**482.** A	**492.** B	**502.** A
453. C	**463.** B	**473.** D	**483.** D	**493.** D	**503.** A
454. A	**464.** C	**474.** A	**484.** C	**494.** C	

Bibliography

1. Principles of Agricultural Engineering. Volume I. A.M. Michael and T.R Ojha. Jain Brothers, New Delhi, 1985.
2. Tractors and their power units. Fourth Edition. John B. Liljedahl, Paul K. Turnquist, David W. Smith and Makoto Hoki, CBS. Publishers and Distributors, New Delhi. 1997.
3. Farm Tractors [Maintenance and Repair]. Second Edition. S.C. Jain and C.R.Rai Standard Publishers Distributors, New Delhi. 1999.
4. Principles of Farm Machinery. Third Edition, R.A. Kepner, Roy Bainer and EL. Barger. CBS Publishers and Distributors, New Delhi, 1997.
5. Farm Machines and Equipment. C.P Nakra. Dhanpat Rai Publishing Company (P) Ltd., New Delhi, 1999.
6. Farm gas engines and tractors. F.R. Jones. Mc-Graw Hill Book Company, New York. 1963.
7. Farm Power. B.D. Mosses and K.R. Frost. John Wiley and Sons, New York. 1967.
8. Electrical Technology. B.L. Theraja. S. Chand and Company Ltd. New Delhi. 1995.
9. Service Manual of Swaraj. Hindustan, International, Massey Fergusson Tractors.
10. A motor vehicle mechanics course. W.J. Burdelt and J.S. Ellis. A Canell Technical Book, London. 1969.
11. Elements of Heat Engines. N.C. Pandya and C.S. Shah. Charter Publishing House, Anand. 1986.
12. Fundamentals of Internal Combustion Engines. Paul W. Gill., H. James, J.R. Smith and E.J. Ziurys. Oxford and IBH Publishing House, New Delhi. 1962.
13. A course in Internal Combustion Engines. ML. Mathur and R.R. Sharma. Dhanpat Rai and Sons, New Delhi, 1989
14. Automobile Engineering. R.E Sharma, Dhanpat Rai & Sons, New Delhi, 1986.
15. I.C. Engine Fundamentals. Jhon B. Heywood. McGraw Hill International Editions. New York. 1989.
16. Automobile Engineering. Fourth Edition. G.B.S. Narang. Khanna Publishers, New Delhi, 1985.

17. Automobile Mechanics: Principles and Practices. Second Edition. Joseph Heitner. Affiliated East-West Press Private Ltd., New Delhi, 1967.

18. A course in Thermodynamics and Heat Engines. Domkundwar and Arora. Dhanpat Rai and Sons New Delhi, 1991.

19. Mechanical Measurement, Third Edition. R.S. Sirohi and H.C. Radhakrishna. New Age International (P) Ltd., Publishers, New Delhi, 1999.

20. An introduction to Energy Conversion. Volume I. Third Reprint. IV Kadambi and Manohar Prasad. Wiley Eastern Ltd., New Delhi, 1984.

21. An introduction to Energy Conversion Volume II IV Kadambi and Manohar Prasad. Wiley Eastern Ltd., New Delhi, 1984.

22. Tractor Performance Data {Volume-ID, April, 2002; Central Farm Machinery Training and Testing Institute, Tractor Nagar, Budni (MP).

Glossary

Accelerator Pump: A piston or diaphragm pump in the carburetor which supplies extra fuel to enrich air-fuel mixture momentarily when the accelerator is suddenly depressed.

Accumulator: The battery of the vehicle which can be recharged.

Additives: Substances added to petrol and oil to improve their quality.

Advanced spark: Act of setting spark timing so that the spark is made to occur before the piston reaches the top dead centre.

Air bleed: An opening into a fuel passage through which air can pass, or bleed into the fuel as it moves through the passage.

Air brake: A braking system that uses compressed air to supply the force required to apply the brakes.

Air cleaner: A device fitted to the engine air intake for filtering and removing dust and other foreign matter from the air before it reaches the engine cylinder.

Air fuel ratio: The proportion of air and fuel (by weight) supplied by the carburetor to the engine.

Alignment: Adjustment of the front wheels of the vehicle.

Alternating current: An electric current that flows first in one direction and then in the other.

Alternator: A device that converts mechanical energy into electric energy for charging the battery and operating electrical accessions in a vehicle. Also known as an ac generator.

Ampere-hour: This represents the capacity of a battery. The product of current in amperes and the hours it can flow till the battery is completely discharged, gives ampere hours.

Anti-freeze: Chemicals like ethylene glycol, added to the engine coolant to lower its freezing point and to raise its boiling point.

Anti-knock quality: Certain chemicals are added to petrol to get anti knock quality. These prevent detonation when the fuel is used in high compression ratio engines.

Arcing: The spark that jumps an air gap between two electrical conductors; for example, the arcing of the distributor contact points.

Atomization: The breaking down of fuel into finely divided particles.

Automatic advance: At higher engine speeds, the ignition spark timing is automatically advanced.

Axle: A shaft which carries the wheels and supports the body of the vehicle through the suspension springs.

Backfire: It is the ignition of mixture in the intake manifold by flame from the cylinder such as might occur from a leaking inlet valve.

Back pressure: The exhaust gases are not released when at atmospheric pressure. This is done at a pressure higher than the atmospheric pressure and is known as back pressure. The resistance to free flow of exhaust gas occurs due to restriction in the muffler.

Battery: An electro-chemical device for storing energy in chemical form so that it can be released as electricity.

BDC: Bottom dead centre. The lowest point where the piston reaches on its downward stroke.

Bead: That part of the tyre which is shaped to fit the rim. It is made of steel wires, wrapped and reinforced by the plies of the tyre.

Bearing: A part which supports a rotating shaft or a moving component. It offers minimum friction.

Bleeding: A process by which air is removed from a hydraulic system (brake or power steering).

Blow by: Leakage of compressed air fuel mixture and burnt gases (from engine cylinder) past the piston rings into the crankcase.

Brake: A device to stop or slow down the vehicle. It changes the kinetic energy of motion into useless and wasted heat energy.

Brake disc: It is a steel or an iron disc which rotates with the wheel. When disc brakes are operated, it gets squeezed between the two friction pads and slows down the vehicle.

Brake drums: The brake drums rotate with the wheels. Brake shoes are brought in contact with the brake drums to slow or stop them.

Brake fluid: A special fluid used in hydraulic brake systems to transmit force through a closed system of tubing known as the brake lines.

Brake light: Lights at the rear of the car which indicate that the brakes are applied.

Brake horse power: Power available from the engine crankshaft to do work such as moving the vehicle.

Breaker points: These are situated in the distributor and consists of one stationary point and one movable point. Such points open and close the ignition primary circuit.

Bore: The diameter of a hole such as cylinder.

By-pass: In alternate path for a flow of fluid (air or liquid).

Butane: A petroleum hydrocarbon compound that remains the liquid state below 0°C at atmospheric pressure. Some times referred to as liquefied petroleum gas. Often combined with propane.

Butterfly: A movable metal disc which governs the flow of air into the carburetor. It serves as a throttle and choke valve.

Bushing: A one piece sleeve placed in a bore to serve as a bearing surface. **Calorific value:** A measure of the heating value of fuel.

Cam: A rotating lobe or eccentric which changes rotary motion into reciprocating motion.

Capacity of engine: The total volume swept by all the pistons of a multi cylinder engine. This is calculated by (number of cylinder x $\pi d^2/4$ x L) where d = diameter of cylinder (or bore dia); L is the length of stroke. The capacity is generally expressed in terms of cc or cm^3 or litres.

Camshaft: The shaft containing lobes or cams to operate the engine valves.

Carburetion: The conversion of liquid fuel to vapour and then mixing it with fresh air for combustion.

Castor angle: The angle between the king-pin axis and a vertical line through the centre of car wheel. It helps the front wheels to return to straight ahead position after cornering.

Centrifugal advance: A rotating-weight mechanism in the distributor that advances and retards ignition timing through the centrifugal force resulting from changes in the rotational speed of the distributor shaft.

Cetane number: It indicates the ignition quality of diesel fuel, or how high a temperature is required to ignite it. The lower the cetane number, the higher the temperature required to ignite a diesel fuel. The high cetane number fuel ignites easily as compared to low cetane number fuel.

Chassis: The assembly of mechanisms that makes up the major operating part of the vehicle.

Check valve: A valve that opens to permit the passage of air or fluid in one direction only or operates to prevent (check) some undesirable action.

Coil: An electrical device which converts the low battery voltage into a high voltage to produce an electric spark.

Combustion: A chemical change through burning process, especially oxidation, accompanied by the production of flame and heat.

Combustion chamber: The space between the cylinder head and piston head when piston is at the top dead centre position. In that chamber, the fuel air mixture burns.

Choke: A device such as a valve placed in a carburetor to restrict the volume of air admitted.

Circuit breaker: A resettable protective device that opens an electric circuit to prevent damage when the circuit is overheated by excess current flow.

Compression: Reduction in the volume of a gas by squeezing it into a smaller space. Increasing the pressure reduces the volume and increases the density and temperature of the gas.

Compression ignition engine: The engine in which fuel is ignited due to the heat of compression.

Compression ratio: The volume of engine cylinder when piston is at BDC, divided by the volume of cylinder when the piston is at TDC.

Condenser: A capacitor in the ignition system which stores electric charge as the contact points remain open. It reduces pitting of contact points.

Connecting rod: In the engine, the rod that connects the crank on the crankshaft with the piston.

Coolant: A mixture of water and antifreeze chemical used to carry away heat from the engine.

Crankcase: The lower portion of the engine which encloses the crankshaft. It also includes the lower part of the engine block and the oil pan.

Crankcase dilution: The unburnt fuel which finds its way past the piston rings into the crankcase oil causing dilution.

Crankshaft: The main shaft of an engine which together with the connecting rods, changes the reciprocating motion of the piston to rotary motion.

Cycle: The series of events *i.e.*, intake, compression, power and exhaust which are completed in four (or two) strokes of the piston. These events are repeated.

Cylinder liner: A replaceable sleeve or liner interposed between the piston and the cylinder wall or water jacket. It is a tube like structure inserted between the piston and the cylinder wall.

Decarborising: During overhaul, carbon is removed mechanically from piston crown, grooves, combustion chamber and valves.

Detonation: It is also referred as spark knock. In the combustion chamber of a spark-ignition engine, an uncontrolled second explosion (after the spark occurs at the spark plug) with spontaneous combustion of the remaining compressed air-fuel mixture, resulting in a pinging sound.

Diaphragm: A flexible sheet of cotton fabric or rubber that separates an area into compartments.

Dieseling: A condition in which a spark ignition engine continues to run after the ignition is off, caused by carbon deposits or hot spots in the combustion chamber glowing sufficiently to furnish heat for combustion.

Differential: A gear assembly between axles that permits one wheel to turn at a different speed from the other, while transmitting power from the drive shaft to the wheel axles.

Disc brakes: A braking system which comprises friction pads, calipers and a metal disc rotating with the wheel. Braking action is caused when the pads held in the calipers press against the metal disc.

Distributor: Any device that distributes. A rotary switch which distributes high voltage surges to engine cylinders in proper sequence. It is generally run by the camshaft.

Drop arm: A lever fixed to the outer end of the steering box.

Drum brake: A brake in which curved brake shoes press against the inner circumference of a metal drum to produce the braking action.

Dry sump lubrication: The oil pump takes up oil from the sump and feeds it to a separate tank. Better oil cooling is obtained in this way.

Dynamometer: A device which absorbs the engine torque and measures it. The brake horse power can be determined from the torque and rotational speed.

Dynamic balance: The balance of an object when it is in motion (For example, the dynamic balance of a rotating wheel).

Earth: The electrical component is earthed to the chassis or body of the vehicle, *i.e.*, it forms a part of earth return system.

Eccentric: A disk or offset section (of a shaft, for example) used to convert rotary to reciprocating motion. Sometimes called a cam.

Earthing strips: The negative of the battery is earthed to the vehicle body by a thick braided wire.

Electronic ignition: An ignition system that uses transistors and other semiconductor devices as an electric switch to turn the primary current on and off. It has no circuit breaker points.

Electronic petrol injection: A device to inject petrol in petrol engines. It controls the amount of fuel injected by sensing the suction pressure, engine speed, throttle position and engine temperature.

Emission control: Any device or modification added on to or designed into a motor vehicle for the purpose of reducing air-polluting emissions.

Exhaust gas: The remains of combustion of fuel-air mixture which are discharged out of an engine. It consists of carbon dioxide, carbon monoxide, unburnt hydrocarbons, oxides of nitrogen and water vapour.

Exhaust manifold: A device with several passages through which exhaust gases leave the engine combustion chambers and enter the exhaust piping system.

Fan: A bladed device in the back of the radiator that rotates to draw cooling air through the radiator or around the engine cylinders. Such a fan is driven by a fan belt from the crank shaft pulley or it may have an independent electric motor.

Feed back carburetor: A carburetor used with an oxygen sensor and a control system to automatically adjust the air-fuel ratio for minimum exhaust emissions.

Feeler gauge: Thin steel metal strips from 0.002 inch to 0.025 inch. These are used to measure clearance between valve and rocker, contact breaker points etc.

Final drive: The final speed reduction gearing in the power train.

Firing order: The order in which the engine cylinders fire or deliver their power strokes, beginning with number 1 cylinder

Float level: The float position of which the needle valve closes the fuel inlet to the carburetor, to prevent further delivery of fuel.

Fluid: Any liquid or gas

Fluid coupling: A device in the power train consisting of driving and driven member transmits power from the engine, through a fluid (or oil)

Fluid flywheel: It is similar to fluid coupling except that it has an additional stator in between the driving and the driven member. It can provide a variable gear ratio.

Flywheel ring gear: A gear, fitted around the fly wheel that is engaged by teeth on the starting motor drive to crank the engine.

Four wheel drive: A transmission system in which the engine power can be transmitted to all the four wheels. Such vehicles are used in rough terrain.

Front wheel drive: A vehicle having its drive wheels located on the front axle. Results in good directional stability and a compact engine and other assemblies.

Fuel: Any combustible substance (Example, petrol or diesel).

Fuel pump: Mechanical or electric pump used to transfer fuel from the tank to the carburetor or fuel-injection system.

Full throttle: When the accelerator pedal of the vehicle is depressed fully, the throttle valve of the carburetor is in wide-open throttle position.

Fuse: A device designed to open an electric circuit when the current is excessive, to protect equipment in the circuit.

Gap: The air space between the two electrodes of the spark or between the contact breaker points.

Gas-turbine Engine: Rotary internal combustion engine in which the gases rotate a curved-vaned rotor attached to a shaft.

Gasket: A thin layer of soft material, such as paper, cork, rubber or copper, placed between two flat surfaces to make a tight seal.

Gasohol: An engine fuel made by mixing 10 per cent ethyl alcohol with 90 per cent unleaded gasoline.

Gasoline: A liquid blend of hydrocarbons, obtained from crude oil, used as the fuel in most automobile engines.

Gears: Mechanical devices that transmit power or turning force from one shaft to another.

Governor: A mechanical device which puts a limit to the speed of the engine. If load increases, the engine speed goes down. A governor operates to increase fuel supply and hence the speed.

Grease: A thick non-flowing lubricant used in places where oil is unsuitable.

Ground: The return path for current in an electric circuit.

Half shaft: The- axle shaft connecting the differential and the road wheel.

Helical gear: Gears having teeth cut at an angle to the axis of the gear. Such gears run silently.

Horse power: The rate at which work is done. One metric horse power is equal to 45,000 mkg of work per minute.

Hub: The centre part of a wheel.

Hydraulics: The use of a liquid under pressure to transfer force or motion, or to increase an applied force.

Hydraulic brakes: A braking system that uses hydraulic pressure to force the brake shoes against the brake drums as the brake pedal is depressed.

Idle: Refers to the engine operating at its slow speed with a machine not in motion.

Idle port: An opening in the carburetor body through which fuel flows when the vehicle is idling.

Idle-mixture screw: The adjustment screw (on some carburetor) that can be turned in or out to lean or enrich the idle mixture.

Ignition: Refers to the burning of air fuel mixture in the combustion chamber by the help of a spark across the electrodes of a spark plug in case of petrol engine and by the help of heat of compression in case of diesel engine.

Ignition advance: At higher engine speeds, the spark is made to jump earlier before the piston reaches TDC position.

Ignition lag: In a diesel engine, the delay in time between the injection of fuel and the start of the combustion.

I-Head engine: An engine having overhead valves *i.e.*, valves in the cylinder head.

Ignition system: The system that furnishes high voltage sparks to the engine cylinders to fire the compressed air-fuel mixture. Consists of the battery, ignition coil, ignition distributor, ignition switch, wiring and spark plugs.

Inertia: The property of an object that causes it to resist any change in its speed or direction of travel.

Intake manifold: A casting with several passages through which air or air-fuel mixture flows from the air intake or carburetor to the ports in the cylinder head or cylinder block.

Idler gear: The reverse idler gear introduced in the transmission to give change of direction.

Iso-octane: A testing fuel which has gnod resistance to knocking. Combined with normal heptane in different proportions, it is used to find out the anti-knock quality of a fuel.

Jet: A calibrated passage in the carburetor through which fuel flows. It is generally made of brass.

Journal: The portion of the shaft in contact with a bearing.

Knock: A heavy metallic engine sound that varies with engine speed. The name also used for detonation, pinging and spark knock.

King-pin: An alloy steel hardened pin in the swivel member to steer the front wheel.

King-pin Inclination: The king-pin is inclined inward to obtain light steering.

Land: The portion of the piston between two ring grooves.

Lead: (Pronounced 'leed'), A cable or conductor that carries electric current. **Lead:** (Pronounced 'Led'), A heavy metal used in lead-acid storage batteries.

Leaded gasoline: Tetra ethyl lead is added to improve anti-knock quality of gasoline (petrol).

L-Head Engine: In this type, valves are located in the cylinder block.

Linkage: A hydraulic system or an assembly of rods or links, used to transmit motion.

Liner: Usually a thin section placed between two parts such as a replaceable cylinder liner in an engine.

Lobe: The nose of the ignition timer cam.

Lubricant: Oil and grease used between two sliding components to reduce friction.

Magneto: An electrical generator giving H.T. current without a battery.

Main bearings: In the engine, the bearings that support the crankshaft.

Main jet: The fuel nozzle or jet, in the carburetor that supplies the fuel when the throttle is partially to fully open.

Master cylinder: The fluid cylinder carrying the piston connected to the brake pedal in the hydraulic braking system.

Mechanical brake: A brake which is operated entirely by mechanical means.

Misfiring: This is caused when any one of the spark plugs does not fire and causes erratic running.

Mono block: An engine which has crankcase and cylinder block cast in one piece.

Muffler: In the engine exhaust system, a device through which the exhaust gases must pass and which reduces the exhaust noise.

Multi grade oil: Lubricant oil, the viscosity of which does not change much with the variation of temperature.

Mutual induction: The condition in which a voltage is induced in one coil by a changing magnetic field caused by a changing current in another coil. The magnitude of the induced voltage depends on the number of turns in two coils.

Needle valve: A small, tapered needle pointed valve that can move into or out of seat to close or open the passage through it. Used to control the fuel level in the carburetor float bowl.

Negative camber: The front wheels are tilted inwards to improve the cornering power of the tires at high speed.

Nozzle: The opening or jet through which fuel or air passes as it is discharged.

Octane rating: A measure of the anti-knock properties of a gasoline. The higher the octane rating, the more resistant the gasoline is to spark knock or detonation.

Oil: A liquid lubricant usually made from crude oil and used for lubrication.

Oil bath cleaner: It traps dirt from intake air which enters into the carburetor.

Oil pan: The detachable lower part of the engine which encloses the crankcase and acts as an oil reservoir.

Oil pump: A gear or rotor pump which circulates lubricating oil from the engine sump to the oil galleries.

Open Circuit: In an electric circuit, a break or opening which prevents the passage of current.

Overhead-camshaft engine: An engine in which the camshaft is mounted over the cylinder head, instead of inside the cylinder block.

Overhead-valve engine: An engine in which the valves are mounted in the cylinder head above the combustion chamber, instead of in the cylinder block. In this type of engine, the cam shaft is mounted in the cylinder block and the valves are actuated by pushrods

Parking brake: Mechanically operated brake that is independent of the foot-operated service brakes on the vehicle, set when the vehicle is parked.

Petrol injection system: A system which does not have a carburetor. It injects a fixed quantity of petrol at proper time into the inlet port.

Petroleum: The crude oil from which gasoline, lubricating oil and other such products are refined.

Pilot bearing: A small bearing, in the centre of the fly wheel and of the crankshaft, which carries the forward end of the clutch shaft.

Pinion gear: The smaller of two meshing gears.

Piston engine: An internal-combustion engine using reciprocating pistons.

Piston crown: The piston head which receives the pressure of the expanding gases.

Piston pin: The cylindrical or tubular metal piece that attaches the piston to the connecting rod. Also called wrist pin or gudgeon pin.

Piston rings: Rings fitted into grooves in the piston. There are two types, compression rings for sealing the compression pressure in the combustion chamber and oil rings to scrape excessive oil off the cylinder wall.

Piston skirts: The portion of the piston below the piston pin.

Pitman arm: In the steering system, the arm that is connected between the steering gear sector shaft and tie rod. It swings back and forth for steering as the steering wheel is turned.

Plies: The layers of cords of rubber or steel fabrics in a tire casing, each of these layers is a ply.

Power brakes: A brake system that uses hydraulic or vacuum and atmospheric pressure to provide most of the force required for braking.

Power steering: A steering system that uses hydraulic pressure from a pump to multiply the driver's steering force.

Pre combustion chamber: In some engines, a separate small combustion chamber where combustion begins.

Pre-ignition: Due to overheated combustion chamber, the mixture burns before the production of actual spark in the spark plug.

Prime mover: A source of mechanical power.

Pressure regulator: A device that operates to prevent excessive pressure from developing. In hydraulic systems, a valve that opens to release oil from a line when the oil pressure reaches a specified maximum,

Pressure-relief valve: A.valve that opens to relieve excessive pressure.

Pressurize: To apply more than atmospheric pressure to a gas or liquid.

Propeller shaft: A revolving shaft which connects the gear box and the differential.

Push rod: Rods which transmit motion from the camshaft to the rockers.

Races: The metal rings on which ball or roller bearings rotate.

Reach: The length of the threaded portion of the spark plug. There are short and long reach spark plugs.

Rebore: A reconditioning process in which the bore is made bigger to accommodate a piston of greater dia than the standard one.

Relay: A relay is connected to the ignition switch so as to avoid heavy current from passing through the switch.

Relief valve: A valve that opens when a preset pressure is reached. This relieves or prevents excessive pressures.

Ring gear: A large gear carried by the differential case, meshes with and is driven by the drive pinion.

Rocker arm: In engines with the valves in the cylinder head, a device that rocks on a shaft (or pivots on a stud) as the cam moves the pushrod, causing a valve to open.

Rotary engine: An engine, such as a gas turbine or a wankel, in which the power is delivered to a spinning rotor (and not to a reciprocating piston as in piston engine).

Rated horsepower: The power recommended by the engine manufacturer to rate its power, allowed for safe loads etc.

Scavenging: Process of removing exhaust gases from the cylinder of an engine by fresh charge.

Scrapper rings: Same as oil control rings.

Semiconductor: A material that acts as an insulator under some conditions and as a conductor under other conditions.

Sensor: Any device that receives and reacts to a signal, such as a change in voltage, temperature or pressure.

Service brake: The foot-operated brake used for retarding, stopping and controlling the vehicle during normal driving conditions.

Side valve engine: Engine in which the valves are placed in the cylinder block. Such a design is not favored because it has lesser volumetric efficiency and lower compression ratio.

Smoke: Air borne particles emitted by the engine. It does not contain water vapour such particles result from combustion may be in sufficient quantity to be visible in atmosphere. Blue color indicates too much of oil in the combustion chamber and black color means too much of fuel in the air-fuel mixture.

Spark-ignition engine: An engine in which fuel is ignited by the heat from an electric spark as it jumps the gap at the end of the spark plug.

Starting motor: The electric motor that cranks the engine, or turns the crankshaft, for starting.

Starting-motor drive: The drive mechanisms mid gear on the end of the starting motor armature shaft, used to cotiple the starting motor to, and disengage it from, the fly wheel ring-gear teeth.

Steering shaft: The shaft extending from the steering gear to the steering wheel.

Steering system: The mechanism that enables the driver to turn the wheels for changing the direction of vehicle movement.

Stoichiometric ratio: In a spark ignition engine, the ideal air fuel mixture ratio of 14.7: 1, which must be maintained on engines.

Super charger: In the intake system of the engine, a pump that pressurizes the ingoing air or air fuel mixture. This increases the amount of fuel that can be burned, increasing engine power. If the supercharger is driven by the engine exhaust gas, it is called a turbocharger.

Suspension: A system which supports the body of a vehicle and minimizes vibrations from the road wheels to the body.

Synchronizer: To cause two events to occur at the same time. That device is used in transmission that synchronizes gears about to be meshed, so that no gear clash will occur.

Tachometer: An instrument to measure the rpm of the engine.

Throat: The narrowest area of the venturi. The petrol is delivered in the form of a jet.

Throttle valve: It is a valve provided at the base of the carburetor to regulate the volume of air-fuel mixture entering the intake manifold, thereby controlling the engine speed.

Throw out bearing: In the clutch, the bearing that can be moved into the release levers by clutch-pedal action, to disengage the engine crankshaft from the transmission,

Thrust bearing: Bearing which resists the thrust.

Tie rod: The part of the track rod assembly which joins the steering arms of the front wheel assembly.

Toe-in: The front wheels are kept slightly inwards, so that they come to correct position when loaded.

Toe-out: In front-wheel driven cars, the front wheels are kept pointing outwards.

Track rod: An alloy steel bar which connects the front wheels so that both wheels move together when steered.

Transverse engine: The engine is bolted to the chassis across the vehicle not lengthwise.

Tread: The part of the tyre that contacts the road. It is the thickest part of tyre and is cut with grooves to provide traction for driving and stopping.

Turning circle: The steering wheel being in its fully locked, position, the diameter of the smallest circle described by the front wheel.

Universal joint: In the power train, a jointed connection in the drive shaft that permits the driving angle to change.

Vacuum: Negative gauge pressure, or a pressure less than atmospheric pressure.

Valve: A device that can be opened or closed to allow or stop the flow of a liquid or gas.

Valve overlap: The number of degrees of crankshaft rotation, during which the intake and exhaust valves are opened together.

Valve seat: The surface against which a valve comes to rest to provide a seal against leaking.

Valve timing: The timing of the opening and closing of the valves in relation to the piston position.

Valve train: The valve operating mechanism of an engine; includes all components from the camshaft to the valve.

Vapour lock: A condition in the fuel system in which gasoline vaporizes in the fuel line; bubbles of gasoline vapour restrict or prevent fuel delivery to the carburetor.

Vent: An opening through which air can leave an enclosed chamber.

Venturi: The reduced area of the carburetor. Here the velocity energy of air-stream increases and therefore pressure energy decreases. This creates a vacuum to suck in petrol.

Viscosity: The property of oil to flow. It decreases with the increase of oil temperature.

Viscosity Index: A measurement by which the viscosity of oil changes with temperature.

Volatility: A measure of the ease with which a liquid vaporizes. Volatility has a direct relationship to the flammability of a fuel.

Volumetric efficiency: A measure of how completely the engine cylinder fills up air or air-fuel mixture on the intake stroke of an engine.

Wankel engine: A rotary type engine in which a triangular rotor revolves eccentrically in an oval chamber to produce power.

Wet sump lubrication: The lubrication system which has oil in the sump positioned underneath the crankcase.

Wheel alignment: Aligning the front wheels relative to each other. Setting of castor, camber, toe in or toe-out and king-pin inclination.

Wheel base: The distance between the centerlines of the front and rear axles.

Some Viva-Voce Questions

Q.1. What is an engine ?

Ans. Engine is a device that converts heat energy to mechanical energy.

Q.2. Classify the engines on the basis of combustion ?

Ans. The engines are classified as internal combustion engines (combustion of fuel takes place internally in the combustion chamber inside the cylinder) and external combustion engines (Combustion of fuel takes place outside the engine cylinder).

Q.3. What is a stroke and a cycle of an internal combustion (1C) engine ?

Ans. The linear distance traveled by the piston from TDC to BDC or BDC to TDC is called a stroke. The four sequential strokes of suction, compression, power and exhaust occurring in a repetitive order is called a cycle.

Q.4. What is negative pressure ?

Ans. The pressure in an enclosed space having less than atmospheric pressure is called negative pressure.

Q.5. How suction stroke takes place in I.C. engine ?

Ans. When piston starts moving from TDC to BDC, pressure decreases or negative pressure is created inside engine cylinder, causing air or air-fuel mixture to enter into it.

Q.6. Differentiate between a 2-stroke and 4-stroke cycle engine.

Ans. An engine which completes four strokes *i.e.*, suction, compression, power and exhaust in 2 strokes of piston or in one revolution of crankshaft is called a 2-stroke engine. It produces one power stroke in each revolution of crankshaft. An engine which completes the above four strokes in four strokes of piston or in two revolutions of crankshaft is called a 4-stroke cycle engine. It produces one power stroke in every two revolutions of crankshaft.

Q.7. What is called idle stroke in *I.e.* engine and how many idle strokes occurring in 2-stroke as well as 4-stroke cycle engine ?

Ans. The stroke or event that does not produce power is called idle stroke. There is one idle stroke is 2-stroke cycle engine and three idle strokes in a 4-stroke cycle engine .

Q.8. Why piston top is deflected in 2-stroke cycle engine ?

Ans. The deflected piston top facilitates the easy movement of incoming charge to the top of the cylinder and slope on the other side deflects the exhaust gases to get out from the exhaust port, minimizing the mixing of fresh charge with exhaust gases.

Q.9. Why crankcase is airtight in 2-stroke cycle engine ?

Ans. The fresh charge enters into the crankcase and is temporarily stored in the crankcase during first stroke (suction and compression). Hence in order to prevent dilution of stored charge, the crankcase is made air tight.

Q.10. What is called scavenging ?

Ans. Displacement or removal of exhaust gases from the cylinder by fresh charge is called scavenging.

Q.11. Why lighter flywheel is provided in 2- stroke cycle engine compared to 4-stroke cycle engine ?

Ans. The size of fly wheel is lighter as it stores less energy for supplying energy to only one idle stroke in 2-stroke cycle engine as compared to three idle strokes in 4-stroke cycle engine. The more heavier the size of flywheel, the more storing of energy in it due to its heavier mass.

Q.12. Why thermal efficiency is poor and mechanical efficiency is higher in case of 2-stroke cycle engine compared to 4-stroke cycle engine ?

Ans. The poor thermal efficiency is due to chances of dilution of fresh charge or loss of fresh charge along with eadiaust gases and higher mechanical efficiency owing to less moving parts like camshaft, valves operating components etc., in case of 2-stroke cycle engine compared to 4-stroke cycle engine.

Q.13. Why diesel engine is called compression ignition (Cl) engine and petrol engine as spark ignition (SI) engine ?

Ans. In diesel engine, the ignition of fuel takes place by the heat of compression and in case of petrol engine, the ignition occurs by electric spark produced by battery and ignition coil.

Q.14. What is a diesel engine ?

Ans. It is an engine, designed to convert chemical energy of heavier fuel like (diesel oil containing around 16 carbon atoms) into mechanical energy.

Q.15. What is a petrol engine ?

Ans. It is an engine, designed to convert chemical energy of lighter fractions of petroleum (petroleum containing around five to eight carbon atoms) into mechanical energy.

Q.16. How a diesel engine differs from petrol engine ?

Ans.

(i) Diesel fuel is ignited by the heat of compression inside the cylinder where as petrol fuel is ignited by electric spark.

(ii) Only air is sucked in diesel engine during suction stroke but air and fuel mixture is sucked into the engine cylinder in case of petrol engine.

(iii) Fuel injector is fitted to the engine cylinder in diesel engine where as spark plug is fitted to the engine cylinder in case of petrol engine.

Q.17. What are the advantages and disadvantages of a C.I. engine ?

Ans. Advantages:

(i) Less wear and tear than petrol engine

(ii) More uniform speed is obtained over a wide range of engine speed.

(iii) Fuel injection equipment is more reliable than carburetor and elec-trical ignition system of petrol engine.

(iv) Thermal efficiency of C.I. engine is approximately 50 % more than that of petrol engine.

Disadvantages

(i) C.I. engine is costlier than petrol engine for the same output.

(ii) CI. engine has more starting difficulties.

(iii) CI. engine is to be more stronger and heavier than petrol engine.

Q.18. Define combustion.

Ans. It is the chemical combination of oxygen with any substance causing it to burn and producing heat energy,

Q.19. What is called a combustion chamber ?

Ans. The volume of space in the engine cylinder above the piston when piston is at TDC.

Q.20. Why air is necessary for burning of fuel ?

Ans. Air supplies oxygen. Combustion cannot take place without oxygen.

Q.21. What are the advantages and disadvantages of two stroke cycle engine.

Ans. The advantages are *(i)* The engines are lighter in weight *(ii)* They are simpler in construction, *(iii)* They have more power strokes, *(iv)* They usually operate in either direction. The disadvantages are *(i)* They have difficulties in controlling the fuel mixture, (ii) More fuel consumption, *(iii)* Cooling and lubrication of engine are not proper, *(iv)* Under fluctu-ating loads, operation is not very satisfactory.

Q.22. Why high compression ratio is used in diesel engine ?

Ans. Diesel gets ignited due to high temperature and pressure, created by high compression. Hence, compression ratio is high *i.e.* (14:1 to 22:1).

Q.23. Why low compression ratio is used in petrol engine ?

Ans. Petrol gets ignited due to electric spark by spark plug provided in the engine. Hence high compression ratio is not required for producing high temperature and pressure which is not desired for ignition of petrol.

Q.24. Why diesel engine is known as constant pressure engine ?

Ans. In diesel engine, the rate of admission of fuel is so regulated that pressure remains almost constant during injection of fuel until combustion is complete, so it is known as constant pressure engine.

Q.25. Why petrol engine is known as constant volume engine or following on Otto cycle?

Ans. The volume remains constant when the heat is added to the cylinder at the end of compression stroke and is rejected from-the cylinder during exhaust stroke.

Q.26. What is the difference between ignition and combustion ?

Ans. Ignition is the process of setting on fire with fuel or air fuel mixture in an I.C. engine where as combustion is the process of uniting fuel chemi-cally with oxygen for producing intense heat *i.e.* in engine cylinder.

Q.27. Give the reason for increased emphasis on diesel engine.

Ans. (i) It is relatively a fuel efficient engine *(ii)* It has no electrical ignition system *(iii)* It has better torque speed characteristics.

Q.28. What is called engine blow by?

Ans. It is the leakage of compressed air-fuel mixture and burnt gases (from combustion) past the piston rings into the crankcase .

Q.29. How backfire occurs in an engine ?

Ans. It is the ignition of air-fuel mixture in the intake manifold of the engine. In this case, a pre-explosion of charge occurs when it passes to the inlet manifold and carburetor through opened inlet valve.

Q.30. What do you mean by compression ratio ?

Ans. It is the ratio of volume of air or air-fuel mixture when piston is at BDC to the volume of air when piston at TDC. The compression ratio of 14:1 means 14 parts of charge in the cylinder at the beginning (piston at BDC) compresses to 1 part of charge at the end (piston at TDC).

Q.31. What are the firing orders in 2,3,4, 5,6 and 8 cylinder, 4-stroke cycle engine?

Ans. 2-cylinder -* 1-2; 3-cylinder -* 1-3-2; 4-cylinder -» 1-2-4-3 or 1-3-4-2, 5-cylinder -^-1-3-5-4-2; 6- cylinder -* 1-5-3-6-2-4 or 1-4-3-6-2-5; 8-cylinder -» 1-5-2-6-8-4-7-3 or 1-6-2-8-4-7-3-5.

Q.32. What is the difference between crankshaft and crank ?

Ans. The crankshaft is that part of the engine which transmits the reciprocat-ing motion of the piston to the driven unit in the form of rotary motion. It is that part to which connecting rods are attached. Crank is a part or arm right angle to the crankshaft axis.

Q.33. What is a combustion chamber ?

Ans. The space between the cylinder head and piston head when piston is at top dead centre. In this space, fuel air mixture burns.

Q.34. What is piston crown and skirt ?

Ans. The crown is the piston head which receives the pressure of the expanding gases and skirt is the portion of piston below the piston pin.

Q.35. What is the difference between a fuel and oil ?

Ans. Fuel is *a* substance whose chemical energy is converted to heat energy and then mechanical energy. Oil is a substance used particularly for lubrica-tion purpose.

Q.36. What is a supercharger ?

Ans. In the intake system of the engine it is a pump that pressurizes the ingoing air or air-fuel mixture. This increases the amount of fuel that can be burned increasing engine power. If the supercharger is driven by the engine exhaust gas, then it is called a turbocharger. Supercharger is driven by crankshaft.

Q.37. What is called vapor lock in fuel line ?

Ans. A condition in the fuel system in which gas-line vaporizes in the fuel line or fuel pump; bubbles of gasoline vapor restrict or prevent fuel delivery to the carburetor.

Q.38. What is a decompression lever ?

Ans. It is a lever to release the compression pressure in combustion chamber during starting or cranking the engine.

Q.39. What is carburetion and function of carburetor ?

Ans. Carburetion is the process of conversion of liquid fuel to vapor and then mixing it with fresh air for combustion. The function of carburetor is to atomize the fuel and to mix air and petrol in correct proportion to get the desired results from the engine.

Q.40. What is called venturi and what is called venturi principle in a carburetor ?

Ans. Venturi is the reduced area of the carburetor. When air passes through venturi, its velocity increases and pressure decreases, causing the dis-charge of fuel from the fuel nozzle. The vacuum created in the venturi helps in sucking in petrol.

Q.41. How ignition quality of a diesel fuel be improved ?

Ans. The improvement can be done by adding small percentages of additives which are more easily ignitable hydrocarbons.

Q.42. What is meant by ignition delay or ignition lag ?

Ans. In a diesel engine, the delay in time between the injection of fuel and the start of combustion.

Q.43. What is the fuel injector in a diesel engine ?

Ans. To inject a small volume of fuel (diesel) in a fine atomized form and to assist in bringing each droplet with highly compressed air for quick and complete combustion.

Q.44. What type of lubrication is done in a 2- stroke cycle engine ?

Ans. Charge lubrication or petroil lubrication is done in 2-stroke cycle engine, in which lubricant is mixed with fuel and is filled in fuel tank.

Q.45. What is the ratio of oil to be added to a litre of petrol in a 2-stroke engine ?

Ans. 25 ml of lubricant is mixed with 1 litre of petrol.

Q.46. Can kerosene fuel be used in petrol engine ?

Ans. No.

Q.47. What is a stoichiometric ratio ?

Ans. It refers to ideal or correct air-fuel mixture that should be maintained in engine. In SI engine, the air fuel ratio is 15:1.

Q.48. What is called a valve train ?

Ans. Valve train refers to the valve operating mechanism of an engine, includ-ing all components from the camshaft to the valve.

Q.49. What is a crank case in an engine ?

Ans. The part of engine which supports and encloses the crankshaft and cam-shaft is called crankcase. It is the lower section of cylinder block and contains oil pan.

Q.50. What is crankcase ventilation ?

Ans. The circulation of air through the crankcase of a running engine to remove water, blow by and other vapors; prevents oil dilution, contami-nation, sludge formation and pressure buildup.

Q.51. What is called a spark advance ?

Ans. Spark advance is an amount of degrees of which crankshaft locks of reaching TDC, when spark occurs in the engine cylinder.

Q.52. What is the purpose of using a choke valve in S.I. engine ?

Ans. It restricts or chokes off air flow to the carburetor. It is used in cold and starting condition to provide rich mixture.

Q.53. Where is the location of choke valve and throttle valve in a carburetor ?

Ans. The choke valve is located at the top and throttle valve at the bottom of the carburetor. .

Q.54. Why the size of flywheel is large in single cylinder engine ?

Ans. Large size fly wheel is able to supply required energy during idle stroke and overcome the friction between moving parts.

Q.55. Why the chamber in the carburetor is called a float chamber ?

Ans. The float chamber includes float bowl, float and needle valve. The float maintains the level of fuel in the float bowl by opening and closing the needle valve for fuel inlet and fuel cut off to the carburetor.

Q.56. What is the function of throttle valve ?

Ans. The throttle valve regulates the quantity of charge to enter into the cylinder. It is operated with the help of accelerator pedal. By controlling the charge, the speed of engine is also controlled.

Q.57. How air-fuel ratio is controlled in a diesel engine ?

Ans. This is controlled by adjusting fuel injector to spray right quantity of fuel to the combustion chamber according to load requirements. The fuel injector is actuated by governor and governor by accelerator pedal; which is operated by the driver as per requirements.

Q.58. How the speed of diesel engine is controlled ?

Ans. The speed is controlled by a speed governing device or governor.

Q.59. What is the basic difference between hit and miss governing system and throttle governing system ?

Ans. In case of hit and miss governing system, number of power strokes is changed according to load requirements but in case of throttle governing system, number of power stroke remains same but amount of air fuel mixture is changed or intensity of explosion is also changed.

Q.60. What is the difference between a centrifugal and pneumatic governor ?

Ans. The main principle of centrifugal governor lies in the fact that when a weight rotates about a point, it tends to fly outward due to centrifugal force. If the weight is constrained by a linkage, it can be made to operate control rod of fuel injection pump. Pneumatic governor uses the suction created in the engine induction pipe to control the fuel injection pump.

Q.61. How the vehicle speed can be changed ?

Ans. Vehicle speed can be changed by using the accelerator pedal within a certain power transmission gear ratio in gear box. The change in speed is small as compared to change of vehicle speed by varying different power transmission gear ratio in gear box.

Q.62. What is the function of control rod in fuel injection pump of diesei engine ?

Ans. The control rod movement in fuel injection pump controls the amount of fuel to be injected to the combustion chamber according to load require-ments.

Q.63. Why counter weights are provided in the crankshaft ?

Ans. To balance or prevent the vibration of the crank shaft.

Q.64. Differentiate between cylinder, cylinder block and cylinder head ?

Ans. Cylinder block is a solid casting or metal casting which includes cylinder, water jacket or cooling fins in air cooling engine. Cylinder is a round hole in which the piston reciprocates. It is also known as engine bore. Cylinder head is a detachable portion of an engine which covers cylinder, includes spark plugs, valves etc.

Q.65. Why camshaft is called a half time gear ?

Ans. Cam shaft gear is bigger in size than that of the crank shaft gear and has twice as many teeth as that of crankshaft gear. The speed of the camshaft is exactly half the speed of the crankshaft in four stroke cycle engine. For this reason, the camshaft is commonly called as half time gear. It also operates the ignition timing mechanism, lubricating oil pump and fuel pump. It is mounted on the crankcase, parallel to the crankshaft.

Q.66. What is called a piston pin and a piston boss ?

Ans. Piston pin is a cylindrical piece of alloy steel which passes through the piston bosses and upper end of the connecting rod, so that the movement of piston is transmitted to the connecting rod. Piston boss is the grooved portion of the piston which supports the piston pin.

Q.67. Why piston skirts are slotted or cut ?

Ans. To accommodate the expansion of piston, subjected to high temperature, to reduce weight and to give space for counter weights on the crankshaft.

Q.68. What are the functions of flywheel ?

Ans. (i) Smooth out the flow of power *(ii)* Serves as the pressure surface for clutch plate *(iii)* Has teeth that mesh with starting motor drive pinion *(iv)* Engine timing marks are indicated for fixing timing of the engine.

Q.69. Why groves are provided in the piston ?

Ans. To fit the piston rings, *i.e.*, compression rings and oil rings. Compression rings are plain and placed nearest to the piston head but oil rings are grooved and slotted and located below the compression ring.

Q.70. Why piston ring gap is provided ?

Ans. The clearance between the ends of the piston ring is called piston ring clearance or gap. The gap is provided to prevent breakage or buckling of ring at the time of expansion due to heat effects.

Q.71. Why big end bearing is split into two shells ?

Ans. It is split into two shells to prevent the back and forth movement of the crankshaft.

Q.72. What Is called a timing gear ?

Ans. Timing gear is the combination of crankshaft gear and camshaft gear. Timing gear controls the timing of ignition, timing of opening and closing of valves as well as fuel injection timing.

Q.73. What is the difference between valve clearance and tappet clearance ?

Ans. Valve clearance is the gap between rocker arm and tip of the valve stem in case of overhead valve type of engine. The tappet clearance is the gap between the cam follower (or tappet) and the tip of the valve stem in case of side valve engine.

Q.74. What is a valve timing diagram ?

Ans. It is the diagram of crankshaft rotation with respect to opening and closing of inlet and exhaust valve. The inlet valve generally opens, about 10° before TDC and closes about 10° after BDC . Similarly exhaust valve opens about 10° before BDC and closes about 10° after TDC. This arrange-ment is provided in order to allow more time for opening of inlet valve and exhaust valve for entry of required amount of air or air fuel mixture in suction stroke and escape of burnt gases in exhaust stroke.

Q.75. What is the function of muffler in a tractor ?

Ans. (*i*) To reduce the temperature of exhaust gases

(*ii*) To damp down the sound of exhaust gases.

(*iii*) To reduce the speed of outgoing gases.

Q.76. What do you mean by cranking ?

Ans. Cranking is the process of allowing flywheel to move at the time of starting the engine. The methods generally adopted are with the help of hand and by an electric motor,

Q.77. What is the difference between a thrust and torque ?

Ans. Thrust is a stress tending to push anything out of alignment. Torque is the twisting or turning effort given to a body.

Q.78. Why engine lubrication is needed ?

Ans. *(i)* To reduce wear of parts *(ii)* To seal the piston ring with cylinder wall, *(iii)* To clean and cool the moving parts *(iv)* To work as a cushion for moving parts.

Q.79. Differentiate between an oil strainer and oil filter ?

Ans. The oil strainer keeps the lubricating oil from coarse particles of carbon, rags, scales etc. The filter removes the fine contaminants from the oil.

Q.80. Why cooling is required for an engine ?

Ans. *(i)* To maintain optimum temperature of engine for efficient operation under all conditions *(ii)* To dissipate surplus heat for protection of engine components like cylinder, cylinder head, piston, piston rings, valves etc. *(iii)* To maintain the lubricating property of oil inside the engine cylinder.

Q.81. What is the effect of over cooling of the engine ?

Ans. *(i)* Engine loses power *(ii)* Uneconomical burning of fuel takes place

Q.82. What is the effect of overheating of the engine ?

Ans. *(i)* Engine may cease *(ii)* Cylinder , head or piston may crack *(iii)* Lubrication will become ineffective.

Q.83. What are the advantages of air cooling of the engine ?

Ans. In this system, no radiator is required and no risk of freezing of the liquid. The engine also becomes lighter as compared to liquid cooling engine.

Q.84. Why water is considered a common coolant ?

Ans. Water is available everywhere and it has high specific heat.

Q.85. What is the function of a radiator in a water cooled engine ?

Ans. In liquid cooling system, radiator removes heat from coolant passing through it; receives hot coolant from the engine and sends the coolant to the engine at a lower temperature.

Q.86. The term radiator used in the literature for engine cooling system is a misnomer. Do you agree ? Why ?

Ans. The radiator used in an engine cooling system is meant for cooling the water circulating round the engine . This water which picks up heat from the engine walls loses the same by passing through copper tubes which are in turn cooled by air.

Thus, the radiator performs the cooling action due to the air convection currents picking up heat from the copper tubes. The radiator exchanges heat with the atmosphere more by convection than by radiation. Hence, the name can be termed as a misnomer.

Q.87. What is a water jacket ?

Ans. Water jacket is placed around the cylinder for cooling. It is cast into the cylinder blocks and heads.

Q.88. What are the commonly used antifreeze materials ?

Ans. The most commonly used antifreeze materials are either alcohol or alcohol base or ethylene glycol.

Q.89. Why a thermostat valve is provided in the radiator ?

Ans. Thermostat valve is a kind of check valve which opens and closes with the effect of temperature. It is fitted between the water outlet of the engine and the top of the radiator. It restricts the coolant flow to radiator to maintain desired engine operating temperature. It is opened when the temperature of circulating water reaches to about 82°C. After opening of thermostat valve., the hot water circulates through the radiator and gets cooled.

Q.90. How a fan and pump gets drive in the cooling system of an engine ?

Ans. The pump gets power from crankshaft through belt. The fan is driven by the same belt that drives the pump.

Q.91. What is the use of pressure cooling system in I.C. engine ?.

Ans. The cooling system works under pressure with the help of a pressure cap provided in the top of the radiator. The cap is equipped with a spring controlled valve, which permits the escape of liquid or steam if the pres-sure becomes too high. This system permits operating the engine at a higher temperature without boiling and losing it by evaporation, resulting in the less consumption of liquid.

Q.92. Why it is dangerous to put cold water in the engine and to open pressure cap while it is very hot ?

Ans. The cooling of hot metal by cold water may develop cracks and damage the engine. The opening of radiator pressure cap in hot condition may cause sudden escape of hot water and steam from the radiator and results in the accident to the operator.

Q.93. What is the basic difference between a wet sump lubrication and a dry sump lubrication?

Ans. The wet sump lubrication system has oil in the sump positioned under-neath the crankcase. The pump delivers oil under pressure to oil gallery.

In dry sump lubrication, there are two sumps and two pumps (one is called scavenge pump and the other is the pressure pump) .The scavenge pump takes up oil from the sump under low pressure and feeds it to a separate tank, positioned outside the engine cylinder. From the separate tank or storage tank, the oil is supplied to oil gallery under pressure through a pressure pump. The storage tank has fins around it, in order to constantly cool the oil by atmospheric air. In case, engine being run upside down, the supply of lubricant is continued .

Q.94. Why clutch is used in automobile vehicle/tractor ?

Ans. Clutch is a device, used to connect and disconnect the engine power with or from the transmission gears. It is operated generally during starting, stopping and changing of gears. It is located in between the engine fly wheel and the gear box.

Q.95. Which ones are driving members and driven members in a clutch system of vehicle ?

Ans. Pressure plate and fly wheel are the driving members and clutch plate is the driven member.

Q.96. Why clutch plate is also called as a friction plate ?

Ans. The power is transmitted through clutch plate due to frictional force between driving and driven members of a clutch.

Q.97. Why anti-rattle spring is provided in the release fingers of clutch?

Ans. To reduce or prevent vibration of release levers or fingers.

Q.98. What is called a power train ?

Ans. The mechanism that carries power from the engine crankshaft to the drive wheel is called a power train. It includes clutch, transmission, propeller shaft, differential, rear axle and rear wheel.

Q.99. With what device the torque and speed of a vehicle change ?

Ans. Torque and speed of a vehicle can be changed with the help of a gear box. When torque of a drive wheel increases in order to improve tractive force, the speed of the vehicle decrease. Because torque multiplication means speed reduction and vice versa.

Q.100. With what relation, one can explain that when torque of drive wheel increases, its speed decreases ?

Ans. HP (horse power = 2rNT/4500

As horse power of engine remains same, then by increasing torque, speed will decrease from the above expression.

Q.101. Differentiate between a gear and pinion.

Ans. Gear and pinion are toothed wheel when mesh with each other, power is transmitted from one shaft to another. They occupy less space compared to chain, belt etc. The smaller of the two meshing gears is called a pinion.

Q.101. How the gear ratio is defined ?

Ans. The gear ratio is defined as $N_A / Ng = Tg / T_A$ where N is the rpm of gear and T is the number of teeth of gear.

Q.102. Differentiate between sliding mesh, constant mesh and synchromesh gear box.

Ans. In sliding mesh gear box, the gears on the main shaft slide along it. By operating the gear shift lever, the gears on the main shaft are allowed to slide on it and are meshed with the respective gears on the countershaft.

In constant mesh gear box, all the gears of the main shaft are in constant mesh with the corresponding gears of the countershaft. The dog clutch or coupling device lock the gears in the main shaft as and when required.

In synchromesh gear box, a transmission is designed to synchronize that is, to equalize the speeds of the matching gears before they are meshed, so that no gear clash occurs. In this case, synchronizer is used in place of dog clutch in constant mesh gear box.

Q.103. Why the component differential is named so ?

Ans. Because differential allows the different speed to the rear drive wheels. It consists of a gear assembly between axles that permits one wheel to turn at a different speed from the other, while transmitting power from drive shaft to the wheel axles.

Q.104. What is the use of differential lock in automobile vehicle/tractor ?

Ans. Differential lock is a device or a coupling sleeve to join both the half axles of drive wheels so that even if one wheel is under more resistance, it comes out from mud, soft ground or loose soil by equalizing the speed among two drive wheels.

Q.105. What is called a power steering ?

Ans. A steering system that uses hydraulic pressure from a pump to multiply the driver's steering force.

Q.106. Why torque is not maximum at maximum engine speed of a diesel engine?

Ans. A diesel engine develops more torque at intermediate speeds than at the maximum speed. This is because the volumetric efficiency is higher at intermediate speeds. Thus there is greater amount of air to burn during power stroke. At higher speeds, the volumetric efficiency and the combus-tion pressures are both lower, causing torque also to be lower.

Q.107. Why rear wheel is bigger than front wheel of tractor ?

Ans. *(i)* Bear wheel is a traction wheel that needs more ground contact, which is possible by taking it bigger in size *(ii)* Front whccl is a steering wheel, which does not require more ground contact as compared to rear wheel.

Q.108. What is the meaning of ballasting ?

Ans. Ballasting is the process of adding weight on tires. The front wheels of tractor is generally ballasted to improve its stability.

Q.109. What would happen if petrol is used in diesel engine ?

Ans. There may be the preignition of petrol due to its low vaporising temperature resulting in loss of power and engine may not start.

Q.110. What would happen if diesel is used in petrol engine ?

Ans. The diesel fuel would not be vaporized while passing through the carburetor and engine may not start.

Q.111. How a Cycle rickshaw takes a turn without a differential ?

Ans. A Cycle rickshaw has one rear wheel as a drive wheel fixed to the shaft and other rear wheel is free on the shaft. That free wheel does the function of a differential as in an automobile vehicle.

Appendix-I

Specifications and Performance Data of Some Tractors

Sl. no.	*Particulars\ Name of tractors*	*Mitsubishi Shakti MT 180D* I	*Mahindra B-275* II	*Escort 450* III	*Swaraj 855* IV	*FORD 5630* V	*TAPE MF-1035* VI	*Sonalika International DI-750* VII	*L&T John Deere 5310* VIII	*HMT 4511* IX
A.	Engine:									
	1. Bore/Stroke (mm)	70/78	88.9/ 101.6	105/ 120	110/ 116	100/ 115	88.9/ 12.7	91.4/ 127	106.5/ 110	102/ 110
	2. No. of cylinders	3	3	3	3	4	3	3	3	3
	3. Cubic capacity (cc)	900	1892	3117	3308	3613	2365	2500	2900	2698
	4. Rated speed (rpm)	2700	2600	2200	2000	2500	2000	2250	2400	2200
	5. Cooling system	Water cooled	Water cooled	Water cooled	Water cooled	Water cooled	Water cooled	Water cooled	Water cooled	Water cooled
B.	Transmission:									
	1. No. of speeds: forward/reverse	6/02	8/2	a/2	8/2	8/2	6/2	8/2	9/3	8/2
	2. Range of speed (kmph); forward/ reverse	1.24-1 4, 02/ 1.58- 6.96	2.S4- 27.4S/ 3.93- 11.45	2.37- 27.4S/ 3.20- 12.96	2.68- 33. SOI 3.75- 14.80	3.26- 32.30/ 4.55- 16.40	2.30- 24.70/ 3.15- 12.50	2.78- 34.74/ 3.89- 15.33	2.25- 31.60/ 3.78- 24,50	2.00- 29.13/ 3.81- 13.11
	3. Clutch	Single	single	single	single	Dual	single	Dual	Dual	Dual
	4. Type of PTO	Engine	Engine	Engine	Engine	Engine	Engine	Engine	Engine	Engine
	5. Standard PTO speed (rpm)	540±10	540±10	1000 ±25	1000 ±25	540±10	540±10	1000±25	540±10	540±10
C.	Type of hydraulic system	MDC	ADDC	ADDC	ADDC	ADDC	ADDC	ADDC	ADDC	ADDC
D.	Minimum radius of Turning space (m)	2.69	3.4	3.46	3.54	3.95	3.19	3.69	3.77	3.63
E.	Construction									
	1. Fuel tank capacity (litre)	18	50	48	50	62	35	44.5	72	63.5
	2. Type of brake	Mechanical shoe	Mechanical disc	Dry disc	Mechanical dry disc	Mechanical dry oil immersed	Mechanical shoe	Mechanical disc	Hydraulic	Hydraulic
	3. Tyre size; Front/Rear	5-12/ 8-18	6-16/ 12.4-28	6-16/ 13.6-28	6-16/ 13.5-28	7.50-16/ 18.4-30	6-16/ 12.4-28	6-16/ 13.6-28	6.50-20/ 16.9-28	6-16/ 13.6-28
	4. Track setting (mm), Front/Rear	785/ 720-870	1230/ 1325	1350/ 1650	1350/ 1650	1360/ 1750	1350/ 1650	1350/ 1680	1450/ 1850	1430/ 1750
	5. Wheel base (mm)	1420	1830	1965	1950	2165	1820	1940	2050	2017
	6. Ground clearance (mm)	190	325	410	415	370	325	385	450	390

Table: (*Contd...*)

Table: *(Contd...)*

Sl. no.	Particulars \ Name of tractors	Mitsubishi Shakti MT 180D	Mahindra B-275	Escort 450	Swaraj 855	FORD 5630	TAPE MF-1035	Sonalika International DI-750	L&T John Deere 5310	HMT 4511
		I	II	III	IV	V	VI	VII	VIII	IX
7.	Overall dimension (mm) length/width/height	2530/930/1465	3060/1680/2220	3300/1705/2080	3385/1750/2072	3890/1995/2585	2985/1630/2172	3545/1715/2100	3600/1865/2320	3557/1857/2100
8.	Unballasted mass (kg) Front/Rear/Total	315/440/755	710/1080/1790	710/1170/1880	755/1160/1915	1190/1810/3000	605/875/1430	790/1180/1970	755/1400/2155	770/1250/2020
9.	Ballasted mass (kg) Front/Rear/Total	3S4/485/870	830/1880/1710	815/2230/3045	815/1840/2655	1324/2552/3876	690/1635/2325	830/2130/2960	1055/2020/3075	870/2100/2970
10	Height of trailer hitch (mm)	267	347-797	692	626-800	538-843	375-555	610-810	547-767	495-735
11	Recommendation for wet land operation	Yes	Yes	Yes	Yes	No	No	No	No	Yes
F.	PTO performance:									
1.	Max. Power, kW (Ps)	12 (16.3)	25.5 (34.6)	31.2 (42.4)	33 (44.9)	43.4 (59.0)	21.6 (29.4)	28.9 (39.3)	37.4 (50.8)	30.1 (40.9)
2.	Engine speed corresponding to max. power (rpm) /SFC at max. power (g/kWh)	2713/355	2528/257	2204-2213/250	2008/256	2495/252	2045/347	2250/264	2373/243	2193/265
3.	Power at standard PTO speed (kw)/ Max. torque (N-m)	11/45.5	21.6/106.2	28.1/159.3	32.5/181.7	38.4/195.2	18.4/120.3	27.8/140.6	37.4/178	28.4/145.1
4.	Torque backup (%) /Smoke level (Bosch No.)	9.2/4.04	9.4/3.0	17.9/1.60	17.6/1.60	17,6/2.20	19.3/3.95	14.7/2.1	18.51/2.36	10.5/22
G.	Drawbar performance, (Unballasted/Ballasted):									
1.	Max. Power (kW)	9.1/10.1 (2W) 10.4/10.9 (4W)	21.3/21.7	25.8/26.5	28.8/28.9	36.6/37.5	18.5/19.1	25.5/24.9	31.8/30.8	25.8/26
2.	Speed at max. power (kmph)	6.68/6.69 (2W) 6.92/6.87 (4W)	8.46/8.87	8.64/8.75	9.76/9.89	8.10/8.16	8.34/8.85	10.08/10.12	10.75/7.07	10.03/7.55
3.	Max. pull (kN)	6.8/7.8 (2W) 7.7/9.4 (4W)	12.6/18.6	16.9/26.0	16.4/20.1	21.3/27.5	10.1/16.1	15.4/22.3	16.4/22.3	17.0/18.3
H.	Hydraulic performance:									
1.	Max. power, (kW)/ Corresponding flow rate (l/min)	2.81/11.25	4.1/19	5.0/25.0	4.0/16.00	8.1/30.5	3.60/14.50	3.8/15.0	7.88/27	5.75/23.0
2.	Lift capacity at standard frame (kN)	4	7	6.92	8.98	9.79	9.62	7.5	13.57	7.8
I.	Brake performance (Wnballasted/Ballasted):									
1.	Minimum stopping distance (m)	1.80/1.50	6.0/5.6	4.70/5.00	6.75/6.40	7.00/5.90	6.95/6.30	6.90/5.60	6.60/5.95	7.40/5.85

Table: *(Contd...)*

Table: *(Contd...)*

SI. no.	*Particulars\ Name of tractors*	*Mitsubishi Shakti MT 180D* I	*Mahindra B-275* II	*Escort 450* III	*Swaraj 855* IV	*FORD 5630* V	*TAPE MF-1035* VI	*Sonalika International DI-750* VII	*L&T John Deere 5310* VIII	*HMT 4511* IX
2.	Force required for deceleration of 2.5 m/square second (N)	225/ 215	225/ 190	210/ 200	190/ 240	230/ 245	130/ 160	170/ 175	165/ 180	275/ 310
3.	Parking performance	Satisfactory	Satisfactory	Satisfactory	Satisfactory	Satisfactory	Satisfactory	Satisfactory	Satisfactory	Satisfactory
J.	Haulage performance (2W/4W trailer):									
1.	Gross mass, tonnes	3.0/NA	4.0/5.5	5.0/6.5	5.0/6.0	5.0/8.0	4.5/5.5	5.0/6.5	5.0/7.0	5.0/6.0
2.	Average travel speed (km/h)	13.53-13.71/ NA	26.2-26.4/ 25.0-25.5	27.65-27.90/ 25.50-26.30	29.20-31.25 29.40-29.60	31.50/ 31.90	25.75/ 25.90	22.50/ 22.90	30.50/ 30.90	27.60/ 27.90
3.	Distance travelled per litre of fuel consumption (km)	5.56-5.80/ NA	S.94-6.0/ 5.50-5.70	4.88-4.90/ 5.00	5.22-5.45/ 5,36-5.55	4.18-4.29/ 3.81-3.82	4.32/ 4.80	5.30-5.50/ 4.80-4.90	5.22/ 5,0	5.22/ 5.0

[1 HP = 1.013 metric HP (Ps); 1 Ps = 735.5 W]
Reference (22)
Abbreviation: MDC-Manual draft control; ADDC- Automatic draft and depth control

Appendix-II

Specificatioms of Some Power Tillers

1. Mitsubishr-Shakti VWH-120 Power Tiller

Type:	Horizontal 4-stroke/single cylinder diesel engine
Bore/Stroke (mm)	92/95
Swept Volume (Cubic centimeter)	631
Compression Ratio	21:1
Maximum Torque	3.0 kg^{-m} at 1900 rpm
Engine output (hp)	9 at 2200 rpm
Rated speed(rpm)	2200
Specific fuel consumption	190 gm/hp-hr
Cooling System	Water cooled radiator type
Lighting System	12 volt/35 watt
Starting System	Hand cranking
Fuel tank capacity	11 litres
Gross weight	387 kg
Over all dimensions (L x W x H)	2320 mm × 810 mm × 1300 mm
Ground clearance(mm)	200
Speed range of Power Tiller	Forward speed (kmph):
Six numbers of forward speed	1st gear; 1.60, 2nd gear; 2.25, 3rd gear; 3.45,
Two numbers of reverse speed	4th gear; 5.55, 5th gear; 8.50, 6th gear; 13.50
	Reverse speed (kmph): R1;1.30, R2;4.60
Rotavator tilling width	540 mm
Rotavator tilling depth	150 mm
No. of tyres of rotavator	16

2. KAMCO Power Tiller

Type: Engine: ER 90; Tiller: KMB 20	Horizontal 4-stroke/single cylinder diesel engine
Engine output (hp)	9-12
Rated speed(rpm)	2000
Fuel consumption	1.5lit/hr
Cooling System	Water cooled radiator type
Lighting System	12 volt/40 watt
Starting System	Hand cranking
Fuel tank capacity	11 litres
Gross weight	485 kg
Over all dimensions (L × W × H)	2250 mm x 820 mm x 1030 mm
Tyre size	6×12
Ground clearance	203mm
Wheel track	Maximum, 930 mm; minimum, 690 mm
Speed range of Power tiller	Forward speed (kmph):
Six numbers of forward speed	1st gear; 1.50, 2nd gear; 2.4, 3rd gear; 3.5,4th gear;
Two numbers of reverse speed	5.5, 5th gear; 8.5, 6th gear; 14.0
	Reverse speed (kmph):R1;1.30, R2;4.9

Rotavator tilling width	600 mm
Rotavator tilling depth	190 mm
No. of tyres of rotavator	20
Rotavator speed	215-315 rpm

3. VST-Shakti 130 DI PowerTiller

Type:	Horizontal 4-stroke/single cylinder diesel engine
Bore/Stroke (mm)	95/95
Swept Volume (Cubic centimeter)	673.4
Compression Ratio	18:1
Maximum Torque	4.2 kg^{-m} at 1900 rpm
Engine output (hp)	13 at 2400 rpm
Rated speed(rpm)	2400
Specific fuel consumption	195 gm/hp-hr
Cooling System	Water cooled radiator type
Lighting System	12 volt/35 watt
Starting System	Hand cranking
Fuel tank capacity	11 litres
Gross weight	395 kg
Ground clearance	200 mm
Over all dimensions (L × W × H)	2320 mm × 810 mm × 1300 mm
Speed range of Power Tiller Six numbers of forward speed Two numbers of reverse speed	Forward speed (kmph): 1st gear; 1.80, 2nd gear; 2.64, 3rd gear; 4.20, 4th gear; 6.42. 5th gear; 9.50 6th gear; 15.00 Reverse speed (kmph):R1;1.55, R2;5.60
Tilling width	600 mm
Tilling depth	150 mm
No. of tynes of rotavator	18

Haulage Test:

The haulage test should be carried out with two wheel or four wheel trailers at the gross trailer loads recommended by the manufacturer. The haulage test is conducted on a level track having about 80 per cent of tarmacadam surface and 20 per cent of earthen surface and having gradients not exceeding 8 per cent at certain short length.

28.4 POWER MEASUREMENT METHODS

To understand power and its measurement, certain terms need to be defined and are as follows:

Power: It is the rate of doing work. Its unit is Newton meter per second (watt).

Indicated power: It is the power available at the piston inside the engine cylinder. It is the power developed in the cylinder without any auxiliary unit.

Brake power: It is the power output of the engine crankshaft.

Friction power: The power required to run the engine at any given speed without production of useful power. It represents the friction and pumping losses of an engine.

Indicated house power = brake power + friction power

Power-take off power: It is the power delivered by a tractor through its pto shaft.

Drawbar power: It is the power of a tractor, measured at the end of the drawbar. This power is used for pulling loads at the drawbar.

Belt power: It is the power of the engine, measured at the end of a suitable belt receiving drive from the pto shaft.

Pulley power: Power measured by coupling the pulley shaft directly to the dynamometer for pulley work.

Observed power: The power obtained at the dynamometer without any correction for atmospheric temperature, pressure or vapour pressure.

Corrected power: Corrected power is obtained by correcting the observed power to standard conditions of sea level pressure and temperature.

Rated engine speed: The speed of engine in rev/min specified by manufacturer for continuous operation at full load.

Rated horsepower: The power stated by the manufacturer that the engine will generate. It may or may not be the maximum power generated by the engine. Generally manufacturer rates the power within a safe limit avoiding the overloading of the engine.

Subject Index

Work done per minute = Work done per stroke x No. of power strokes per minute = P x A x L x N. This is usually written as (PLAN). Horse power (HP) = 4500 m. kg/min. Hence I.H.P. = PLAN/4500. In case of four-stroke engine, N = revolution per minute/2, as there is only one power stroke in every two revolutions of the crank shaft. In case of two-stroke engine, N = revolution per minute as there are one power stroke per revolution of crankshaft.

23.4.5 Measurement of I.H.P. of Multi Cylinder Engine (Morse Test)

This method is used in multi-cylinder engines to measure I.H.P without the use of indicator. The BHP of the engine is measured by making each cylinder inoperative in turn. If the engine consists of let 4- cylinder, then the BHP of the engine should be measured four times by making each cylinder inoperative turn by turn. This is applicable to petrol as well as for diesel engine. The cylinder of a petrol engine is made inoperative by shorting the spark plug where as in case of diesel engine, fuel supply is cut-off to the required cylinder.

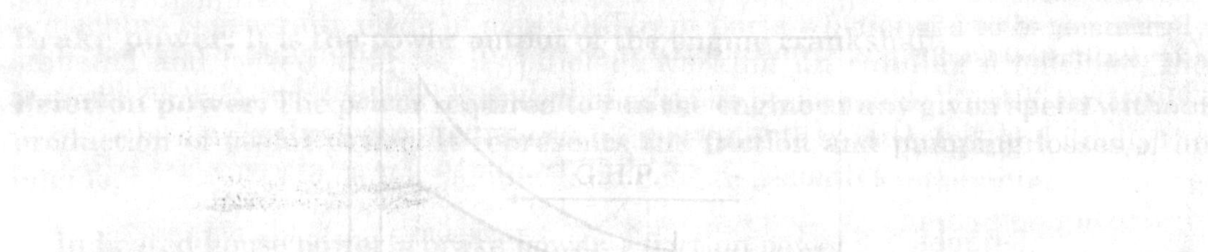

Fig. 23.4: Morse Test

In Fig. 23.4, the top area of the indicator diagram is a measure of the gross horsepower (G. H.P) developed by the engine, the bottom area the pumping horsepower (P.H.P). The net IHP per cylinder = (GHP-PHP). The net B.H.P per cylinder = (IHP - FHP) where F.H.P. stands for frictional horsepower with all the 'n' cylinders operating if there are 'n' cylinders in an engine. The total BHP *i.e.*, $(BHP)_n - (GPH)_n - (PHP)_B - (FHP)_n$. If one cylinder is inoperative, then the power developed by that cylinder (IHP) is lost and the speed of the engine will fall as the load on the engine remains the same. The engine speed can be maintained its original value by reducing the load on the engine. This is necessary to maintain the FHP constant, because it is assumed that the FHP is independent of load and depends only on the speed of the engine. Hence by maintaining rpm of the engine constant, the FHP remains constant. By making one cylinder inoperative, the new BHP represents that of an engine having (n-1) cylinders minus the pumping and friction losses of the inoperative cylinder.

When one cylinder is cut-off, then $(BHP)_{n-1} = (GHP)_{n-1} - (PHP)_{n-1} - (FHP)_{n-1} - (PHP + FHP)$ of idle cylinder. Then the reduction in BHP by cutting-off one cylinder is

$[(BHP)_n - (BHP)_{n-1}] = [(GHP)_n - (PHP)_n - (FHP)_n] - [(GHP)_{n-1} - (PHP)_{n-1} - (FHP)_{n-1} - (FHP)_{n-1} - (PH...$